Recent Developments in Cell Biology

Volume I

Recent Developments in Cell Biology
Volume I

Edited by **Samantha Granger**

R Callisto Reference

New York

Published by Callisto Reference,
106 Park Avenue, Suite 200,
New York, NY 10016, USA
www.callistoreference.com

Recent Developments in Cell Biology: Volume I
Edited by Samantha Granger

International Standard Book Number: 978-1-63239-530-6 (Hardback)

Printed in the United States of America.

Contents

Preface

Cell biology is that part of biology that focuses on the cell, which is the basic unit of life. It studies cells along with many defining characteristics like their physiological properties, their life cycle, their structure, interactions with their environment, the organelles they contain, division and even death. It seeks to understand life and cellular processes at the molecular level. Cell biology is an interdisciplinary field and is closely related to other areas of biology such as molecular biology, genetics and biochemistry. Cell biology includes a great diversity of research, from single-celled organisms like bacteria and protozoa to the many specialized cells in multi-cellular organisms such as humans, plants, and sponges. There are several main sub divisions within cell biology itself - the study of cell energy, the study of biochemical mechanisms that support cell metabolism and even a branch that focuses solely on the structure of cell components. It is becoming more and more important to discover and research the basic systems and mechanisms that allow cells to have distinct properties and coordinate the activities that form the essential systems that constitute any living cell. This is critical for gaining an understanding of human disease. Thus there is a demand for highly skilled researchers in this discipline.

This book is an attempt to compile and collate all available research on cell biology under one umbrella. I am grateful to those who put their hard work, effort and expertise into these researches as well as those who were supportive in this endeavor.

Editor

Regulation of Tissue Fibrosis by the Biomechanical Environment

Wayne Carver and Edie C. Goldsmith

Department of Cell Biology and Anatomy, University of South Carolina, School of Medicine, Columbia, SC 29209, USA

Correspondence should be addressed to Wayne Carver; wayne.carver@uscmed.sc.edu

Academic Editor: Mauro S. G. Pavao

The biomechanical environment plays a fundamental role in embryonic development, tissue maintenance, and pathogenesis. Mechanical forces play particularly important roles in the regulation of connective tissues including not only bone and cartilage but also the interstitial tissues of most organs. *In vivo* studies have correlated changes in mechanical load to modulation of the extracellular matrix and have indicated that increased mechanical force contributes to the enhanced expression and deposition of extracellular matrix components or fibrosis. Pathological fibrosis contributes to dysfunction of many organ systems. A variety of *in vitro* models have been utilized to evaluate the effects of mechanical force on extracellular matrix-producing cells. In general, application of mechanical stretch, fluid flow, and compression results in increased expression of extracellular matrix components. More recent studies have indicated that tissue rigidity also provides profibrotic signals to cells. The mechanisms whereby cells detect mechanical signals and transduce them into biochemical responses have received considerable attention. Cell surface receptors for extracellular matrix components and intracellular signaling pathways are instrumental in the mechanotransduction process. Understanding how mechanical signals are transmitted from the microenvironment will identify novel therapeutic targets for fibrosis and other pathological conditions.

1. Introduction

Mechanical forces play integral roles in embryonic development, homeostasis, and pathogenesis. All cells in multicellular organisms are exposed to mechanical forces of varying degrees. Endothelial cells, for instance, are exposed to shear stress due to the passage of fluid through the cardiovascular system. Chondrocytes and other cells in joints are exposed to repetitive compressive forces. The effects of mechanical forces on cells and tissues have received greater attention as models have been developed to systematically analyze these effects. Many of the early studies in this regard were focused on cells and tissues that are influenced by obvious mechanical force including the cardiovascular and musculoskeletal systems. Early investigations in the mechanobiology field relied on relatively simple and imprecise systems. For instance, studies have utilized a hanging-drop culture system to examine the effects of tensile forces on connective tissue cells [1]. As interest grew in the mechanobiology field, innovative systems were developed to apply tensile strain to rat calvarial cells cultured on ribbons of collagen [2] and compressive forces to chick long bones [3].

The mechanobiology field began to move forward rapidly as *in vitro* model systems were developed to more precisely isolate the effects of mechanical forces on cellular processes. Various systems were engineered to apply uniaxial or multiaxial distension or stretch to cells grown on deformable substrata. These systems date back several decades to studies conducted on smooth muscle cells that were cultured on deformable elastin matrices [4, 5]. Among other responses, these studies illustrated a role for mechanical force in the growth and maintenance of skeletal and cardiovascular cells [6–9]. It has become increasingly clear that many aspects of cell behavior can be modulated by mechanical force including cell proliferation, differentiation, migration, and gene expression. The realization that most cells respond to mechanical stimuli has resulted in enhanced interest in the contribution of these forces to pathogenesis including tissue fibrosis and in the mechanisms whereby cells detect and respond to these forces.

Studies by Leung et al. [5] were among the first to illustrate that cyclic mechanical loading promotes the production of extracellular matrix (ECM) components by vascular smooth muscle cells. The ECM is a dynamic network composed primarily of collagens, noncollagenous glycoproteins, and proteoglycans. The ECM was historically appreciated for its function as a three-dimensional scaffold that played an essential role in tissue development and function. Alterations in ECM composition, organization, and accumulation can deleteriously impact embryonic development and organ homeostasis in adults. For instance, deficits in collagen production result in vascular weakness and aneurysms [10]. On the other extreme, increased accumulation of ECM components or fibrosis results in dysfunction of many organs.

The expression of ECM components is regulated by diverse biochemical factors including growth factors, cytokines, and hormones (see [11, 12] for recent reviews). In addition, ECM production can be modulated by electrical and mechanical stimuli. Until relatively recently, the role of mechanical forces in regulating gene expression and cell behavior has received little attention. This has changed as it has been realized that all cells are exposed to mechanical forces, and with the advent of *in vitro* testing systems the effects of these forces and the mechanisms of their actions have been and continue to be investigated.

2. Mechanical Stretch and Promotion of Tissue Fibrosis

Cells can be exposed to diverse types of extrinsic mechanical forces including mechanical stretch (tension), compression, and shear stress. A number of early studies utilized cells cultured on deformable membranes to examine the cellular effects of mechanical stretch. These studies illustrated that mechanical stretch of isolated cells mimicked many of the responses that had been characterized to increased load *in vivo*. For instance, mechanical stretch of skeletal myotubes elicited a hypertrophic response that included increased general protein synthesis and enhanced accumulation of contractile proteins [13].

Alterations in mechanical load *in vivo* had been known for some time to impact synthesis and deposition of the ECM. For instance, increased cardiovascular load has for some time been correlated to increased deposition of ECM components. The period immediately after birth is associated with increased cardiovascular load and rapid growth of the heart [14]. This period of "physiological hypertrophy" is also associated with rapid deposition and organization of ECM components, particularly interstitial collagens [15–18]. Increased mechanical load as seen during aortic constriction or stenosis also promotes myocardial hypertrophy and fibrosis in the adult heart [19, 20]. While a number of mechanical stretch devices have been utilized to mimic changes in mechanical forces seen *in vivo*, all have generally illustrated that mechanical stretch of matrix-producing cells (largely fibroblasts and smooth muscle cells) results in increased production of ECM components or a profibrotic response [21–24].

To more accurately mimic the *in vivo* environment, apparatuses are being developed to investigate mechanical forces in three-dimensional *in vitro* systems. Several recent studies have applied mechanical loads to cells cultured in three-dimensional scaffolds [25, 26]. The use of three-dimensional constructs provides important insight into the effects of complex mechanical forces on tissue properties, and the development of systems to apply and analyze mechanical load to these constructs will also be advantageous to efforts to engineer functional tissue constructs.

3. Effects of Tissue Stiffness on Fibrosis

While two-dimensional *in vitro* systems have been invaluable in elucidating the effects of mechanical forces on cells and the mechanisms of mechanotransduction, cells function within a three-dimensional environment whose mechanical properties can change during development [27] or various pathological conditions including fibrosis [28, 29], cancer [30–34], and atherosclerosis [35]. Due to accumulation of ECM components and cross-linking of these components, alterations in tissue stiffness are a common feature of fibrosis. For instance, pathological scars are stiffer relative to unwounded normal skin and typically consist of thicker collagen bundles [36]. Accumulation of ECM components alters the tissue mechanical properties, which in turn can deleteriously impact organ function [37]. Component cells sense and respond to ECM rigidity, which can regulate cell growth [38], shape [39], migration [40, 41], and differentiation [42, 43].

Seminal studies by Mauch et al. [44] were among the first to evaluate the effects of the biomechanical microenvironment on the expression of ECM components. The expression of ECM components and ECM-modifying enzymes was compared between cells cultured on tissue culture plastic, a rigid substratum, and three-dimensional collagen gels, a more flexible substratum. These studies illustrated that collagen expression is markedly decreased in fibroblasts cultured in three-dimensional collagen scaffolds compared to cells grown on tissue culture plastic. This effect was, at least in part, regulated at the mRNA level as $\alpha1(I)$, $\alpha2(I)$, and $\alpha1(III)$ collagen mRNAs were diminished in cells cultured in the three-dimensional scaffolds. Further studies by this group of investigators illustrated that collagenase activity is enhanced by culture in three-dimensional scaffolds promoting a collagenolytic phenotype in the less rigid environment of the collagen gels [45]. A number of studies have subsequently supported the concept that matrix rigidity propagates the profibrotic response. Culture of human colon fibroblasts on matrices that mimic the mechanical properties of the normal colon or the pathologically stiff colon of Crohn's disease patients demonstrated enhanced expression of ECM components and increased proliferation of fibroblasts on the stiffer matrix [46]. Similarly, culture of human dermal fibroblasts in collagen gels that were made stiffer by prestraining resulted in enhanced expression of collagen by dermal fibroblasts relative to that in unstrained scaffolds [47]. Liu et al. [48] have utilized a novel photopolymerization approach to generate polyacrylamide scaffolds with stiffness gradients that span the range of normal and fibrotic lung tissue (0.1 to 50 kPa). In this system, proliferation of lung fibroblasts was induced by increased scaffold stiffness. In contrast, matrix stiffness protected cells from apoptosis in response to serum starvation. The patterns of collagen α1(I) and α1(III) mRNA expression paralleled proliferation with increasing expression in stiffer regions of the scaffold. The expression of prostaglandin, which is an endogenous antifibrotic factor, was opposite to that of the collagens with increased levels in the less rigid portions of the construct. These studies and others indicate that the biomechanical properties of the microenvironment can direct the expression of ECM components and ECM-modifying enzymes with stiffer tissue properties contributing to enhanced ECM production. Less rigid matrices appear to promote an anti-fibrotic environment that includes increased production of matrix-degrading proteases and anti-fibrotic agents like prostaglandin.

Matrix rigidity impacts not only the expression of ECM components but also other parameters associated with fibrosis including the deposition and organization of these components. Studies by Halliday and Tomasek [49] illustrated that fibroblasts cultured in stabilized three-dimensional collagen gels generate stress that is transmitted throughout the collagen scaffold. These cells develop large actin microfilament bundles and organize fibronectin into extracellular fibrils. Fibroblasts cultured in free-floating collagen gels generate less stress and lack fibronectin-containing fibrils. More recently, Carraher and Schwarzbauer [50] utilized a polyacrylamide model to evaluate the role of matrix stiffness on fibronectin organization. Polyacrylamide scaffolds have become popular three-dimensional models as their rigidity can be modulated by altering the ratios of the components contributing to polymerization of the scaffold. Similar to previous studies, this work illustrated that growth of cells on more rigid substrates promoted fibronectin assembly and activation of focal adhesion kinase (FAK). Furthermore, activation of ECM receptors of the integrin family by Mn^{2+} on softer substrates stimulated fibronectin assembly illustrating that integrin activity is an important mediator of this process (discussed further below). Previous studies have illustrated that the conformation of fibronectin on more rigid substrata is extended, which exposes additional binding sites for cells to fibronectin [51]. This is consistent with other studies illustrating that multiple proteins that are involved in mechanotransduction become extended in response to mechanical force thus revealing cryptic interaction sites that mediate activity of the proteins. Indeed, providing exogenous unfolded fibronectin to cells in "soft" polyacrylamide gels increases FAK activation to a similar degree as culture in more rigid gels [50].

4. ECM Density and Myofibroblast Formation

An important step in tissue fibrosis of many organs is the formation of myofibroblasts or myofibroblast-like cells. These cells are characterized by enhanced contractile activity, formation of stress fibers, and expression of α-smooth muscle actin. Myofibroblasts are responsible for alterations to connective tissues including increased synthesis of ECM components. In addition, these cells produce cytokines and growth factors that promote the fibrotic response in an autocrine/paracrine manner. Myofibroblasts are derived from a variety of cells in response to tissue damage and stress including quiescent fibroblasts, blood-derived fibrocytes, mesenchymal stem cells, stellate cells of the liver, and others [52, 53]. Regardless of their origin, myofibroblasts likely arise as an acute and beneficial response to repair damaged tissue. Continued myofibroblast contraction and production of ECM components become deleterious and in many cases yield to stiff fibrotic tissue that obstructs and destroys organ function [54]. Stiffened tissue further promotes myofibroblast formation perpetuating scar formation.

Studies using a three-dimensional collagen scaffold system illustrated that collagen deformability or compliance is inversely related to the transformation of cells into a myofibroblast phenotype [55]. Culture of cells on plastic coated with thin films of collagen (minimal compliance and maximal generation of intracellular tension) resulted in the highest levels of α-smooth muscle actin expression, routinely used as a marker for myofibroblast formation. Culture of cells in free-floating collagen gels (maximal compliance and least generation of intracellular tension) yielded the lowest relative level of α-smooth muscle actin expression. Similar results have been obtained in experiments examining matrix rigidity and differentiation of bronchial fibroblasts to a myofibroblast phenotype [56]. Culture of bronchial fibroblasts on polydimethylsiloxane substrates of variable stiffnesses (1–50 kPa) was performed to evaluate the effects of matrix mechanical properties on myofibroblast formation [56]. Increased scaffold stiffness promoted myofibroblast formation and increased α-smooth muscle actin and interstitial collagen expression. In the former studies, the expression of the α1 and α2 integrins, which are collagen receptors, correlated to enhanced myofibroblast formation on collagen-coated plastic [55]. Incubation of cells with function-blocking antibodies to these integrins attenuated myofibroblast formation indicating that generation of intracellular tension via integrin-ECM interactions is critical to the transformation process. More recent studies have illustrated an interaction between the mechanical properties of three-dimensional collagen gels and the biochemical environment [57]. In these studies, there was no difference in α-smooth muscle actin expression between cells in free-floating and constrained collagen gels cultured in low serum (5%); however, enhanced α-smooth muscle actin expression was seen in constrained gels at higher serum levels (10%). These studies and others illustrate integration of mechanical and biochemical signals by cells.

The conversion of hepatic stellate cells to a myofibroblast phenotype is a critical step in liver fibrosis and is part of the pathway to cirrhosis in chronic liver disease. Culture of hepatic stellate cells on tissue culture plastic and in high levels of serum results in their spontaneous conversion to a myofibroblast phenotype [58]. Culture of hepatic cells on Matrigel, a relatively soft basement membrane-like matrix, retains the quiescent nature of hepatic stellate cells [59]. Furthermore, culture of differentiated hepatic myofibroblasts on Matrigel results in loss of myofibroblast characteristics [60]. The mechanisms of the dedifferentiation of these cells are not well understood. Recent studies by Olsen et al. [61] to evaluate the role of substrate stiffness on differentiation of hepatic stellate cells utilized polyacrylamide scaffolds coated with various ECM substrates. These studies illustrated that increased matrix stiffness is capable of promoting myofibroblast formation independent of growth factor or cytokine stimulation. However, addition of TGF-β to the culture medium enhanced differentiation on stiff scaffolds, again indicating interactions between the mechanical and biochemical environments. These studies also illustrated that interactions between the cells and the surrounding ECM and generation of mechanical tension are critical to the conversion to a myofibroblast phenotype. That is, coating of polyacrylamide scaffolds with collagen or fibronectin promoted myofibroblast formation to a much greater degree than polyacrylamide scaffolds coated with poly-L-lysine. Cell adhesion to poly-L-lysine is through electrostatic charges and not via specific integrin receptors. Studies with foreskin fibroblasts have illustrated that alterations in integrin expression accompany changes in substrate rigidity and myofibroblast formation [62]. In these studies, cells cultured on less rigid polyacrylamide gels expressed little α-smooth muscle actin and primarily the $\alpha2\beta1$ integrin. Culture of cells on more rigid substrata resulted in enhanced expression of α-smooth muscle actin and a switch to expression primarily of $\alpha v\beta3$ integrin.

Fibroblasts isolated from diseased patients or animal models typically retain characteristics of their altered phenotype *in vitro* [63]. Indeed, comparison of fibroblasts from normal individuals and individuals with idiopathic pulmonary fibrosis illustrated differences in proliferation and contractile activity on rigid substrates [64]. However, the fibroblasts from idiopathic pulmonary fibrosis patients remained responsive to alterations in matrix rigidity with decreased proliferation and contractile properties when plated in soft matrices. This suggests that the myofibroblast phenotype is not a permanent state but can be reversed by alterations in the matrix properties. In contrast to this, studies culturing fibroblasts for prolonged periods on matrices of different mechanical properties suggest the conversion to a myofibroblast phenotype is a more "permanent" condition [65]. Culture of cells on a rigid matrix for three weeks resulted in sustained fibrotic activity, even after moving the cells to softer matrices. Understanding the plasticity of the fibrotic phenotype is critical to development of novel therapeutic approaches to fibrosis.

Recent studies have been carried out utilizing a novel photodegradable cross-linker-polyethylene glycol scaffold in which exposure to ultraviolet light can modulate the mechanical properties of the substratum to evaluate the effects on myofibroblast conversion of heart valve interstitial cells [66]. Similar to studies with other cell types, increased elastic modulus of the scaffold yielded an enhanced proportion of α-smooth muscle actin-containing cells. Interestingly, and of potential therapeutic significance, the proportion of myofibroblasts in the scaffolds decreased by approximately half when the elastic modulus was decreased by photodegradation. This coincided with a reduction in connective tissue growth factor and in proliferation. The classic dogma has been that once fibrosis has begun, it cannot be reversed; however, recent studies have illustrated that fibrosis can be halted or even reversed depending upon the extent of its progression [67]. The above studies suggest that alteration in the ECM biomechanical properties may be an important therapeutic target that is able to modulate myofibroblast formation and fibrosis.

Recent studies with gold nanoparticles have shown that they can be used for both measuring cell-induced deformation of the ECM as well as modulating matrix stiffness and formation of myofibroblasts. Stone et al. [68] described a method using the light scattering properties of gold nanorods as a pattern marker to track cardiac fibroblast deformation of a two-dimensional collagen matrix using digital image correlation. This study detected areas of both tensile and compressive strain within the collagen films and displacements on the order of 18 μm [68]. Recently this method was applied to examine age-dependent differences in cellular mechanical behavior. Cardiac fibroblasts isolated from neonatal and adult rats were examined for their ability to deform a two-dimensional collagen film and three-dimensional collagen gels [69]. While no significant differences in strain were detected between the cell populations on the two-dimensional films, neonatal fibroblasts were significantly more contractile in three-dimensional collagen gels and expressed higher levels of α-smooth muscle actin compared to adult fibroblasts. Inclusion of negatively charged, polyelectrolyte-coated gold nanorods within three-dimensional collagen gels significantly reduced the ability of neonatal cardiac fibroblasts to contract these gels and was accompanied by a significant decrease in both the expression of α-smooth muscle actin and type I collagen [70]. This study suggested that the presence of the surface-modified nanorods impaired the ability of the fibroblasts to transform into myofibroblasts. In addition, it has been shown that negatively charged nanorods accelerated the *in vitro* assembly to type I collagen, and rheological characterization of the mechanical properties of these constructs revealed that these gels were stiffer and more elastic than controls or gels containing positively charged gold nanorods [71]. These latter studies would suggest that nanomaterials may hold promise as a means to both alter the mechanical properties of the ECM and the formation of the myofibroblast phenotype associated with pathological fibrosis.

Another mechanism to take advantage of matrix mechanical properties therapeutically is in targeting death of cells via alterations in matrix rigidity. It has long been known that interactions with the ECM are necessary for survival of normal cells. However, the effects of the mechanical properties of the ECM on cell survival are only recently being addressed. Using polyacrylamide gels of varying rigidity coated with type I collagen, Wang et al. [72] illustrated that proliferation of NIH 3T3 cells is enhanced on stiffer scaffolds. These studies also illustrated that apoptosis of NIH 3T3 cells was increased by almost two fold on less rigid collagen-coated polyacrylamide gels. The effect of matrix stiffness on apoptosis was absent in H-ras-transformed cells. A similar increase in apoptosis was seen in cells from the rat annulus fibrosis when cultured on softer polyacrylamide scaffolds [73]. These studies suggest that decreasing local matrix stiffness will result in apoptosis, potentially of matrix-producing myofibroblasts or other cells.

The ability of matrix mechanical properties to direct cell behavior is also being integrated into novel tissue engineering approaches, particularly in attempting to develop vascularized tissue constructs [74]. Examination of the invasive activity of endothelial cells plated onto the surface of collagen scaffolds has been used as an angiogenic model. Increasing the stiffness of the collagen scaffolds by cross-linking with microbial transglutaminase resulted in increased numbers of angiogenic sprouts and enhanced cell invasion independent of ECM pore size or density [75]. Under the appropriate biochemical and mechanical conditions, endothelial cells are able to form three-dimensional networks. Utilizing polyacrylamide gels functionalized with peptide sequences derived from cell adhesion sequences, the effect of scaffold mechanical properties on network formation was evaluated [76]. Endothelial cells formed stable networks on relatively soft functionalized polyacrylamide gels (Young's modulus of 140 Pa) in the absence of angiogenic biochemical factors (bFGF or VEGF). On stiffer polyacrylamide scaffolds (2500 Pa), endothelial cells failed to assemble into networks in the presence or absence of angiogenic factors. Thus, the elastic modulus of hydrogels is able to direct the migration and organization of vascular cells [74].

5. Transduction of Mechanical Signals

Studies utilizing *in vitro* systems have provided fundamental information regarding the molecular mechanisms whereby cells detect and respond to mechanical forces. During the past two decades, extensive progress has been made in understanding "mechanotransduction" or the mechanisms whereby physical stimuli are converted into chemical signals by cells [77, 78]. Despite the fact that the types of mechanical forces cells experience are variable, including externally applied forces (stretch, shear stress, compression, etc.) and forces generated by cells themselves, the molecular mechanisms whereby this information is transduced appear to have similarities. Alterations in the three-dimensional conformation of mechanosensitive proteins or adhesion structures are often at the foundation of this process. Studies utilizing mechanical stretch systems were fundamental in implicating cell surface integrins as central components of cell adhesion complexes and fundamental to mechanotransduction [79]. Integrins are heterodimers composed of an alpha and a beta chain that serve as the primary family of receptors for ECM components [80–82]. There are over twenty different α/β heterodimer combinations, and specific α/β heterodimers serve as receptors for particular ECM ligand(s). The response of cells to mechanical stretch varies depending upon the ECM substratum suggesting a role for specific integrin heterodimers [79, 83]. Utilizing function-blocking antibodies to specific integrins ($\alpha4$ and $\alpha5$ chains) or arginine-glycine-aspartic acid (RGD) peptides to prevent integrin-ECM interactions, MacKenna et al. [79] were among

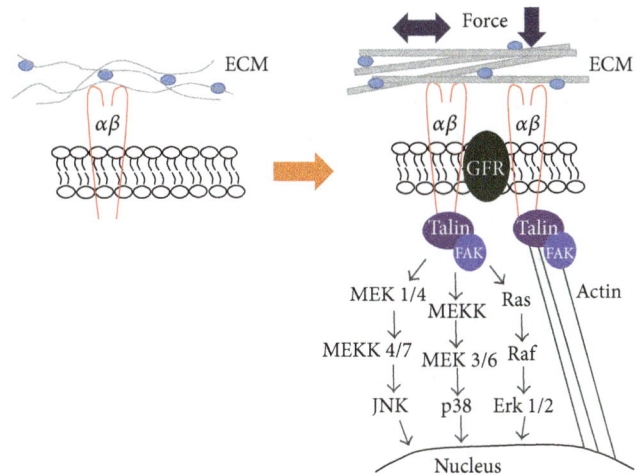

FIGURE 1: This schematic illustrates the transduction of mechanical force from the microenvironment to the cell. Extrinsically applied force results in alteration in the three-dimensional structure of the ECM and activation of integrin-associated signaling and transmission of signals via the actin cytoskeleton. These forces subsequently result in accumulation of ECM components and a stiffer ECM, which exacerbates the fibrotic response.

the first to show roles for specific integrins in the response of fibroblasts to mechanical stretch.

These early studies set the stage for extensive research focused on the mechanisms whereby cells detect mechanical changes in the microenvironment and transduce these into biochemical and molecular alterations in the cytoplasm and nucleus. The cell-ECM linkage involving integrins and a myriad of associated proteins is a critical component of this process (Figure 1). It has become increasingly clear that integrin-based adhesions are dynamic and complex structures that transmit information from the ECM to the cell and vice versa [84]. Integrins, which lack intrinsic enzyme activity, provide a physical linkage from the ECM to the actin cytoskeleton and to a wide array of signaling proteins. In fact, integrin complexes can contain over a hundred different proteins, many that bind in a force-dependent manner [85, 86]. The characterization of the ECM-integrin-cytoskeletal linkage has contributed to the concept of tensegrity in which signals can be transmitted from the ECM to the cytoplasm and nucleus via these physical connections [87, 88]. Several proteins can simultaneously bind integrins and actin and are thus thought to participate in mechanotransduction via the physical ECM-integrin-cytoskeleton linkage including vinculin, talin, and α-actinin [89, 90].

A number of signaling molecules associate directly or indirectly with the integrin cytoplasmic domain including focal adhesion kinase (FAK). FAK was initially identified as a Src kinase substrate [91, 92]. As integrins do not have intrinsic enzyme activity, FAK is a critical mediator of integrin-induced signaling events. The activation of FAK is initiated by autophosphorylation of tyrosine at position 397 and can be induced by clustering of integrins [93, 94]. In turn, FAK can activate integrins, which strengthens cell adhesions with the ECM [95]. Activated FAK can act independently or as part of a Src-containing complex to phosphorylate other signaling proteins or act as a scaffold in the recruitment of additional proteins to cell adhesions.

Exposure of cells to mechanical force results in activation of numerous intracellular signaling pathways including protein kinases such as protein kinase C, c-Jun N-terminal kinases (JNK), extracellular signal-regulated kinases (Erk), and others (see [96] for recent review). Activation of these pathways ultimately leads to activation of transcription factors and cell activities that comprise the response of a given cell to mechanical events.

While there appear commonalities in signaling pathways induced by various types of mechanical forces, *in vitro* studies illustrate that cells respond differently to diverse types of mechanical perturbations. The type of mechanical force can modulate differentiation of connective tissue cells. The ratio between tensile and compression type forces can promote either differentiation into cartilage or bone [97]. Exposing heart fibroblasts to constant versus cyclic mechanical stretch resulted in differences in collagen gene expression [98]. Similarly, exposing vascular endothelial cells to cyclic stretch resulted in differences in growth factor expression and branch formation compared to constant stretch [99]. Application of steady mechanical force on aortas resulted in more pronounced FAK activation compared to pulsatile stretch [100]. These studies suggest that while generalities may be developed regarding the response of cells to mechanical force, the details of this response likely vary depending on the type of force and in a cell- or tissue-specific manner.

6. YAP/TAZ as Mechanotransducers

Recent studies have illustrated that signals from the ECM and cell adhesion sites converge on two components of the Hippo pathway, Yes-associated protein (YAP) and transcriptional coactivator with PDZ-binding motif (TAZ) [101, 102]. Analysis of the expression of YAP and TAZ illustrated that the levels of these proteins were enhanced in endothelial cells cultured on stiff fibronectin-containing polyacrylamide hydrogels (10–40 kPa) compared to cells growing on soft hydrogels (0.7–1.0 kPa) [101]. The expression of YAP and TAZ on stiff hydrogels was similar to that seen in cells cultured on plastic culture dishes. In addition, the subcellular localizations of YAP and TAZ are altered by the ECM mechanical environment. These proteins are predominantly located in the cytoplasm of cells grown in softer matrices but are translocated to the nucleus in cells cultured in stiff substrates. YAP and TAZ modulate the activity of transcription factors, including LEAD, RUNx, and Smads in the nucleus. Among the transcriptional targets of the YAP and TAZ system are connective tissue growth factor and TGF-β, two important biochemical factors that promote fibrosis, and transglutaminase-2, an important component of ECM deposition and turnover [103].

Several recent studies have begun to evaluate the functional roles of YAP and TAZ in mediating the response of cells to mechanical forces. In humans, the trabecular meshwork of the eye is approximately twentyfold stiffer in individuals with glaucoma than in normal individuals [104]. Cells from the trabecular meshwork have been cultured on hydrogels of varying stiffness representing normal and glaucomatous conditions (5 kPa and 75 kPa, resp.) to evaluate the role of the YAP/TAZ system in the progression of fibrosis associated with glaucoma. Similar to the above studies, culture of trabecular meshwork cells on stiffer ECM resulted in enhanced expression of TAZ and transglutaminase-2. Interestingly, YAP expression was decreased relative to that on softer scaffolds suggesting that there may be cell-specific regulation of YAP and TAZ in response to altered mechanical properties of the microenvironment.

7. Conclusions and Future Directions

It has become increasingly clear that most cells in the vertebrate body are exposed to varying degrees of mechanical forces. These forces impact embryonic development, homeostasis, and pathological conditions including fibrosis. Historically most of the studies that focused on mechanical force as a profibrotic stimulus utilized two-dimensional stretch or compression models with isolated matrix-producing cells. These studies have provided substantial knowledge regarding the responses of cells to mechanical force and the underlying mechanisms of this response. However, these systems do not adequately mimic the *in vivo* three-dimensional environment. This has led to development of three-dimensional models to evaluate the effects of mechanical forces in a more *in vivo*-like environment. The realization that the biomechanical properties of the microenvironment can promote fibrosis and other responses has led to renewed interest in the effects of mechanical forces on cell and tissue behavior.

While extensive knowledge has been gained regarding the effects of the mechanical environment on cells and tissues, many questions remain regarding the molecular mechanisms of these effects. Identification of novel mechanoresponsive proteins such as YAP and TAZ will provide new therapeutic targets to modulate the deleterious effects of increased mechanical force. As it is becomingly increasing clear that tissue stiffness may precede fibrosis or at least contribute to ongoing fibrosis, identifying methods to modulate the mechanical properties of the microenvironment may also yield novel therapeutic approaches. Along these lines, specific nanomaterials may provide such reagents. However, the mechanisms whereby these materials regulate tissue properties have not been elucidated.

References

[1] C. A. L. Bassett and I. Herrmann, "Influence of oxygen concentration and mechanical factors on differentiation of connective tissues in vitro," *Nature*, vol. 190, no. 4774, pp. 460–461, 1961.

[2] C. K. Yeh and G. A. Rodan, "Tensile forces enhance prostaglandin E synthesis in osteoblastic cells grown on collagen ribbons," *Calcified Tissue International*, vol. 36, supplement 1, pp. S67–S71, 1984.

[3] G. A. Rodan, T. Mensi, and A. Harvey, "A quantitative method for the application of compressive forces to bone in tissue

culture," *Calcified Tissue International*, vol. 18, no. 2, pp. 125–131, 1975.

[4] D. Y. M. Leung, S. Glagov, and M. B. Matthews, "Cyclic stretching stimulates synthesis of matrix components by arterial smooth muscle cells in vitro," *Science*, vol. 191, no. 4226, pp. 475–477, 1976.

[5] D. Y. M. Leung, S. Glagov, and M. B. Mathews, "A new in vitro system for studying cell response to mechanical stimulation. Different effects of cyclic stretching and agitation on smooth muscle cell biosynthesis," *Experimental Cell Research*, vol. 109, no. 2, pp. 285–298, 1977.

[6] H. H. Vandenburgh, "Motion into mass: how does tension stimulate muscle growth?" *Medicine and Science in Sports and Exercise*, vol. 19, no. 5, pp. S142–S149, 1987.

[7] J. L. Samuel, I. Dubus, F. Contard, K. Schwartz, and L. Rappaport, "Biological signals of cardiac hypertrophy," *European Heart Journal*, vol. 11, pp. 1–7, 1990.

[8] V. J. Dzau, "Local contractile and growth modulators in the myocardium," *Clinical Cardiology*, vol. 16, no. 5, pp. II5–II9, 1993.

[9] T. Yamazaki, I. Komuro, and Y. Yazaki, "Molecular aspects of mechanical stress-induced cardiac hypertrophy," *Molecular and Cellular Biochemistry*, vol. 163-164, pp. 197–201, 1996.

[10] J. Löhler, R. Timpl, and R. Jaenisch, "Embryonic lethal mutation in mouse collagen I gene causes rupture of blood vessels and is associated with erythropoietic and mesenchymal cell death," *Cell*, vol. 38, no. 2, pp. 597–607, 1984.

[11] T. Bowen, R. H. Jenkins, and D. J. Fraser, "MicroRNAs, transforming growth factor beta-1 and tissue fibrosis," *The Journal of Pathology*, vol. 229, pp. 274–285, 2013.

[12] K. Lee and C. M. Nelson, "New insights into the regulation of epithelial-mesenchymal transition and tissue fibrosis," *International Review of Cell and Molecular Biology*, vol. 294, pp. 171–221, 2012.

[13] H. Vandenburgh and S. Kaufman, "In vitro model for stretch-induced hypertrophy of skeletal muscle," *Science*, vol. 203, no. 4377, pp. 265–268, 1979.

[14] S. F. Hopkins Jr., E. P. McCutcheon, and D. R. Wekstein, "Postnatal changes in rat ventricular function," *Circulation Research*, vol. 32, no. 6, pp. 685–691, 1973.

[15] T. K. Borg and J. B. Caulfield, "Collagen in the heart," *Texas Reports on Biology and Medicine*, vol. 39, pp. 321–333, 1979.

[16] T. K. Borg, "Development of the connective tissue network in the neonatal hamster heart," *American Journal of Anatomy*, vol. 165, no. 4, pp. 435–443, 1982.

[17] W. Carver, L. Terracio, and T. K. Borg, "Expression and accumulation of interstitial collagen in the neonatal rat heart," *Anatomical Record*, vol. 236, no. 3, pp. 511–520, 1993.

[18] G. L. Engelmann, "Coordinate gene expression during neonatal rat heart development. A possible role for the myocyte in extracellular matrix biogenesis and capillary angiogenesis," *Cardiovascular Research*, vol. 27, no. 9, pp. 1598–1605, 1993.

[19] K. T. Weber, J. S. Janicki, S. G. Shroff, R. Pick, R. M. Chen, and R. I. Bashey, "Collagen remodeling of the pressure-overloaded, hypertrophied nonhuman primate myocardium," *Circulation Research*, vol. 62, no. 4, pp. 757–765, 1988.

[20] J. E. Jalil, C. W. Doering, J. S. Janicki, R. Pick, S. G. Shroff, and K. T. Weber, "Fibrillar collagen and myocardial stiffness in the intact hypertrophied rat left ventricle," *Circulation Research*, vol. 64, no. 6, pp. 1041–1050, 1989.

[21] P. R. Kollros, S. R. Bates, M. B. Mathews, A. L. Horwitz, and S. Glagov, "Cyclic AMP inhibits increased collagen production by cyclically stretched smooth muscle cells," *Laboratory Investigation*, vol. 56, no. 4, pp. 410–417, 1987.

[22] W. Carver, M. L. Nagpal, M. Nachtigal, T. K. Borg, and L. Terracio, "Collagen expression in mechanically stimulated cardiac fibroblasts," *Circulation Research*, vol. 69, no. 1, pp. 116–122, 1991.

[23] A. A. Lee, T. Delhaas, L. K. Waldman, D. A. Mackenna, F. J. Villarreal, and A. D. McCulloch, "An equibiaxial strain system for cultured cells," *American Journal of Physiology*, vol. 271, no. 4, pp. C1400–C1408, 1996.

[24] R. P. Butt and J. E. Bishop, "Mechanical load enhances the stimulatory effect of serum growth factors on cardiac fibroblast procollagen synthesis," *Journal of Molecular and Cellular Cardiology*, vol. 29, no. 4, pp. 1141–1151, 1997.

[25] A. Auluck, V. Mudera, N. P. Hunt, and M. P. Lewis, "A three-dimensional in vitro model system to study the adaptation of craniofacial skeletal muscle following mechanostimulation," *European Journal of Oral Sciences*, vol. 113, no. 3, pp. 218–224, 2005.

[26] R. K. Birla, Y. C. Huang, and R. G. Dennis, "Development of a novel bioreactor for the mechanical loading of tissue-engineered heart muscle," *Tissue Engineering*, vol. 13, no. 9, pp. 2239–2248, 2007.

[27] T. Mammoto and D. E. Ingber, "Mechanical control of tissue and organ development," *Development*, vol. 137, no. 9, pp. 1407–1420, 2010.

[28] M. Yin, L. Lian, D. Piao, and J. Nan, "Tetrandrine stimulates the apoptosis of hepatic stellate cells and ameliorates development of fibrosis in a thioacetamide rat model," *World Journal of Gastroenterology*, vol. 13, no. 8, pp. 1214–1220, 2007.

[29] W. Tomeno, M. Yoneda, K. Imajo et al., "Evaluation of the liver fibrosis index calculated by using real-time tissue elastography for the non-invasive assessment of liver fibrosis in chronic liver diseases," *Hepatology Research*, 2012.

[30] M. J. Paszek, N. Zahir, K. R. Johnson et al., "Tensional homeostasis and the malignant phenotype," *Cancer Cell*, vol. 8, no. 3, pp. 241–254, 2005.

[31] T. A. Ulrich, E. M. De Juan Pardo, and S. Kumar, "The mechanical rigidity of the extracellular matrix regulates the structure, motility, and proliferation of glioma cells," *Cancer Research*, vol. 69, no. 10, pp. 4167–4174, 2009.

[32] W. A. Lam, L. Cao, V. Umesh, A. J. Keung, S. Sen, and S. Kumar, "Extracellular matrix rigidity modulates neuroblastoma cell differentiation and N-myc expression," *Molecular Cancer*, vol. 9, article 35, 2010.

[33] P. Schedin and P. J. Keely, "Mammary gland ECM remodeling, stiffness, and mechanosignaling in normal development and tumor progression," *Cold Spring Harbor Perspectives in Biology*, vol. 3, no. 1, p. a003228, 2011.

[34] A. Pathak and S. Kumar, "Independent regulation of tumor cell migration by matrix stiffness and confinement," *Proceedings of the National Academy of Sciences of USA*, vol. 109, pp. 10334–10339, 2012.

[35] T. Y. Choi, N. Ahmadi, S. Sourayanezhad, I. Zeb, and M. J. Budoff, "Relation of vascular stiffness with epicardial and pericardial adipose tissues and coronary atherosclerosis," *Atherosclerosis*, 2013.

[36] R. A. F. Clark, G. S. Ashcroft, M. J. Spencer, H. Larjava, and M. W. J. Ferguson, "Re-epithelialization of normal human

excisional wounds is associated with a switch from $\alpha v\beta 5$ to $\alpha v\beta 6$ integrins," *British Journal of Dermatology*, vol. 135, no. 1, pp. 46–51, 1996.

[37] B. Hinz, "Tissue stiffness, latent TGF-β1 Activation, and mechanical signal transduction: implications for the pathogenesis and treatment of fibrosis," *Current Rheumatology Reports*, vol. 11, no. 2, pp. 120–126, 2009.

[38] H. Wang, M. Dembo, and Y. Wang, "Substrate flexibility regulates growth and apoptosis of normal but not transformed cells," *American Journal of Physiology*, vol. 279, no. 5, pp. C1345–C1350, 2000.

[39] T. Yeung, P. C. Georges, L. A. Flanagan et al., "Effects of substrate stiffness on cell morphology, cytoskeletal structure, and adhesion," *Cell Motility and the Cytoskeleton*, vol. 60, no. 1, pp. 24–34, 2005.

[40] M. P. Sheetz, D. P. Felsenfeld, and C. G. Galbraith, "Cell migration: regulation of force on extracellular-matrix-integrin complexes," *Trends in Cell Biology*, vol. 8, no. 2, pp. 51–54, 1998.

[41] S. R. Peyton and A. J. Putnam, "Extracellular matrix rigidity governs smooth muscle cell motility in a biphasic fashion," *Journal of Cellular Physiology*, vol. 204, no. 1, pp. 198–209, 2005.

[42] A. J. Engler, F. Rehfeldt, S. Sen, and D. E. Discher, "Microtissue elasticity: measurements by atomic force microscopy and its influence on cell differentiation," *Methods in Cell Biology*, vol. 83, pp. 521–545, 2007.

[43] A. J. Engler, M. A. Griffin, S. Sen, C. G. Bönnemann, H. L. Sweeney, and D. E. Discher, "Myotubes differentiate optimally on substrates with tissue-like stiffness: pathological implications for soft or stiff microenvironments," *Journal of Cell Biology*, vol. 166, no. 6, pp. 877–887, 2004.

[44] C. Mauch, A. Hatamochi, K. Scharffetter, and T. Krieg, "Regulation of collagen synthesis in fibroblasts within a three-dimensional collagen gel," *Experimental Cell Research*, vol. 178, no. 2, pp. 493–503, 1988.

[45] C. Mauch, B. Adelmann-Grill, A. Hatamochi, and T. Krieg, "Collagenase gene expression in fibroblasts is regulated by a three-dimensional contact with collagen," *FEBS Letters*, vol. 250, no. 2, pp. 301–305, 1989.

[46] L. A. Johnson, E. S. Rodansky, K. L. Sauder et al., "Matrix stiffness corresponding to strictured bowel induces a fibrogenic response in human colonic fibroblasts," *Inflammatory Bowel Diseases*, vol. 19, pp. 891–903, 2013.

[47] D. Karamichos, N. Lakshman, and W. M. Petroll, "Regulation of corneal fibroblast morphology and collagen reorganization by extracellular matrix mechanical properties," *Investigative Ophthalmology and Visual Science*, vol. 48, no. 11, pp. 5030–5037, 2007.

[48] F. Liu, J. D. Mih, B. S. Shea et al., "Feedback amplification of fibrosis through matrix stiffening and COX-2 suppression," *Journal of Cell Biology*, vol. 190, no. 4, pp. 693–706, 2010.

[49] N. L. Halliday and J. J. Tomasek, "Mechanical properties of the extracellular matrix influence fibronectin fibril assembly in vitro," *Experimental Cell Research*, vol. 217, no. 1, pp. 109–117, 1995.

[50] C. L. Carraher and J. E. Schwarzbauer, "Regulation of matrix assembly through rigidity-dependent fibronectin conformational changes," *The Journal of Biological Chemistry*, 2013.

[51] E. Klotzsch, M. L. Smith, K. E. Kubow et al., "Fibronectin forms the most extensible biological fibers displaying switchable force-exposed cryptic binding sites," *Proceedings of the National Academy of Sciences of the United States of America*, vol. 106, no. 43, pp. 18267–18272, 2009.

[52] D. A. Brenner, T. Kisseleva, D. Scholten et al., "Origin of myofibroblasts in liver fibrosis," *Fibrogenesis & Tissue Repair*, vol. 5, supplement 1, article S17, 2012.

[53] B. Hinz, S. H. Phan, V. J. Thannickal et al., "Recent developments in myofibroblast biology: paradigms for connective tissue remodeling," *American Journal of Pathology*, vol. 180, no. 4, pp. 1340–1355, 2012.

[54] B. Hinz, "Mechanical aspects of lung fibrosis: a spotlight on themyofibroblast," *Proceedings of the American Thoracic Society*, vol. 9, pp. 137–147, 2012.

[55] P. D. Arora, N. Narani, and C. A. G. McCulloch, "The compliance of collagen gels regulates transforming growth factor-β induction of α-smooth muscle actin in fibroblasts," *American Journal of Pathology*, vol. 154, no. 3, pp. 871–882, 1999.

[56] Y. Shi, Y. Dong, Y. Duan, X. Jiang, C. Chen, and L. Deng, "Substrate stiffness influences TGF-ß,1-induced differentiation of bronchial fibroblasts into myofibroblasts in airway remodeling," *Molecular Medicine Reports*, vol. 7, pp. 419–424, 2013.

[57] P. A. Galie, M. V. Westfall, and J. P. Stegemann, "Reduced serum content and increased matrix stiffness promote the cardiac myofibroblast transition in 3D collagen matrices," *Cardiovascular Pathology*, vol. 20, no. 6, pp. 325–333, 2011.

[58] S. L. Friedman, "Hepatic stellate cells: protean, multifunctional, and enigmatic cells of the liver," *Physiological Reviews*, vol. 88, no. 1, pp. 125–172, 2008.

[59] S. L. Friedman, F. J. Roll, J. Boules, D. M. Arenson, and D. M. Bissel, "Maintenance of differentiated phenotype of cultured rat hepatic lipocytes by basement membrane matrix," *Journal of Biological Chemistry*, vol. 264, no. 18, pp. 10756–10762, 1989.

[60] M. D. A. Gaça, X. Zhou, R. Issa, K. Kiriella, J. P. Iredale, and R. C. Benyon, "Basement membrane-like matrix inhibits proliferation and collagen synthesis by activated rat hepatic stellate cells: evidence for matrix-dependent deactivation of stellate cells," *Matrix Biology*, vol. 22, no. 3, pp. 229–239, 2003.

[61] A. L. Olsen, S. A. Bloomer, E. P. Chan et al., "Hepatic stellate cells require a stiff environment for myofibroblastic differentiation," *American Journal of Physiology*, vol. 301, no. 1, pp. G110–G118, 2011.

[62] C. Jones and H. P. Ehrlich, "Fibroblast expression of β-smooth muscle actin, $\beta 2\beta 1$ integrin and $\beta v\beta 3$ integrin: influence of surface rigidity," *Experimental and Molecular Pathology*, vol. 91, no. 1, pp. 394–399, 2011.

[63] M. L. Burgess, L. Terracio, T. Hirozane, and T. K. Borg, "Differential integrin expression by cardiac fibroblasts from hypertensive and exercise-trained rat hearts," *Cardiovascular Pathology*, vol. 11, no. 2, pp. 78–87, 2002.

[64] A. Marinkovic, F. Liu, and D. J. Tschumperlin, "Matrices of physiological stiffnesspotently inactivate idiopathic pulmonary fibrosis fibroblasts," *American Journal of Respiratory Cell and Molecular Biology*, vol. 48, pp. 422–430, 2013.

[65] J. L. Balestrini, S. Chaudhry, V. Sarrazy, A. Koehler, and B. Hinz, "The mechanical memory of lung myofibroblasts," *Integrative Biology*, vol. 4, no. 4, pp. 410–421, 2012.

[66] H. Wang, S. M. Haeger, A. M. Kloxin, L. A. Leinwand, and K. S. Anseth, "Redirecting valvular myofibroblasts into dormant fibroblasts through light-mediated reduction in substrate modulus," *PLoS One*, vol. 7, article e39969, 2012.

[67] G. Garrison, S. K. Huang, K. Okunishi et al., "Reversal of myofibroblast differentiation by prostaglandin e2," *American Journal of Respiratory Cell and Molecular Biology*, vol. 48, pp. 550–558, 2013.

[68] J. W. Stone, P. N. Sisco, E. C. Goldsmith, S. C. Baxter, and C. J. Murphy, "Using gold nanorods to probe cell-induced collagen deformation," *Nano Letters*, vol. 7, no. 1, pp. 116–119, 2007.

[69] C. G. Wilson, J. W. Stone, V. Fowlkes et al., "Age-dependent expression of collagen receptors and deformation of type i collagen substrates by rat cardiac fibroblasts," *Microscopy and Microanalysis*, vol. 17, no. 4, pp. 555–562, 2011.

[70] P. N. Sisco, C. G. Wilson, E. Mironova, S. C. Baxter, C. J. Murphy, and E. C. Goldsmith, "The effect of gold nanorods on cell-mediated collagen remodeling," *Nano Letters*, vol. 8, no. 10, pp. 3409–3412, 2008.

[71] C. G. Wilson, P. N. Sisco, F. A. Gadala-Maria, C. J. Murphy, and E. C. Goldsmith, "Polyelectrolyte-coated gold nanorods and their interactions with type I collagen," *Biomaterials*, vol. 30, no. 29, pp. 5639–5648, 2009.

[72] S. Wang, E. Cukierman, W. D. Swaim, K. M. Yamada, and B. J. Baum, "Extracellular matrix protein-induced changes in human salivary epithelial cell organization and proliferation on a model biological substratum," *Biomaterials*, vol. 20, no. 11, pp. 1043–1049, 1999.

[73] Y. Zhang, C. Zhao, L. Jiang, and L. Dai, "Substrate stiffness regulates apoptosis and the mRNA expression of extracellular matrix regulatory genes in the rat annular cells," *Matrix Biology*, vol. 30, no. 2, pp. 135–144, 2011.

[74] M. V. Turturro, S. Sokic, J. C. Larson, and G. Papavasiliou, "Effective tuning of ligand incorporation and mechanical properties in visible light photopolymerized poly(ethylene glycol) diacrylate hydrogels dictates cell adhesion and proliferation," *Biomedical Materials*, vol. 8, no. 2, article 025001, 2013.

[75] P. F. Lee, Y. Bai, R. L. Smith, K. J. Bayless, and A. T. Yeh, "Angiogenic responses are enhanced in mechanically and microscopically characterized, microbial transglutaminase crosslinked collagen matrices with increased stiffness," *Acta Biomaterialia*, 2013.

[76] R. L. Saunders and D. A. Hammer, "Assembly of human umbilical vein endothelial cells on compliant hydrogels," *Cellular and Molecular Bioengineering*, vol. 3, no. 1, pp. 60–67, 2010.

[77] C. C. Dufort, M. J. Paszek, and V. M. Weaver, "Balancing forces: architectural control of mechanotransduction," *Nature Reviews Molecular Cell Biology*, vol. 12, no. 5, pp. 308–319, 2011.

[78] H. Zhang and M. Labouesse, "Signaling through mechanical inputs: a coordinated process," *Journal of Cell Science*, vol. 125, pp. 3039–3049, 2012.

[79] D. A. MacKenna, F. Dolfi, K. Vuori, and E. Ruoslahti, "Extracellular signal-regulated kinase and c-Jun NH2-terminal kinase activation by mechanical stretch is integrin-dependent and matrix-specific in rat cardiac fibroblasts," *Journal of Clinical Investigation*, vol. 101, no. 2, pp. 301–310, 1998.

[80] C. A. Buck and A. F. Horwitz, "Cell surface receptors for extracellular matrix molecules," *Annual Review of Cell Biology*, vol. 3, pp. 179–205, 1987.

[81] M. J. Humphries, Y. Yasuda, K. Olden, and K. M. Yamada, "The cell interaction sites of fibronectin in tumour metastasis," *Ciba Foundation symposium*, vol. 141, pp. 75–93, 1988.

[82] E. Ruoslahti, "Fibronectin and its receptors," *Annual Review of Biochemistry*, vol. 57, pp. 375–413, 1988.

[83] J. Atance, M. J. Yost, and W. Carver, "Influence of the extracellular matrix on the regulation of cardiac fibroblast behavior by mechanical stretch," *Journal of Cellular Physiology*, vol. 200, no. 3, pp. 377–386, 2004.

[84] P. Roca-Cusachs, T. Iskratsch, and M. P. Sheetz, "Finding the wekest link: exploring integrin-mediated mechanical molecular pathways," *Journal of Cell Science*, vol. 125, pp. 3025–3038, 2012.

[85] R. Zaidel-Bar, S. Itzkovitz, A. Ma'ayan, R. Iyengar, and B. Geiger, "Functional atlas of the integrin adhesome," *Nature Cell Biology*, vol. 9, no. 8, pp. 858–867, 2007.

[86] A. M. Pasapera, I. C. Schneider, E. Rericha, D. D. Schlaepfer, and C. M. Waterman, "Myosin II activity regulates vinculin recruitment to focal adhesions through FAK-mediated paxillin phosphorylation," *Journal of Cell Biology*, vol. 188, no. 6, pp. 877–890, 2010.

[87] D. E. Ingber, "Control of capillary growth and differentiation by extracellular matrix: use of a tensegrity (tensional integrity) mechanism for signal processing," *Chest*, vol. 99, no. 3, pp. 34S–40S, 1991.

[88] D. E. Ingber, "Integrins, tensegrity, and mechanotransduction," *Gravitational and Space Biology Bulletin*, vol. 10, no. 2, pp. 49–55, 1997.

[89] K. Burridge and P. Mangeat, "An interaction between vinculin and talin," *Nature*, vol. 308, no. 5961, pp. 744–746, 1984.

[90] D. R. Critchley, "Biochemical and structural properties of the integrin-associated cytoskeletal protein talin," *Annual Review of Biophysics*, vol. 38, no. 1, pp. 235–254, 2009.

[91] M. D. Schaller, C. A. Borgman, B. S. Cobb, R. R. Vines, A. B. Reynolds, and J. T. Parsons, "pp125(FAK), a structurally distinctive protein-tyrosine kinase associated with focal adhesions," *Proceedings of the National Academy of Sciences of the United States of America*, vol. 89, no. 11, pp. 5192–5196, 1992.

[92] T. M. Weiner, E. T. Liu, R. J. Craven, and W. G. Cance, "Expression of focal adhesion kinase gene and invasive cancer," *The Lancet*, vol. 342, no. 8878, pp. 1024–1025, 1993.

[93] A. J. Pelletier, T. Kunicki, Z. M. Ruggeri, and V. Quaranta, "The activation state of the integrin α(IIb)β3 affects outside-in signals leading to cell spreading and focal adhesion kinase phosphorylation," *Journal of Biological Chemistry*, vol. 270, no. 30, pp. 18133–18140, 1995.

[94] H. Chen, P. A. Appeddu, H. Isoda, and J. Guan, "Phosphorylation of tyrosine 397 in focal adhesion kinase is required for binding phosphatidylinositol 3-kinase," *Journal of Biological Chemistry*, vol. 271, no. 42, pp. 26329–26334, 1996.

[95] K. E. Michael, D. W. Dumbauld, K. L. Burns, S. K. Hanks, and A. J. García, "Focal adhesion kinase modulates cell adhesion strengthening via integrin activation," *Molecular Biology of the Cell*, vol. 20, no. 9, pp. 2508–2519, 2009.

[96] A. Mammoto, T. Mammoto, and D. E. Ingber, "Mechanosensitive mechanisms in transcriptional regulation," *Journal of Cell Science*, vol. 125, pp. 3061–3073, 2012.

[97] D. R. Carter, G. S. Beaupré, N. J. Giori, and J. A. Helms, "Mechanobiology of skeletal regeneration," *Clinical Orthopaedics and Related Research*, no. 355, pp. S41–S55, 1998.

[98] W. Carver, M. L. Nagpal, M. Nachtigal, T. K. Borg, and L. Terracio, "Collagen expression in mechanically stimulated cardiac fibroblasts," *Circulation Research*, vol. 69, no. 1, pp. 116–122, 1991.

[99] W. Zheng, L. P. Christensen, and R. J. Tomanek, "Differential effects of cyclic and static stretch on coronary microvascular endothelial cell receptors and vasculogenic/angiogenic responses," *American Journal of Physiology*, vol. 295, no. 2, pp. H794–H800, 2008.

[100] S. Lehoux, B. Esposito, R. Merval, and A. Tedgui, "Differential regulation of vascular focal adhesion kinase by steady stretch and pulsatility," *Circulation*, vol. 111, no. 5, pp. 643–649, 2005.

[101] S. Dupont, L. Morsut, M. Aragona et al., "Role of YAP/TAZ in mechanotransduction," *Nature*, vol. 474, no. 7350, pp. 179–184, 2011.

[102] G. Halder, S. Dupont, and S. Piccolo, "Tranduction of mechanical and cytoskeletal cues by YAP and TAZ," *Nature Reviews Molecular Cell Biology*, vol. 13, pp. 591–600, 2012.

[103] V. Raghunathan, C. T. McKee, W. Cheung et al., "Influence of extracellular matrix proteins and substratum topography on corneal epithelial alignment and migration," *Tissue Engineering A*, 2013.

[104] S. M. Thomasy, J. A. Wood, P. H. Kass, C. J. Murphy, and P. Russell, "Substratum stiffness and latrunculin B regulate matrix gene and protein expression in human trabecular meshwork cells," *Investigative Ophthalmology & Visual Science*, vol. 53, no. 2, pp. 952–958, 2012.

Prostate Stem Cells in the Development of Benign Prostate Hyperplasia and Prostate Cancer: Emerging Role and Concepts

Akhilesh Prajapati,[1] **Sharad Gupta,**[2] **Bhavesh Mistry,**[1] **and Sarita Gupta**[1]

[1] *Department of Biochemistry, Faculty of Science, The Maharaja Sayajirao University of Baroda, Vadodara, Gujarat 390005, India*
[2] *Ex-assistant Professor karamsad medical college and Gupta Pathological laboratory, Vadodara, Gujarat 390001, India*

Correspondence should be addressed to Sarita Gupta; sglmescrl@gmail.com

Academic Editor: Mauro S. G. Pavao

Benign Prostate hyperplasia (BPH) and prostate cancer (PCa) are the most common prostatic disorders affecting elderly men. Multiple factors including hormonal imbalance, disruption of cell proliferation, apoptosis, chronic inflammation, and aging are thought to be responsible for the pathophysiology of these diseases. Both BPH and PCa are considered to be arisen from aberrant proliferation of prostate stem cells. Recent studies on BPH and PCa have provided significant evidence for the origin of these diseases from stem cells that share characteristics with normal prostate stem cells. Aberrant changes in prostate stem cell regulatory factors may contribute to the development of BPH or PCa. Understanding these regulatory factors may provide insight into the mechanisms that convert quiescent adult prostate cells into proliferating compartments and lead to BPH or carcinoma. Ultimately, the knowledge of the unique prostate stem or stem-like cells in the pathogenesis and development of hyperplasia will facilitate the development of new therapeutic targets for BPH and PCa. In this review, we address recent progress towards understanding the putative role and complexities of stem cells in the development of BPH and PCa.

1. Introduction

Prostate gland is a male accessory reproductive endocrine organ, which expels proteolytic solution in the urethra during ejaculation. In humans, the prostate is located immediately below the base of the bladder surrounding the neck region of the urethra. It is mainly associated with three types of disorders, namely, benign prostate hyperplasia (BPH), prostate cancer (PCa), and prostatitis. BPH and PCa are the most common pathophysiological conditions of prostate gland in elderly men. These diseases already represent significant challenges for health-care systems in most parts of the world. Epidemiologically, BPH is more prevalent in Asian population [1, 2]. Whereas, PCa is more common in the western world [3, 4]. Both the diseases are complex and multifactorial. Factors predisposing to the development of BPH or PCa include hormonal imbalance, oxidative stress, environmental pollutants, inflammation, hereditary, aging, and, more particularly, stromal to epithelial cells crosstalk [5–7]. So far, variety of growth factors and hormonal factors, including androgens and estrogens, has been described in the hyperplastic development of the prostate gland [8–10]. However, the cellular and molecular processes underlying the pathogenesis and development of BPH or PCa are poorly understood.

Stem cells have an extensive capacity to propagate themselves by self-renewal and to differentiate into tissue-specific progeny. It is well know that stem cells are required to maintain and repair tissues throughout the lifetime. The requirement to understand the biology of stem cells derived from the prostate is increasing, as new evidence suggests that BPH and PCa may arise from the stem or stem-like cell compartments [11–13]. This review summarises the biology of prostate stem or stem-like cells and their contribution in pathogenesis and development of BPH and PCa.

2. Prostatic Cellular Compartments

The prostate is a hormonally regulated glandular organ whose growth accelerates at sexual maturity due to androgen action on both stromal and epithelial cells [14, 15]. The human

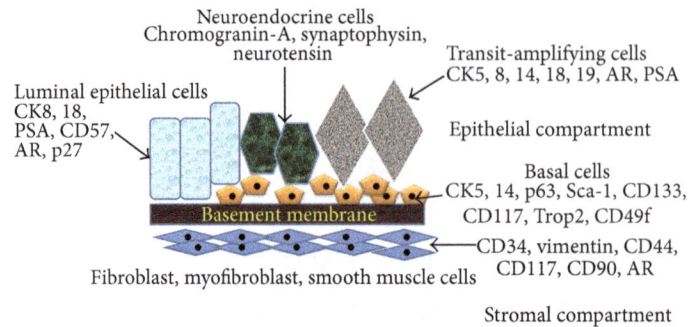

FIGURE 1: Prostatic cellular compartments and stem cell identity markers. Pictorial representation of different prostatic cells and their respective cellular markers.

prostate is a complex ductal-acinar gland that is divided into three anatomically distinct zones: peripheral, transitional, and central zones, which are surrounded by a dense and continuous fibromuscular stroma [16–18]. BPH, a nonmalignant overgrowth found in older men, mainly, develops in the transitional zone, while PCa arises primarily in the peripheral zone [19].

At histological level, human prostate contains mainly two types of cells that are called epithelial and stromal cells. The stromal to epithelial ratio in normal prostate of human is 2 : 1 [18, 20]. The epithelial cell layer is composed of four differentiated cell types known as basal, secretory luminal, neuroendocrine (NE), and transit-amplifying (TA) cells that are identified by their morphology, location, and distinct marker expression (Figure 1). The basal cells form a layer of flattened to cuboidal shaped cells above the basement membrane and express p63 (a homolog of the tumor suppressor gene *p53*), Bcl-2 (an anti-apoptotic factor), Cluster designation (CD) 44, hepatocyte growth factor (HGF), and the high molecular weight cytokeratins (CK) 5 and 14. The expression of androgen receptor (AR) is low or undetectable in the basal cells, which makes the basal cells independent of androgens for their survival [21–23]. The luminal cells are the major cell type of the prostate that form a layer of columnar-shaped cells above the basal layer and constitute the exocrine compartment of the prostate, secreting prostate-specific antigen (PSA) and prostatic acid phosphatase (PAP) into the lumen. They are terminally differentiated, androgen dependent, and nonproliferating cells, expressing low molecular weight CK8 and 18, CD57 and p27^{Kip1} (a cell cycle inhibitor) [22–24] along with high levels of AR. NE cells are rare cells scattered in the basal and luminal layers of the prostate. They are terminally differentiated and androgen-insensitive cells, expressing chromogranin A, synaptophysin, and neuron-specific enolase (NSF) [23, 25, 26]. The NE cells also produce and secrete neuropeptides such as bombesin, calcitonin, and neurotensin that are believed to support epithelial cell growth and differentiation [19, 27, 28]. Additionally, there is a small group of intermediate cells referred to as TA cells that express both basal as well as luminal cell markers (CK5, CK8, CK14, CK18, AR, and PSA) [29–32]. The epithelial layer is surrounded by a stromal layer, which forms a peripheral boundary of the prostate gland. The stromal cell layer consists of several types of cells that include smooth muscle cells (the most abundant

cell type in stroma), fibroblasts, and myofibroblasts. Stromal cells express mesenchymal markers like CD34, vimentin, CD44, CD117, and CD90 [33].

3. Stem Cell in Normal Prostate

Prostatic epithelium is, structurally and functionally, a highly complex tissue composed of multiple differentiated cell types, including basal, luminal, and neuroendocrine cells, along with small population of relatively undifferentiated cells generally known as "stem cells" that are endowed with self-renewal and differentiation capacities [26]. If the stem cells are key target for mutagenic changes and tumourigenesis in human prostate, we need to understand more about stem cell status in normal prostate tissue.

As the adult prostate is relatively slow-growing organ with limited cycles of cell proliferation and apoptosis, the possible existence of adult prostate stem cells (PSCs) was controversial for many years. Several investigations based on stem cell models have elegantly defined role of stem cells in cellular turnover and morphogenesis of normal prostate [30, 34]. Evidence for the existence of the stem cells in normal prostate came from the studies which demonstrated that adult rodent prostate can undergo multiple rounds of castration-induced regression and testosterone-induced regrowth [35–37]. Adult PSCs were believed to reside within the basal cell layer because of the ability of the basal cells to survive and undergo regression and regeneration following repeated castration and androgen replacement [38–40]. Adult mouse prostate epithelial cells, when transplanted along with the urogenital sinus mesenchymal cells under the renal capsule, generated normal murine prostate like structures [41]. Prostate glands were also regenerated when dissociated cells were implanted in Matrigel subcutaneously into immunodeficient mice [42]. Studies, including 5-bromo-2-deoxyuridine (BrdU) retention analysis, showed that the enriched population of BrdU-labelled cells possessing stem cell features (quiescent, high proliferation potential) are localized at the proximal region of mouse prostate duct [43] and are programmed to regenerate proximal-distal ductal axis [44]. The proximal region of the prostatic duct is surrounded by a thick band of smooth muscle cells [45] that are known to produce high level of transforming growth factor-beta (TGF-β) [46], which is known to play a critical role in maintaining the relative dormancy of the PSCs [47]. Independent study by Burger et al. also identified a candidate population of PSCs in the proximal region of mouse prostatic ducts, using stem cell surface marker known as stem cell antigen 1 (Sca-1, also known as Ly6a) [48]. In addition to high expression of Sca-1, these cells were shown to coexpress integrin $\alpha 6$ (CD49f) and Bcl-2. The cells with these properties showed a higher efficiency to generate prostatic tissue in an *in vivo* reconstitution assay [48]. Lawson et al. showed that sorting prostatic cells for CD45(−)CD31(−)Ter119(−)Sca-1(+)CD49f(+) antigenic profile results in a 60-fold enrichment for colony and sphere-forming cells that can self-renew and expand to form spheres for many generations [49]. Leong and colleagues identified CD117 (c-Kit, stem cell factor receptor) as a new marker of a rare adult mouse PSC population that showed all the functional characteristics of stem cells including self-renewal and full differentiation potential. The CD117(+) single stem cell defined by the phenotype Lin(−)Sca-1(+)CD133(+)CD44(+)CD117(+) regenerated functional, secretion-producing prostate after transplantation *in vivo*. Moreover, CD117(+) PSCs showed long-term self renewal capacity after serial isolation and transplantation *in vivo*. CD117 expression was predominantly localized to the proximal region of the mouse prostate and was upregulated after castration-induced prostate involution, consistent with prostate stem cell identity and function [50].

Stem cells in the human prostate have been identified and isolated using the cell surface markers such as integrin $\alpha 2\beta 1$ [51], CD133 (Prominin-1) [52], and CK6a (cytokeratin 6a) [53]. Based on high expression of $\alpha 2\beta 1$ integrin, Collins and colleagues identified PSCs in the basal layer and showed that the $\alpha 2\beta 1^{\text{high}}$ integrin cells represent ~1% of basal cell population in the human prostate [51]. This selected PSC population was enriched through rapid adherence to the type I collagen and showed higher colony-forming efficiency *in vitro*. Furthermore, when the $\alpha 2\beta 1^{\text{high}}$ integrin cells were grafted subcutaneously together with stromal cells in Matrigel into nude mice, they formed prostatic gland structures *in vivo*. Nevertheless, these glandular-like structures, although containing basal cytokeratin positive as well as AR, PAP, and PSA positive cells, lack well-defined basal and luminal organizations [51]. However, recent studies by Missol-Kolka et al. have reported that the overall expression of CD133 in human prostate is not strictly limited to the rare basal stem and progenitor cells, but it is also expressed in some of the secretory luminal cells [54]. Furthermore, it has been shown that CD133 is downregulated in prostate cancer tissues and upregulated in the luminal cells in the vicinity of cancer area. In contrast to the human CD133, the mouse CD133 has been shown to express widely in prostate [54]. Several other surface markers, such as aldehyde dehydrogenase (ALDH), tumor-associated calcium signal transducer 2 (Trop-2), ATP-binding cassette transporter family membrane efflux pump (ABCG2), p63, and CD44, have also been reported for identification and isolation of the PSCs from the prostate tissues of human and mouse [49, 55–60]. Moreover, Trop2(+)CD44(+)CD49f(+) were used as the markers to identify basal stem cells with enhanced prostasphere-forming

and tissue-regenerating abilities [61]. Unlike the murine PSCs, the human PSCs are randomly distributed within the basal epithelial layer throughout the acini and ductal regions of the prostate [51, 52]. In addition to the expression of stem-cell-specific markers, different studies have also shown that PSCs express both basal and luminal cell-specific markers in fetal and adult stages of prostate development [13, 22, 31, 62, 63]. Several studies have proposed the existence of different cell compartments based on stem-cell-driven differentiation hierarchical arrangements within the prostate epithelium [24, 29, 30, 64].

In addition to prostate epithelial stem cells, stromal stem cells (SSCs) have also been reported to exist in the prostate, where they are postulated to carry out function of replacing and regenerating local cells that are lost to normal tissue turnover, injury, or aging [65–67]. These subpopulation of SSCs expressed mesenchymal stem cell (MSC) markers such as CD34 and Sca-1, showed a high proliferative activity and ability to differentiate into fibroblastic, myogenic, adipogenic, and osteogenic lineages [68]. Of all these potential lineages, the most characteristic cell type derived from prostate stromal stem cell is fibroblast or smooth muscle cells [68, 69]. Growth factors that have regulatory effects on SSCs include members of TGF-β superfamily, the insulin-like growth factors, the fibroblast growth factors, the platelet-derived growth factor, and Wnts [70]. It is believed that the differentiation of stromal stem cells to smooth muscle cells is due to paracrine effects of prostrate epithelial cells, which permanently commit the stromal stem cells to mature into androgen receptor (AR) expressing smooth muscle cells [68].

4. Stem Cell in Benign Prostate Hyperplasia (BPH)

BPH is a slow progressive enlargement of the prostate gland which can lead to lower urinary tract symptoms (LUTS) in elderly men. It is characterized by hyperproliferation of epithelial and stromal cells in the transition zone of the prostate gland, which can be observed histopathologically [71]. Despite of its obvious importance as a major health problem, little is known in terms of biological processes that contribute to the development of BPH. To explain the etiology behind the pathogenesis of BPH, several theories, including stem cell, hormonal imbalance, apoptosis, epithelial-mesenchymal transition, embryonic awakening, and inflammation, have been proposed in recent years, and all of them seem to contribute together to some extent in the pathogenesis of BPH [12, 72]. According to stem cell theory, the stem cell population residing in the prostate gland is increased due to abnormal proliferation and apoptosis of stem cells, which may eventually contribute to BPH pathogenesis. Earlier, it was reported by Berry et al. that stem cell population is responsible for prostate gland maintenance [73]. Changes in tissue consistency and cellular hyperplasia are accompanied by downregulation of apoptotic factors and increased level of antiapoptotic factors that decrease the rate of prostatic cell death and, thus, contributing to hyperproliferation of prostatic tissue [74]. It has been reported that stromal to epithelial ratio is altered in BPH, where the ratio increases from 2:1 in normal glands to 5:1 in BPH [75]. Because stromal hyperproliferative activity is thought to promote the development of BPH, the existence of adult stem cells in the prostate stromal compartment is speculated to expand the stroma in response to stimuli during the pathogenesis of BPH [68]. Lin et al. showed that primary culture of prostate cells from BPH patients possessed many common stem cell markers, including CD30, CD44, CD54, neuronspecific enolase (NSE), CD34, vascular endothelial growth factor receptor-1 (Flt-1), and stem cell factor (SCF, also known as KIT ligand or steel factor) [68]. Compared to CD30, CD44, CD54, and NSE, the CD34, Flt-1, and SCF markers were expressed at low level. These stem cells were negative for CD11b, stem cell antigen-1 (SCA-1), SH2, AA4.1, and c-Kit. Furthermore, among this stem cell population only a fraction (5%) of the stem cells was positive for CD133 [68]. Although the origin of these stem cells is not known, the CD49(+)CD54(+)NSE(+)SCF(+) cell marker profile of these cells suggests that they are in a lineage closely related to MSCs. The stem cell population with the above profile possessed ability to differentiate or transdifferentiate into myogenic, adipogenic, and osteogenic lineages [68, 76]. Ceder et al. reported the possible existence of prostate stromal stem/progenitor cells in the adult human prostate [76]. This stromal population expressed vimentin (a mesenchymal marker), CD133, c-Kit, and SCF, with expression profiles similar to those observed in the Cajal cells of gastrointestinal tract, which represent a subset of stem cell-like cells. Several studies have identified c-Kit-expressing interstitial cells in the stromal compartment of human prostate [77–79]. Altered patterns of c-Kit expression have been reported in benign lesions of prostate and breast tissues [80, 81]. It has been shown that the c-Kit expression and number of c-Kit(+) interstitial cells were significantly higher in BPH than those of the normal prostate. Furthermore, it has been suggested that c-Kit regulates cell proliferation in prostate and plays a crucial role in the pathophysiology of BPH via altering the expression of JAK2 and STAT1 [77].

Stem cells from the BPH samples expressing CD49f, CD44, or CD133 markers have been shown to possess monolayer- and spheroid-colony-forming ability, where the highest (98%) recovery of colony-forming cells (CFCs) was achieved by CD49f(+) cells as compared to CD44(+) (17%) or CD133(+) (3%) cells [82]. These CFCs showed the capacity to undergo clonal proliferation, generates branching ductal structures, and they expressed both basal and luminal lineage markers. Further characterization of CD49f(+) cells revealed that they are comprised of two cell types: CK5(+) basal epithelial cells and CD31(+) endothelial cells [82]. Sca-1- and CD34-expressing cells isolated from BPH tissue showed a high proliferative capacity and increased plasticity, as these cells were able to differentiate into fibroblastic, myogenic, adipogenic and osteogenic lineages, similar to that of MSCs [68, 83]. Furthermore, Burger and colleagues found that cells with high Sca-1 expression had considerably more growth potential, and proliferative capabilities than cells expressing low or no Sca-1 antigen [48]. Expression of pluripotency markers such as *Oct4A*, *Sox2*, *c-Myc*, and *Klf4* might represent a stemness-specific gene signature. A very recent study has demonstrated a relatively high expression of stemness-associated genes, including *Oct4A*, *Sox2*, *c-Myc*, *Nanog*, and *Klf4*, in BPH as compared to normal prostate tissue [84]. Thus, several studies have revealed the presence of stem cells that express pluripotency-associated markers and are hyperproliferative and capable of differentiation into different cell lineages within the hyperplastic prostate tissue. The presence of these high proliferative and plastic stem cells in the BPH tissue samples suggests that BPH could occur as a result of changes in the stem cell properties that could ultimately give rise to a clonal expansion of cell populations.

5. Stem Cell in Prostate Cancer (PCa)

PCa is the most prevalent and is the second most frequently diagnosed cancer and sixth leading cause of cancer-related deaths among men in the world [85]. Its etiology, although not clear, is partly attributed to multigenic and epigenetic mechanisms and the heterogeneous nature of this disease [4, 86–88]. Gleason and others described that when the transition of normal gland into adenocarcinoma of prostate takes place, its normal histological structure is disrupted and results in abnormal proliferation of the glandular structure, destruction of basement membrane, and progressive loss of basal cells (<1%) [87, 89]. In addition, AR(+) luminal cells increase and contribute in bulk of prostate mass (>99%) in PCa [90]. It is hypothesised that prostate cancer arises from AR(+) luminal cells and dramatic loss of basal cells. To support this hypothesis several investigations have been conducted [4, 91–93]. In addition, mouse basal population expressing Lin(−)Sca-1(+)CD49f^{high} cells can differentiate into luminal cells in xenograft [49]. Lin(−)Sca-1(+)CD49f^{high} cells from a Pten−/− mouse model display cancer stem cells phenotypes, which gave rise to adenocarcinoma after transplantation [94]. It has been reported that basal cells are the possible cells of prostate cancer origin [95]. When Goldstein et al., especially injected the mixture of urogenital sinus mesenchyme (UGSM) with human prostate basal (expressing CD49f^{high} and Trop2^{high}) or luminal cells (expressing CD49f^{dow} and Trop2^{high}) into the subcutaneous space of immunodeficient NOD(−)SCID(−)IL(−)2Rg−/− mice, only basal cells formed prostatic duct after 16 week, whereas no prostatic duct or adenocarcinoma developed when using luminal cells [91, 95]. Luminal derived grafts lack epithelial structures and mimicked transplantation of UGSM cell alone [95]. Collins et al. reported basal cancer stem cells isolated from human prostate cancer biopsies expressing Cd44(+), α2β1^{high}, and Cd133(+) and cell surface markers were of self renewal *in vitro* [96]. ALDH^{high} is another marker used for cancer stem cells in human prostate cancer cell lines. Cells expressing ALDH^{high}α2(+)/α6(+)/αv(+)-integrin CD44(+) showed increased tumourigenicity and metastasis *in vivo* and enhanced invasiveness *in vitro* [97]. Prostate cancer stem cells isolated from LNCaP and DU145 cell lines also showed expression of CD44(+), α2β1^{high}, and CD133(+) markers [98, 99]. In addition, CD44(+) population isolated from xenograft human tumour and cell lines displayed high tumour initiating ability and metastasis *in vitro* [100]. Recently, Rajasekhar and his group isolated a small cell population expressing TRA-1-60(+)CD151(+)CD166(+) markers that displayed stem cell like features with increased NF-kB signalling along with basal cell markers, and this recapitulates the cellular hierarchy of the tumour origin from basal cells [101].

Over all data from several investigators indicated that origin of prostate cancer can be from basal stem cell population, which expresses CD44(+), $\alpha 2\beta 1^{high}$, CD133(+), ALDH^{high}, and other normal basal stem cell markers.

6. Stem Cell Niche and Plasticity

Stem cells are localized in a defined microenvironment, which is known as their "niche." The main function of a niche probably is to provide specific factors necessary for the maintenance of the stem cell properties via a combination of intracellular and intercellular signalling. These factors include a complex array of growth factors, cytokines, chemokines, and adhesive molecules known to be capable of altering the balance between proliferation, differentiation, and quiescence in stem cell populations [102, 103]. One can probably assume that this is equally true for prostatic stem cells as it is for other stem cell populations.

PSCs reside in niche areas within the basal layer of the epithelial compartment at a low percentage of approximately 0.5–1% [34]. PSCs population in the prostate undergoes a series of phenotype changes. Specifically, the basal SCs do not express the AR or the p63 protein. They have extended proliferative potential by slow cycling. According to these studies, it is postulated that, in addition to the reserve stem-cell population, there is a "TA" cell type, which is characterized by the expression of p63, as well as other basal markers such as CK5 and 14, Jagged-1, and Notch-1 [64, 104, 105]. A TA cell does not express AR protein and it is dependent, for proliferation, but not for survival, on andromedans secreted by stromal cells [105]. Under normal conditions a PSC is slow cycling in that it divides occasionally, undergoing asymmetric division to give rise to a new PSC along with a more differentiated TA daughter cell. TA cell undergoes a limited number of rapidly amplifying cell division cycles to increase the cell population derived from a single PSC before leaving the proliferative compartment to produce intermediate cell [106]. This intermediate cell expresses both epithelial specific (CK5 and 14) and luminal specific (CK8 and 18) cytokines, AR mRNA (but not protein), and prostate stem cell antigen (PSCA) [105, 107]. As an intermediate cell migrates through the basal layer, it differentiates into various terminally differentiated cell lineages of prostate epithelium.

7. Is BPH/PCa a Stem Cells Disease?

Numerous investigators demonstrated presence of stem cell in prostate tissue by using various high-end techniques that may contribute to local invasive to metastatic disease in human and research animals. In normal tissue-development, homeostasis is maintained by differentiation of stem cells and

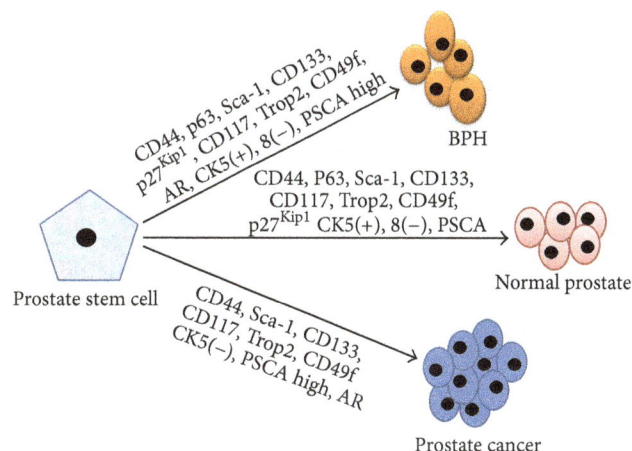

FIGURE 2: Cellular identity of stem cells in prostate. Stem cell model of normal tissue renewal, BPH and PCa.

programmed cells death in regular cell cycle. This mechanism is established through interactions with tissue specific environmental factors such as growth factors and steroid hormones. Many signalling molecules and factors involvement have been reported in stem cell self-renewal and implication in cancer stem cells (CSCs) regulation (Figure 2).

Although the precise role of stem cells in tumourigenesis is still in debate, it is widely accepted that cancers can arise from normal stem cells which may accumulate mutation, genetic changes, and molecular pathway alterations that disrupt self-renewal control capacity (Table 1). It has been reported that, in prostate, putative stem/progenitor cells can reside in CK5(+) 8(−) basal cells. A diagnostic feature of human prostate cancer is the loss of basal cells [108], indicating cancer origin cells as basal cells. In BPH, CD133(+) cells expressed genes related to undifferentiated cells such as TDGF1 (teratocarcinoma-derived growth factor 1) and targets of the Wnt and Hedgehog developmental pathways, whereas CD133(−) cells showed upregulation of genes related to proliferation and metabolism. In cancer, CD133(+) cells specifically displayed more TA population phenotype with increased metabolic activity and proliferation, possibly explaining the transition from a relatively quiescent state to an active growing tumour phenotype. This reflects that CD133 isolates from benign and malignant tissues show biologically distinct characteristics [109]. CSCs exploit many of the signal pathways such as notch, hedgehog- and TGF-β, which play, important role in proliferation and differentiation in prostate stem cell [110, 111]. The sonic hedgehog signalling element receptor PTCH1 and glioma-associated oncogene homolog-1 (GL1) transcription factor were especially reported to be colocalized with p63 basal marker in BPH and PCa cells, expressing CD44/CK8/14. This suggests that hedgehog pathway may induce differentiation of prostate stem/progenitor cells into CD44(+)/P63(+/−) hyperplasia basal cells [112]. Other studies on DNA damage and proliferation markers p27^{Kip1}, cyclin D3, and Ki-67, revealed interesting findings. It has been shown that p27^{Kip1} is significantly upregulated in BPH,

TABLE 1: Molecular alterations in BPH and PCa.

Factors	Normal prostate	BPH	PCa
Prostate-specific factors			
5 α reductase	Normal	Upregulated	Upregulated
Androgen receptor (AR)	Normal	Upregulated	Upregulated
AR coactivator	Normal	Upregulated	Upregulated
Androgen corepressor	Normal	Upregulated	Upregulated
PSA level in serum	(0–4 ng/mL)	(2–8 ng/mL)	(4–10 ng/mL)
Growth factors	FGF-2,7,9 IGF 1,2 IGFBP-2	FGF 1,2,9 IGF-2 high IGFBP-3	FGF-1,2,6,8 IGF-1 high IGFBP-2 high IGFBP-3 high
NE cells	Normal	Number decrease	Number increase
Luminal cell factors	Vimentin Intracellular space normal PMSA normal	Vimentin increase Intracellular space increase PMSA decrease	Vimentin over exp Intracellular space decrease PMSA increase
Basal cells	Present	Present	Absent
Stromal cell factor	Fibroblast content normal NMMHC Elastin SMMHC	Fibroblast content increase NMMHC increase Elastin decrease SMMHC decrease	Fibroblast content increase NMMHC Elastin increase SMMHC decrease
Stem cell markers	CD44, P63, Sca-1, CD133, CD117, Trop2, CD49f, p27^{Kip1}, CK5(+), 8(−), PSCA	CD44, p63, Sca-1, CD133, p27^{Kip1}, CD117, Trop2, CD49f, AR, CK5(+), 8(−), PSCA high	CD44, Sca-1, CD133, CD117, Trop2, CD49f, CK5(−), PSCA high, AR

whereas it is downregulated in PCa. In addition to downregulation of p27^{Kip1}, there is also up regulation of Ki-67 and cyclin D3 in PCa [113].

Several lines of evidence have been indicated that CSCs exhibit both stem cells and cancer cells characteristics. CSCs have the ability to form tumors when transplanted into an animal host. CSCs can be distinguished from other cells within the tumor by cell division and alterations in their gene expression profile [114].

Advanced prostate cancer is androgen independent and basal cells can be phenotypically identified in the majority of metastases [115]. Studies from several investigators revealed that tumor-initiating cells are negative for AR and p63 and expressed the stem cell markers Oct-4, Nanog, Sox-2, Nestin, CD44, CD133, and CD117. Moreover, Sca-1-positive cells having the ability with prostate-regeneration activity, showed evidence of a basal and luminal lineage [96, 100, 116, 117]. Gu et al. demonstrated human telomerase reverse transcriptase-(hTERT-) positive epithelial cells could regenerate tumor in mice that resembled the original tumor in patients [118]. These finding may be indicative of CSC role in prostate cancer.

The growing understanding of the prostate stem cell biology provides the rationale for acute approaches. But without a clear definition of stem cells in normal prostate and BPH/PCa, it is difficult to determine whether the cancer cell of origin in prostate is a stem cell, multipotent progenitor/TA cells, or a more differentiated progeny. Nonetheless, evidence exists that the cellular origin can include both basal and luminal cells.

8. Conclusion

The prostate stem cells are a key role player in prostate tumourigenesis and enlargement disorders. But their precise role in disease pathogenesis remains unknown. The prostate stromal and epithelial compartments and their reciprocal paracrine and autocrine interactions are crucial regulators of prostatic tissue homeostasis. The combination of the prostatic cell surface markers, such as Sca-1, CD133, p63, and CD49f, can aid in the identification of prostate stem cell populations. However, a prostate-specific stem cell marker has yet to be identified. The study of CSCs is still in its early stages. No standard treatments have yet been developed as a result of research on CSCs. The isolation and characterization of epithelial, stromal stem cells and cancer stem cells in the prostate will lead to understanding normal stem cells and CSCs activity to identify new strategies for the control of prostate diseases without harming normal cells milieu.

References

[1] M. L. Gaynor, "Isoflavones and the prevention and treatment of prostate disease: is there a role?" *Cleveland Clinic Journal of Medicine*, vol. 70, no. 3, pp. 203–204, 2003.

[2] L. Denis, M. S. Morton, and K. Griffiths, "Diet and its preventive role in prostatic disease," *European Urology*, vol. 35, no. 5-6, pp. 377–387, 1999.

[3] A. Jemal, R. Siegel, E. Ward, Y. Hao, J. Xu, and M. J. Thun, "Cancer statistics, 2009," *CA Cancer Journal for Clinicians*, vol. 59, no. 4, pp. 225–249, 2009.

[4] M. M. Shen and C. Abate-Shen, "Molecular genetics of prostate cancer: new prospects for old challenges," *Genes and Development*, vol. 24, no. 18, pp. 1967–2000, 2010.

[5] W. W. Barclay, R. D. Woodruff, M. C. Hall, and S. D. Cramer, "A system for studying epithelial-stromal interactions reveals distinct inductive abilities of stromal cells from benign prostatic hyperplasia and prostate cancer," *Endocrinology*, vol. 146, no. 1, pp. 13–18, 2005.

[6] S. M. Harman, E. J. Metter, J. D. Tobin, J. Pearson, and M. R. Blackman, "Longitudinal effects of aging on serum total and free testosterone levels in healthy men," *Journal of Clinical Endocrinology and Metabolism*, vol. 86, no. 2, pp. 724–731, 2001.

[7] J. C. Nickel, J. Downey, I. Young, and S. Boag, "Asymptomatic inflammation and/or infection in benign prostatic hyperplasia," *BJU International*, vol. 84, no. 9, pp. 976–981, 1999.

[8] J. D. McConnell, "The pathophysiology of benign prostatic hyperplasia," *Journal of Andrology*, vol. 12, no. 6, pp. 356–363, 1991.

[9] M. Mimeault and S. K. Batra, "Recent advances on multiple tumorigenic cascades involved in prostatic cancer progression and targeting therapies," *Carcinogenesis*, vol. 27, no. 1, pp. 1–22, 2006.

[10] M. Marcelli and G. R. Cunningham, "Hormonal signaling in prostatic hyperplasia and neoplasia," *Journal of Clinical Endocrinology and Metabolism*, vol. 84, no. 10, pp. 3463–3468, 1999.

[11] A. Y. Nikitin, A. Matoso, and P. Roy-Burman, "Prostate stem cells and cancer," *Histology and histopathology*, vol. 22, no. 9, pp. 1043–1049, 2007.

[12] M. Notara and A. Ahmed, "Benign prostate hyperplasia and stem cells: a new therapeutic opportunity," *Cell Biology and Toxicology*, vol. 28, no. 6, pp. 435–442, 2012.

[13] G. L. Powers and P. C. Marker, "Recent advances in prostate development and links to prostatic diseases," *Wiley Interdisciplinary Reviews*, vol. 5, no. 2, pp. 243–256, 2013.

[14] G. S. Prins and O. Putz, "Molecular signaling pathways that regulate prostate gland development," *Differentiation*, vol. 76, no. 6, pp. 641–659, 2008.

[15] Y. Sugimura, G. R. Cunha, and A. A. Donjacour, "Morphogenesis of ductal networks in the mouse prostate," *Biology of Reproduction*, vol. 34, no. 5, pp. 961–971, 1986.

[16] J. E. McNeal, "Anatomy of the prostate and morphogenesis of BPH," *Progress in Clinical and Biological Research*, vol. 145, pp. 27–53, 1984.

[17] J. E. McNeal and D. G. Bostwick, "Anatomy of the prostatic urethra," *Journal of the American Medical Association*, vol. 251, no. 7, pp. 890–891, 1984.

[18] B. G. Timms, "Prostate development: a historical perspective," *Differentiation*, vol. 76, no. 6, pp. 565–577, 2008.

[19] C. Abate-Shen and M. M. Shen, "Molecular genetics of prostate cancer," *Genes and Development*, vol. 14, no. 19, pp. 2410–2434, 2000.

[20] G. Bartsch and H. P. Rohr, "Comparative light and electron microscopic study of the human, dog and rat prostate: an approach to an experimental model for human benign prostatic hyperplasia (light and electron microscopic analysis): a review," *Urologia Internationalis*, vol. 35, no. 2, pp. 91–104, 1980.

[21] H. Bonkhoff and K. Remberger, "Widespread distribution of nuclear androgen receptors in the basal cell layer of the normal and hyperplastic human prostate," *Virchows Archiv*, vol. 422, no. 1, pp. 35–38, 1993.

[22] Y. Wang, S. W. Hayward, M. Cao, K. A. Thayer, and G. R. Cunha, "Cell differentiation lineage in the prostate," *Differentiation*, vol. 68, no. 4-5, pp. 270–279, 2001.

[23] R. M. Long, C. Morrissey, J. M. Fitzpatrick, and R. W. G. Watson, "Prostate epithelial cell differentiation and its relevance to the understanding of prostate cancer therapies," *Clinical Science*, vol. 108, no. 1, pp. 1–11, 2005.

[24] A. M. De Marzo, A. K. Meeker, J. I. Epstein, and D. S. Coffey, "Prostate stem cell compartments: expression of the cell cycle inhibitor p27(Kip1) in normal, hyperplastic, and neoplastic cells," *The American Journal of Pathology*, vol. 153, no. 3, pp. 911–919, 1998.

[25] H. Bonkhoff, U. Stein, and K. Remberger, "Endocrine-paracrine cell types in the prostate and prostatic adenocarcinoma are postmitotic cells," *Human Pathology*, vol. 26, no. 2, pp. 167–170, 1995.

[26] J. A. Schalken and G. van Leenders, "Cellular and molecular biology of the prostate: stem cell biology," *Urology*, vol. 62, supplement 1, no. 5, pp. 11–20, 2003.

[27] G. P. Amorino and S. J. Parsons, "Neuroendocrine cells in prostate cancer," *Critical Reviews in Eukaryotic Gene Expression*, vol. 14, no. 4, pp. 287–300, 2004.

[28] P. A. Abrahamsson, "Neuroendocrine differentiation in prostatic carcinoma," *Prostate*, vol. 39, no. 2, pp. 135–148, 1999.

[29] H. Bonkhoff, U. Stein, and K. Remberger, "Multidirectional differentiation in the normal, hyperplastic, and neoplastic human prostate: simultaneous demonstration of cell-specific epithelial markers," *Human Pathology*, vol. 25, no. 1, pp. 42–46, 1994.

[30] H. Bonkhoff and K. Remberger, "Differentiation pathways and histogenetic aspects of normal and abnormal prostatic growth: a stem cell model," *Prostate*, vol. 28, no. 2, pp. 98–106, 1996.

[31] Y. Xue, F. Smedts, F. M. Debruyne, J. J. de la Rosette, and J. A. Schalken, "Identification of intermediate cell types by keratin expression in the developing human prostate," *Prostate*, vol. 34, no. 4, pp. 292–301, 1998.

[32] D. L. Hudson, M. O'Hare, F. M. Watt, and J. R. W. Masters, "Proliferative heterogeneity in the human prostate: evidence for epithelial stem cells," *Laboratory Investigation*, vol. 80, no. 8, pp. 1243–1250, 2000.

[33] T. Takao and A. Tsujimura, "Prostate stem cells: the niche and cell markers," *International Journal of Urology*, vol. 15, no. 4, pp. 289–294, 2008.

[34] J. T. Isaacs and D. S. Coffey, "Etiology and disease process of benign prostatic hyperplasia," *Prostate*, vol. 2, pp. 33–50, 1989.

[35] H. F. English, R. J. Santen, and J. T. Isaacs, "Response of glandular versus basal rat ventral prostatic epithelial cells to androgen withdrawal and replacement," *Prostate*, vol. 11, no. 3, pp. 229–242, 1987.

[36] G. S. Evans and J. A. Chandler, "Cell proliferation studies in the rat prostate: II. The effects of castration and androgen-induced regeneration upon basal and secretory cell proliferation," *Prostate*, vol. 11, no. 4, pp. 339–351, 1987.

[37] A. P. M. Verhagen, T. W. Aalders, F. C. S. Ramaekers, F. M. J. Debruyne, and J. A. Schalken, "Differential expression of keratins in the basal and luminal compartments of rat prostatic epithelium during degeneration and regeneration," *Prostate*, vol. 13, no. 1, pp. 25–38, 1988.

[38] D. P. DeKlerk and D. S. Coffey, "Quantitative determination of prostatic epithelial and stromal hyperplasia by a new technique. Biomorphometrics," *Investigative Urology*, vol. 16, no. 3, pp. 240–245, 1978.

[39] N. Kyprianou and J. T. Isaacs, "Identification of a cellular receptor for transforming growth factor-β in rat ventral prostate and its negative regulation by androgens," *Endocrinology*, vol. 123, no. 4, pp. 2124–2131, 1988.

[40] M. Montpetit, P. Abrahams, A. F. Clark, and M. Tenniswood, "Androgen-independent epithelial cells of the rat ventral prostate," *Prostate*, vol. 12, no. 1, pp. 13–28, 1988.

[41] L. Xin, H. Ide, Y. Kim, P. Dubey, and O. N. Witte, "In vivo regeneration of murine prostate from dissociated cell populations of postnatal epithelia and urogenital sinus mesenchyme," *Proceedings of the National Academy of Sciences of the United States of America*, vol. 100, supplement 1, pp. 11896–11903, 2003.

[42] M. Azuma, A. Hirao, K. Takubo, I. Hamaguchi, T. Kitamura, and T. Suda, "A quantitative matrigel assay for assessing repopulating capacity of prostate stem cells," *Biochemical and Biophysical Research Communications*, vol. 338, no. 2, pp. 1164–1170, 2005.

[43] A. Tsujimura, Y. Koikawa, S. Salm et al., "Proximal location of mouse prostate epithelial stem cells: a model of prostatic homeostasis," *Journal of Cell Biology*, vol. 157, no. 7, pp. 1257–1265, 2002.

[44] K. Goto, S. N. Salm, S. Coetzee et al., "Proximal prostatic stem cells are programmed to regenerate a proximal-distal ductal axis," *Stem Cells*, vol. 24, no. 8, pp. 1859–1868, 2006.

[45] J. A. Nemeth and C. Lee, "Prostatic ductal system in rats: regional variation in stromal organization," *Prostate*, vol. 28, no. 2, pp. 124–128, 1996.

[46] J. A. Nemeth, J. A. Sensibar, R. R. White, D. J. Zelner, I. Y. Kim, and C. Lee, "Prostatic ductal system in rats: tissue-specific expression and regional variation in stromal distribution of transforming growth factor-beta 1," *Prostate*, vol. 33, no. 1, pp. 64–71, 1997.

[47] S. N. Salm, P. E. Burger, S. Coetzee, K. Goto, D. Moscatelli, and E. L. Wilson, "TGF-β maintains dormancy of prostatic stem cells in the proximal region of ducts," *Journal of Cell Biology*, vol. 170, no. 1, pp. 81–90, 2005.

[48] P. E. Burger, X. Xiong, S. Coetzee et al., "Sca-1 expression identifies stem cells in the proximal region of prostatic ducts with high capacity to reconstitute prostatic tissue," *Proceedings of the National Academy of Sciences of the United States of America*, vol. 102, no. 20, pp. 7180–7185, 2005.

[49] D. A. Lawson, L. Xin, R. U. Lukacs, D. Cheng, and O. N. Witte, "Isolation and functional characterization of murine prostate stem cells," *Proceedings of the National Academy of Sciences of the United States of America*, vol. 104, no. 1, pp. 181–186, 2007.

[50] K. G. Leong, B.-E. Wang, L. Johnson, and W.-Q. Gao, "Generation of a prostate from a single adult stem cell," *Nature*, vol. 456, no. 7223, pp. 804–810, 2008.

[51] A. T. Collins, F. K. Habib, N. J. Maitland, and D. E. Neal, "Identification and isolation of human prostate epithelial stem cells based on $\alpha2\beta1$-integrin expression," *Journal of Cell Science*, vol. 114, no. 21, pp. 3865–3872, 2001.

[52] G. D. Richardson, C. N. Robson, S. H. Lang, D. E. Neal, N. J. Maitland, and A. T. Collins, "CD133, a novel marker for human prostatic epithelial stem cells," *Journal of Cell Science*, vol. 117, no. 16, pp. 3539–3545, 2004.

[53] M. Schmelz, R. Moll, U. Hesse et al., "Identification of a stem cell candidate in the normal human prostate gland," *European Journal of Cell Biology*, vol. 84, no. 2-3, pp. 341–354, 2005.

[54] E. Missol-Kolka, J. Karbanová, P. Janich et al., "Prominin-1 (CD133) is not restricted to stem cells located in the basal compartment of murine and human prostate," *Prostate*, vol. 71, no. 3, pp. 254–267, 2011.

[55] R. I. Bhatt, M. D. Brown, C. A. Hart et al., "Novel method for the isolation and characterisation of the putative prostatic stem cell," *Cytometry A*, vol. 54, no. 2, pp. 89–99, 2003.

[56] P. E. Burger, R. Gupta, X. Xiong et al., "High aldehyde dehydrogenase activity: a novel functional marker of murine prostate stem/progenitor cells," *Stem Cells*, vol. 27, no. 9, pp. 2220–2228, 2009.

[57] A. S. Goldstein, D. A. Lawson, D. Cheng, W. Sun, I. P. Garraway, and O. N. Witte, "Trop2 identifies a subpopulation of murine and human prostate basal cells with stem cell characteristics," *Proceedings of the National Academy of Sciences of the United States of America*, vol. 105, no. 52, pp. 20882–20887, 2008.

[58] A. Y. Liu, L. D. True, L. Latray et al., "Cell-cell interaction in prostate gene regulation and cytodifferentiation," *Proceedings of the National Academy of Sciences of the United States of America*, vol. 94, no. 20, pp. 10705–10710, 1997.

[59] J. C. Pignon, C. Grisanzio, Y. Geng, J. Song, R. A. Shivdasani, and S. Signoretti, "p63-expressing cells are the stem cells of developing prostate, bladder, and colorectal epithelia," *Proceedings of the National Academy of Sciences of the United States of America*, vol. 110, no. 20, pp. 8105–8110, 2013.

[60] M. Yao, R. A. Taylor, M. G. Richards et al., "Prostate-regenerating capacity of cultured human adult prostate epithelial cells," *Cells Tissues Organs*, vol. 191, no. 3, pp. 203–212, 2010.

[61] I. P. Garraway, W. Sun, C. P. Tran et al., "Human prostate sphere-forming cells represent a subset of basal epithelial cells capable of glandular regeneration in vivo," *Prostate*, vol. 70, no. 5, pp. 491–501, 2010.

[62] G. van Leenders, H. Dijkman, C. Hulsbergen-van de Kaa, D. Ruiter, and J. Schalken, "Demonstration of intermediate cells during human prostate epithelial differentiation in situ and in vitro using triple-staining confocal scanning microscopy," *Laboratory Investigation*, vol. 80, no. 8, pp. 1251–1258, 2000.

[63] W. W. Barclay, L. S. Axanova, W. Chen et al., "Characterization of adult prostatic progenitor/stem cells exhibiting self-renewal and multilineage differentiation," *Stem Cells*, vol. 26, no. 3, pp. 600–610, 2008.

[64] R. A. Taylor, R. Toivanen, and G. P. Risbridger, "Stem cells in prostate cancer: treating the root of the problem," *Endocrine-Related Cancer*, vol. 17, no. 4, pp. R273–R285, 2010.

[65] C. G. Roehrborn, J. D. McConnell, M. Lieber et al., "Serum prostate-specific antigen concentration is a powerful predictor of acute urinary retention and need for surgery in men with clinical benign prostatic hyperplasia," *Urology*, vol. 53, no. 3, pp. 473–480, 1999.

[66] M. J. Naslund and M. Miner, "A review of the clinical efficacy and safety of 5α-reductase inhibitors for the enlarged prostate," *Clinical Therapeutics*, vol. 29, no. 1, pp. 17–25, 2007.

[67] P. Boyle, C. Roehrborn, R. Harkaway, J. Logie, J. de La Rosette, and M. Emberton, "5-alpha reductase inhibition provides superior benefits to alpha blockade by preventing AUR and BPH-related surgery," *European Urology*, vol. 45, no. 5, pp. 620–627, 2004.

[68] V. K. Lin, S.-Y. Wang, D. V. Vazquez, C. C. Xu, S. Zhang, and L. Tang, "Prostatic stromal cells derived from benign prostatic hyperplasia specimens possess stem cell like property," *Prostate*, vol. 67, no. 12, pp. 1265–1276, 2007.

[69] G. R. Cunha, S. W. Hayward, R. Dahiya, and B. A. Foster, "Smooth muscle-epithelial interactions in normal and neoplastic prostatic development," *Acta Anatomica*, vol. 155, no. 1, pp. 63–72, 1996.

[70] C. Richard, G. Kim, Y. Koikawa et al., "Androgens modulate the balance between VEGF and angiopoietin expression in prostate epithelial and smooth muscle cells," *Prostate*, vol. 50, no. 2, pp. 83–91, 2002.

[71] G. A. Schuster and T. G. Schuster, "The relative amount of epithelium, muscle, connective tissue and lumen in prostatic hyperplasia as a function of the mass of tissue resected," *Journal of Urology*, vol. 161, no. 4, pp. 1168–1173, 1999.

[72] J. Tang and J. Yang, "Etiopathogenesis of benign prostatic hyprplasia," *Indian Journal of Urology*, vol. 25, no. 3, pp. 312–317, 2009.

[73] S. J. Berry, D. S. Coffey, J. D. Strandberg, and L. L. Ewing, "Effect of age, castration, and testosterone replacement on the development and restoration of canine benign prostatic hyperplasia," *Prostate*, vol. 9, no. 3, pp. 295–302, 1986.

[74] N. Kyprianou, H. Tu, and S. C. Jacobs, "Apoptotic versus proliferative activities in human benign prostatic hyperplasia," *Human Pathology*, vol. 27, no. 7, pp. 668–675, 1996.

[75] E. Shapiro, M. J. Becich, V. Hartanto, and H. Lepor, "The relative proportion of stromal and epithelial hyperplasia is related to the development of symptomatic benign prostate hyperplasia," *Journal of Urology*, vol. 147, no. 5, pp. 1293–1297, 1992.

[76] J. A. Ceder, L. Jansson, R. A. Ehrnström, L. Rönnstrand, and P.-A. Abrahamsson, "The characterization of epithelial and stromal subsets of candidate stem/progenitor cells in the human adult prostate," *European Urology*, vol. 53, no. 3, pp. 524–532, 2008.

[77] M. Imura, Y. Kojima, Y. Kubota et al., "Regulation of cell proliferation through a KIT-mediated mechanism in benign prostatic hyperplasia," *Prostate*, vol. 72, no. 14, pp. 1506–1513, 2012.

[78] A. Lammie, M. Drobnjak, W. Gerald, A. Saad, R. Cote, and C. Cordon-Cardo, "Expression of c-kit and kit ligand proteins in normal human tissues," *Journal of Histochemistry and Cytochemistry*, vol. 42, no. 11, pp. 1417–1425, 1994.

[79] A. Shafik, I. Shafik, and O. El-Sibai, "Identification of c-kit-positive cells in the human prostate: the interstitial cells of Cajal," *Archives of Andrology*, vol. 51, no. 5, pp. 345–351, 2005.

[80] A. Kondi-Pafiti, N. Arkadopoulos, C. Gennatas, V. Michalaki, M. Frangou-Plegmenou, and P. Chatzipantelis, "Expression of c-kit in common benign and malignant breast lesions," *Tumori*, vol. 96, no. 6, pp. 978–984, 2010.

[81] R. Simak, P. Capodieci, D. W. Cohen et al., "Expression of c-kit and kit-ligand in benign and malignant prostatic tissues," *Histology and Histopathology*, vol. 15, no. 2, pp. 365–374, 2000.

[82] H. Yamamoto, J. R. Masters, P. Dasgupta et al., "CD49f is an efficient marker of monolayer- and spheroid colony-forming cells of the benign and malignant human prostate," *PLoS ONE*, vol. 7, no. 10, Article ID e46979, 2012.

[83] M. F. Pittenger, A. M. Mackay, S. C. Beck et al., "Multilineage potential of adult human mesenchymal stem cells," *Science*, vol. 284, no. 5411, pp. 143–147, 1999.

[84] C. Le Magnen, L. Bubendorf, C. Ruiz et al., "Klf4 transcription factor is expressed in the cytoplasm of prostate cancer cells," *European Journal of Cancer*, vol. 49, no. 4, pp. 955–963, 2013.

[85] A. Jemal, F. Bray, M. M. Center, J. Ferlay, E. Ward, and D. Forman, "Global cancer statistics," *CA Cancer Journal for Clinicians*, vol. 61, no. 2, pp. 69–90, 2011.

[86] N. J. Maitland and A. Collins, "A tumour stem cell hypothesis for the origins of prostate cancer," *BJU International*, vol. 96, no. 9, pp. 1219–1223, 2005.

[87] N. J. Maitland, F. M. Frame, E. S. Polson, J. L. Lewis, and A. T. Collins, "Prostate cancer stem cells: do they have a basal or luminal phenotype?" *Hormones and Cancer*, vol. 2, no. 1, pp. 47–61, 2011.

[88] E. E. Oldridge, D. Pellacani, A. T. Collins, and N. J. Maitland, "Prostate cancer stem cells: are they androgen-responsive?" *Molecular and Cellular Endocrinology*, vol. 360, no. 1-2, pp. 14–24, 2011.

[89] D. F. Gleason, "Classification of prostatic carcinomas," *Cancer Chemotherapy Reports*, vol. 50, no. 3, pp. 125–128, 1966.

[90] C. Grisanzio and S. Signoretti, "p63 in prostate biology and pathology," *Journal of Cellular Biochemistry*, vol. 103, no. 5, pp. 1354–1368, 2008.

[91] Z. A. Wang and M. M. Shen, "Revisiting the concept of cancer stem cells in prostate cancer," *Oncogene*, vol. 30, no. 11, pp. 1261–1271, 2011.

[92] H. Korsten, A. Ziel-van der Made, X. Ma, T. van der Kwast, and J. Trapman, "Accumulating progenitor cells in the luminal epithelial cell layer are candidate tumor initiating cells in a Pten knockout mouse prostate cancer model," *PLoS ONE*, vol. 4, no. 5, Article ID e5662, 2009.

[93] X. Ma, A. C. Ziel-van der Made, B. Autar et al., "Targeted biallelic inactivation of Pten in the mouse prostate leads to prostate cancer accompanied by increased epithelial cell proliferation but not by reduced apoptosis," *Cancer Research*, vol. 65, no. 13, pp. 5730–5739, 2005.

[94] D. J. Mulholland, L. Xin, A. Morim, D. Lawson, O. Witte, and H. Wu, "Lin-Sca-1+CD49fhigh stem/progenitors are tumor-initiating cells in the Pten-null prostate cancer model," *Cancer Research*, vol. 69, no. 22, pp. 8555–8562, 2009.

[95] A. S. Goldstein, J. Huang, C. Guo, I. P. Garraway, and O. N. Witte, "Identification of a cell of origin for human prostate cancer," *Science*, vol. 329, no. 5991, pp. 568–571, 2010.

[96] A. T. Collins, P. A. Berry, C. Hyde, M. J. Stower, and N. J. Maitland, "Prospective identification of tumorigenic prostate cancer stem cells," *Cancer Research*, vol. 65, no. 23, pp. 10946–10951, 2005.

[97] C. van den Hoogen, G. van der Horst, H. Cheung et al., "High aldehyde dehydrogenase activity identifies tumor-initiating and metastasis-initiating cells in human prostate cancer," *Cancer Research*, vol. 70, no. 12, pp. 5163–5173, 2010.

[98] E. M. Hurt, B. T. Kawasaki, G. J. Klarmann, S. B. Thomas, and W. L. Farrar, "CD44+CD24- prostate cells are early cancer progenitor/stem cells that provide a model for patients with poor prognosis," *British Journal of Cancer*, vol. 98, no. 4, pp. 756–765, 2008.

[99] C. Wei, W. Guomin, L. Yujun, and Q. Ruizhe, "Cancer stem-like cells in human prostate carcinoma cells DU145: the seeds of the cell line?" *Cancer Biology and Therapy*, vol. 6, no. 5, pp. 763–768, 2007.

[100] L. Patrawala, T. Calhoun, R. Schneider-Broussard et al., "Highly purified CD44+ prostate cancer cells from xenograft human tumors are enriched in tumorigenic and metastatic progenitor cells," *Oncogene*, vol. 25, no. 12, pp. 1696–1708, 2006.

[101] V. K. Rajasekhar, L. Studer, W. Gerald, N. D. Socci, and H. I. Scher, "Tumour-initiating stem-like cells in human prostate cancer exhibit increased NF-κB signalling," *Nature Communications*, vol. 2, article 162, 2011.

[102] A. D. Whetton and G. J. Graham, "Homing and mobilization in the stem cell niche," *Trends in Cell Biology*, vol. 9, no. 6, pp. 233–238, 1999.

[103] A. Spradling, D. Drummond-Barbosa, and T. Kai, "Stem cells find their niche," *Nature*, vol. 414, no. 6859, pp. 98–104, 2001.

[104] I. V. Litvinov, D. J. Vander Griend, Y. Xu, L. Antony, S. L. Dalrymple, and J. T. Isaacs, "Low-calcium serum-free defined medium selects for growth of normal prostatic epithelial stem cells," *Cancer Research*, vol. 66, no. 17, pp. 8598–8607, 2006.

[105] J. T. Isaacs, "Prostate stem cells and benign prostatic hyperplasia," *Prostate*, vol. 68, no. 9, pp. 1025–1034, 2008.

[106] D. L. Hudson, "Epithelial stem cells in human prostate growth and disease," *Prostate Cancer and Prostatic Diseases*, vol. 7, no. 3, pp. 188–194, 2004.

[107] C. P. Tran, C. Lin, J. Yamashiro, and R. E. Reiter, "Prostate stem cell antigen is a marker of late intermediate prostate epithelial cells," *Molecular Cancer Research*, vol. 1, no. 2, pp. 113–121, 2002.

[108] P. A. Humphrey, "Diagnosis of adenocarcinoma in prostate needle biopsy tissue," *Journal of Clinical Pathology*, vol. 60, no. 1, pp. 35–42, 2007.

[109] C. J. Shepherd, S. Rizzo, I. Ledaki et al., "Expression profiling of CD133+ and CD133- epithelial cells from human prostate," *Prostate*, vol. 68, no. 9, pp. 1007–1024, 2008.

[110] A. T. Collins and N. J. Maitland, "Prostate cancer stem cells," *European Journal of Cancer*, vol. 42, no. 9, pp. 1213–1218, 2006.

[111] R. Blum, R. Gupta, P. E. Burger et al., "Molecular signatures of prostate stem cells reveal novel signaling pathways and provide insights into prostate cancer," *PLoS ONE*, vol. 4, no. 5, Article ID e5722, 2009.

[112] B.-Y. Chen, J.-Y. Liu, H.-H. Chang et al., "Hedgehog is involved in prostate basal cell hyperplasia formation and its progressing towards tumorigenesis," *Biochemical and Biophysical Research Communications*, vol. 357, no. 4, pp. 1084–1089, 2007.

[113] D. Nikoleishvili, A. Pertia, O. Trsintsadze, N. Gogokhia, L. Managadze, and A. Chkhotua, "Expression of p27(Kip1), cyclin D3 and Ki67 in BPH, prostate cancer and hormone-treated prostate cancer cells," *International Urology and Nephrology*, vol. 40, no. 4, pp. 953–959, 2008.

[114] J. M. Rosen and C. T. Jordan, "The increasing complexity of the cancer stem cell paradigm," *Science*, vol. 324, no. 5935, pp. 1670–1673, 2009.

[115] A. Y. Liu, P. S. Nelson, G. D. van Engh, and L. Hood, "Human prostate epithelial cell-type cDNA libraries and prostate expression patterns," *Prostate*, vol. 50, no. 2, pp. 92–103, 2002.

[116] G. J. L. H. van Leenders and J. A. Schalken, "Stem cell differentiation within the human prostate epithelium: implications for prostate carcinogenesis," *BJU International*, vol. 88, Supplement, no. 2, pp. 35–42, 2001.

[117] N. Craft, C. Chhor, C. Tran et al., "Evidence for clonal outgrowth of androgen-independent prostate cancer cells from androgen-dependent tumors through a two-step process," *Cancer Research*, vol. 59, no. 19, pp. 5030–5036, 1999.

[118] A. Gu, J. Yuan, M. Wills, and S. Kasper, "Prostate cancer cells with stem cell characteristics reconstitute the original human tumor in vivo," *Cancer Research*, vol. 67, no. 10, pp. 4807–4815, 2007.

3

MicroRNAs in Kidney Fibrosis and Diabetic Nephropathy: Roles on EMT and EndMT

Swayam Prakash Srivastava, Daisuke Koya, and Keizo Kanasaki

Department of Diabetology & Endocrinology, Kanazawa Medical University, Uchinada, Ishikawa 920-0293, Japan

Correspondence should be addressed to Keizo Kanasaki; kkanasak@kanazawa-med.ac.jp

Academic Editor: Achilleas D. Theocharis

MicroRNAs (miRNAs) are a family of small, noncoding RNAs that regulate gene expression in diverse biological and pathological processes, including cell proliferation, differentiation, apoptosis, and carcinogenesis. As a result, miRNAs emerged as major area of biomedical research with relevance to kidney fibrosis. Fibrosis is characterized by the excess deposition of extracellular matrix (ECM) components, which is the end result of an imbalance of metabolism of the ECM molecule. Recent evidence suggests that miRNAs participate in the fibrotic process in a number of organs including the heart, kidney, liver, and lung. Epithelial mesenchymal transition (EMT) and endothelial mesenchymal transition (EndMT) programs play vital roles in the development of fibrosis in the kidney. A growing number of the extracellular and intracellular molecules that control EMT and EndMT have been identified and could be exploited in developing therapeutics for fibrosis. This review highlights recent advances on the role of miRNAs in the kidney diseases; diabetic nephropathy especially focused on EMT and EndMT program responsible for the development of kidney fibrosis. These miRNAs can be utilized as a potential novel drug target for the studying of underlying mechanism and treatment of kidney fibrosis.

1. Introduction

MicroRNAs (miRNAs) are short noncoding RNAs that modulate fundamental cellular processes such as differentiation, proliferation, death, metabolism, and pathophysiology of many diseases by inhibiting target gene expression via inhibition of protein translation or by inducing mRNA degradation. By recent estimates, nearly 1000 human miRNAs target and downregulate at least 60% of human protein coding genes expressed in the genome [1]. The understandings of miRNAs in molecular mechanisms on various disease processes are now expanding day by day. In the current scenario, miRNAs play the role of conductors in the pathogenesis of fibrosis diseases. There are many literatures that organ-specific miRNAs alterations cause fibrotic disorders [2]. Fibrosis is the leading cause of organ dysfunction in diseases, either as outcome of an uncontrolled reaction to chronic tissue injury or as the primary disease itself in predisposed individuals [3]. Fibrosis of the kidney is caused by prolonged injury and dysregulation of normal wound healing process in association with an excess deposition of extracellular matrix. In such fibrotic process, kidney fibroblasts play important roles but the origin of fibroblasts remains elusive. In addition to the activation of residential fibroblasts, other important sources of fibroblasts have been proposed such as pericytes, fibrocytes, and fibroblasts originated from epithelial mesenchymal transition, endothelial mesenchymal transition. The two main loci for fibrosis in the kidney are the tubulointerstitial space and the glomerulus. Recent studies using transgenic mice have demonstrated that primary changes in glomeruli can lead to progressive glomerulosclerosis and renal failure [4]. For these reasons and knowing the multitude of pathways that miRNAs can affect, it is envisaged that investigating the roles of miRNAs in fibrosis could not only advance our understanding of the pathogenesis of this common condition but might also provide new targets for therapeutic intervention. In this review we focused on roles of miRNA biology in the kidney disease especially in epithelial mesenchymal transition (EMT) and endothelial mesenchymal transition (EndMT) programs.

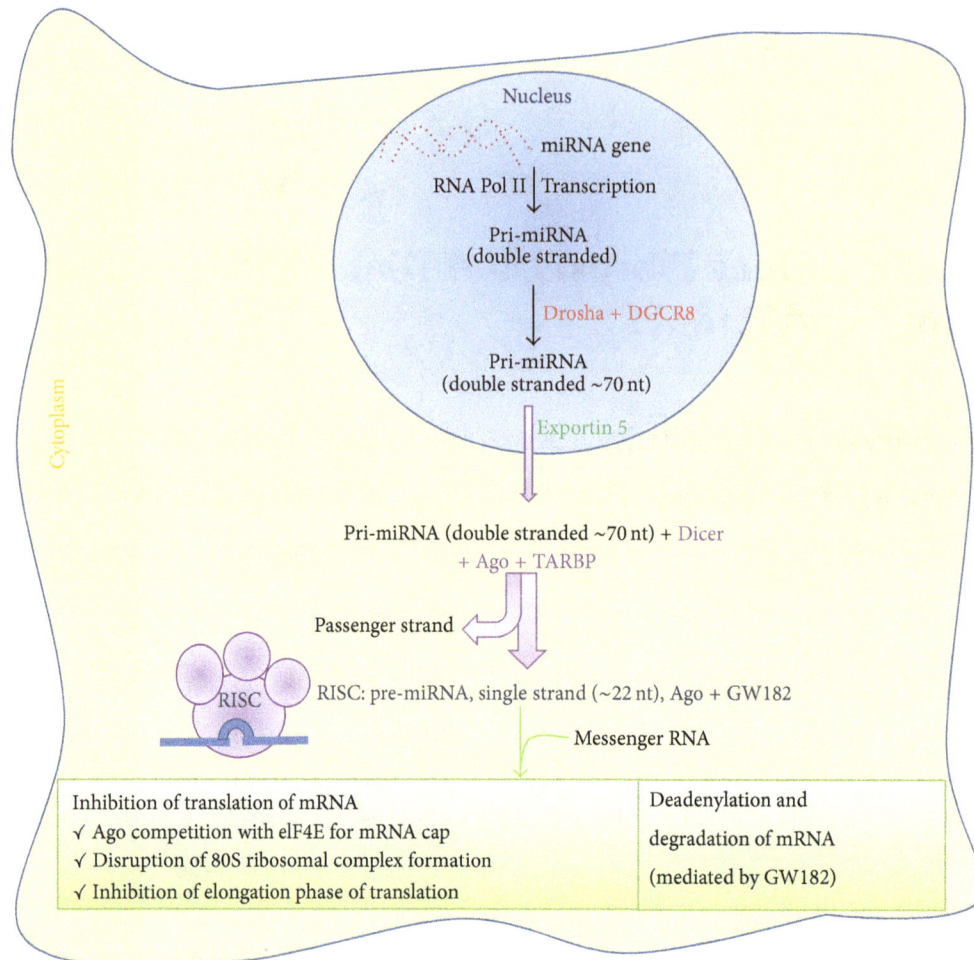

FIGURE 1: Schematic presentation of biogenesis and action of miRNAs. Ago: Argonaute; DGCR8: DiGeorge syndrome critical region 8; elF4E: eukaryotic initiation factor 4E; GW182: glycine-tryptophan protein-182; nt: nucleotides; RISC: miRNA-induced silencing complex; TARBP: transactivation-responsive RNA-binding protein.

2. miRNA Gene and Transcription

miRNAs are single-stranded RNAs (ssRNAs) of ~22 nt in length that are generated from endogenous hairpin-shaped transcripts [5]. miRNAs function as guide molecules in posttranscriptional gene regulation by base-pairing with the target mRNAs, usually in the $3'$ untranslated region (UTR). Binding of a miRNA to the target mRNA typically leads to translational repression and exonucleolytic mRNA decay, although highly complementary targets can be cleaved endonucleolytically. Over one-third of human genes are predicted to be directly targeted by miRNAs.

The 1st step in miRNAs biogenesis is nuclear processing by Drosha; the primary transcripts (pri-miRNAs) that are generated by Pol II are usually several kilobases long and contain local stem-loop structures (Figure 1). The first step of miRNA maturation is cleavage at the stem of the hairpin structure, which releases a small hairpin that is termed a pre-miRNA. This reaction takes place in the nucleus by the nuclear RNase III-type protein Drosha. Drosha requires a cofactor, the DiGeorge syndrome critical region gene 8 (DGCR8) protein in humans (Pasha in *D. melanogaster* and

C. elegans) [6]. Together with DGCR8 (or Pasha), Drosha forms a large complex known as the microprocessor complex, which is ~500 kDa in *D. melanogaster* and ~650 kDa in humans [6]. Drosha and DGCR8 are conserved only in animals. The 2nd step in biogenesis is the nuclear export by the exportin 5. The trimmed precursor (pre-miRNA) hairpins from both canonical and noncanonical miRNA pathways are then transported by an exportin 5 (EXP 5, member of nuclear transport family). As with the other nuclear transport receptor, EXP 5 binds cooperatively to its cargo and the GTP-bound form of the cofactor Ran in the nucleus and releases the cargo following the hydrolysis of GTP in the cytoplasm. EXP 5 recognizes the >14 bp dsRNA stem along with a short $3'$ overhang (1–8 nt) [7]. The 3rd step is cytoplasmic processing by the Dicer, pre-miRNA in the cytoplasm is typically further processed by the Dicer and transactivation-response RNA-binding protein (TRBP) RNase III enzyme complex to form the mature double-stranded ~22-nucleotide miRNA. Finally, the 4th step is argonaute loading, Argonaute proteins then unwind the miRNA duplex and facilitate incorporation of the miRNA-targeting strand (also known as the guide strand) into the AGO-containing RNA-induced silencing complex

(RISC). The RISC-miRNA assembly is then guided to specific target sequences in mRNAs. The initial recognition of mRNAs by the RISC-miRNA complex is driven primarily by Watson-Crick base-pairing of nucleotides 2 to 8 in the mature miRNA (seed sequence) with specific mRNA target sequences chiefly located in the 3' untranslated region, and additional base-pairing affords greater affinity and targeting efficiency [8].

3. Regulation of miRNAs Biogenesis

Precise control of miRNA levels is crucial to maintain normal cellular functions, and dysregulation of miRNA is often associated with human diseases, such as cancer [9].

3.1. Regulation at Transcriptional Level. Transcription is a major point of regulation in miRNA biogenesis. Numerous Pol II-associated transcription factors are involved in transcriptional control of miRNA genes. For instance, myogenic transcription factors, such as myogenin and myoblast determination 1 (MyoD1), bind upstream of miR-1 and miR-133 loci and induce the transcription of these miRNAs during myogenesis. Some miRNAs are under the control of tumour-suppressive or oncogenic transcription factors. The tumour suppressor p53 activates the miR-34 family of miRNAs [10], whereas the oncogenic protein MyC transactivates or represses a number of miRNAs that are involved in the cell cycle and apoptosis [11].

3.2. Regulation at Posttranscriptional Level. Drosha processing is also another important point of regulation. miR-21 is induced in response to bone morphogenetic protein (BMP)/transforming growth factor-β (TGF-β) signaling without transcriptional activation [12]. It was proposed that SMAD proteins activated by BMP/TGF-β interact with Drosha and DDX5 (also known as p68) to stimulate Drosha processing, although the detailed mechanism for this remains unclear. The let-7 miRNAs show interesting expression patterns. The primary transcript of let-7 (pri-let-7) is expressed in both undifferentiated and differentiated ES cells, whereas mature let-7 is detected only in differentiated cells, indicating that let-7 might be posttranscriptionally controlled [13]. Recent studies show that an RNA-binding protein, LIN28, is responsible for the suppression of let-7 biogenesis. Several different mechanisms of LIN28 action have been proposed: blockage of Drosha processing interference [14] with Dicer processing and terminal uridylation of pre-let-7 [14]. RNA editing is another possible way of regulating miRNA biogenesis. The alteration of adenines to inosines, a process that is mediated by adenine deaminases (ADARs), has been observed in miR-142 [15] and miR-151 [16]. Because the modified pri-miRNAs or pre-miRNAs become poor substrates of RNase III proteins, editing of the precursor can interfere with miRNA processing. Editing can also change the target specificity of the miRNA if it occurs in miRNA sequences [17].

3.3. Feedback Circuits in miRNA Networks. Two types of feedback circuits are frequently observed: single-negative feedback and double-negative feedback. The levels of Drosha and Dicer are controlled by single-negative feedback to maintain the homeostasis of miRNA production [18]. Drosha constitutes a regulatory circuit together with DGCR8; Drosha downregulates DGCR8 by cleaving *DGCR8* mRNA, whereas DGCR8 upregulates Drosha through protein stabilization. Double-negative feedback control is also often used as an effective genetic switch of specific miRNAs during differentiation. An interesting example is the conserved loop that involves let-7 and LIN28. miRNA let-7 suppresses LIN28 protein synthesis, whereas LIN28 blocks let-7 maturation. The miR-200 family and the transcriptional repressors Zeb1 and Zeb2 also constitute a double-negative feedback loop that functions in epithelial-mesenchymal transition program (EMT) [19].

4. EMT in Renal Fibrosis

EMT involves a series of changes through which epithelial cells lose their epithelial characteristics and acquire properties typical of mesenchymal cells. EMT facilitates cell movement and the generation of new tissue types during development and also contributes to the pathogenesis of disease. Earlier the role of EMT in renal fibrosis was discussed in the review [20]. Figure 2 displayed unique phenotypes of epithelial and mesenchymal cells. Epithelial cells are normally associated tightly with their neighbors, which inhibit their potential for movement and dissociation from the epithelial layer. Epithelia contour the cavities and surfaces of organs throughout the body and also form many glands. In contrast, mesenchymal cells do not form a regular layer of cells or specialized intercellular adhesion complexes. Mesenchymal cells are elongated in shape relative to epithelial cells and exhibit end-to-end polarity and focal adhesions, allowing for increased migratory capacity. Although mesenchymal cells may be polarized when migrating or interacting with neighboring cells, they lack the typical apical-basal polarity seen in epithelia. Moreover, mesenchymal cells migrate easily within tissues individually or collectively by forming a chain of migrating cells. Mesenchymal cells are essential for development as they can migrate large distances across the embryo to give rise to a particular organ. In the adult, the main function of fibroblasts, prototypical mesenchymal cells that exist in many tissues, is to maintain structural integrity by secreting extracellular matrix (ECM). Dr. Kalluri proposed classification of EMT into following three subtypes based on context [21]. Type 1 EMT involves the transition of primordial epithelial cells into motile mesenchymal cells and is associated with the generation of diverse cell types during embryonic development and organogenesis. These type 1 EMTs neither cause fibrosis nor induce invasion, and, in many cases, the mesenchymal cells that are generated later undergo MET to give rise to secondary epithelia. Type 2 EMT involves transition of secondary epithelial cells to tissue fibroblasts and is associated with wound healing, tissue regeneration, and organ fibrosis. In contrast to type 1, type 2 EMT is induced in response to inflammation but stops once inflammation is attenuated, especially during wound healing and tissue

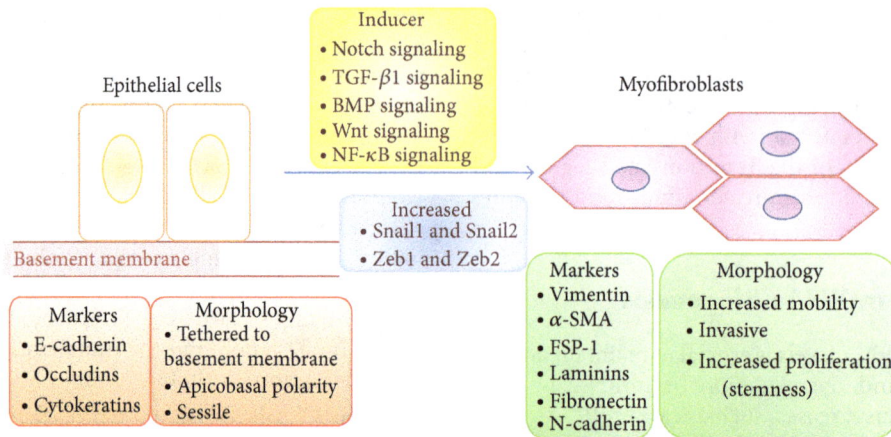

FIGURE 2: Biochemical changes during EMT in fibrosis. Repression of the transcription factors Snail1, Snail2, Zeb1, and Zeb2 is important for the maintenance of epithelial morphology. Several factors that are upregulated in the context of inflammation, including nuclear factor-κB (NF-κB), TGF-β1, bone morphogenetic proteins (BMPs), Wnt, and Notch signaling proteins, can activate the Snail-Zeb pathway, leading to mesenchymal differentiation in these cells. FSP-1: fibroblast-specific protein-1.

regeneration [22]. During organ fibrosis, type 2 EMT continues to respond to persistent inflammation, resulting in tissue destruction [22]. Type 3 EMT occurs in carcinoma cells that have formed solid tumors and is associated with their transition to metastatic tumor cells that have the potential to migrate through the bloodstream and, in some cases, form secondary tumors at other sites through mesenchymal epithelial transition (MET) [23]. Fibroblast-specific protein 1 (FSP-1; also known as S100A4 and MTS-1), an S100 class of cytoskeletal protein, α-SMA, and collagen I have provided reliable markers to characterize the mesenchymal products generated by the EMTs that occur during the development of fibrosis in various organs [21]. These markers, along with discoidin domain receptor tyrosine kinase 2 (DDR2), vimentin, and desmin, have been used to identify epithelial cells of the kidney, liver, lung, and intestine that are in the midst of undergoing an EMT associated with chronic inflammation. Such cells continued to exhibit epithelial-specific morphology and molecular markers, such as cytokeratin and E-cadherin, but showed concomitant expression of the FSP-1 mesenchymal marker and α-SMA. Such cells are likely to represent the intermediate stages of EMT, when epithelial markers continue to be expressed, but new mesenchymal markers have already been acquired. The behavior of these cells provided one of the first indications that epithelial cells under inflammatory stresses can advance to various extents through an EMT, creating the notion of "partial EMTs." Eventually, these cells leave the epithelial layer, negotiate their way through the underlying basement membrane, and accumulate in the interstitium of the tissue, where they ultimately shed all of their epithelial markers and gain a fully fibroblastic phenotype. Inflammatory injury to the mouse kidney can result in the recruitment of a diverse array of cells that can trigger an EMT through their release of growth factors, such as TGF-β, PDGF, EGF, and FGF-2 [21]. Most prominent among these cells are macrophages and activated resident fibroblasts that accumulate at the site of injury and release these growth factors. In addition, these cells release

chemokines and MMPs, notably MMP-2, MMP-3, and MMP-9. The significance of TGF-β-induced EMT for progression of organ fibrosis has been demonstrated in studies using BMP-7, an antagonist of TGF-β signaling, in mouse models of kidney, liver, billiard tract, lung, and intestinal fibrosis [24]. BMP-7 functions as an endogenous inhibitor of TGF-β-induced EMT [24]. Among other effects, it reverses the TGF-β-induced loss of the key epithelial protein, E-cadherin. Restoration of E-cadherin levels by BMP-7 is mediated via its cognate receptors, activin like kinase-2/-3/-6 (ALK-2/-3/-6), and downstream transcription factors smads [24]. Systemic administration of recombinant BMP-7 to mice with severe fibrosis resulted in reversal of EMT and repair of damaged epithelial structures, with repopulation of healthy epithelial cells, all presumably mediated via an MET. This reversal was also associated with restoration of organ function, a substantial decrease in FSP-1+ and α-SMA+ interstitial fibroblasts, and the de novo activation of BMP-7 signaling [24]. However, these different EMT programs may be induced and regulated by a common set of stimuli, signal transduction pathways, transcription factors, and posttranslational regulations [22].

4.1. miRNAs in EMT. Genome-wide analysis for miRNAs has revealed that the miR200 family and miR205 are highly associated with EMT [25]. This change is reflected in a strong correlation between the expression of the miR200 family and E-cadherin across numerous cell lines and epithelial tissues [25, 26]. The miR200 family binds to the $3'$ UTRs of RNA and suppresses the expression of Zeb1 and SIP1, which repress E-cadherin. The miR200 family is thereby capable of enforcing epithelial phenotypes. Additional EMT-related downstream targets of the miR200 family have been identified: miR141 inhibits TGF-β2 [26] and miR200a suppresses β-catenin (CTNNB1) [27]. miRNAs are also associated with the TGF-β signaling pathway. The expression of miR155 increases during TGF-β-induced EMT in mammary epithelial cells through smad4-mediated transcriptional upregulation and facilitates loss of cell polarity and tight junctions [28, 29]. Moreover, epithelial cells expressing miR155 responded more rapidly to TGF-β. A key downstream target of miR155 is RhoA, which plays a role in the formation and stabilization of cell junctions. RhoA contains three conserved regions that may serve as binding sites for miR155 [28]. These data suggest that miR155 may provide further inhibitory effects on RhoA during EMT, in addition to TGF-β-mediated ubiquitination and degradation. The expression levels of miR29a and miR21 also are increased upon TGF-β-induced EMT in mammary epithelial cells [28], although their role in EMT has not been completely elucidated. Overexpression of miR29a suppresses the expression of tristetraprolin (known as zinc finger protein 36 homolog, ZFP36) and leads to EMT in cooperation with the Ras signaling pathway [29].

4.2. Regulation of EMT. It was recently shown that miR9 directly targets the mRNA encoding E-cadherin [30]. Ectopic expression of miR9 led to EMT in human mammary epithelial cells [31]. Moreover, a significant number of breast carcinoma cells located at the edge of miR9-expressing tumors expressed mesenchymal markers including vimentin, whereas few cells located in intratumoral regions were vimentin-positive, suggesting that miR9 may sensitize cells to EMT-inducing signals from the tumor microenvironment [30]. The EMT-inducing transcription factors have recently emerged as transcriptional regulators of miRNAs. miR21 is highly expressed in various tumors and known to induce metastasis through EMT. The promoter regions of miR21 include consensus E-box sequences that serve as binding sites for Zeb1 [31]. Binding of Zeb1 induces transcription of miR21 and also blocks bone morphogenetic protein- (BMP-)6-mediated inhibition of EMT in breast cancer cells [31].

5. EndMT in Renal Fibrosis

Vascular endothelial cells share several common traits with epithelial cells and can generate fibroblasts by undergoing a phenotypic transition similar to EMT, referred to as endothelial-mesenchymal transition (EndMT). EndMT is characterized by the loss of endothelial markers including CD31 and vascular endothelial cadherin (VE-cadherin) and the expression of mesenchymal proteins including α-smooth muscle actin (αSMA) [32]. EndMT contributes to cardiac fibrogenesis which results in progressive stiffening of the ventricular walls, loss of contractility, and abnormalities in cardiac conductance [32]. EndMT is also involved in pulmonary fibrosis, idiopathic hypertension [33], and corneal fibrosis [34]. Many growth factors and signaling pathways that govern EMT also regulate EndMT in the embryonic heart and during cardiac fibrosis. However, as compared to EMT, relatively little is known about EndMT. Earlier role of EndMT in renal fibrosis was discussed and reviewed by many researchers [35, 36]. In the adult organism, pathological conditions such as injury, inflammation, or aging can awaken EndMT and induce the fibrosis of the involved organs. The EndMT program has also been suggested to contribute to the development and progression of cardiac fibrosis, pulmonary

fibrosis, hepatic fibrosis, corneal fibrosis, intestinal fibrosis, and wound healing in addition to renal fibrosis [37]. Zeisberg et al., 2008, designed and conducted a landmark experiment that first confirmed the contribution of EndMT in renal fibrosis in three mouse models, unilateral ureteral obstruction (UUO; a model used to study progressive tubulointerstitial fibrosis), streptozotocin- (STZ-) induced diabetic nephropathy, and α3 chain of collagen type 4 (COL4A3) knockout mice (a mouse model for Alport syndrome). They found that a considerable proportion of myofibroblasts coexpress the endothelium marker CD31, also known as platelet endothelial cell adhesion molecule-1, and the (myo) fibroblast markers αSMA and fibroblast-specific protein-1 (FSP-1, also known as S100A4) in all three models [38]. Furthermore, they analyzed the kidneys 6 months after a single injection of STZ in CD1 mice, which exhibited progressive glomerular sclerosis and tubulointerstitial fibrosis. A double-immunolabeling experiment demonstrated that approximately 40% of all FSP-1 (+) and 50% of αSMA (+) stromal cells in STZ kidneys were also CD31-positive [38]. In the kidneys of 22-week-old COL4A3 knockout mice, 45% of all αSMA-positive fibroblasts and 60% of all FSP-1-positive fibroblasts were CD31-positive, suggesting that these fibroblasts are likely of endothelial origin and that EndMT may contribute substantially to the accumulation of fibroblasts in the development and progression of renal fibrosis [38]. Li et al., 2009, also confirmed that EndMT occurs and contributes to the generation of myofibroblasts in early diabetic renal fibrosis. Using endothelial-lineage tracing with Tie2-cre, LoxP-enhanced green fluorescent protein (EGFP) transgenic mice, they identified a significant number of interstitial αSMA-positive cells (myofibroblasts) of an endothelial origin in the fibrotic kidneys from mice with STZ-induced diabetic nephropathy [39]. ECs line the entire circulatory and lymphatic system, forming the inner lining of blood vessels and lymphatic vessels. These cells, which are anatomically similar to squamous epithelium, express apical-basal polarity and are tightly bound by adherens junctions and tight junctions [40]. These cells demonstrate a disparate set of biomarkers including VE-cadherin, CD31, TIE1, TIE2, von Willebrand factor (vWF), and cytokeratins. Similar to EMT, during EndMT, ECs lose their adhesion and apical-basal polarity to form highly invasive, migratory, spindle-shaped, elongated mesenchymal cells. Biochemical changes accompany these distinct changes in cell polarity and morphology, including the decreased expression of endothelial markers and the acquisition of mesenchymal markers (FSP-1, αSMA, SM22α, N-cadherin, fibronectin, vimentin, types I and III collagen, nestin, CD73, MMP-2, and MMP-9) [40] (Figure 3).

5.1. miRNAs in EndMT. Indirectly, miRNAs also upregulate many genes via suppression of their repressor molecules. The expression of both primary and mature miRNA-21 (miR-21), especially the latter, was upregulated by TGF-β by silencing phosphatase and tensin homolog (PTEN) and activating the Akt pathway in EndMT *in vitro* (HUVECs) and *in vivo* (heart) [41]. smad3 signaling increases the expression of miR-21 in the kidney to promote renal fibrosis in response to TGF-β whereas smad2 negatively regulates the posttranscriptional

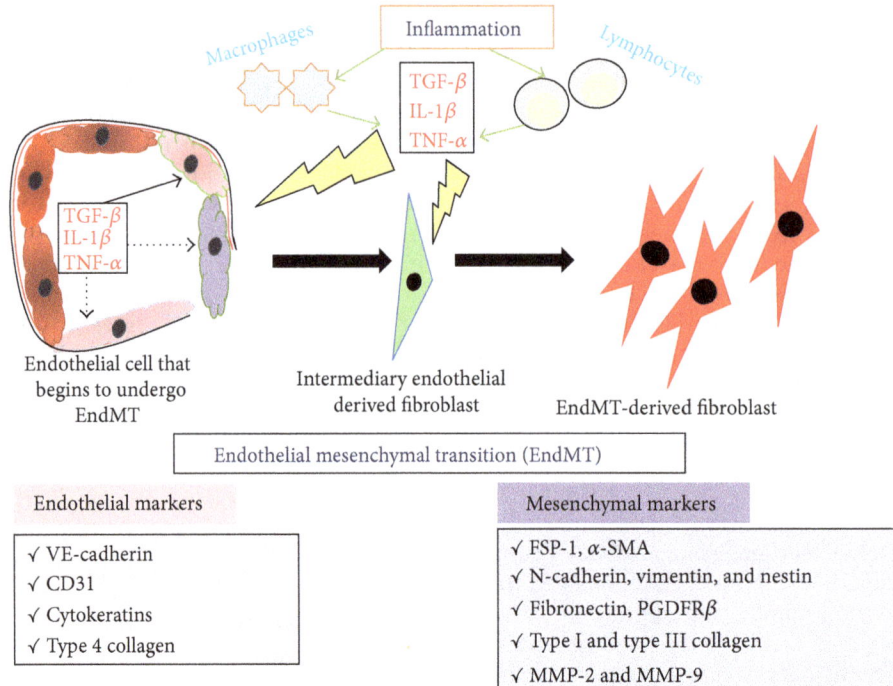

FIGURE 3: Biochemical changes during EndMT program. The EndMT program causes decreased expression of endothelial markers VE-cadherin, CD31, cytokeratins, and type 4 collagen and a gain of mesenchymal markers FSP-1, αSMA, N-cadherin, vimentin, fibronectin, type I and type III collagen, and MMP-2 and MMP-9. FSP-1: fibroblast-specific protein-1; α-SMA: α-smooth muscle actin; and MMP: matrix metalloproteinase.

modification of miR-21 [42]. miR-23 inhibits TGF-β-induced EndMT in mouse ECs, and miR-23 in the embryonic heart is required to restrict endocardial cushion formation by inhibiting hyaluronic acid synthase 2 (Has2) expression and extracellular hyaluronic acid production in Zebrafish dicer mutants [43]. Using the miRNA array analysis, Ghosh et al. [44] found that although miR-125b, let-7c, let-7g, miR-21, miR-30b, and miR-195 were significantly elevated during EndMT, the levels of miR-122a, miR-127, miR-196, and miR-375 were significantly downregulated. MiR-125b is approximately 4-fold higher in EndMT-derived fibroblast compared with MCECs [44]. The level of cellular p53, the major target of miR-125b and a known negative modulator of TGF-β-induced profibrotic signaling, was significantly suppressed with an elevated level of αSMA [44]. Blockade of FGF signaling induced EndMT program can be mimicked by the let-7b or let-7c miRNA inhibition. Although these studies were mostly performed in the heart, MCECs, or HUVECs, they still suggest that the specific suppression of upregulated miRNAs, such as miR-21, or the specific overexpression of downregulated miRNAs, such as miR-125b, may be a viable approach to blocking the induction of EndMT in a wide variety of organs.

6. miRNAs in Kidney Disease and Diabetic Nephropathy

Diabetic nephropathy is a progressive kidney disease and a major debilitating complication of both type 1 and type 2

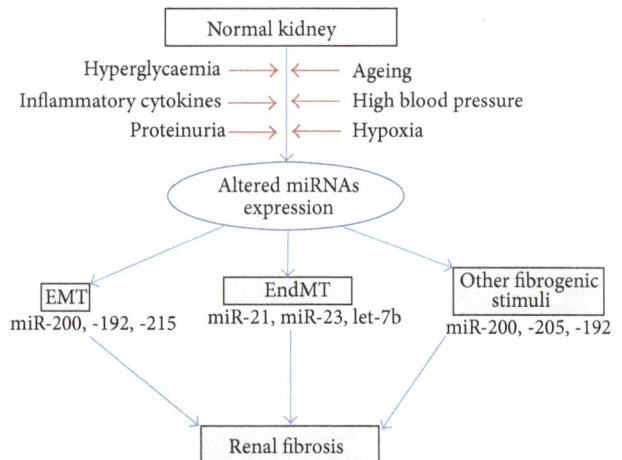

FIGURE 4: Implications of miRNAs in renal fibrosis.

diabetes that can lead to end-stage renal disease (ESRD) and related cardiovascular disorders. Absence or lower levels of particular miRNAs in the kidney compared with other organs may permit renal specific expression of target proteins that are important for kidney functions [45]. Figure 4 depicts the connection between the role of miRNAs and kidney fibrosis. Altered expression of miRNAs causes renal fibrosis by inducing EMT, EndMT, and other fibrogenic stimuli. The accumulative effects of hyperglycaemia, inflammatory cytokines, proteinuria, ageing, high blood pressure, and hypoxia result into alteration of miRNAs expression profiles. The altered miRNAs level causes the initiation of such transition program in normal kidney, finally fibrosis. Some of the miRNAs that are more abundant in the kidney compared with other organs include miR-192, miR-194, miR-204, miR-215, and miR-216. A critical role of miRNA regulation in the progression of glomerular and tubular damage and the development of proteinuria been suggested by studies in mice with podocyte-specific deletion of Dicer [46]. There was a rapid progression of renal disease with initial development of albuminuria followed by pathological features of glomerulosclerosis and tubulointerstitial fibrosis. It is likely that these phenotypes are due to the global loss of miRNAs because of Dicer deletion, but, given multiple miRNAs and their myriad targets, the precise pathways responsible require identification. These investigators also identified specific miRNA changes, for example, the downregulation of the miR-30 family when Dicer was deleted. Of relevance, the miR-30 family was found to target connective tissue growth factor, a profibrotic molecule that is also downstream of transforming growth factor (TGF)-β [47]. Thus, the targets of these miRNAs may regulate critical glomerular and podocyte functions. These findings have also been complemented by an elegant study revealing a developmental role for the miR-30 family during pronephric kidney development in *Xenopus* [48]. Sun et al. [49] identified five miRNAs (-192, -194, -204, -215, and -216) that were highly expressed in human and mouse kidney using miRNA microarray. A recent report using new proteomic approaches to profile and identify miRNA targets demonstrated that miRNAs repress their targets at both the mRNA and translational levels and that the effects are mostly relatively mild [50]. The role of miR-192 remains controversial and highlights the complex nature of miRNA research. Kato et al. [51] observed increased renal expression of miR-192 in streptozotocin-(STZ-) induced diabetes and in the db/db mouse and demonstrated that transforming growth factor (TGF-β1) upregulated miR-192 in mesangial cells (MCs). miR-192 repressed the translation of Zeb2, a transcriptional repressor that binds to the E-box in the collagen 1α2 (col1α2) gene. They proposed that miR-192 repressed Zeb2 and resulted in increased col1α2 expression *in vitro* and contributed to increased collagen deposition *in vivo*. These data suggest a role for miR-192 in the development of the matrix accumulation observed in DN. It is interesting that the expression of miR-192 was increased by TGF-β in mouse MCs (mesangial cells), whereas, conversely, the expression of its target, Zeb2, was decreased [51]. This also paralleled the increased Col1 α2 and TGF-β expression [51]. These results suggested that the increase in TGF-β *in vivo* in diabetic glomeruli and *in vitro* in MCs can induce miR-192 expression, which can target and downregulate Zeb2 thereby to increase Col1 α2. This is supported by the report showing that miR-192 is upregulated in human MCs treated with high glucose [51]. TGF-β induced downregulation of Zeb2 (*via* miR-192) and Zeb1 (*via* potentially another miRNA) can cooperate to enhance Col1 α2 expression *via* de-repression at E-box elements [51]. In contrast to the above, other reports suggest the relationship between miR-192 and renal fibrosis may be more complicated. Krupa et al. [52] identified two miRNAs in human renal biopsies, the

expression of which differed by more than twofold between progressors and nonprogressors with respect to DN, the greatest change occurring in miR-192 which was significantly lower in patients with advanced DN, correlating with tubulointerstitial fibrosis and low glomerular filtration rate. They also reported, in contrast to the Kato et al. [51] study in MCs, that TGF-β1 decreased expression of miR-192 in cultured proximal tubular cells (PTCs). These investigators concluded that a decrease in miR-192 is associated with increased renal fibrosis *in vivo*. Interestingly, connective tissue growth factor (CTGF) treatment also resulted in fibrogenesis but caused the induction of miR-192/215 and, consequently, decreased Zeb2 and increased E-cadherin. The contrasting findings above highlight the complex nature of miRNA research. Some of the differences may relate to models and/or experimental conditions; however, one often overlooked explanation is that some effects of miRNAs and inhibitors are likely to be indirect in nature. A recent report also showed that BMP6-induced miR-192 decreases the expression of Zeb1 in breast cancer cells [53]. Thus, TGF-β induced increase in the expression of key miRNAs (miR-192 and miR-200 family members) might coordinately downregulate E-box repressors Zeb1 and Zeb2 to increase Col1α2 expression in MCs related to the pathogenesis of DN. The proximal promoter of the *Col1a2* gene responds to TGF-β *via* smads and SP1. Conversely, the downregulation of Zeb1 and Zeb2 by TGF-β *via* miR-200 family and miR-192 can affect upstream E-box regions. Because E-boxes are present in the upstream genomic regions of the miR-200 family, miR-200 family members may themselves be regulated by Zeb1 and Zeb2 [54]. It is possible that the miR-200 family upregulated by TGF-β or in diabetic glomeruli under early stages of the disease can also regulate collagen expression related to diabetic kidney disease by targeting and downregulating E-box repressors. miR-192 might initiate signaling from TGF-β to upregulate miR-200 family members, which subsequently could amplify the signaling by further regulating themselves through down regulation of E-box repressors. Such events could lead to progressive renal dysfunction under pathologic conditions such as diabetes, in which TGF-β levels are enhanced. Conversely, there are several reports that miR-200 family members and miR-192 can be suppressed by TGF-β, and this promotes epithelial-to-mesenchymal transition (EMT) in cancer and other kidney-derived epithelial cell lines *via* subsequent upregulation of targets Zeb1 and Zeb2 to repress E-cadherin [54, 55].

7. Prospective

The discovery of miRNAs in 1993 in the nematode made the tremendous revolution in the field of RNA world. There are several major challenges in exploring the role of miRNAs in kidney diseases. Now miRNA-based therapeutics has already entered Phase 2 clinical trials. miR-122 antagonists are the indicator of hepatitis C virus and now in the Phase 2 clinical trials [56]. miR-208/499 antagonists are the indicator for chronic failure and now in the preclinical development. Likewise miR-195 antagonists are also in preclinical development, used for the indicator of postmyocardial infarction remodeling. Some of the miRNAs (miR-34 and miR-7) are in preclinical development for the miRNA replacement therapy of cancer [57, 58]. This rapid progress from discovery to development reflects the importance of miRNAs as critical regulators in human disease and holds the promise of yielding a new class of therapeutics that could represent an attractive addition to the current drug pipeline of Big Pharma. Most importantly many fundamental questions remain regarding miRNA biology. The mechanism of regulation of miRNA is not completely clear. While many miRNAs are located within the intron of the host gene, their expression does not correlate perfectly with that of host genes suggesting further, posttranscriptional regulation [59]. Furthermore, the use of the miRNAs as therapeutic agents is attractive but faces considerable challenges, including development of safe and reliable organ and cell-specific delivery system, avoidance of toxicity derived from off-target effects and from activation of the innate and adaptive immune response. Current health statistics suggest that nearly 45% of all deaths in the western world can be attributed to some types of chronic fibro-proliferative disease [60]. EMT and EndMT have become a key topic in the study of organ fibrosis, since stressed and injured epithelium can give rise to myofibroblasts and thereby contribute to fibrogenesis therapeutics for fibrosis. The participation of EMT and EndMT in the pathogenesis of various fibrotic disorders requires confirmation and validation from further studies of human clinical pathological conditions. Future efforts should also be devoted to further understanding of the molecular mechanisms and the regulatory controls involved in these processes including miRNA regulation. These efforts would eventually lead to the development of novel therapeutic approaches for these incurable and often devastating disorders by targeting miRNAs.

Abbreviations

ADARs:	Adenine deaminases
αSMA:	α-Smooth muscle actin
COL4A3:	$\alpha3$ chain of collagen type 4
BMP:	Bone morphogenetic protein
CTGF:	Connective tissue growth factor
DN:	Diabetic nephropathy
DGCR8:	DiGeorge syndrome critical region gene 8 protein
EndMT:	Endothelial mesenchymal transition
EMT:	Epithelial mesenchymal transition
ECM:	Extracellular matrix
EXP 5:	Exportin 5 (member of nuclear transport family)
EGFP:	Enhanced green fluorescent protein (EGFP)
ESRD:	End-stage renal disease
FSP-1:	Fibroblast-specific protein-1
FGF:	Fibroblast growth factor
microRNAs:	miRNAs
MyoD1:	myoblast determination 1
MET:	Mesenchymal epithelial transition
MCs:	Mesangial cells
MMP:	Matrix metalloproteinase
PTEN:	Phosphatase and tensin homolog
PTCs:	Proximal tubular cells
RISC:	RNA-induced silencing complex
TRBP:	Transactivation-response RNA-binding protein
TGF-β:	Transforming growth factor-β
UTRs:	Untranslated regions
VE-cadherin:	Vascular endothelial cadherin.

Conflict of Interests

The authors declare there is no conflict of interests in this work.

Acknowledgments

The authors' laboratory is supported by a Grant from the Japan Society for the Promotion of Science to Keizo Kanasaki (23790381), and Daisuke Koya (25670414, 25282028) and research grants from the Japan Research Foundation for Clinical Pharmacology to Keizo Kanasaki In addition, this work was partially supported by Grants for Promoted Research to Keizo Kanasaki (S2012-5, S2013-13) from Kanazawa Medical University. Keizo Kanasaki was currently also supported by several foundation grants from the following: the Novartis Foundation (Japan) for the Promotion of Science, the Takeda Science Foundation, and the Banyu Foundation. The authors declare that no financial conflict of interests exists. Swayam Prakash Srivastava is supported by the Japanese Government MEXT (Ministry of Education, Culture, Sports, Science, and Technology) Fellowship Program.

References

[1] D. P. Bartel, "MicroRNAs: target recognition and regulatory functions," *Cell*, vol. 136, no. 2, pp. 215–233, 2009.

[2] X. Jiang, E. Tsitsiou, S. E. Herrick et al., "MiRNAs and the regulation of fibrosis," *FEBS Journal*, vol. 277, no. 9, pp. 2015–2021, 2010.

[3] T.A. Wynn and T.R. Ramalingam, "Mechanisms of fibrosis: therapeutic translation for fibrotic disease," *Nature Medicine*, vol. 18, no. 7, pp. 1028–1040, 2012.

[4] P. Boor, T. Ostendorf, and J. Floege, "Renal fibrosis: novel insights into mechanisms and therapeutic targets," *Nature Reviews Nephrology*, vol. 6, no. 11, pp. 643–656, 2010.

[5] V. N. Kim, "MicroRNA biogenesis: coordinated cropping and dicing," *Nature Reviews Molecular Cell Biology*, vol. 6, no. 5, pp. 376–385, 2005.

[6] J. Han, Y. Lee, K.-H. Yeom et al., "The Drosha-DGCR8 complex in primary microRNA processing," *Genes and Development*, vol. 18, no. 24, pp. 3016–3027, 2004.

[7] C. Gwizdek, B. Ossareh-Nazari, A. M. Brownawell et al., "Exportin-5 mediates nuclear export of minihelix-containing RNAs," *Journal of Biological Chemistry*, vol. 278, no. 8, pp. 5505–5508, 2003.

[8] D. P. Bartel, "MicroRNAs: target recognition and regulatory functions," *Cell*, vol. 136, no. 2, pp. 215–233, 2009.

[9] Q. Jiang, Y. Wang, Y. Hao et al., "miR2Disease: a manually curated database for microRNA deregulation in human disease," *Nucleic Acids Research*, vol. 37, no. 1, pp. D98–D104, 2009.

[10] L. He, X. He, S. W. Lowe, and G. J. Hannon, "microRNAs join the p53 network—another piece in the tumour-suppression puzzle," *Nature Reviews Cancer*, vol. 7, no. 11, pp. 819–822, 2007.

[11] T.-C. Chang, D. Yu, Y.-S. Lee et al., "Widespread microRNA repression by Myc contributes to tumorigenesis," *Nature Genetics*, vol. 40, no. 1, pp. 43–50, 2008.

[12] B. N. Davis, A. C. Hilyard, G. Lagna et al., "SMAD proteins control DROSHA-mediated microRNA maturation," *Nature*, vol. 454, no. 7200, pp. 56–61, 2008.

[13] F. G. Wulczyn, L. Smirnova, A. Rybak et al., "Posttranscriptional regulation of the let-7 microRNA during neural cell specification," *FASEB Journal*, vol. 21, no. 2, pp. 415–426, 2007.

[14] S. R. Viswanathan, G. Q. Daley, and R. I. Gregory, "Selective blockade of microRNA processing by Lin28," *Science*, vol. 320, no. 5872, pp. 97–100, 2008.

[15] W. Yang, T. P. Chendrimada, Q. Wang et al., "Modulation of microRNA processing and expression through RNA editing by ADAR deaminases," *Nature Structural and Molecular Biology*, vol. 13, no. 1, pp. 13–21, 2006.

[16] Y. Kawahara, B. Zinshteyn, T. P. Chendrimada, R. Shiekhattar, and K. Nishikura, "RNA editing of the microRNA-151 precursor blocks cleavage by the Dicer—TRBP complex," *EMBO Reports*, vol. 8, no. 8, pp. 763–769, 2007.

[17] Y. Kawahara, M. Megraw, E. Kreider et al., "Frequency and fate of microRNA editing in human brain," *Nucleic Acids Research*, vol. 36, no. 16, pp. 5270–5280, 2008.

[18] S. Tokumaru, M. Suzuki, H. Yamada et al., "*let-7* regulates Dicer expression and constitutes a negative feedback loop," *Carcinogenesis*, vol. 29, no. 11, pp. 2073–2077, 2008.

[19] C. P. Bracken, P. A. Gregory, N. Kolesnikoff et al., "A double-negative feedback loop between ZEB1-SIP1 and the microRNA-200 family regulates epithelial-mesenchymal transition," *Cancer Research*, vol. 68, no. 19, pp. 7846–7854, 2008.

[20] R. M. Carew, B. Wang, and P. Kantharidis, "The role of EMT in renal fibrosis," *Cell and Tissue Research*, vol. 347, no. 1, pp. 103–116, 2012.

[21] R. Kalluri and R. A. Weinberg, "The basics of epithelial-mesenchymal transition," *Journal of Clinical Investigation*, vol. 119, no. 6, pp. 1420–1428, 2009.

[22] M. Zeisberg, C. Bottiglio, N. Kumar et al., "Bone morphogenic protein-7 inhibits progression of chronic renal fibrosis associated with two genetic mouse models," *American Journal of Physiology*, vol. 285, no. 6, pp. F1060–F1067, 2003.

[23] J. M. López-Nouoa and M. A. Nieto, "Inflammation and EMT: an alliance towards organ fibrosis and cancer progression," *EMBO Molecular Medicine*, vol. 1, no. 6-7, pp. 303–314, 2009.

[24] J. P. Thiery, "Epithelial-mesenchymal transitions in tumour progression," *Nature Reviews Cancer*, vol. 2, pp. 442–454, 2002.

[25] S.-M. Park, A. B. Gaur, E. Lengyel et al., "The miR-200 family determines the epithelial phenotype of cancer cells by targeting the E-cadherin repressors ZEB1 and ZEB2," *Genes and Development*, vol. 22, no. 7, pp. 894–907, 2008.

[26] U. Burk, J. Schubert, U. Wellner et al., "A reciprocal repression between ZEB1 and members of the miR-200 family promotes EMT and invasion in cancer cells," *EMBO Reports*, vol. 9, no. 6, pp. 582–589, 2008.

[27] H. Xia, S. S. Ng, S. Jiang et al., "miR-200a-mediated downregulation of ZEB2 and CTNNB1 differentially inhibits nasopharyngeal carcinoma cell growth, migration and invasion," *Biochemical and Biophysical Research Communications*, vol. 391, no. 1, pp. 535–541, 2010.

[28] W. Kong, H. Yang, L. He et al., "MicroRNA-155 is regulated by the transforming growth factor β/Smad pathway and contributes to epithelial cell plasticity by targeting RhoA," *Molecular and Cellular Biology*, vol. 28, no. 22, pp. 6773–6784, 2008.

[29] C. A. Gebeshuber, K. Zatloukal, and J. Martinez, "miR-29a suppresses tristetraprolin, which is a regulator of epithelial polarity and metastasis," *EMBO Reports*, vol. 10, no. 4, pp. 400–405, 2009.

[30] L. Ma, J. Young, H. Prabhala et al., "MiR-9, a MYC/MYCN-activated microRNA, regulates E-cadherin and cancer metastasis," *Nature Cell Biology*, vol. 12, no. 3, pp. 247–256, 2010.

[31] J. Du, S. Yang, D. An et al., "BMP-6 inhibits microRNA-21 expression in breast cancer through repressing δEF1 and AP-1," *Cell Research*, vol. 19, no. 4, pp. 487–496, 2009.

[32] E. M. Zeisberg, O. Tarnavski, M. Zeisberg et al., "Endothelial-to-mesenchymal transition contributes to cardiac fibrosis," *Nature Medicine*, vol. 13, no. 8, pp. 952–961, 2007.

[33] A. Kitao, Y. Sato, S. Sawada-Kitamura et al., "Endothelial to mesenchymal transition via transforming growth factor-β1/smad activation is associated with portal venous stenosis in idiopathic portal hypertension," *American Journal of Pathology*, vol. 175, no. 2, pp. 616–626, 2009.

[34] Y. Nakano, M. Oyamada, P. Dai, T. Nakagami, S. Kinoshita, and T. Takamatsu, "Connexin43 knockdown accelerates wound healing but inhibits mesenchymal transition after corneal endothelial injury in vivo," *Investigative Ophthalmology and Visual Science*, vol. 49, no. 1, pp. 93–104, 2008.

[35] J. He, Y. Xu, D. Koya, and G. J. Hannon, "Role of the endothelial-to mesenchymal transition in renal fibrosis of chronic kidney disease," *Clinical and Experimental Nephrology*, 2013.

[36] J. Li and J. F. Bertram, "Review: Endothelial-myofibroblast transition, a new player in diabetic renal fibrosis," *Nephrology*, vol. 15, no. 5, pp. 507–512, 2010.

[37] S. Piera-Velazquez, Z. Li, and S. A. Jimenez, "Role of endothelial-mesenchymal transition (EndoMT) in the pathogenesis of fibrotic disorders," *American Journal of Pathology*, vol. 179, no. 3, pp. 1074–1080, 2011.

[38] E. M. Zeisberg, S. E. Potenta, H. Sugimoto et al., "Fibroblasts in kidney fibrosis emerge via endothelial-to-mesenchymal transition," *Journal of the American Society of Nephrology*, vol. 19, no. 12, pp. 2282–2287, 2008.

[39] J. Li, X. Qu, and J. F. Bertram, "Endothelial-myofibroblast transition contributes to the early development of diabetic renal interstitial fibrosis in streptozotocin-induced diabetic mice," *American Journal of Pathology*, vol. 175, no. 4, pp. 1380–1388, 2009.

[40] D. Medici and R. Kalluri, "Endothelial-mesenchymal transition and its contribution to the emergence of stem cell phenotype," *Seminars in Cancer Biology*, vol. 22, pp. 379–384, 2012.

[41] R. Kumarswamy, I. Volkmann, V. Jazbutyte, S. Dangwal, D.-H. Park, and T. Thum, "Transforming growth factor-β-induced endothelial-to-mesenchymal transition is partly mediated by MicroRNA-21," *Arteriosclerosis, Thrombosis, and Vascular Biology*, vol. 32, no. 2, pp. 361–369, 2012.

[42] X. Zhong, A. C. K. Chung, H.-Y. Chen et al., "Smad3-mediated upregulation of miR-21 promotes renal fibrosis," *Journal of the American Society of Nephrology*, vol. 22, no. 9, pp. 1668–1681, 2011.

[43] A. K. Lagendijk, M. J. Goumans, S. B. Burkhard et al., "MicroRNA-23 restricts cardiac valve formation by inhibiting

has2 and extracellular hyaluronic acid production," *Circulation Research*, vol. 109, no. 6, pp. 649–657, 2011.

[44] A. K. Ghosh, V. Nagpal, J. W. Covington et al., "Molecular basis of cardiac endothelial-to-mesenchymal transition (EndMT): differential expression of microRNAs during EndMT," *Cellular Signalling*, vol. 24, no. 5, pp. 1031–1036, 2012.

[45] C.-G. Liu, G. A. Calin, B. Meloon et al., "An oligonucleotide microchip for genome-wide microRNA profiling in human and mouse tissues," *Proceedings of the National Academy of Sciences of the United States of America*, vol. 101, no. 26, pp. 9740–9744, 2004.

[46] S. Shi, L. Yu, C. Chiu et al., "Podocyte-selective deletion of dicer induces proteinuria and glomerulosclerosis," *Journal of the American Society of Nephrology*, vol. 19, no. 11, pp. 2159–2169, 2008.

[47] R. F. Duisters, A. J. Tijsen, B. Schroen et al., "MiR-133 and miR-30 Regulate connective tissue growth factor: implications for a role of micrornas in myocardial matrix remodeling," *Circulation Research*, vol. 104, no. 2, pp. 170–178, 2009.

[48] R. Agrawal, U. Tran, and O. Wessely, "The miR-30 miRNA family regulates *Xenopus* pronephros development and targets the transcription factor Xlim1/Lhx1," *Development*, vol. 136, no. 23, pp. 3927–3936, 2009.

[49] Y. Sun, S. Koo, N. White et al., "Development of a micro-array to detect human and mouse microRNAs and characterization of expression in human organs," *Nucleic Acids Research*, vol. 32, no. 22, p. e188, 2004.

[50] D. Baek, J. Villén, C. Shin, F. D. Camargo, S. P. Gygi, and D. P. Bartel, "The impact of microRNAs on protein output," *Nature*, vol. 455, no. 7209, pp. 64–71, 2008.

[51] M. Kato, J. Zhang, M. Wang et al., "MicroRNA-192 in diabetic kidney glomeruli and its function in TGF-β-induced collagen expression via inhibition of E-box repressors," *Proceedings of the National Academy of Sciences of the United States of America*, vol. 104, no. 9, pp. 3432–3437, 2007.

[52] A. Krupa, R. Jenkins, D.D. Luo, A. Lewis, A. Phillips, and D. Fraser, "Loss of microRNA-192 promotes fibrogenesis in diabetic nephropathy," *Journal of the American Society of Nephrology*, vol. 21, pp. 438–447, 2010.

[53] S. Yang, J. Du, and Z. Wang, "Dual mechanism of deltaEF1 expression regulated by bone morphogenetic protein-6 in breast cancer," *The International Journal of Biochemistry & Cell Biology*, vol. 41, pp. 853–861, 2009.

[54] U. Burk, J. Schubert, U. Wellner et al., "A reciprocal repression between ZEB1 and members of the miR-200 family promotes EMT and invasion in cancer cells," *EMBO Reports*, vol. 9, no. 6, pp. 582–589, 2008.

[55] M. Korpal, E. S. Lee, G. Hu et al., "The miR-200 family inhibits epithelial-mesenchymal transition and cancer cell migration by direct targeting of E-cadherin transcriptional repressors ZEB1 and ZEB2," *Journal of Biological Chemistry*, vol. 283, no. 22, pp. 14910–14914, 2008.

[56] R. E. Lanford, E. S. Hildebrandt-Eriksen, A. Petri et al., "Therapeutic silencing of microRNA-122 in primates with chronic hepatitis C virus infection," *Science*, vol. 327, no. 5962, pp. 198–201, 2010.

[57] P. Trang, P. P. Medina, J. F. Wiggins et al., "Regression of murine lung tumors by the let-7 microRNA," *Oncogene*, vol. 29, no. 11, pp. 1580–1587, 2010.

[58] C. Liu, K. Kelnar, B. Liu et al., "The microRNA miR-34a inhibits prostate cancer stem cells and metastasis by directly repressing CD44," *Nature Medicine*, vol. 17, no. 2, pp. 211–216, 2011.

[59] J. M. Thomson, M. Newman, J. S. Parker et al., "Extensive post-transcriptional regulation of microRNAs and its implications for cancer," *Genes and Development*, vol. 20, no. 16, pp. 2202–2207, 2006.

[60] T. A. Wynn, "Common and unique mechanisms regulate fibrosis in various fibroproliferative diseases," *Journal of Clinical Investigation*, vol. 117, no. 3, pp. 524–529, 2007.

4

In Vitro Culture-Induced Pluripotency of Human Spermatogonial Stem Cells

Jung Jin Lim,[1] **Hyung Joon Kim,**[1] **Kye-Seong Kim,**[2] **Jae Yup Hong,**[3] **and Dong Ryul Lee**[1, 4]

[1] *Fertility Center of CHA Gangnam Medical Center, College of Medicine, CHA University, 606-5 Yeoksam-dong, Gangnam-gu, Seoul 135-081, Republic of Korea*
[2] *Department of Anatomy and Cell Biology, College of Medicine, Hanyang University, Seoul 133-791, Republic of Korea*
[3] *Department of Urology, CHA Bundang Medical Center, CHA University, Seongnam 463-712, Republic of Korea*
[4] *Department of Biomedical Science, College of Life Science, CHA University, Seoul 135-081, Republic of Korea*

Correspondence should be addressed to Dong Ryul Lee; drleedr@cha.ac.kr

Academic Editor: Thomas Skutella

Unipotent spermatogonial stem cells (SSCs) can be transformed into ESC-like cells that exhibit pluripotency *in vitro*. However, except for mouse models, their characterization and their origins have remained controversies in other models including humans. This controversy has arisen primarily from the lack of the direct induction of ESC-like cells from well-characterized SSCs. Thus, the aim of the present study was to find and characterize pluripotent human SSCs in *in vitro* cultures of characterized SSCs. Human testicular tissues were dissociated and plated onto gelatin/laminin-coated dishes to isolate SSCs. In the presence of growth factors SSCs formed multicellular clumps after 2–4 weeks of culture. At passages 1 and 5, the clumps were dissociated and were then analyzed using markers of pluripotent cells. The number of SSEA-4-positive cells was extremely low but increased gradually up to ~ 10% in the SSC clumps during culture. Most of the SSEA-4-negative cells expressed markers for SSCs, and some cells coexpressed markers of both pluripotent and germ cells. The pluripotent cells formed embryoid bodies and teratomas that contained derivatives of the three germ layers in SCID mice. These results suggest that the pluripotent cells present within the clumps were derived directly from SSCs during *in vitro* culture.

1. Introduction

Embryonic germ cells (EGCs), which can be derived from fetal unipotent primordial germ cells (PGCs), are pluripotent and have expression patterns of cell surface and gene markers similar to those of embryonic stem cells (ESCs). These markers include alkaline phosphatase, OCT-4, SSEA-4, NANOG, TRA-1-60, and REX-1. Other important characteristics, such as multicellular colony formation, maintaining normal and stable karyotypes, the ability to proliferate continuously, and the ability to differentiate into all three embryonic germ layers, can be acquired during *in vitro* induction [1]. Although it has been suggested that PGCs are typically unipotent and are able to produce only germ cells [2], several studies have shown that a small number of PGCs express OCT4 and NANOG during various stages of prenatal development. These results provide evidence that there exists a population of multipotent PGC. In contrast to the induction of pluripotent stem cells (iPSCs), this type of induction was a solely culture-induced procedure and did not rely on the introduction of exogenous transcription factors.

Spermatogonial stem cells (SSCs) are derived from PGCs during the neonatal period and can self-renew and produce large numbers of differentiating germ cells that become spermatozoa throughout adult life. Recently, some groups reported that SSCs obtained from neonatal and adult mouse testes can be induced to form multipotent SSCs (mSSCs) or multipotent germline stem cells (mGSCs) during *in vitro* culture, and these cells may have a pluripotency similar to that of ESCs [3, 4]. In mice, mSSCs (mGSCs) are phenotypically similar to ESC/EG cells except with respect to their genomic imprinting pattern. These stem cells can differentiate into various types of somatic cells *in vitro* and can produce teratomas *in vivo* [3]. These multipotent cells were isolated and established from adult mouse testes using genetic selection, and the rate of establishment of cell lines was about 27% [4]. Additionally, another group established a similar type of multipotent cells derived from GPR125+ spermatogonial progenitor cells, and derivatives of the three germ layers (contractile cardiac tissues *in vitro* and formed functional blood vessels *in vivo*) have been generated [5]. These results suggest that stem cells in the germline lineage may retain the ability to generate multipotent cells [4].

Conrad et al. reported that human adult germ-line stem cells (haGSCs) from testicular tissue can be induced to form ESC-like cells that display multipotency *in vitro* [6]. Several researchers have also reported the establishment of human mSSC lines with different morphologies in previous papers [7–9]. Those studies were performed with various culture systems, and the findings for human adult testicular tissue remain questionable [10–12]. In 2010, Ko et al. reported that clusters of cells from human testicular fibroblasts (hTFCs) can be easily established from human testicular cultures. Those cells that were not pluripotent were found to be morphologically similar to haGSCs [13]. It was believed that the controversy regarding the characterization of human mSSCs was primarily due to a lack of a protocol for direct induction from well-characterized SSCs. Recently, our group reported that highly pure human SSCs were isolated using a gelatin/laminin-coated dish. These cells proliferated under exogenous feeder-free culture conditions, and then their functions were characterized [14]. The aim of the present study was to identify and isolate human pluripotent SSCs derived from the long-term *in vitro* culture of well-characterized human SSCs.

2. Materials and Methods

2.1. Patient Samples. Testicular tissues were donated from obstructive azoospermic (OA) patients subjected to multiple testicular sperm extraction (TESE)-intracytoplasmic sperm injection (ICSI) treatment. When sperm were found in the dissected samples, the testicular material remaining after clinical requirements was donated for this study after obtaining the patient's consent. This study was approved by the Institutional Review Board of the CHA Gangnam Medical Center, Seoul, Korea.

2.2. Isolation and In Vitro Culture of SSCs. The isolation and culturing of human SSCs was performed as described in our previous report [14]. Briefly, the testicular tissues of 18 OA patients were placed in 10 mL of enzyme solution A containing 0.5 mg/mL type I collagenase (Sigma-Aldrich, St. Louis, MO), 10 μg/mL DNase I (Sigma-Aldrich), 1 μg/mL soybean trypsin inhibitor (Gibco/Invitrogen, Grand Island, NY), and 1 mg/mL hyaluronidase (Sigma-Aldrich) in Ca^{++}/Mg^{++}-free PBS and incubated for 20 min at room temperature (~25°C). After the peritubular cells were removed in the washing step, the seminiferous tubules were re-dissociated in 10 mL of Enzyme Solution B containing 5 mg/mL collagenase, 10 μg/mL DNase I, 1 μg/mL soybean trypsin inhibitor, and 1 mg/mL hyaluronidase in

Ca^{++}/Mg^{++}-free PBS and incubated for 30 min at 37°C. After incubation, the sperm in the dissociated testicular cell samples were removed using a modified two-gradient (35%–70%) Percoll method. The recovered testicular cells were then plated and incubated on uncoated dishes in Germ Cell Culture Medium I, which consisted of DMEM (Gibco/Invitrogen) containing 20% FBS (Gibco/Invitrogen), 10 μmol/L 2-mercaptoethanol (Gibco/Invitrogen), 1% non-essential amino acids (Gibco/Invitrogen), and 10 ng/mL rat GDNF (R&D systems), at 37°C under a humidified atmosphere of 5% CO_2 in air. Over the following 48 hours, unattached cells were harvested and replated on laminin-coated dishes in Germ Cell Culture Medium II, which consisted of StemPro-34 SFM (Invitrogen) supplemented with 6 mg/mL D(+)glucose, 5×10^{-5} M β-mercaptoethanol, 1 μM d(L)-lactic acid, 2 mM L-glutamine, 30 μM pyruvic acid, 10^{-4} M ascorbic acid, 60 ng/mL progesterone, 30 ng/mL β-estradiol (Sigma-Aldrich), 0.2% BSA (ICN Chemicals), 100 U/mL penicillin, 100 μg/mL streptomycin, 1 × insulin-transferrin-selenium (ITS) supplement, 1 × MEM vitamin solution, 1 × MEM non-essential amino acids, 20 ng/mL mouse EGF, 10 ng/mL human bFGF (Invitrogen), 1% KSR (Invitrogen), 10 ng/mL rat GDNF (R&D systems), and 10^3 U/mL LIF (Chemicon, Billerica, MA) (modified from [14]). To obtain highly pure SSCs, un-attached cells collected after a 4-h incubation were re-sorted by magnetic activated cell separation (MACS) using an anti-CD9 antibody. The isolated SSCs slowly proliferated and then formed slightly attached clump-like structures ($\geq$10 cells) on the bottom of dishes 2–4 weeks after seeding. During culture, approximately 80% of the medium was changed carefully every other day under a stereomicroscope to avoid the loss of floating clumps. Only clumps were collected. The clumps were dissociated by trypsinization and then re-plated every 2 weeks using the same medium. After every passage, cells clumps were divided into two groups. One group was fixed or sampled for characterization, and the other group was passaged using the method previously described.

2.3. Characterization of SSCs from In Vitro Culture. To characterize isolated highly pure SSCs and to investigate the relative expression levels of multipotent markers in the SSC clumps, we performed immunocytochemistry using the SSC markers GFR α1 (Chemicon International) and CD9 (Chemicon international) and the pluripotent stem cell markers OCT-4 (Santa Cruz), SSEA-4 (Chemicon international), TRA-1-60 (Chemicon international), and TRA-1-81 (Chemicon international). The samples were washed three times in DPBS with 5% FBS and were then fixed in paraformaldehyde (4% v/v in DPBS) for 24 hours. For permeabilization, the cell clumps were incubated in 0.1% Triton X-100 in DPBS for 1 hour. After washing three times with DPBS, the nonspecific binding of antibodies was suppressed by incubating the cells in blocking solution (4% normal goat serum in DPBS) for 30 min at room temperature. After washing three times with PBS, immunocytochemical staining was performed by incubating the fixed samples with primary antibody diluted 1 : 200–1 : 500 with DPBS containing 0.1%

Tween-20 and 1% BSA for 60 min at room temperature or overnight at 4°C. Immunoreactive proteins were then detected using CY3- or FITC-conjugated secondary antibodies diluted 1 : 500 with DPBS for 60 min at room temperature. Finally, samples were counterstained with 1 μg/mL 4′,6′-diamidino-2-phenylindole (DAPI; Sigma). Following multiple washes, samples were mounted in Vectashield mounting medium (Vector laboratories, Burlingame, CA). The staining was viewed using an inverted confocal laser scanning microscope (LSM 510; Carl Zeiss, Oberkochen, Germany) with fluorescence at a 400x magnification. Micrographs were stored in LSM (Zeiss LSM Image Browser version 2.30.011; Carl Zeiss Jena GmbH, Jena, Germany).

Alkaline phosphatase activity was assessed by histochemical staining. Cells were fixed in 4% paraformaldehyde at room temperature for 1 min, washed twice with PBS and stained with an alkaline phosphatase substrate solution (10 mL FRV-Alkaline Solution, 10 mL Naphthol AS-BI Alkaline Solution; Alkaline Phosphatase kit, Sigma-Aldrich) for 30 min at room temperature. Alkaline phosphatase activity was detected colorimetrically (red) by light microscopy.

2.3.1. RT-PCR. RT-PCR was performed to assess the expression of multipotent marker genes, specifically, *OCT4*, *NANOG*, and *Integrin* $\alpha 6$, in clumps from SSCs. Total RNA was extracted from 100 colonies using the TRIzol method (Gibco). Amplification was performed in a 20 μL reaction mixture containing 10 mmol/L Tris-HCl (pH 8.3), 2 mmol/L MgCl$_2$, 50 mmol/L KCl, 0.25 mmol/L dNTP, 3–5 pmol of each primer, and 1.25 IU Taq polymerase (Gibco). The following genes were amplified using the primers indicated in parentheses: *OCT-4* (F: 5′-GGA AAG GCT TCC CCC TCA GGG AAA GG-3′, R: 5′-AAG AACA TGT GTA AGC TGC GGC CC-3′, 460 bp, GenBank accession number NM002701); *NANOG* (F: 5′-CCC ATC CAG TCA ATC TCA-3′, R: 5′-CCT CCC AAT CCC AAA CAA-3′, 565 bp, GenBank accession number NM024865); *Integrin* $\alpha 6$ (F: 5′-GGG AGC CTC TTC GGC TTC TC-3′, R: CAC ATG TCA CGA CCT TGC CC-3′, 286 bp, GenBank accession number NM000210) and 18S ribosomal RNA (F: 5′-TAC CTA CCT GGT TGA TCC TG-3′, R: 5′-GGG TTG GTT TTG ATC TGA TA-3′, 255 bp, GenBank accession number K03432). PCR was initiated with a denaturation step at 94°C for 5 min, followed by 35–40 cycles of 30 s at 94°C, 30 s at 55–60°C, and 30 s at 72°C. A final extension step for 10 min at 72°C completed the amplification reaction, after which the products were separated by 1.5% agarose-gel electrophoresis. Negative controls included mock transcription without mRNA and PCR with distilled deionized water.

2.4. Flow Cytometry. SSC clumps were dissociated in trypsin-EDTA and resuspended in PBS containing 2% FBS. Then, the cells were incubated with APC-conjugated antibody to SSEA-4 (BD/Pharmingen) for 60 min at 4C. Finally, the cells were placed in the flow cytometer (Becton Dickinson FACS IV San Jose, CA, USA) for analysis. Cells without antibody staining were used as negative controls.

2.5. Karyotype Analysis. Chromosome spreads were prepared as described [15]. Briefly, SSCs were treated with 0.06 μg/mL colcemid (Invitrogen) for 2–4 h, trypsinized, incubated in 0.075 M KCl for 10 min, and fixed in Carnoy's fixative. The chromosome number and banding patterns were analyzed with a 300–500 band resolution.

2.6. EB Formation from SSCs. After 5 passages, over 200–400 SSC clumps cultured in HEPES-buffered DMEM/F-12 (Gibco) supplemented with 10 μg/mL ITS (Gibco), 10^{-4} mol/L vitamin C (Sigma), 10 μg/mL vitamin E (Sigma), 3.3×10^{-7} mol/L retinoic acid (Sigma), 3.3×10^{-7} mol/L retinol (Sigma), 1 mmol/L pyruvate (Sigma), 2.5×10^{-5} IU recombinant human FSH (Gonal-F; Serono), 10^{-7} mol/L testosterone (Sigma), 1 × antibiotic-antimycotic (ABAM, containing penicillin, streptomycin and amphotericin B; Gibco), and 10% bovine calf serum (Hyclone), for spontaneous *in vitro* differentiation [6]. SSCs clumps were transferred to 1.0 mL of differentiation culture medium in a 24-well dish and were cultured for up to 4 weeks at 37°C in a humidified atmosphere of 5% CO^2 in air. The medium was replaced on alternate days. After culturing, the EBs were fixed in 10% neutral buffered formalin, embedded in paraffin, stained with hematoxylin and eosin (H&E) and examined immunocytochemically. The endoderm marker α-fetoprotein (Chemicon international), the ectoderm marker nestin (Chemicon international) and the mesoderm marker cardiac troponin I (Chemicon international) were used.

2.7. Teratoma Formation from Pluriotent SSCs. Pluripotency was determined by harvesting ~2,000 SSC clumps (~2×10^5 cells) and injecting them subcutaneously into the back of 4- to 8-week-old severe combined immunodeficient (SCID) mice (CB 17 strain; Jackson Laboratory, Bar Harbor, ME) using a sterile 26 G needle. After 12 weeks, the resulting tumors were fixed in 10% neutral buffered formalin, embedded in paraffin, cut into 5-μm serial sections, H&E-stained and immunocytochemically examined. The human specific markers, α-fetoprotein (Chemicon international), MAP-2 (Chemicon international) and STEM 121 (Stem-121, Stem Cells Inc., Cambridge, UK) were used.

3. Results

3.1. Morphology and Karyotype of In Vitro Cultured SSCs. Significant staining for pluripotent marker (SSEA-4) was detected in hESCs. But testicular tissue did not express this marker (Figure 1(a)). In the primary culture after enzyme treatment, seeding cells exhibited positive signal of GFR $\alpha 1$. However, SSEA-4, a pluripotent marker, was not detected in those cells (Figure 1(b)). After the selection procedure, the isolated and cultivated SSCs exhibited high expression levels of SSC marker, as described in our previous report [14]. Significant staining for SSC markers (CD9 and GFR $\alpha 1$) was detected at a high level in the SSC clumps (Figure 1(c)). SSCs were well maintained and proliferated in culture, ultimately forming small clumps (>10 cells), and were passaged by

(a)

(b)

Primary seeding cells

(c)

(d)

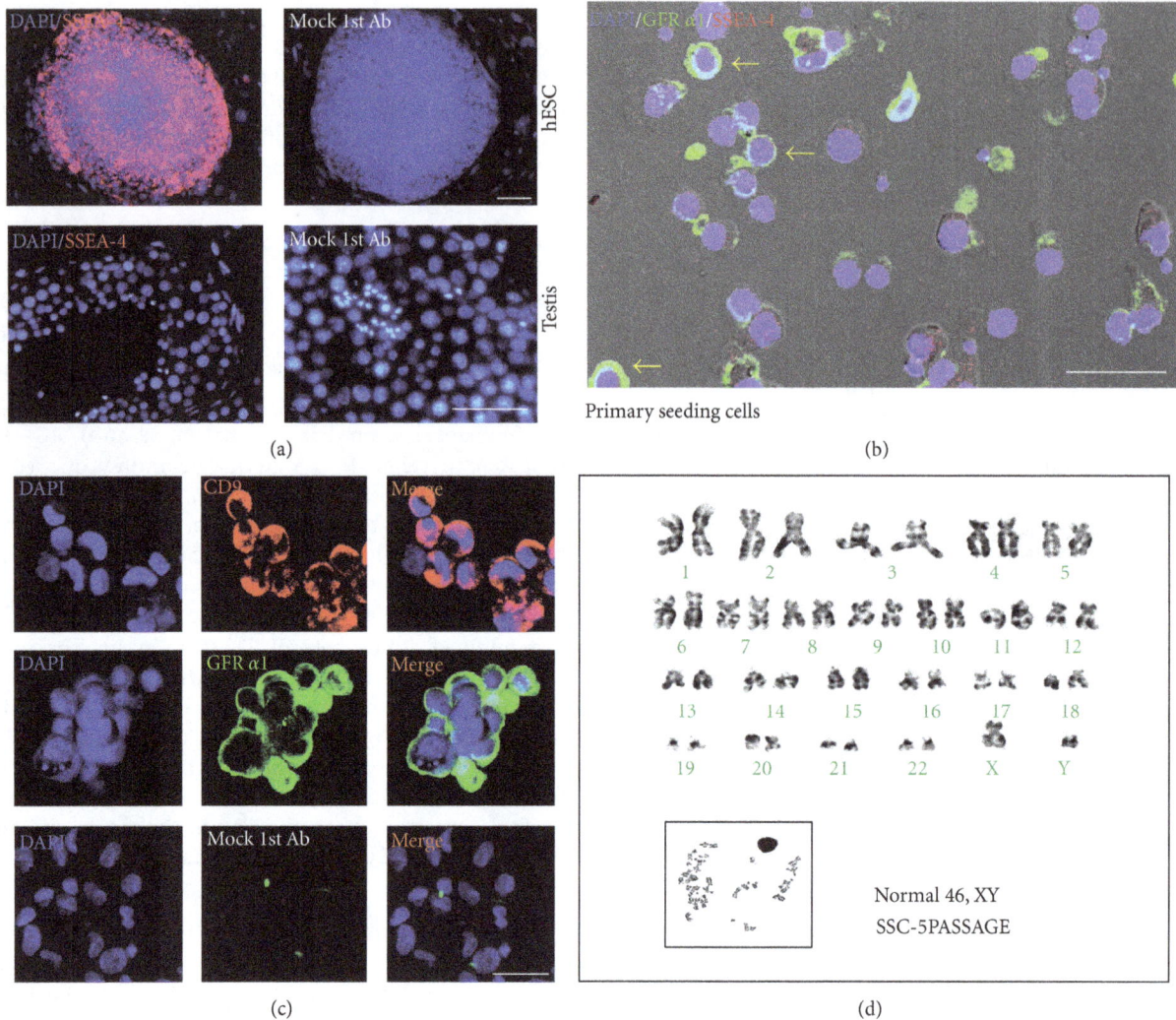

FIGURE 1: Characterization of spermatogonial stem cells (SSCs). (a) Expression of pluripotent stem cells marker (SSEA-4) in the hESCs and testis. (b) Expression of pluripotent stem cells (SSEA-4) and SSCs marker (GFR α1) in the primary seeding cells. The yellow arrows indicate a GFR α1-positive signals which were expressed in SSCs. (c) Localization of specific markers for SSCs (CD9, red, and GFR α1, green) in the cultured cell clumps (at passage 1). (d) Karyotyping of SSCs performed at passage 5. Note: Mock 1st Ab; cultured cells were stained with secondary antibody only as a negative control. Scale bars = 100 μm.

trypsin-dissociation and plating on new culture dishes with fresh medium containing GDNF. SSCs attached to the plate after incubation for 2-3 days and then proliferated by re-forming floating clumps. Somatic cells and differentiated cells attached to the dish. The passaging of floating clumps was repeated every two weeks. Dissociated SSCs continued to proliferate for more than 5 passages (>10 weeks) and re-formed floating clumps. Using this method, we successfully isolated SSCs and maintained proliferating SSC cultures from more than 83.3% (15/18) of OA patients.

To determine the chromosome stability of SSC clumps, karyotyping analysis was performed at passage 5. The results demonstrated that the SSC clumps had a normal karyotype (46, XY), and no indications of other cytogenetic abnormalities were detected (Figure 1(d)).

3.2. Immunocytochemical Staining. The morphology of SSC clumps was flattened and loosely associated at first (upper panel of Figure 2(a)) and then changed to tightly associated clumps (lower panel of Figure 2(a)). High levels of AP activity were associated with multicellular clumps in vitro (Figure 2(b)). Figure 2(c) summarizes the expression of pluripotent stem cell markers in the SSC clumps. At passages 1 to 5, immunostaining analysis showed that SSC clumps expressed pluripotent stem cell markers (Oct-4, SSEA-4, TRA-1-60, and TRA-1-81). The expression levels of these markers were slightly increased up to passage 5, but their relative amounts were still low. Additionally, in SSC clumps, some cells expressed the pluripotent stem cell marker SSEA-4, and most expressed a SSC marker, GFR α1. Interestingly, a few SSCs co-expressed SSEA-4 and GFR α1 (Figure 2(d)).

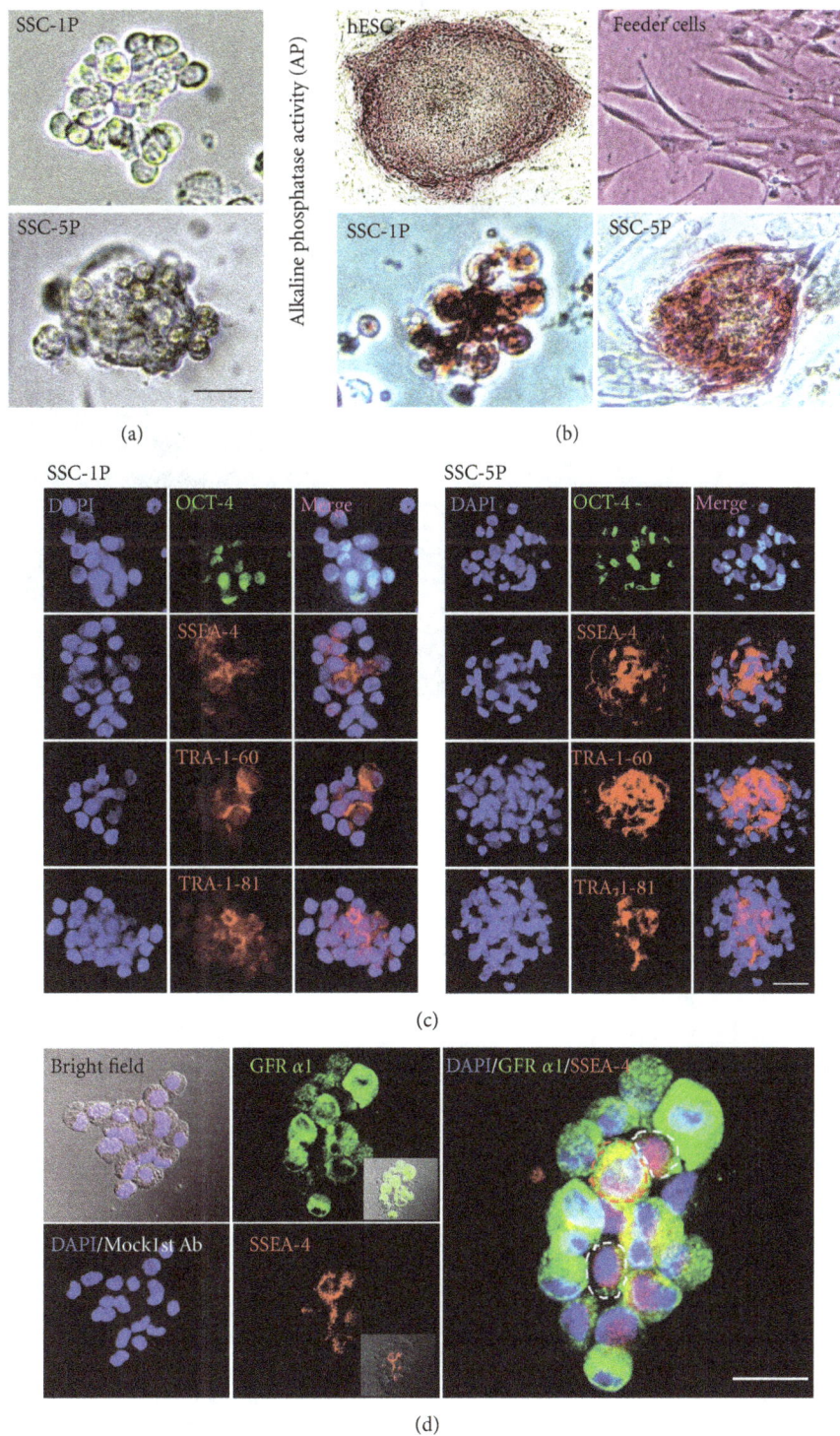

FIGURE 2: Characterization of pluripotent stem cells within spermatogonial stem cell (SSC) clumps. (a) Morphology of spermatogonial stem cell (SSC) clumps after *in vitro* culture (upper: passage 1, lower: passage 5). (b) Alkaline phosphatase activity in SSC colonies after culture (passage 1 and passage 5). CHA-hES4 cells (human embryonic stem cell line, hESCs) were used as a positive control, and feeder cells were used as a negative control. (c) Immunocytochemical analysis of pluripotent stem cell markers (OCT4, SSEA-4, TRA 1–60 and TRA 1–81) was performed with SSC clumps at passage 1 (left panel) and passage 5 (right panel). (d) Colocalization of specific markers for pluripotent stem cells (SSEA-4, red color) and SSCs (GFR α1, green color) in the cultured SSC clumps (at passage 1). The red circle indicates a SSC in which both markers were co-expressed. The yellow circles indicates a mSSC in which pluripotent stem cell-marker were only expressed. Note: Mock 1st Ab; cultured cells stained with secondary antibody only as a negative control. Scale bars = 50 μm.

(a)

(b)

(c)

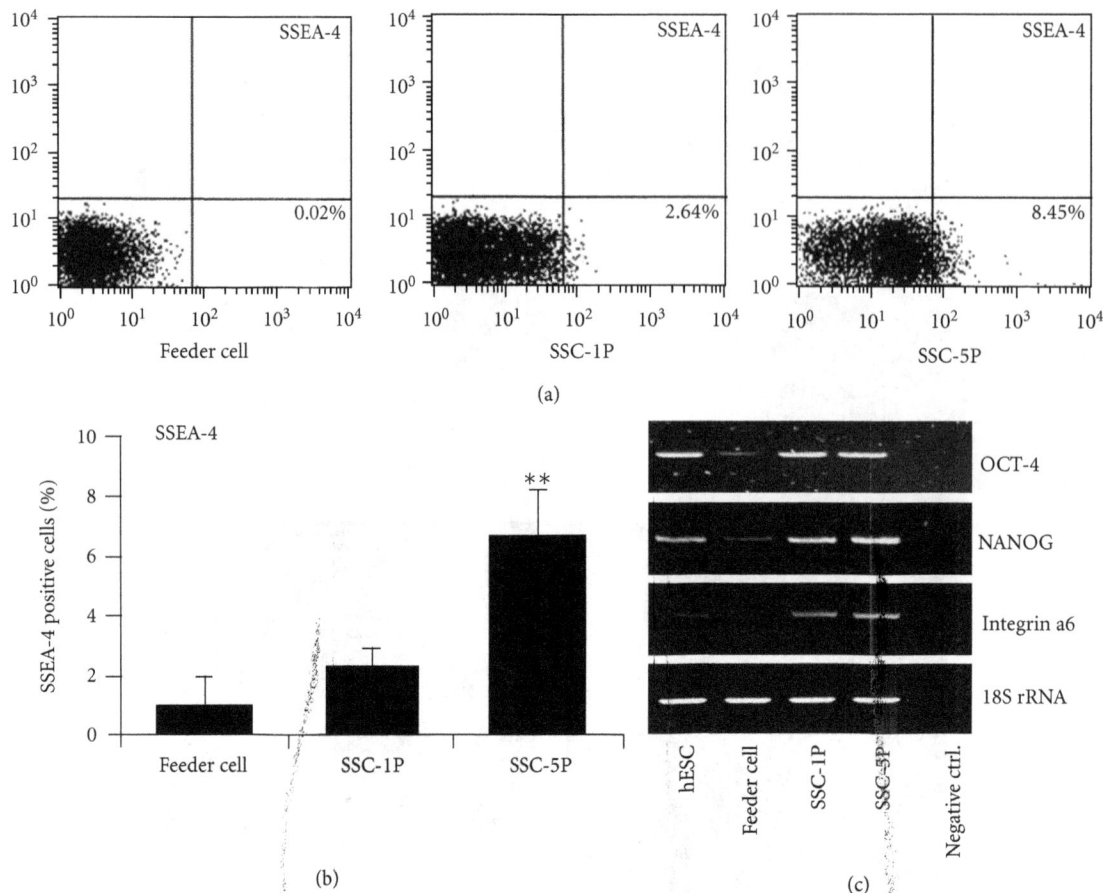

FIGURE 3: Identification of pluripotent stem cells within spermatogonial stem cell (SSC) clumps using flow cytometry and RT-PCR. (a) Flow cytometric analysis of the cultured SSC clumps using a pluripotent stem cell marker (SSEA-4). (b, c) Immunocytochemical analysis and RT-PCR analysis of pluripotent stem cell markers (*OCT-4, NANOG,* and *Integrin α6*) in the cultured SSC clumps. 18S ribosomal RNA was used as an experimental control.

3.3. Flow Cytometric and Gene Expression Analysis of Pluripotent Stem Cell Markers in SSC Clumps.

The flow cytometric analysis indicated that the number of SSEA-4-positive cells in SSC clumps was greater at passage 5 (8.45%) than at passage 1 (2.64%). These results were similar to the immunostaining results (Figures 3(a) and 3(b)). The gene expression levels of the markers *OCT-4, NANOG,* and *Integrin α6* were confirmed by RT-PCR. ESC cells and SSC clumps (from passage 1 and 5) strongly expressed the mRNAs for *OCT-4, NANOG* and *Integrin α6*. In contrast, testicular feeder cells did not express or weakly expressed these genes (Figure 3(c)).

3.4. Spontaneously Differentiation In Vitro.

To investigate the differentiation potential of the SSCs clumps *in vitro*, SSC clumps at passage 5 were spontaneously differentiated using the suspension EB-formation method in the absence of growth factors. After 10–14 days under these culture conditions, the SSC clumps consistently aggregated and formed EB-like structures (Figure 4(a)). Markers of ectodermal progenitor cells (nestin, which is present in neuro-epithelial cells) and endodermal lineage cells (α-fetoprotein, expressed in early and late hepatocytes) were detected in the EB-like

structures [16, 17]. We also found that marker of mesoderm cells (cardiac protein, widely localized in cardiac muscle cells) was expressed in the EB-like structures and the expression of integrin α6, a marker of pluripotent stem cells, was remained after *in vitro* differentiation (Figure 4(b)).

3.5. Teratoma Formation Potential of SSC Clumps.

To confirm the *in vivo* differentiation potential, we subcutaneously injected SSC clumps derived from OA patients into the dorsal skin of immune-deficient mice and examined the ability of these cells to form teratomas *in vivo*. Injected SSC clumps gave rise to teratomas in recipients 12 weeks after transplantation. However, wide-scale expansion to large teratomas was not observed, as is typically observed with hESCs (Figure 5(a)). However, the small teratomas contained derivatives of all three embryonic germ layers. Histological analysis revealed that a variety of cell types was present (Figure 5(b)). Using this assay, teratomas were formed in three of ten (30%) injected mouse. In our previous report, 80–90% density SSC-positive cells did not form teratomas or other tumours after transplanatation into SCID mice (data not shown) [14]. Teratomas from the cultured SSC were distinguished from

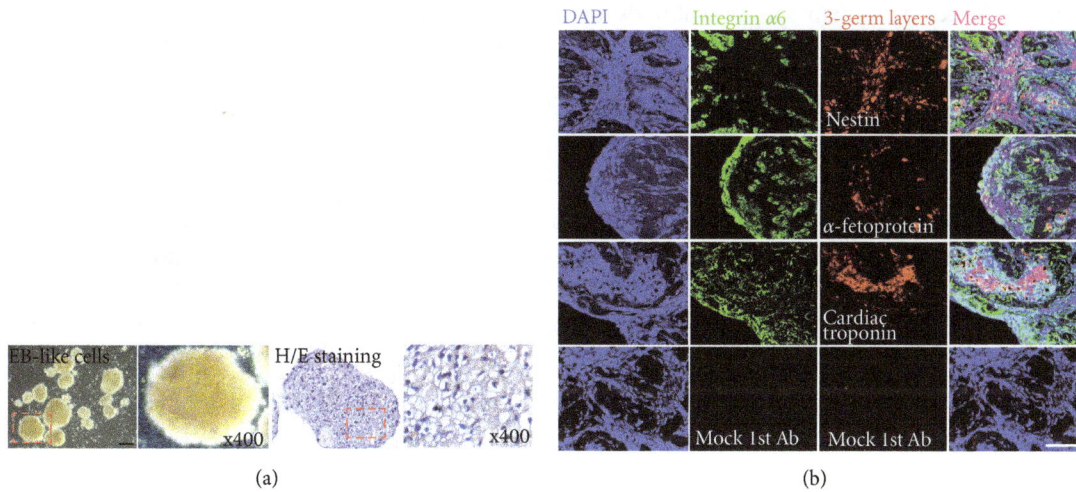

FIGURE 4: Spontaneous differentiation of spermatogonial stem cell (SSC) clumps *in vitro*. (a) Morphology of differentiated EB-like structured cells after plating onto non-coated dishes, and hematoxylin and eosin staining of EB-like structured cells. (b) Expression of three germ layer-specific markers in EB-like structured cells derived from SSC clumps. Markers of ectodermal progenitor cells (nestin), endodermal lineage cells (α-fetoprotein) and mesodermal cells (cardiac protein). Note: Mock 1st Ab; cultured cells stained with secondary antibody only as a negative control. Scale bars = 100 μm.

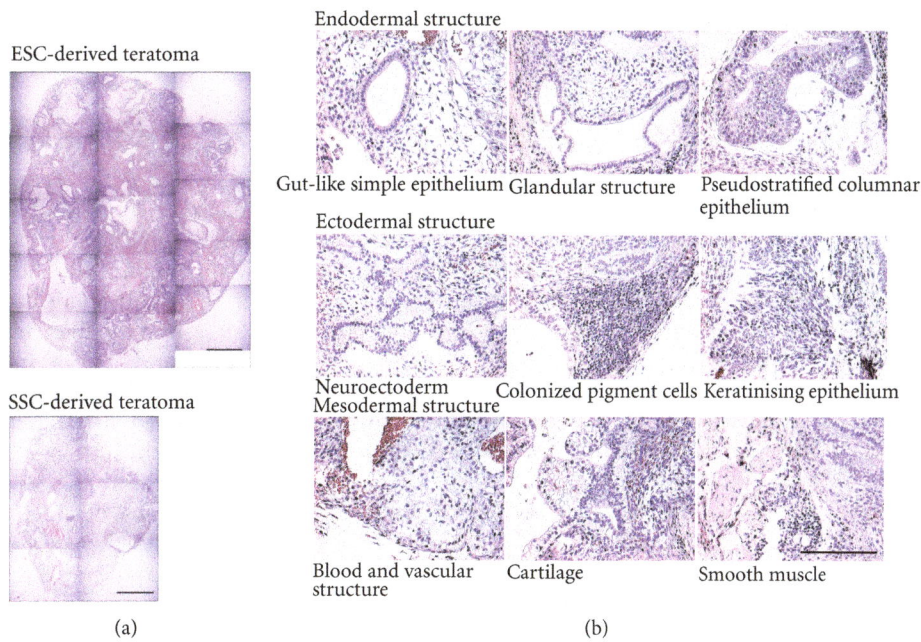

FIGURE 5: Teratoma formation from transplanted pluripotent stem cells within spermatogonial stem cells (SSCs) (a) Teratoma obtained from hESCs (CHA-hES4) and multipotent stem cells among SSCs. (b) Morphology of three germ layer-like structures obtained from multipotent stem cells within SSCs. (c) Immunohistochemical studies for human specific antibody. α-fetoprotein (ectoderm marker reacts with human), MAP-2 (mesoderm marker reacts with human) and STEM121 (cytoplasm marker reacts with human). Note: Mock 1st Ab; cultured cells stained with secondary antibody only as a negative control. Scale bars = 100 μm.

host SCID mouse tissues by human specific antibody (α-fetoprotein, MAP-2, and STEM121) (Figure 5(c)).

4. Discussion

Although the existence of pluripotent stem cells in the testis has been reported in human models [6–9, 18], the characteristics and origin of these cells have remained uncertain [10–13]. In contrast to mouse mSSCs [3], human pluripotent SSCs were isolated from heterogeneous testicular cells including SSCs and somatic cells that were not well-characterized. In the present study, we confirmed that human pluripotent SSCs with ESC-like characteristics can be derived from highly pure SSCs. These cells underwent culture-induced reprogramming without any genetic manipulation.

In 2003, Kanatsu-Shinohara and his colleagues succeeded in culturing SSCs obtained from postnatal mouse testes. SSCs formed uniquely shaped colonies when cultured in the presence of GDNF [19]. Even after 2 years of culture *in vitro*, the cells had the potential to produce normal offspring [20]. Recent studies also indicate that the self-renewal of these SSCs in rodents (and possibly all mammals) is dependent on bFGF and LIF [21, 22]. Based on these properties, we isolated human SSCs from testes and propagated these cells in a modified *in vitro* culture system based on that for mouse SSCs [14]. In the present study, the generation of pluripotent stem cells was observed in long-term cultures of purified human SSCs with a normal chromosome status, but immunocytochemistry and flow cytometric analysis revealed that the relative amount of these cells was disappointingly low (~10%). It has been suggested that the current culture conditions may not fully support the stable maintenance or propagation of pluripotent SSCs, and other factors provided by ESC culture medium or additional feeder cells may be required [18]. Generally, pluripotent stem cells such as ESCs are plated and cultured on mouse embryonic fibroblast feeder cell layers, and following a brief period of attachment and expansion, the resulting outgrowth is disaggregated and replated onto another feeder cell layer [23]. In our culture system, the feeder cells, mouse embryonic fibroblasts, did not affect the proliferation of human pluripotent SSCs (data not shown). Hence, it is important to identify suitable feeder cells for pluripotent SSCs.

The clump shape of SSCs was similar to that of ES cell colonies during long-term culture (Figure 2(a)). Additionally, very small numbers of cells co-expressed GFR α1 and SSEA-4, and some cells expressed SSEA-4 (Figure 2(d)). These results are very similar to our previous results in mice indicating that the intermediate state of SSCs has an expression profile more similar to that of pluripotent stem cells than to that of mSSCs, yet the expression of germ cell markers is preserved [24]. These results suggest that the morphological transformation of SSCs and the start of the expression of specific markers indicate that SSCs can be reprogramed during long-term culture under specific conditions.

Kerr et al. have provided some findings regarding the development and differentiation of human germ cells in the fetal testis, including a very small population of PGCs with a molecular signature including OCT4, NANOG, c-Kit, SSEA-1, SSEA-4, and alkaline phosphatase [25]. Our initial characterization showed that a small number of isolated SSCs expressed markers of undifferentiated stem cells such as OCT4, SSEA-4, Tra-1-60 and Tra-1-81 [26, 27]. The population of these cells was slightly larger after passaging (Figures 2(b) and 3) and was then maintained under defined culture conditions. Pluripotent stem cells have the potential to differentiate into nearly all cell types in the human body [28]. *In vitro* and *in vivo*, these cells are able to generate embryoid bodies or teratomas that express marker genes of all three germ layers and develop into different cell types. The capacity of pluripotent stem cells to differentiate into almost all of the cell types of the human body highlights the potentially promising role of these cells in cell replacement therapies for the treatment of human diseases [29]. If the pluripotent stem cells can be derived from existed/reprogrammed human testis cells, these cells will generate embryoid bodies in culture. In our system, pluripotent cells among SSCs were able to generate EB-like structures and exhibited some characteristics of pluripotency, differentiating into cells of the three germ layers (Figure 4). Additionally, these cells formed teratomas after injection into SCID mice (Figure 5).

5. Conclusion

This result revealed that pluripotent cells were induced from SSCs during *in vitro* culture and were present within the SSC clumps isolated from human adult testicular tissues. Additionally, no genetic modification was needed for this procedure, which can be compared to the generation of iPSCs from somatic cells [11]. The isolation and long-term proliferation of mSSCs from human testicular tissue may allow the development and use of individual cell-based therapies without ethical and immunological problems. In this study, we demonstrated that adult SSCs are able to develop into pluripotent stem cells in *in vitro* culture, which differentiate into cells of the three germ layers. However, before pluripotent stem cells derived from SSCs can be used in treatments, a highly efficient system to induce and propagate these cells on a large scale is required.

Acknowledgments

This work was partly supported by a Grant from the Stem Cell Research Program (2006-2004127) and from the Priority Research Centers Program (2009-0093821) of the Ministry of Education, Science, and Technology, Republic of Korea, and Korea Science and Engineering Foundation. The authors also thank thier laboratory for technical assistance and advice during experiments.

References

[1] M. J. Shamblott, J. Axelman, S. Wang et al., "Derivation of pluripotent stem cells from cultured human primordial germ cells," *Proceedings of the National Academy of Sciences of the United States of America*, vol. 95, no. 23, pp. 13726–13731, 1998.

[2] C. L. Kerr, C. M. Hill, P. D. Blumenthal, and J. D. Gearhart, "Expression of pluripotent stem cell markers in the human fetal testis," *Stem Cells*, vol. 26, no. 2, pp. 412–421, 2008.

[3] M. Kanatsu-Shinohara, K. Inoue, J. Lee et al., "Generation of pluripotent stem cells from neonatal mouse testis," *Cell*, vol. 119, no. 7, pp. 1001–1012, 2004.

[4] K. Guan, K. Nayernia, L. S. Maier et al., "Pluripotency of spermatogonial stem cells from adult mouse testis," *Nature*, vol. 440, no. 7088, pp. 1199–1203, 2006.

[5] M. Seandel, D. James, S. V. Shmelkov et al., "Generation of functional multipotent adult stem cells from GPR125^{+} germline progenitors," *Nature*, vol. 449, no. 7160, pp. 346–350, 2007.

[6] S. Conrad, M. Renninger, J. Hennenlotter et al., "Generation of pluripotent stem cells from adult human testis," *Nature*, vol. 456, no. 7220, pp. 344–349, 2008.

[7] N. Kossack, J. Meneses, S. Shefi et al., "Isolation and characterization of pluripotent human spermatogonial stem cell-derived cells," *Stem Cells*, vol. 27, no. 1, pp. 138–149, 2009.

[8] N. Golestaneh, M. Kokkinaki, D. Pant et al., "Pluripotent stem cells derived from adult human testes," *Stem Cells and Development*, vol. 18, no. 8, pp. 1115–1125, 2009.

[9] S. C. Mizrak, J. V. Chikhovskaya, H. Sadri-Ardekani et al., "Embryonic stem cell-like cells derived from adult human testis," *Human Reproduction*, vol. 25, no. 1, pp. 158–167, 2010.

[10] J. V. Chikhovskaya, M. J. Jonker, A. Meissner, T. M. Breit, S. Repping, and A. M. van Pelt, "Human testis-derived embryonic stem cell-like cells are not pluripotent, but possess potential of mesenchymal progenitors," *Human Reproduction*, vol. 27, no. 1, pp. 210–221, 2012.

[11] K. Takahashi, K. Tanabe, M. Ohnuki et al., "Induction of pluripotent stem cells from adult human fibroblasts by defined factors," *Cell*, vol. 131, no. 5, pp. 861–872, 2007.

[12] N. Tapia, M. J. Araúzo-Bravo, K. Ko, and H. R. Schöler, "Concise review: challenging the pluripotency of human testis-derived ESC-like cells," *Stem Cells*, vol. 29, no. 8, pp. 1165–1169, 2011.

[13] K. Ko, M. J. Araúzo-Bravo, N. Tapia et al., "Human adult germline stem cells in question," *Nature*, vol. 465, no. 7301, pp. E1–E2, 2010.

[14] J. J. Lim, S. Y. Sung, H. J. Kim et al., "Long-term proliferation and characterization of human spermatogonial stem cells obtained from obstructive and non-obstructive azoospermia under exogenous feeder-free culture conditions," *Cell Proliferation*, vol. 43, no. 4, pp. 405–417, 2010.

[15] M. G. A. Nagy K and K. Vinetrstein, *Isolation and Culture of Blastocyst-Derived Stem Cell Lines. Manipulating the Mouse Embryo*, Cold Spring Harbor Laboratory Press, Harbor, NY, USA, third edition, 2001.

[16] Y. Yan, J. Yang, W. Bian, and N. Jing, "Mouse nestin protein localizes in growth cones of P19 neurons and cerebellar granule cells," *Neuroscience Letters*, vol. 302, no. 2-3, pp. 89–92, 2001.

[17] H. Araki, H. Ueda, and S. Fujimoto, "Immunocytochemical localization of alpha-fetoprotein in the developing and carbon tetrachloride-treated rat liver," *Acta Anatomica*, vol. 143, no. 3, pp. 169–177, 1992.

[18] M. Stimpfel, T. Skutella, M. Kubista, E. Malicev, S. Conrad, and I. Virant-Klun, "Potential stemness of frozen-thawed testicular biopsies without sperm in infertile men included into the *in vitro* fertilization programme," *Journal of Biomedicine and Biotechnology*, vol. 2012, Article ID 291038, 15 pages, 2012.

[19] M. Kanatsu-Shinohara, N. Ogonuki, K. Inoue et al., "Long-term proliferation in culture and germline transmission of mouse male germline stem cells," *Biology of Reproduction*, vol. 69, no. 2, pp. 612–616, 2003.

[20] M. Kanatsu-Shinohara, N. Ogonuki, T. Iwano et al., "Genetic and epigenetic properties of mouse male germline stem cells during long-term culture," *Development*, vol. 132, no. 18, pp. 4155–4163, 2005.

[21] F. K. Hamra, K. M. Chapman, D. M. Nguyen et al., "Self renewal, expansion, and transfection of rat spermatogonial stem cells in culture," *PNAS*, vol. 102, pp. 17430–17435, 2005.

[22] B. Y. Ryu, H. Kubota, M. R. Avarbock, and R. L. Brinster, "Conservation of spermatogonial stem cell self-renewal signaling between mouse and rat," *Proceedings of the National Academy of Sciences of the United States of America*, vol. 102, no. 40, pp. 14302–14307, 2005.

[23] M. F. Pera, B. Reubinoff, and A. Trounson, "Human embryonic stem cells," *Journal of Cell Science*, vol. 113, no. 1, pp. 5–10, 2000.

[24] H. J. Kim, H. J. Lee, J. J. Lim et al., "Identification of an intermediate state as spermatogonial stem cells reprogram to multipotent cells," *Molecules and Cells*, vol. 29, no. 5, pp. 519–526, 2010.

[25] C. L. Kerr, C. M. Hill, P. D. Blumenthal, and J. D. Gearhart, "Expression of pluripotent stem cell markers in the human fetal testis," *Stem Cells*, vol. 26, no. 2, pp. 412–421, 2008.

[26] Y. Q. Shi, Q. Z. Wang, S. Y. Liao, Y. Zhang, Y. X. Liu, and C. S. Han, "*In vitro* propagation of spermatogonial stem cells from KM mice," *Frontiers in Bioscience*, vol. 11, no. 2, pp. 2614–2622, 2006.

[27] M. C. Hofmann, L. Braydich-Stolle, and M. Dym, "Isolation of male germ-line stem cells; Influence of GDNF," *Developmental Biology*, vol. 279, no. 1, pp. 114–124, 2005.

[28] M. Schuldiner, O. Yanuka, J. Itskovitz-Eldor, D. A. Melton, and N. Benvenisty, "Effects of eight growth factors on the differentiation of cells derived from human embryonic stem cells," *Proceedings of the National Academy of Sciences of the United States of America*, vol. 97, no. 21, pp. 11307–11312, 2000.

[29] M. Stojkovic, M. Lako, T. Strachan, and A. Murdoch, "Derivation, growth and applications of human embryonic stem cells," *Reproduction*, vol. 128, no. 3, pp. 259–267, 2004.

5

The Exposure of Breast Cancer Cells to Fulvestrant and Tamoxifen Modulates Cell Migration Differently

Dionysia Lymperatou, Efstathia Giannopoulou,
Angelos K. Koutras, and Haralabos P. Kalofonos

Clinical Oncology Laboratory, Division of Oncology, Department of Medicine, University of Patras,
Patras Medical School, 26504 Rio, Greece

Correspondence should be addressed to Haralabos P. Kalofonos; kalofonos@upatras.gr

Academic Editor: Davide Vigetti

There is no doubt that there are increased benefits of hormonal therapy to breast cancer patients; however, current evidence suggests that estrogen receptor (ER) blockage using antiestrogens is associated with a small induction of invasiveness *in vitro*. The mechanism by which epithelial tumor cells escape from the primary tumor and colonize to a distant site is not entirely understood. This study investigates the effect of two selective antagonists of the ER, Fulvestrant (Fulv) and Tamoxifen (Tam), on the invasive ability of breast cancer cells. We found that 17β-estradiol (E_2) demonstrated a protective role regarding cell migration and invasion. Fulv did not alter this effect while Tam stimulated active cell migration according to an increase in Snail and a decrease in E-cadherin protein expression. Furthermore, both tested agents increased expression of matrix metalloproteinases (MMPs) and enhanced invasive potential of breast cancer cells. These changes were in line with focal adhesion kinase (FAK) rearrangement. Our data indicate that the anti-estrogens counteracted the protective role of E_2 concerning migration and invasion since their effect was not limited to antiproliferative events. Although Fulv caused a less aggressive result compared to Tam, the benefits of hormonal therapy concerning invasion and metastasis yet remain to be investigated.

1. Background

Breast cancer is the most frequent malignancy cancer in women. It is estimated that approximately 75% of breast tumors are estrogen receptor (ER) positive, and their growth is stimulated by estrogens [1]. Estrogen-based therapies represent the mainstay in the treatment of hormone-dependent breast cancer with the ER modulator Tamoxifen (Tam) improving significantly the clinical outcome of patients with both early and advanced breast cancer [2]. Furthermore, Fulvestrant (Fulv) that belongs to a recently developed group of antiestrogens (selective estrogen receptor downregulators—SERDs) has extended the therapeutic options in the management of breast cancer patients [2, 3].

Invasion is considered as the hallmark of malignancy and is the first in the cascade of events leading to tumor development and metastasis. During invasion, the tumor cells penetrate into tissues breaking the basement membrane and allowing tumor growth. The invading tumor cells are able to enter the circulation so as distant metastasis occurs [4, 5]. Both invasion and metastasis require cell migration. The cell type and tissue microenvironment define the way of cell movement that is generally categorized as single and collective cell migration. During single cell migration, cells disseminate from the primary tumor as individual using either amoeboid or mesenchymal type movement, while in collective migration cells move as cell sheets or clusters [6, 7].

Degradation of the extracellular matrix (ECM) is one of the most important events in the spread of malignant cells, and it is well documented that it plays an essential role in tumor prognosis [8]. Matrix metalloproteinases (MMPs), zinc finger dependent enzymes, promote invasion, metastasis, and angiogenesis through the digestion of ECM components as well as surface factors' receptor and junctional proteins involved in cell-cell and cell-ECM interactions. MMPs consist of 23 members, which are classified into different groups, including gelatinases. MMP-2 and MMP-9 are gelatinases that are related to tumor invasion and metastasis by their capacity for tissue remodeling via ECM, as well as their involvement in epithelial mesenchymal transition (EMT) [8, 9]. EMT is the key mechanism by which tumor cells gain invasive and metastatic ability, as EMT enables separation of individual cells from the primary tumor mass and promotes cell migration. During EMT, epithelial cells lose polarity and cell-cell contacts and undergo a complete remodeling of the cytoskeleton that leads to the acquisition of the mesenchymal features such as motility, invasiveness, and resistance to apoptosis [10–12]. One of the most pivotal steps in this process is the loss of E-cadherin, a cell-adhesion protein that maintains the cell-cell contacts [13]. However, the expression of E-cadherin is regulated by several transcription factors including Snail, Slug, and Twist. Furthermore, the nonreceptor tyrosine kinase focal adhesion kinase (FAK) is associated with highly invasive breast cancers, and it mediates several pathways leading to proliferation, migration, and adhesion [14]. Phosphorylation is required for FAK activation, and it has been shown that estrogens are able to promote rapid phosphorylation of FAK at tyrosines residues [15].

Despite the undoubted benefits that estrogen-based therapies offer to ER⁺ breast cancer patients, *de novo* and acquired resistance to such therapies presents a major clinical problem [16]. The aim of the current study is to evaluate the effect of antiestrogens Fulv and Tam as well as the active metabolites of Tam, Endoxifen (End), and 4-OH-Tamoxifen (4-OH-T) on migration of 17β-estradiol- (E_2-) stimulated breast cancer cells. We focused on single and collective cell migration since these are the main ways for cells to migrate. To understand the effect of estrogen receptors' inhibition on cell migration, we assessed the effect of the antiestrogens on MMPs levels, on protein levels as well as on localization of E-cadherin and Snail and colocalization of FAK phosphorylated form with actin fibers.

2. Methods

2.1. Cell Culture and Reagents.
In the current study, the human hormone-dependent breast cancer cell lines MCF-7 and T47D were purchased from the American Type Culture Collection (ATCC, USA). The adenocarcinoma cell line MCF-7 was cultured in EMEM supplemented with 2 mM L-glutamine, 0.1 mM nonessential amino acids, and 10% fetal bovine serum (FBS). The ductal carcinoma cell line T47D was cultured in RPMI 1640 supplemented with 4.5 g/L glucose (Sigma-Aldrich, Inc., USA) and 10% FBS. Both mediums were supplemented with 0.01 mg/mL insulin (Sigma-Aldrich, Inc., USA), 1 mM sodium pyruvate, 1.5 g/L sodium bicarbonate, 100 μg/mL penicillin G/streptomycin, 2.5 μg/mL amphotericin B, and 50 μg/mL gentamycin. All mediums and supplements were purchased from Biochrom (Berlin, Germany) unless otherwise indicated. Cells were cultured at 37°C, 5% CO_2, and 100% humidity.

E_2, Fulv, Tam, End, and 4-OH-T were purchased from Sigma-Aldrich (Sigma-Aldrich, Inc., USA). All experiments were performed according to the following conditions: after reaching 70% confluence, cells were washed with phosphate buffer saline (PBS) and incubated with phenol red-free RPMI (rf-RPMI) (Biochrom, Berlin, Germany) with 1% charcoal-stripped serum (CSS) for 24 h to deplete estrogen [17]. Thereafter, cells were treated with E_2 and the tested agents at the indicated time points and doses according to appropriate assay.

2.2. Cell Proliferation Assay.
The effect of E_2 and the tested agents on proliferation of cells was determined using the 3-(4,5-dimethylthiazol-2-yl)-2,5-dephenyltetrazolium-bromide (MTT) assay, as previously described [18]. Briefly, both MCF-7 and T47D cells, were seeded at a density of 2 × 10^4 cells/well in 24-well plates with rf-RPMI supplemented with 1% CSS. Cells were treated with E_2 10 nM alone or in combination with the tested agents: Fulv + E_2, Tam + E_2, End + E_2, and 4-OH-T + E_2 for 48 h. The tested agents were added at two different concentrations: 100 nM and 1 μM. MTT solution (5 mg/mL in PBS) was prepared and a volume equal to 1/10 was added to each well and incubated for 2 h, at 37°C. Medium was removed and 100 μL acidified isopropanol (0.33 mL HCl in 100 mL isopropanol) was added in each well in order solubilise the dark blue formazan crystals. The solution was transferred to 96-well plates and immediately read in a microplate reader (Tecan, Sunrise, Magellan 2) at a wavelength of 570 nm using reference wavelength 620 nm.

2.3. Migration Assay.
Migration assay was performed using boyden chambers (Costar, Avon, France) containing uncoated polycarbonate membranes with 8 μm pores. Briefly, cells were treated with E_2 and the tested agents for 24 h with rf-RPMI supplemented with 1% CSS. Cells were trypsinized and resuspended at 2 × 10^4 cells/0.1 mL in the same medium in presence of E_2 and the tested agents. The bottom chamber was filled with 0.6 mL of rf-RPMI with 10% CSS. The upper chamber was loaded with the solution of 2 × 10^4 cells and incubated for 36 h. After incubation, the membrane was fixed with saline-buffered formalin and stained in 1% toluidine blue solution. Images of cells that have migrated through the filter were captured using an inverted microscope of Nikon (Eclipse TE 2000-U) at magnification of 10X.

2.4. Invasion Assay.
To evaluate the effect of E_2 and tested agents on capacity of cell to invade, a Boyden chamber containing matrigel-coated polycarbonate membranes with 8 μm (Invasion Chambers, BD Biosciences, Oxford, UK) was used. Briefly, cells were treated with E_2 and the tested agents for 24 h with rf-RPMI supplemented with 1% CSS. Cells were trypsinized and resuspended at 1.25 × 10^5/mL in the same medium in presence of E_2 and the tested agents. The bottom chamber was filled with 0.7 mL of rf-RPMI with 10% CSS. The upper chamber was loaded with the solution of 1.25 × 10^5 cells and incubated for 72 h at 37°C. After the incubation, the noninvading cells were removed from the upper compartment using a cotton swab. Transwell filters were fixed with saline-buffered formalin for 10 min and then in 100% methanol for 20 min. Cells were stained

in toluidine blue solution for 10 min and washed twice in 1% PBS. Images of cells that have migrated through the matrigel-coated filter were captured using an inverted microscope of Nikon (Eclipse TE 2000-U) at magnification of 10X.

2.5. Scratch-Wound Assay. The effect of E_2 and the tested agents on collective cell migration was evaluated using 2D scratch-wound assay. Briefly, cells were seeded in 6-well plates at a density of 10^5 cells/well. After reaching 100% of confluence, cells were treated with E_2 and the tested agents in the appropriate medium rf-RPMI with 10% CSS for 24 h. In the confluent cells' monolayer an artificial gap was created with a yellow pipette tip. Then cells were rinsed several times with the appropriate medium to remove dislodged cells. Images of living cells were captured at the indicated time points of 0, 24, and 48 h at magnification of 4X using an inverted microscope (Nikon Eclipse TE 2000-U).

2.6. Zymography. Zymography was used to evaluate the expression both of pro- and active forms of MMP-2 and MMP-9. Supernatants from both cell lines were collected in 48 h, concentrated 80-fold to 50 μL, and analyzed as previously described [18].

2.7. Immunoblotting. E-cadherin and Snail were studied using western blot analysis. Briefly, MCF-7 and T47D cells were treated with E_2 and the tested agents for 24 and 48 h, and then cells were lysed in buffer containing 0.5% NP-40, 0.5% NaDOC, 0.1% SDS, 50 mM Tris (pH 7.0), 150 mM NaCl, 1 mM EDTA (pH 8.0), 1 mM NaF, and a protease inhibitor cocktail (Sigma-Aldrich, Inc., USA), as previously described [19]. Cell extracts were incubated on ice for 30 min, with vortexing every 10 min and centrifuged at 13000 rpm for 30 min. Supernatants were collected and protein concentration was determined with Bradford (Sigma-Aldrich, Inc, USA) assay. Specific protein amount was analyzed using the standard procedure of western blot analysis. A mouse anti-E-cadherin (1:1000, Invitrogen Corporation, Camarillo, CA, USA), a rat anti-Snail (1:1000, Cell Signaling Technology, Inc., Boston, USA), and a mouse antiactin (1:1000, Chemicon, Millipore, Temecula, CA, USA) were used. Detection of the immunoreactive proteins was performed by chemiluminescence using horseradish peroxidase substrate SuperSignal (Pierce, Rockford, IL, USA), according to the manufacturer's instructions.

2.8. Immunofluorescence. Cells were grown in 4-well coverslips (15×10^3 cells/well) in the presence or absence of E_2 and the tested agents for 48 h. Cells were fixed with saline-buffered formalin for 15 min and permeabilized with 0.1% Triton for 5 min. Blocking was performed with 3% bovine serum albumin (BSA) in phosphate buffer saline (PBS) containing 10% FBS for 1 h at 37°C. After the incubation, cells were rinsed once with PBS for 5 min and then incubated with a mouse anti-E-cadherin (1:1000, Invitrogen Corporation, Camarillo, CA, USA), a rat anti-Snail (1:500, Cell Signaling Technology, Inc., Boston, MA, USA), a rabbit anti-Tyr^{397}-FAK antibody (dilution 1:200, R&D Systems, Deutschland, Germany), a mouse anti-ER-α antibody

(dilution 1:500, Chemicon International Inc., Temecula, CA, USA), and phalloidin-fluorescein isothiocyanate labeled (Sigma-Aldrich, Inc., USA) for 1 h at 37°C. Cells were rinsed 3×5 min with PBS and then a chicken anti-mouse Alexa Fluor 488, a chicken anti-rat Alexa Fluor 568, or a donkey anti-rabbit antibody Alexa Fluor 594, (1:1000, molecular probe, Invitrogen Corporation, Camarillo, CA, USA) diluted in blocking solution and an incubation for 30 min at 37°C was followed. Cells were rinsed 2×5 min with PBS; then incubation for 5 min with 5 μM Draq 5 (Biostatus Limited, Shepshed, UK) or DAPI (Vectashield, Vector Laboratories, Inc., US) diluted in PBS was followed for nucleus staining and cells mounted on glass sides. Fluorescence was visualized using a Leica microscope at 63X magnification.

2.9. Statistical Analysis. Differences between groups and controls were tested by one-way ANOVA. Each experiment included at least triplicate measurements. All results are expressed as mean ± SEM from at least three independent experiments.

3. Results

3.1. Fulv, Tam, and the Metabolites End and 4-OH-T Partially Decrease E_2-Induced Cell Proliferation. In the current study, MCF-7 and T47D breast cancer cells were treated with Fulv, Tam, and its metabolites End and 4-OH-T, so as to determine the optimum concentration regarding their effect on cell proliferation. E_2 was used at a concentration of 0.01 μM as previously described [20]. Fulv, Tam, End, and 4-OH-T were tested at the concentrations of 0.1 and 1 μM, as previously described [21–24]. Both cell lines were treated with Fulv and Tam as well as the metabolites of the latter with simultaneous addition of E_2. We showed that E_2 induced cell proliferation in both cell lines 48 h after its addition (Figure 1(a)), as previously described [25]. All the tested agents demonstrated an antiproliferative effect in both concentrations in a dose-dependent manner compared to untreated cells in both cell lines 48 h after their addition, as was expected (Figures 1(b) and 1(c)). Thereafter, all the experiments were performed using 0.01 μM E_2 and 0.1 μM of the tested agent.

3.2. Tam but Not Fulv Stimulates Single Cell Migration. Migration is a pivotal process for both invasion and metastasis allowing cells to change position into tissues or metastasize to distant organs [5, 26]. Cancer cells utilize different ways to migrate, either individual or multicellular [4]. To assess the effect of the tested agents on single cell migration, we used the boyden chamber assay in both cell lines. Cells were pretreated with E_2 and the tested agents for 24 h, and then we observed their ability to migrate through the membrane after 36 h incubation. MCF-7 cells showed greater ability to pass through the membrane compared to T47D cells (Figure 2). E_2 alone or in combination with Fulv did not affect MCF-7 cell migration compared to untreated cells. In contrast the treatment of MCF-7 cells with the combination of E_2 with Tam and its metabolites significantly promotes the motility of cells to migrate through the pores of the membrane

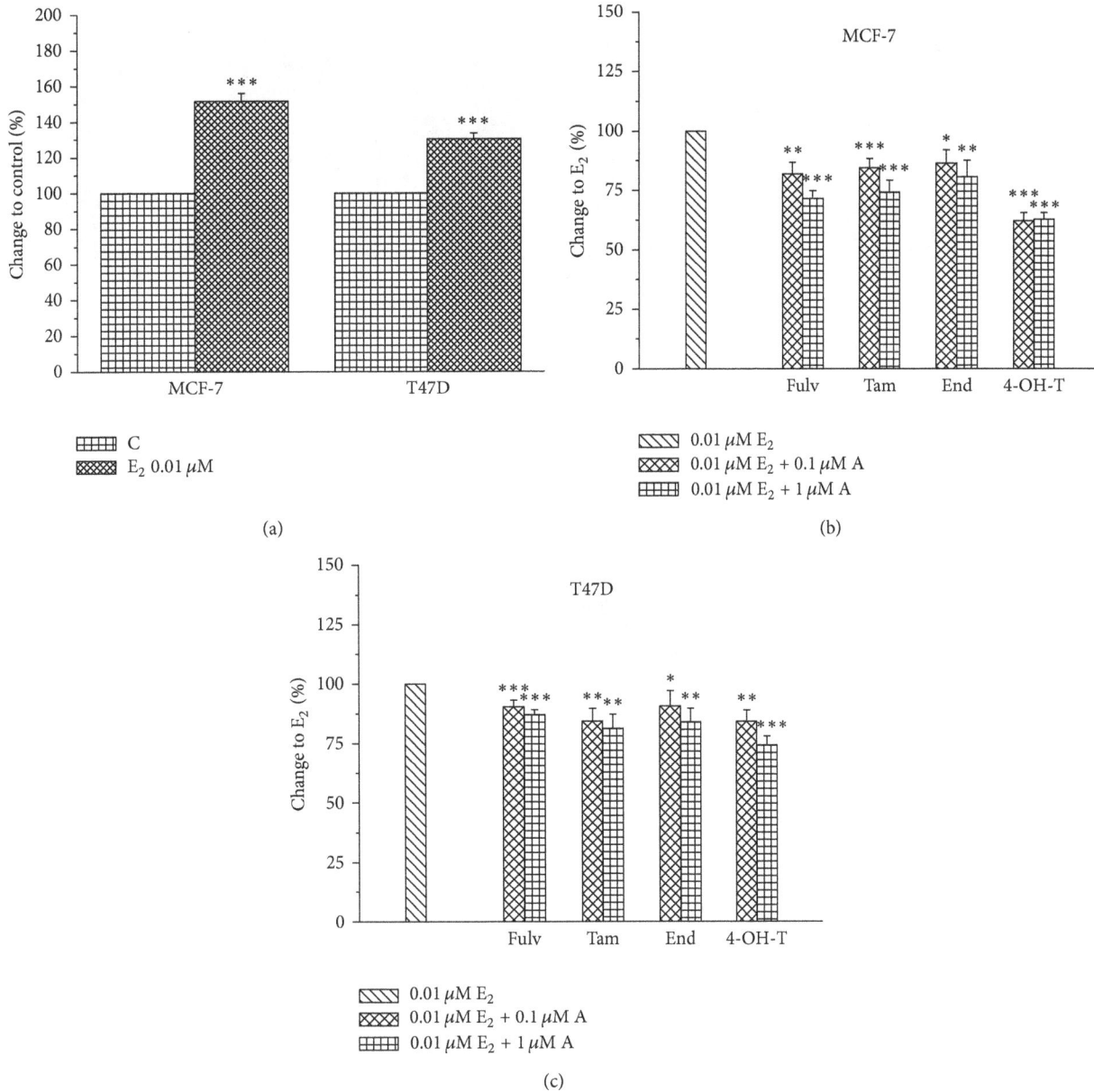

(a)

(b)

(c)

FIGURE 1: The effect of E_2 and the tested agents on cell proliferation. E_2 alone induces cell proliferation of MCF-7 and T47D (a). Cells were pretreated with E_2 (0.01 μM), and the tested agents (A) were added at the concentrations of 0.1 and 1 μM at MCF-7 (b) and T47D (c). Results are expressed as mean $\pm$ SEM of the % change compared to the untreated cells and/or E_2. Asterisks denote a statistically significant difference compared to control (untreated) cells. $^*P < 0.05$, $^{**}P < 0.01$, and $^{***}P < 0.001$.

(Figure 2). In T47D cells the effect of E_2 and the tested agents on cell migration is not reliable since very low number of cells passed through the membrane. The difference in the ratio of ERα/ERβ might contribute to low metastatic ability of T47D cells. MCF-7 cells express very low levels of ERβ compared to T47D cells [27]. According to recent data, ERβ exerts a protective role for the cell by inhibiting the invasiveness and promoting the adhesion [28]. Further, a previous study demonstrated that treatment of MCF-7 cells with E_2 caused a degradation of ERα and an increase of ERβ [29]. This might explain the absence of any effect on MCF-7 cell migration

after their treatment with E_2 alone or in combination with Fulv since Fulv exerts its effect through ERα degradation.

3.3. Collective Cell Migration Is Not Affected by Fulv but It Is Reduced by Tam.
Since E_2 alone or in combination with Fulv did not affect single cell migration, we studied the effect of tested agents on collective cell migration using the scratch wound assay [30]. Both cell lines were treated with E_2 and the tested agents for 24 and 48 h. In MCF-7 cells we found that E_2 alone increased cell migration compared to untreated cells up

(a)

(b)

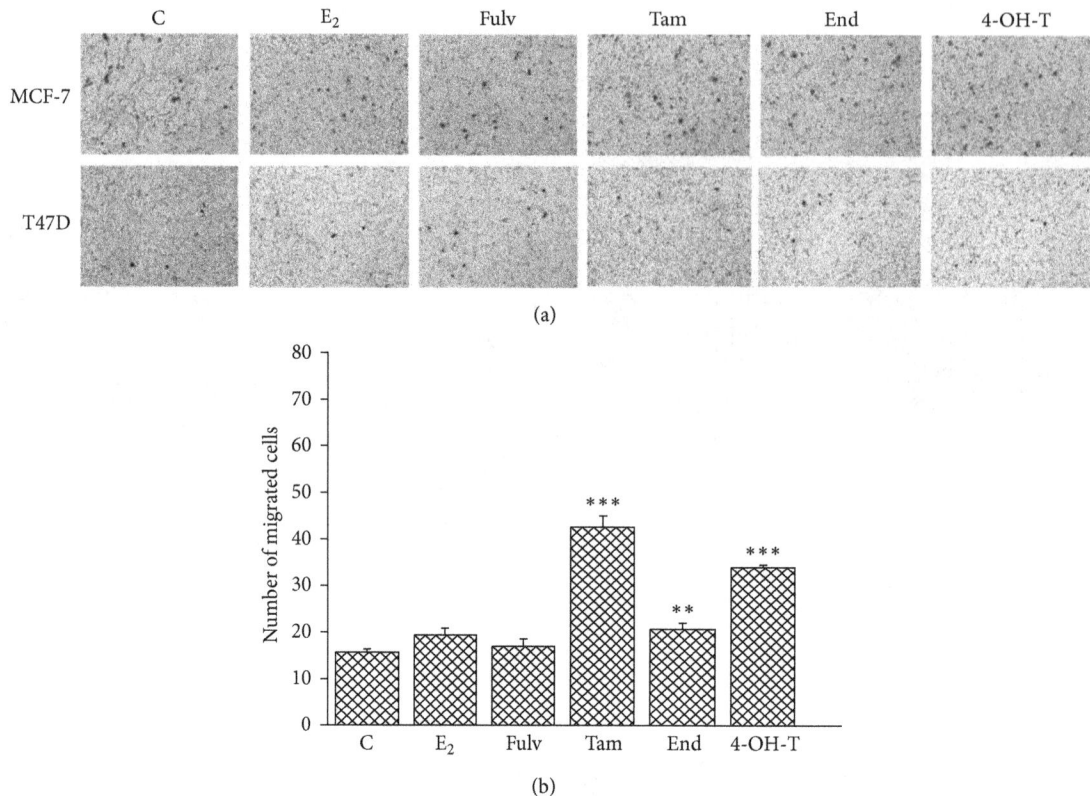

FIGURE 2: Single cell migration in MCF-7 and T47D cells after their treatment with E_2 and antiestrogens. C: control (untreated cells); E_2: cells treated with 17β-estradiol; Fulv: cells treated with E_2 + 100 nM Fulv; Tam: cells treated with E_2 + 100 nM Tam; End: cells treated with E_2 + 100 nM End; and 4-OH-T: cells treated with E_2 + 100 nM 4-OH-T. The image is representative of three independent experiments using a magnification of 10X (a). Quantification of images from boyden chamber assay in MCF-7 cells (b). Results are expressed as mean ± SEM of the % change compared to the untreated cells. Asterisks denote a statistically significant difference compared to E_2 treated cells. $^{**}P < 0.01$ and $^{***}P < 0.001$.

to 48 h (Figure 3). The combination of E_2 with Fulv reversed slightly the effect of E_2 alone. This reversal was more potent when E_2 combined with Tam, End, and 4-OT-T as shown in Figure 3. The same effect of E_2 and tested agents was observed in T47D (data not shown).

3.4. Fulv and Tam Totally Reverse the Protective Effect of E_2 in Cell Invasion.

Because migration plays a crucial role during tumor invasion, we evaluated the influence of Fulv, Tam, and its active metabolites on the invasive capacity of breast cancer cells lines. Cell invasion was studied using a modified boyden chamber assay with a membrane coated with matrigel. Cells were treated with E_2 and tested agents, and the invasion was observed 72 h later. In MCF-7 cells, we found that E_2 alone reduced cell ability to invade and this effect was partially reversed by the combination of E_2 and the tested agents (Figure 4). Fulv and 4-OH-T exerted a better inhibitory effect than Tam and End (Figure 4). T47D cells were not used in this set of experiments, because of the low capacity to migrate the membrane in typical boyden chamber assay. Although, MCF-7 cells are also characterized by low invasive capacity compared to other breast cancer lines, we showed that the treatment with E_2 and the tested agents altered their motility and this prompted us to investigate it further.

3.5. Fulv and Tam Facilitate Invasion through MMPs' Modulation.

MMPs are key players in invasion and metastasis since they promote the invasive potential through digestion of the ECM components [5, 31, 32]. In ER$^+$ breast tumors E_2 exerts a protective role since it regulates the expression both of MMP-2 and MMP-9 as well as syndecan-4 [29] and, therefore, limits the ability of cells to invade the adjacent tissues. By contrast, antiestrogens seem to reverse this effect increasing the level of MMPs [33]. We evaluated the influence of E_2 alone and/or in combination with the tested agents on MMP-2 and MMP-9 levels 24 and 48 h after treatment of cells. Zymography analysis in MCF-7 cells demonstrated a slight decrease on the expression of both MMP-2 and MMP-9 followed the treatment with E_2 up to 48 h. In addition, the combination of cells with E_2 and tested agents reversed the effect of E_2 inducing MMPs levels 24 h after treatment of cells (Figure 5). This phenomenon was preserved for Fulv and End up to 48 h after cells treatment. At the same time point, when E_2 combined with Tam, MMPs levels were not changed compared to E_2 alone while the combination of E_2 with 4-OH-T reduced the levels of MMPs and particularly MMP-9 (Figure 5). In T47D cells any change in MMPs levels was not found after cells treatment with E_2 and the tested agents at any time point tested (data not shown).

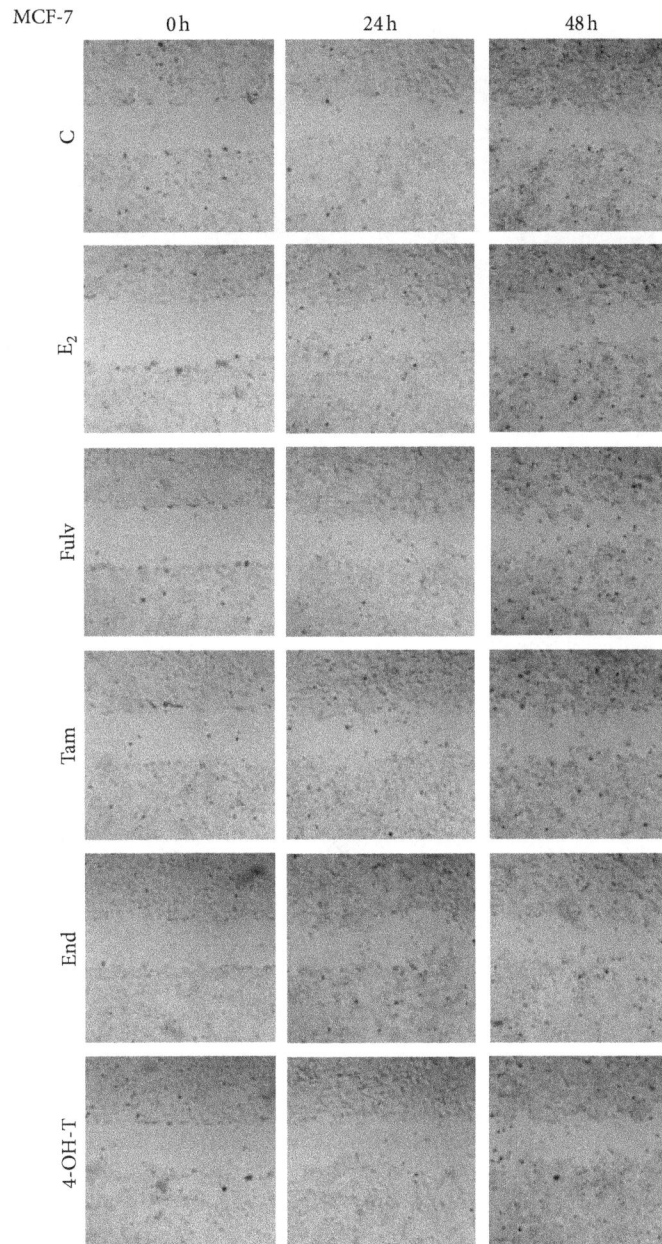

FIGURE 3: Collective cell migration in MCF-7 cells treated with E_2 and antiestrogens. C: control (untreated cells); E_2: cells treated with 17β-estradiol; Fulv: cells treated with E_2 + 100 nM Fulv; Tam: cells treated with E_2 + 100 nM Tam; End: cells treated with E_2 + 100 nM End; and 4-OH-T: cells treated with E_2 + 100 nM 4-OH-T. The image is representative of three independent experiments using a magnification of 4X.

3.6. Tam and End Stimulate EMT-A Different Role for Snail.
At the leading edge of invasiveness and metastasis, epithelial cells undergo EMT. Two major partners of EMT are E-cadherin and Snail. E-cadherin is reversibly downregulated in EMT, and this reduction is associated with increased levels of Snail, a repressor of E-cadherin [28, 34, 35]. Regarding E-cadherin protein levels, we found that E_2 alone and/or in combination with 4-OH-T did not alter protein status 48 h after their addition to MCF-7 cells. The combinations of E_2 with Fulv, Tam, and End caused a decrease in E-cadherin protein levels (Figure 6(a)). Further, regarding Snail protein levels, we found that E_2 alone increased Snail protein status at the same time point. The combination of E_2 with Fulv and/or with 4-OH-T decreased Snail protein. This phenomenon was more potent in the case of E_2 with 4-OH-T. The combination of E_2 with Tam and End increased Snail levels (Figure 6(a)). In T47D cells, E_2 alone as well as its combinations with Tam, End and 4-OH-T did not alter E-cadherin protein levels. The treatment of cells with E_2 and Fulv caused a slight decrease in protein levels (Figure 6(b)). Furthermore Snail protein was decreased only when E_2 combined with Fulv and 4-OH-T (Figure 6(b)).

(a)

(b)

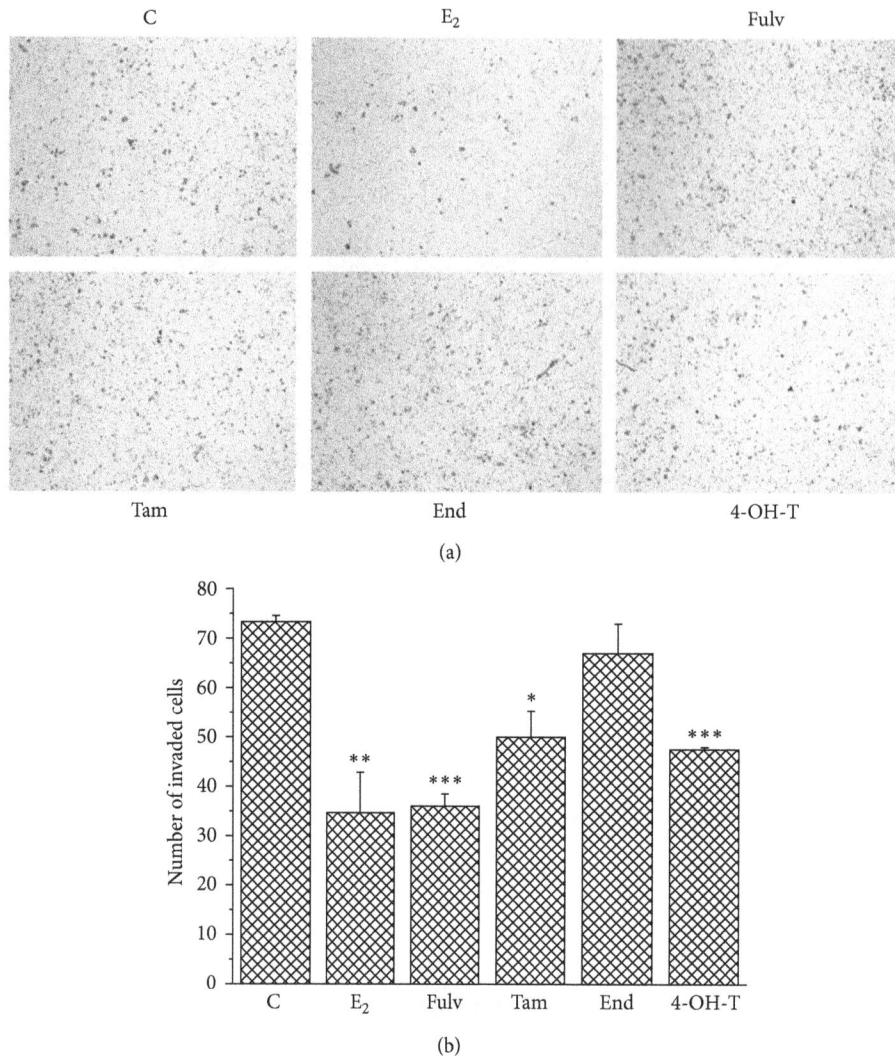

FIGURE 4: The effect of E_2 and the tested agents on MCF-7 cell invasion. C: control (untreated cells); E_2: cells treated with 17β-estradiol; Fulv: cells treated with E_2 + 100 nM Fulv; Tam: cells treated with E_2 + 100 nM Tam; End: cells treated with E_2 + 100 nM End; and 4-OH-T: cells treated with E_2 + 100 nM 4-OH-T. The image is representative of three independent experiments using a magnification of 10X (a). Quantification of images from boyden chamber assay in MCF-7 cells (b). Results are expressed as mean ± SEM of the % change compared to the untreated cells. Asterisks denote a statistically significant difference compared to untreated cells. $^*P < 0.05$, $^{**}P < 0.01$, and $^{***}P < 0.001$.

The complicated results from western blot analysis revealed that, in both cell lines, the protein changes of E-cadherin did not follow the changes of Snail protein levels in order for an EMT phenomenon to be observed. Only in the case that E_2 combined with Tam or End, a decrease of E-cadherin levels followed an increase of Snail levels. In addition, the most important changes were observed at Snail protein after cell treatment with the combinations of E_2 with Fulv and 4-OH-T where Snail levels were decreased (Figure 6). Besides in EMT, the role of Snail is also very important for cell survival. Previous studies have shown that a decrease in Snail protein sensitizes cell to death [34, 36].

3.7. The Antiestrogens on Localization of E-Cadherin and Snail.
In order for the transcription factor Snail to act as repressor of E-cadherin, its nuclear translocation is required. Since western blot analysis did not reveal any significant connection between EMT proteins' expression and treatment of cells with the E_2 and the tested agents, we studied the effect of antiestrogens on these proteins' localization, 48 h after cell treatment. We found that E-cadherin is located in cell membrane and cell-cell junctions in untreated MCF-7 cells as well as in cells treated with E_2 and the tested agents (Figure 7). Snail was localized at both nucleus and cytoplasm in untreated cells or cells treated with E_2 (Figure 7). The combinations of E_2 with Fulv and 4-OH-T retained the cytoplasmic localization and enhanced the nuclear localization. The combinations of E_2 with Tam and End retained the cytoplasmic localization of Snail. Similar effects of E_2 and the tested agents were observed at T47D cells (data not shown).

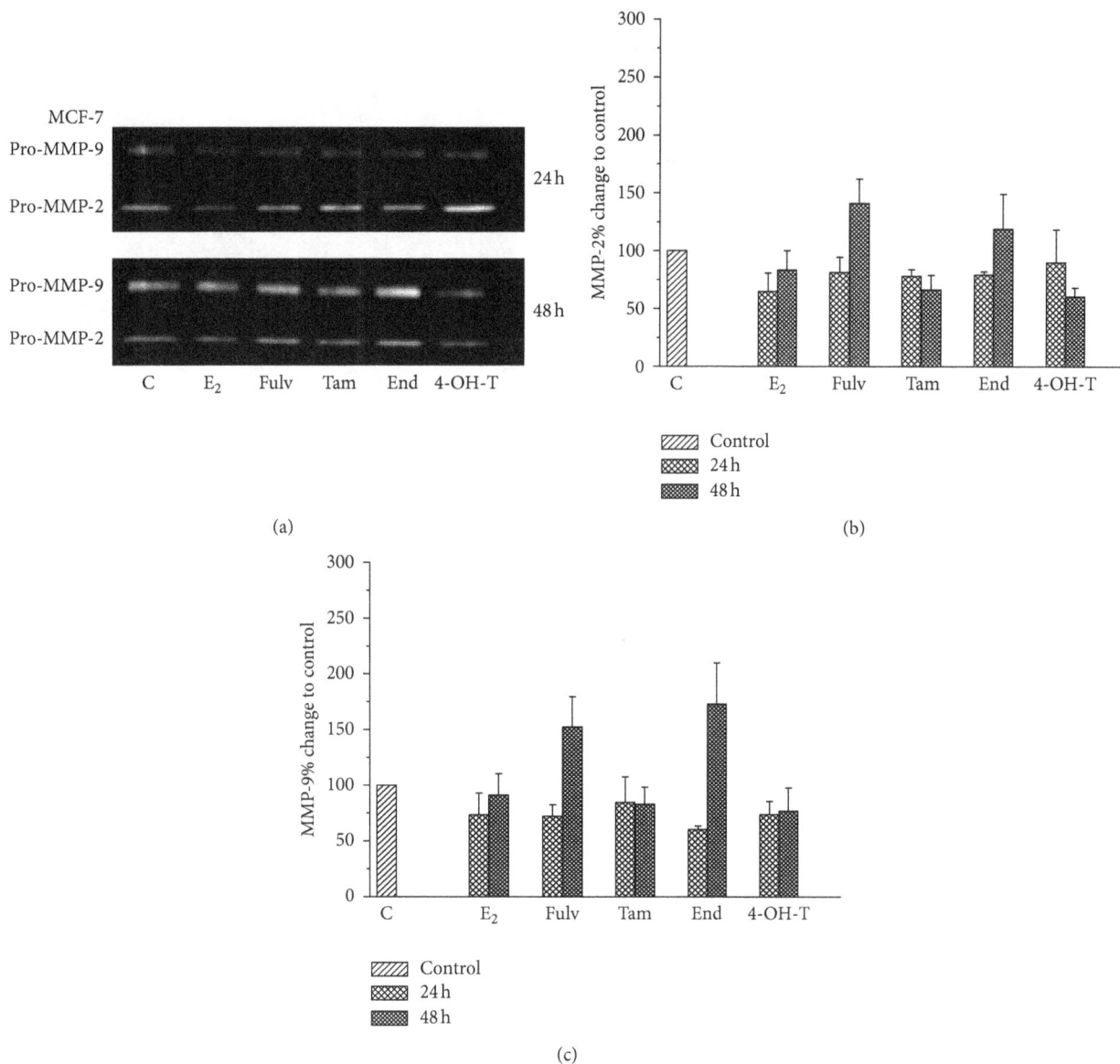

FIGURE 5: MMP-9 and MMP-2 enzyme expression after treatment of MCF-7 cells with E_2 and the tested agents. (a) A representative image of three independent experiments. A quantitative analysis of images for (b) MMP-2 and (c) MMP-9 expression using appropriate software. Results are expressed as mean ± SEM of the % change compared to the untreated cells.

3.8. Fulv and Tam Affect Migration through FAK Phosphorylation and F-Actin Rearrangement. FAK exerts a central role on cell migration and invasion, and its activation is correlated with malignant transformation [37, 38]. In addition, a specific phosphorylation at Tyr^{397} residue is correlated with Tam-resistance [22]. In MCF-7 cells, E_2 exposure resulted in autophosphorylation of FAK in Tyr^{397} residue, which entails activation of FAK. This phenomenon was time dependent, and the highest phosphorylation was observed in 10 min (Figure 8). Thereafter, the phosphorylated signal was down-regulated.

At the time point of 10 min, when the maximum FAK phosphorylation was found, we investigated the impact of Fulv, Tam, and its metabolites in spatial organization of actin fibers. The main finding to emerge was that the treatment of

cells with E_2 combined with Fulv either Tam or End resulted in a less round-like morphology with more leading edges than the other groups (Figure 8). The colocalisation of F-actin with Tyr^{397} FAK appeared mainly at the leading edges. In untreated cells as well as in cells treated with E_2 alone or in combination with 4-OH-T, the spots of Tyr^{397} FAK are scattered all around the cell membrane which is attributed to increased stability (Figure 8). Similar effects of E_2 and the tested agents were observed at T47D cells (data not shown).

4. Discussion

Hormonal therapy has been established for the treatment of ER^+ breast cancer patients. Several clinical trials [39–41] have demonstrated the benefits of this type of treatment,

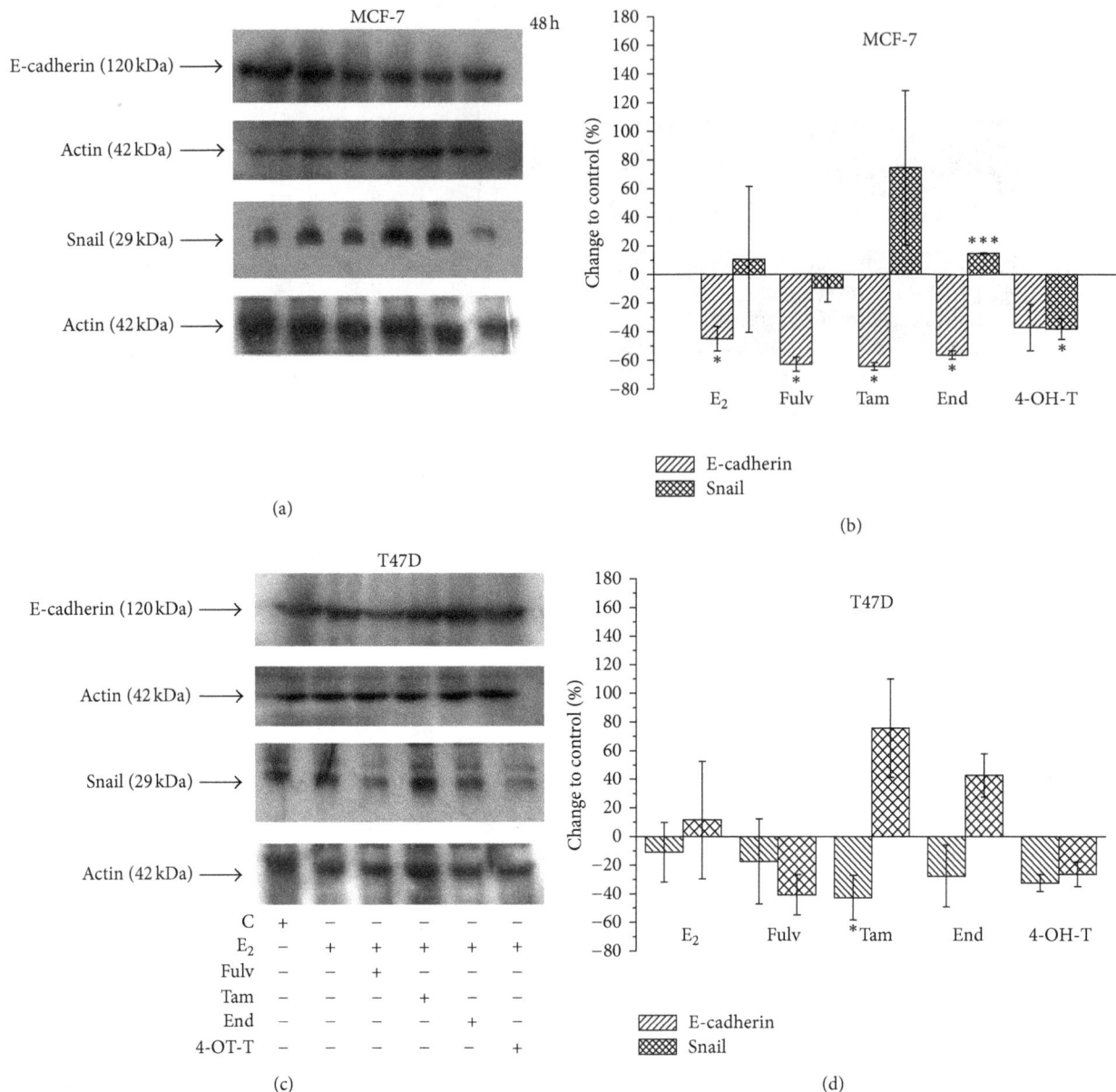

(a)

(b)

(c)

(d)

FIGURE 6: E-cadherin and Snail protein expression in MCF-7 and T47D cells 48 h after treatment of cells with E_2 and the tested agents. A representative image of three independent experiments for both cell lines using western blot analysis, (a) and (c). Quantification of images from western blot analysis in both cell lines, (b) and (d). Results are expressed as % change compared to the untreated cells ± SEM. Asterisks denote a statistically significant difference compared to untreated cells. $^{*}P < 0.05$ and $^{***}P < 0.001$.

and it is generally acceptable that it has contributed to the decrease in breast cancer mortality. Despite the benefits of hormonal therapy, the disease often relapses and secondary tumors develop due to their metastatic potential [42, 43]. *In vitro* studies have assessed the impact of antiestrogens on breast cancer cell invasiveness and MMPs expression [16, 33, 44, 45]. In the present study we evaluated the effect of the antiestrogens Fulv and Tam from a different standpoint, namely, migration that leads to tumor growth, invasion, and metastasis.

There are many types of cell movement that lead to cell migration and invasion according to cell type and

microenvironment [4]. Epithelial cells undergoing EMT can migrate individually. On the other hand, basal- and squamous-originated epithelial cells following EMT or moderately differentiated epithelial cells lacking EMT can migrate collectively [4]. In order to evaluate the effect of E_2 on single and collective cell migration, we applied 2 typical assays: boyden chamber and wound healing, respectively. We found that in MCF-7 cells, E_2 alone failed to stimulate single cell migration while promoting collective cell migration in both cell lines. The failure of E_2 to stimulate single cell migration is in line with the unclear results of western blot analysis for the interaction of EMT proteins, E-cadherin, and Snail as well

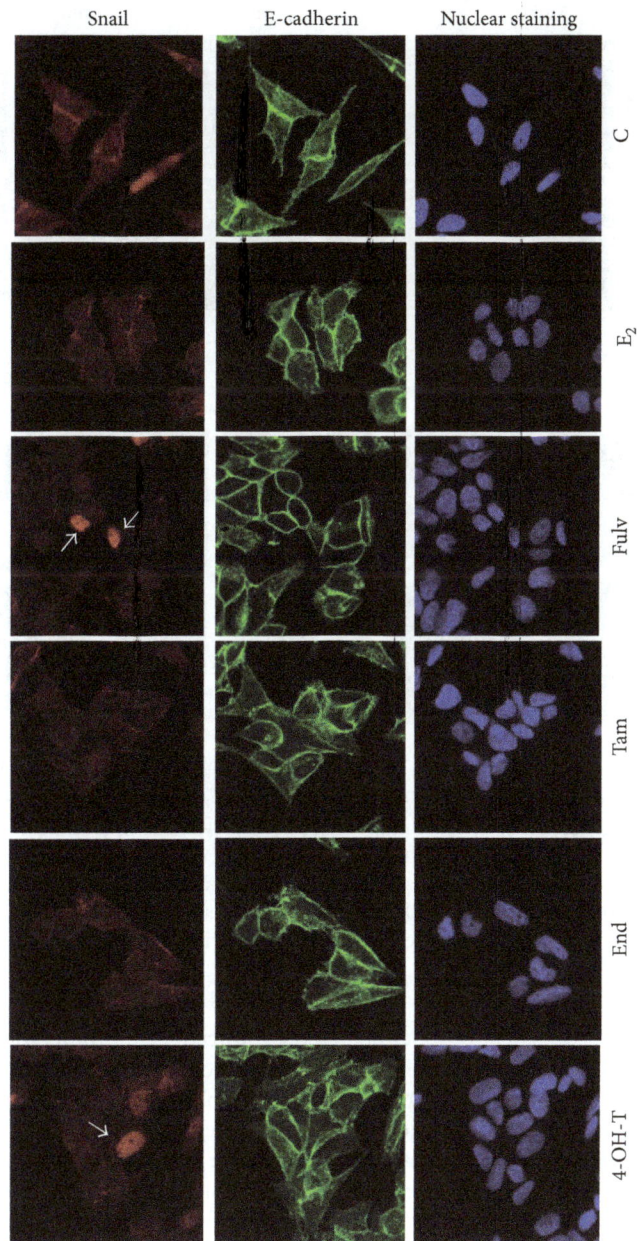

FIGURE 7: E-cadherin and Snail protein localization in MCF-7 cells 48 h after treatment of cells with E_2 and the tested agents. C: control (untreated cells); E_2: cells treated with 17β-estradiol; Fulv: cells treated with E_2 + 100 nM Fulv; Tam: cells treated with E_2 + 100 nM Tam; End: cells treated with E_2 + 100 nM End; and 4-OH-T: cells treated with E_2 + 100 nM 4-OH-T. The image is representative of three independent experiments using a magnification of 63X.

as with the absence of Snail import to the nucleus. Snail is a highly unstable protein and is dually regulated by protein stability and cellular localization. In order for Snail to exert its effect, a nuclear translocation is required [34]. The increase in collective cell migration after treatment of cells with E_2 is in line with the increase in cell proliferation of both cell lines since these are indications of expansive growth with the absence of active migration [46]. In contrast to the increase in cell proliferation and collective cell migration, we found that E_2 decreased the capacity of cells to invade. The decrease in invasiveness was associated with decrease in MMPs. This

is not the first time that a protective role of E_2 is described. Previous studies have shown that E_2 may inhibit breast cancer cell invasion by affecting proteins that modulate cell-cell interactions or increasing the number of desmosomes [47]. The reduced invasiveness of E_2-stimulated cells is also supported by the findings from immunofluorescence assay, where cells demonstrated a more spherical morphology with focal adhesions all a round the cell membrane, which is associated with increased stability.

Using Fulv, the mitogenic effect of E_2 was partially reversed with a decrease in Snail protein levels associated

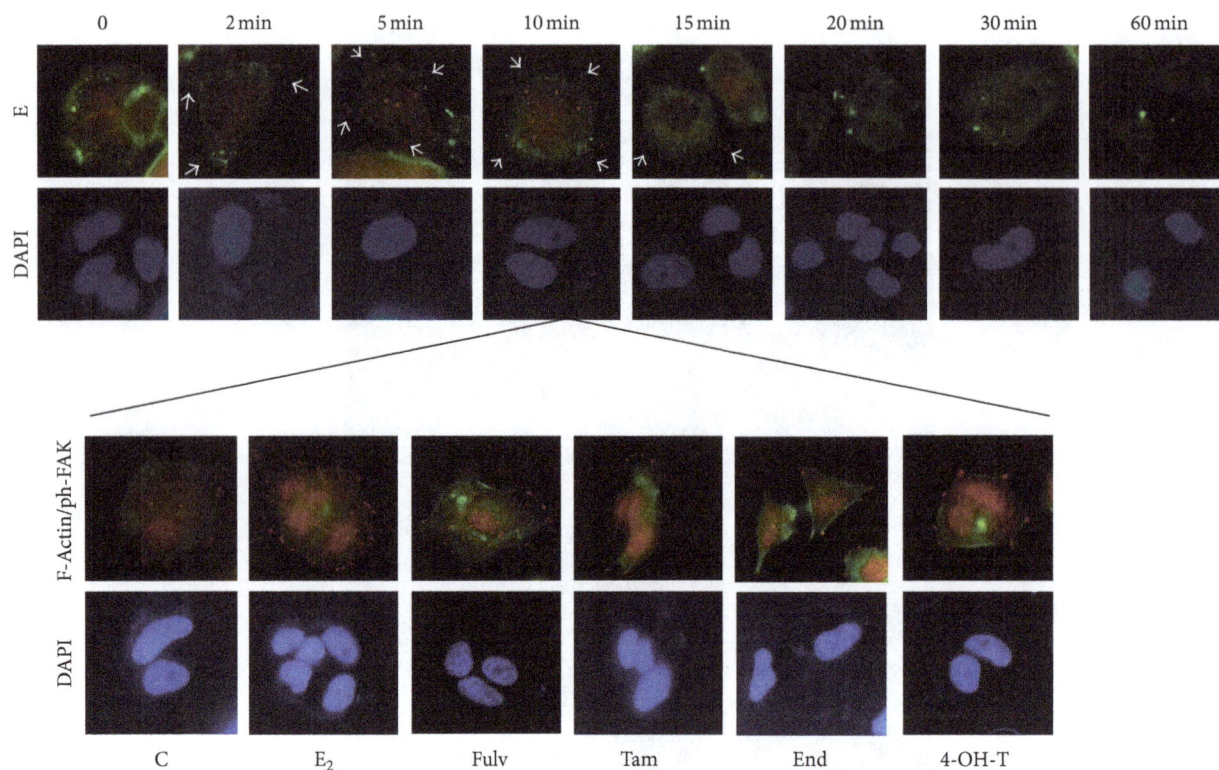

FIGURE 8: The impact of E_2 and the tested agents on Tyr^{397} FAK phosphorylation and F-actin rearrangement. MCF-7 cells exposed to E_2 in a time course manner up to 60 min for detection of the maximum FAK phosphorylation. ERα and Tyr^{397} FAK localisation is indicated with green and red fluorescence, respectively. At the time point of 10 min, F-actin and Tyr^{397} FAK colocalisation (green and red fluorescence, resp.) was observed after the exposure of MCF-7 cells to the tested agents. C: control (untreated cells); E_2: cells treated with 17β-estradiol; Fulv: cells treated with E_2 + 100 nM Fulv; Tam: cells treated with E_2 + 100 nM Tam; End: cells treated with E_2 + 100 nM End; and 4-OH-T: cells treated with E_2 + 100 nM 4-OH-T. The image is representative of three independent experiments using a magnification of 60X.

with its import to nucleus. However, the effect of E_2 either on single or collective cell migration was not altered. Fulv is a selective estrogen downregulator that binds to ER forming an unstable ER-Fulv complex, which is rapidly degraded resulting in ER reduction. Fulv may exert genomic as well as non genomic effects on target cells [16, 48]. A recent publication by Song et al. [48] shows that Fulv at the concentration of 0, 1 μM shuttles ERα from the nucleus to the cytosol and plasma membrane. When Fulv is extranuclear acts as an estrogen agonist but after its entrance to the nucleus blocks the genomic effects of estrogens in transcription and cell proliferation. This might explain the effect on cell proliferation but not on cell migration. Previous data have shown that functional ERα is associated with E-cadherin expression, and this expression as well as cell-cell adhesion may be modulated by antiestrogens resulting in an invasive phenotype [16]. Indeed, we found that Fulv decreased E-cadherin protein expression and increased cell invasion and MMPs expression versus E_2. These data were confirmed by immunofluorescence assay where cells exhibited a less round-like morphology, indication of increased invasiveness.

Tam is a prodrug that is metabolized to End and 4-OH-T so as to exert its therapeutic effect. Although both metabolites are equivalent regarding ERα binding and inhibition of

E_2-induced cell proliferation, it is proposed that End is the principal antiestrogenic metabolite for the antitumour activity observed in breast cancer patients [49]. In the current study we used both metabolites to verify that they act in the same way. Tam and its metabolites stimulated single cell migration and reduced collective cell migration. Regarding Tam and End, the stimulation of single cell migration is in concordance with the E-cadherin protein decrease and Snail protein increase. This might be an indication that an active migration through EMT induction occurs after Tam and End treatment. Although Snail was not detected to the nucleus at the same time point we cannot exclude a positive role of cytosolic Snail in cell migration [50]. These data are also in agreement with the less round-like shape of cells as well as with the scattering of focal adhesions at the leading edges of F-actin revealing a more invasive and potent phenotype. An increase in both MMPs expression and cell invasiveness might facilitate EMT induction. In the case of 4-OH-T, it seems that an EMT phenomenon did not occur because no decrease in E-cadherin or increase in Snail protein levels was detected. In contrast, a reduction in Snail protein in association with a nuclear localization was detected. So far, our data indicate that 4-OH-T promoted single cell migration without EMT. A detailed review of Friedl and Alexander [4], related to the types of cancer cell movement, referred

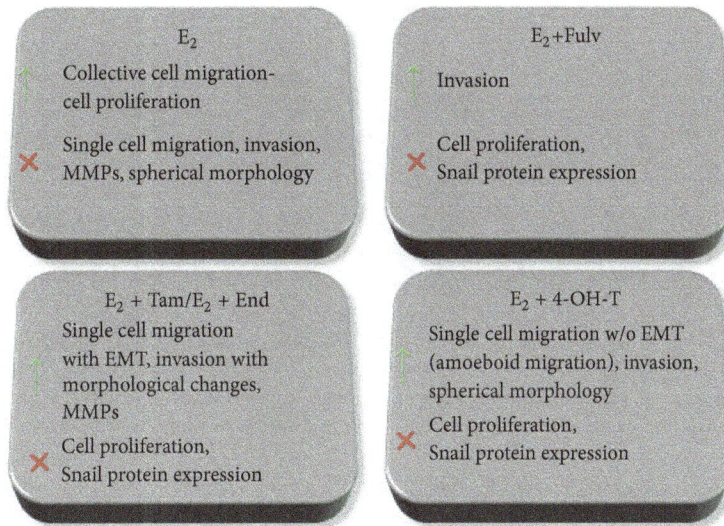

FIGURE 9: An overview of the effect of Fulv, Tam, and metabolites of Tam in migration and invasion of MCF-7 and T47D cells.

to a single cell movement currently known as amoeboid migration. In this type of migration, cells adopt a more spherical shape and migrate without ECM proteolysis. The decrease in MMPs levels and the spherical shape of cells found in our study after treatment with 4-OH-T using zymography and immunofluorescence, respectively, support this type of migration. The decrease of Snail protein and its nuclear location after 4-OH-T treatment seem to correlate with the inhibition of cell proliferation rather than migration. This is compatible with the decrease of cell proliferation that we found after 4-OH-T treatment. This decrease was more potent for 4-OH-T compared to the other agents which did not reduce Snail protein. Regarding invasion, it seems that the active single cell migration with or without EMT was associated with increased invasiveness.

5. Conclusions

Our working hypothesis was that different approaches of estrogen inhibition affected differently breast cancer cell migration and invasion. Summarizing our data, we may conclude that in breast cancer cells after serum E_2 withdrawal (i) E_2 stimulated expansive growth of cells with the absence of EMT but exerted a protective effect by reducing invasiveness, MMPs expression and preserving a more stable phenotype with focal adhesions all around the cell membrane; (ii) the antiestrogens partially counteracted the E_2-induced effect; (iii) Fulv did not affect the expansive growth stimulated by E_2 and promoted cell invasion; (iv) Tam and its metabolites stimulated active single cell migration and increased cell invasiveness. An overview of Fulv and Tam effect is observed in Figure 9.

Although Fulv might result in a less aggressive behaviour of cells compared to Tam, the benefits of hormonal therapy concerning invasion and metastasis yet remain under question.

Abbreviations

ER:	Estrogen receptor
Fulv:	Fulvestrant
Tam:	Tamoxifen
E2:	17β-Estradiol
End:	Endoxifen
4OHT:	4-OH-Tamoxifen
MMPs:	Matrix metalloproteinases
SERDs:	Selective estrogen receptor downregulators
ECM:	Extracellular matrix
EMT:	Epithelial mesenchymal transition
FAK:	Focal adhesion kinase
FBS:	Fetal bovine serum
rf-RPMI:	Phenol red-free RPMI
CSS:	Charcoal-stripped serum
MTT:	3-(4,5-Dimethylthiazol-2-yl)-2, 5-dephenyltetrazolium-bromide.

Conflict of Interests

The authors declare that they have no conflict of interests.

Authors' Contribution

Dionysia Lymperatou participated in acquisition of data as well as data analysis and was involved in drafting the paper; Efstathia Giannopoulou has made substantial contribution to conception and design of the project, participated in acquisition and analysis of data, and was involved in drafting the paper, Angelos K. Koutras has made substantial contribution to conception of the project, was involved in drafting the paper, and revised it critically for important intellectual content; Haralabos P. Kalofonos has made substantial contribution to conception and design of the project, was involved in drafting the paper, and revised it critically for important intellectual content. All authors have given final approval of the version to be published.

Acknowledgments

The authors would like to thank the Medical School, University of Patras, Greece, for providing them with the Advanced Light Microscopy facility and EOGE Oncological Research Fund for financial support.

References

[1] D. L. Hertz, H. L. McLeod, and J. M. Hoskins, "Pharmacogenetics of breast cancer therapies," *Breast*, vol. 18, 3, pp. S59–S63, 2009.

[2] M. J. Higgins and J. Baselga, "Targeted therapies for breast cancer," *Journal of Clinical Investigation*, vol. 121, no. 10, pp. 3797–3803, 2011.

[3] J. F. R. Robertson, A. Llombart-Cussac, J. Rolski et al., "Activity of fulvestrant 500 mg versus anastrozole 1 mg as first-line treatment for advanced breast cancer: results from the FIRST study," *Journal of Clinical Oncology*, vol. 27, no. 27, pp. 4530–4535, 2009.

[4] P. Friedl and S. Alexander, "Cancer invasion and the microenvironment: plasticity and reciprocity," *Cell*, vol. 147, no. 5, pp. 992–1009, 2011.

[5] P. Friedl and K. Wolf, "Tumour-cell invasion and migration: diversity and escape mechanisms," *Nature Reviews Cancer*, vol. 3, no. 5, pp. 362–374, 2003.

[6] S. Schmidt and P. Friedl, "Interstitial cell migration: integrin-dependent and alternative adhesion mechanisms," *Cell and Tissue Research*, vol. 339, no. 1, pp. 83–92, 2010.

[7] P. Friedl and D. Gilmour, "Collective cell migration in morphogenesis, regeneration and cancer," *Nature Reviews Molecular Cell Biology*, vol. 10, no. 7, pp. 445–457, 2009.

[8] A. Jezierska and T. Motyl, "Matrix metalloproteinase-2 involvement in breast cancer progression: a mini-review," *Medical Science Monitor*, vol. 15, no. 2, pp. RA32–RA40, 2009.

[9] Y. Sullu, G. G. Demirag, A. Yildirim, F. Karagoz, and B. Kandemir, "Matrix metalloproteinase-2 (MMP-2) and MMP-9 expression in invasive ductal carcinoma of the breast," *Pathology Research and Practice*, vol. 207, no. 12, pp. 747–753, 2011.

[10] H. Dang, W. Ding, D. Emerson, and C. B. Rountree, "Snail1 induces epithelial-to-mesenchymal transition and tumor initiating stem cell characteristics," *BMC Cancer*, vol. 11, article 396, 2011.

[11] J. P. Thiery, H. Acloque, R. Y. J. Huang, and M. A. Nieto, "Epithelial-mesenchymal transitions in development and disease," *Cell*, vol. 139, no. 5, pp. 871–890, 2009.

[12] J. P. Their, "Epithelial-mesenchymal transitions in tumor progression," *Nature Reviews Cancer*, vol. 2, no. 6, pp. 442–454, 2002.

[13] P. J. Kowalski, M. A. Rubin, and C. G. Kleer, "E-cadherin expression in primary carcinomas of the breast and its distant metastases," *Breast Cancer Research*, vol. 5, no. 6, pp. R217–R222, 2003.

[14] S. K. Mitra, D. A. Hanson, and D. D. Schlaepfer, "Focal adhesion kinase: in command and control of cell motility," *Nature Reviews Molecular Cell Biology*, vol. 6, no. 1, pp. 56–68, 2005.

[15] A. Sanchez and J. Villanueva, "PI3K-based molecular signatures link high PI3K pathway activity with low ER levels in ER+ breast cancer," *Expert Review of Proteomics*, vol. 7, no. 6, pp. 819–821, 2010.

[16] A. C. Borley, S. Hiscox, J. Gee et al., "Anti-oestrogens but not oestrogen deprivation promote cellular invasion in intercellular adhesion-deficient breast cancer cells," *Breast Cancer Research*, vol. 10, no. 6, article R103, 2008.

[17] O. K. Weinberg, D. C. Marquez-Garban, M. C. Fishbein et al., "Aromatase inhibitors in human lung cancer therapy," *Cancer Research*, vol. 65, no. 24, pp. 11287–11291, 2005.

[18] E. Giannopoulou, K. Dimitropoulos, A. A. Argyriou, A. K. Koutras, F. Dimitrakopoulos, and H. P. Kalofonos, "An *in vitro* study, evaluating the effect of sunitinib and/or lapatinib on two glioma cell lines," *Investigational New Drugs*, vol. 28, no. 5, pp. 554–560, 2010.

[19] K. Lundgren, B. Nordenskjöld, and G. Landberg, "Hypoxia, Snail and incomplete epithelial-mesenchymal transition in breast cancer," *British Journal of Cancer*, vol. 101, no. 10, pp. 1769–1781, 2009.

[20] S. Oesterreich, W. Deng, S. Jiang et al., "Estrogen-mediated down-regulation of E-cadherin in breast cancer cells," *Cancer Research*, vol. 63, no. 17, pp. 5203–5208, 2003.

[21] I. R. Hutcheson, L. Goddard, D. Barrow et al., "Fulvestrant-induced expression of ErbB3 and ErbB4 receptors sensitizes oestrogen receptor-positive breast cancer cells to heregulin β1," *Breast Cancer Research*, vol. 13, no. 2, article R29, 2011.

[22] S. Hiscox, W. G. Jiang, K. Obermeier et al., "Tamoxifen resistance in MCF7 cells promotes EMT-like behaviour and involves modulation of β-catenin phosphorylation," *International Journal of Cancer*, vol. 118, no. 2, pp. 290–301, 2006.

[23] R. Gopalakrishna, U. Gundimeda, J. A. Fontana, and R. Clarke, "Differential distribution of protein phosphatase 2A in human breast carcinoma cell lines and its relation to estrogen receptor status," *Cancer Letters*, vol. 136, no. 2, pp. 143–151, 1999.

[24] X. Wu, M. Subramaniam, S. B. Grygo et al., "Estrogen receptor-beta sensitizes breast cancer cells to the anti-estrogenic actions of endoxifen," *Breast Cancer Research*, vol. 13, no. 2, article R27, 2011.

[25] T. N. Mitropoulou, G. N. Tzanakakis, D. Kletsas, H. P. Kalofonos, and N. K. Karamanos, "Letrozole as a potent inhibitor of cell proliferation and expression of metalloproteinases (MMP-2 and MMP-9) by human epithelial breast cancer cells," *International Journal of Cancer*, vol. 104, no. 2, pp. 155–160, 2003.

[26] S. H. Ngalim, A. Magenau, G. Le Saux, J. J. Gooding, and K. Gaus, "How do cells make decisions: engineering micro- and nanoenvironments for cell migration," *Journal of Oncology*, vol. 2012, Article ID 363106, 2010.

[27] J. Sastre-Serra, M. Nadal-Serrano, D. G. Pons, A. Valle, J. Oliver, and P. Roca, "The effects of 17β-estradiol on mitochondrial biogenesis and function in breast cancer cell lines are dependent on the ERα/ERβ Ratio," *Cellular Physiology and Biochemistry*, vol. 29, no. 1-2, pp. 261–268, 2012.

[28] S.-H. Park, L. W. T. Cheung, A. S. T. Wong, and P. C. K. Leung, "Estrogen regulates snail and slug in the down-regulation of E-cadherin and induces metastatic potential of ovarian cancer cells through estrogen receptor α," *Molecular Endocrinology*, vol. 22, no. 9, pp. 2085–2098, 2008.

[29] O. C. Kousidou, A. Berdiaki, D. Kletsas et al., "Estradiol-estrogen receptor: a key interplay of the expression of syndecan-2 and metalloproteinase-9 in breast cancer cells," *Molecular Oncology*, vol. 2, no. 3, pp. 223–232, 2008.

[30] G. Cory, "Scratch-wound assay," *Methods in Molecular Biology*, vol. 769, pp. 25–30, 2011.

[31] E. S. Radisky and D. C. Radisky, "Matrix metalloproteinase-induced epithelial-mesenchymal transition in breast cancer," *Journal of Mammary Gland Biology and Neoplasia*, vol. 15, no. 2, pp. 201–212, 2010.

[32] C. Gialeli, A. D. Theocharis, and N. K. Karamanos, "Roles of matrix metalloproteinases in cancer progression and their pharmacological targeting," *The FEBS Journal*, vol. 278, no. 1, pp. 16–27, 2011.

[33] U. W. Nilsson, S. Garvin, and C. Dabrosin, "MMP-2 and MMP-9 activity is regulated by estradiol and tamoxifen in cultured human breast cancer cells," *Breast Cancer Research and Treatment*, vol. 102, no. 3, pp. 253–261, 2007.

[34] Y. Wu and B. P. Zhou, "Snail: more than EMT," *Cell Adhesion and Migration*, vol. 4, no. 2, pp. 199–203, 2010.

[35] M. Guarino, B. Rubino, and G. Ballabio, "The role of epithelial-mesenchymal transition in cancer pathology," *Pathology*, vol. 39, no. 3, pp. 305–318, 2007.

[36] S. Vega, A. V. Morales, O. H. Ocaña, F. Valdés, I. Fabregat, and M. A. Nieto, "Snail blocks the cell cycle and confers resistance to cell death," *Genes and Development*, vol. 18, no. 10, pp. 1131–1143, 2004.

[37] M. Luo and J.-L. Guan, "Focal adhesion kinase: a prominent determinant in breast cancer initiation, progression and metastasis," *Cancer Letters*, vol. 289, no. 2, pp. 127–139, 2010.

[38] S. Hiscox, P. Barnfather, E. Hayes et al., "Inhibition of focal adhesion kinase suppresses the adverse phenotype of endocrine-resistant breast cancer cells and improves endocrine response in endocrine-sensitive cells," *Breast Cancer Research and Treatment*, vol. 125, no. 3, pp. 659–669, 2011.

[39] S. Chia and W. Gradishar, "Fulvestrant: expanding the endocrine treatment options for patients with hormone receptor-positive advanced breast cancer," *Breast*, vol. 17, no. 3, pp. S16–S21, 2008.

[40] S. Chia, W. Gradishar, L. Mauriac et al., "Double-blind, randomized placebo controlled trial of fulvestrant compared with exemestane after prior nonsteroidal aromatase inhibitor therapy in postmenopausal women with hormone receptor-positive, advanced breast cancer: rsults from EFECT," *Journal of Clinical Oncology*, vol. 26, no. 10, pp. 1664–1670, 2008.

[41] A. Howell, J. F. R. Robertson, P. Abram et al., "Comparison of fulvestrant versus tamoxifen for the treatment of advanced breast cancer in postmenopausal women previously untreated with endocrine therapy: a multinational, double-blind, randomized trial," *Journal of Clinical Oncology*, vol. 22, no. 9, pp. 1605–1613, 2004.

[42] N. Normanno, M. Di Maio, E. de Maio et al., "Mechanisms of endocrine resistance and novel therapeutic strategies in breast cancer," *Endocrine-Related Cancer*, vol. 12, no. 4, pp. 721–747, 2005.

[43] S. Hiscox, L. Morgan, D. Barrow, C. Dutkowski, A. Wakeling, and R. I. Nicholson, "Tamoxifen resistance in breast cancer cells is accompanied by an enhanced motile and invasive phenotype: inhibition by gefitinib ('Iressa', ZD1839)," *Clinical and Experimental Metastasis*, vol. 21, no. 3, pp. 201–212, 2004.

[44] N. Goto, H. Hiyoshi, I. Ito, M. Tsuchiya, Y. Nakajima, and J. Yanagisawa, "Estrogen and antiestrogens alter breast cancer invasiveness by modulating the transforming growth factor-β signaling pathway," *Cancer Science*, vol. 102, no. 8, pp. 1501–1508, 2011.

[45] S. M. A. Abidi, E. W. Howard, J. J. Dmytryk, and J. T. Pento, "Differential influence of antiestrogens on the *in vitro* release of gelatinases (type IV collagenases) by invasive and non-invasive breast cancer cells," *Clinical and Experimental Metastasis*, vol. 15, no. 4, pp. 432–439, 1997.

[46] O. Ilina, G.-J. Bakker, A. Vasaturo, R. M. Hofmann, and P. Friedl, "Two-photon laser-generated microtracks in 3D collagen lattices: Principles of MMP-dependent and -independent collective cancer cell invasion," *Physical Biology*, vol. 8, no. 1, Article ID 015010, 2011.

[47] M. Maynadier, P. Nirdé, J.-M. Ramirez et al., "Role of estrogens and their receptors in adhesion and invasiveness of breast cancer cells," *Advances in Experimental Medicine and Biology*, vol. 617, pp. 485–491, 2008.

[48] R. X.-D. Song, Y. Chen, Z. Zhang et al., "Estrogen utilization of IGF-1-R and EGF-R to signal in breast cancer cells," *Journal of Steroid Biochemistry and Molecular Biology*, vol. 118, no. 4-5, pp. 219–230, 2010.

[49] H. Brauch and V. C. Jordan, "Targeting of tamoxifen to enhance antitumour action for the treatment and prevention of breast cancer: the 'personalised' approach?" *European Journal of Cancer*, vol. 45, no. 13, pp. 2274–2283, 2009.

[50] D. Domínguez, B. Montserrat-Sentís, A. Virgós-Soler et al., "Phosphorylation regulates the subcellular location and activity of the snail transcriptional repressor," *Molecular and Cellular Biology*, vol. 23, no. 14, pp. 5078–5089, 2003.

6

Use of Insulin to Increase Epiblast Cell Number: Towards a New Approach for Improving ESC Isolation from Human Embryos

Jared M. Campbell,[1,2] **Michelle Lane,**[1,3] **Ivan Vassiliev,**[2] **and Mark B. Nottle**[2]

[1] *Discipline of Obstetrics and Gynaecology, School of Paediatrics and Reproductive Health, University of Adelaide, Medical School South, Level 3, Frome Road, Adelaide, SA 5005, Australia*

[2] *Centre for Stem Cell Research, University of Adelaide, Medical School South, Level 3, Frome Road, Adelaide, SA 5005, Australia*

[3] *Repromed, 180 Fullarton Road, Dulwich, SA 5065, Australia*

Correspondence should be addressed to Jared M. Campbell; jared.campbell@adelaide.edu.au

Academic Editor: Deepa Bhartiya

Human embryos donated for embryonic stem cell (ESC) derivation have often been cryopreserved for 5–10 years. As a consequence, many of these embryos have been cultured in media now known to affect embryo viability and the number of ESC progenitor epiblast cells. Historically, these conditions supported only low levels of blastocyst development necessitating their transfer or cryopreservation at the 4–8-cell stage. As such, these embryos are donated at the cleavage stage and require further culture to the blastocyst stage before hESC derivation can be attempted. These are generally of poor quality, and, consequently, the efficiency of hESC derivation is low. Recent work using a mouse model has shown that the culture of embryos from the cleavage stage with insulin to day 6 increases the blastocyst epiblast cell number, which in turn increases the number of pluripotent cells in outgrowths following plating, and results in an increased capacity to give rise to ESCs. These findings suggest that culture with insulin may provide a strategy to improve the efficiency with which hESCs are derived from embryos donated at the cleavage stage.

1. Introduction

The embryo begins as a single totipotent cell it then undergoes multiple rounds of division coupled with differentiation until it forms a blastocyst and has the potential to implant in the uterus. The pluripotent epiblast of the inner cell mass (ICM) then undergoes further division and differentiation to develop into the fetus and eventually a fully developed organism. Epiblast cells can be isolated and cultured in conditions which allow embryonic stem cell (ESC) lines to be derived. ESC lines, especially hESC lines, hold considerable promise in the fields of drug discovery, developmental biology and regenerative medicine. However, the efficiency of hESC derivation is low. This paper brings together work by our group which used a mouse model to develop a strategy for improving the efficiency of hESC derivation. As most embryos donated for human ESC derivation were cultured in relatively simple media, now known to perturb development, before being frozen at the precompaction stage up to 10 years

earlier, we examined the hypothesis that the period where they are subsequently cultured to the blastocyst stage could be exploited to improve the efficiency with which cell lines could be derived. In particular, our work has focussed on adding insulin to culture media to increase epiblast cell number.

In initial studies [1], we showed that in vitro culture of embryos during the precompaction stage in a simple medium that was designed to model the culture conditions that human embryos available for hESC derivation were previously exposed to reduces embryo quality; as highlighted by reduced developmental rates, decreased epiblast cell number and altered gene expression in outgrowths compared with that seen for embryos culture in modern G1 medium. Some aspects of this reduction in quality could be restored after compaction by culture in a medium designed to support postcompaction embryo development in vitro (G2), including epiblast cell number. However, epiblast cell number was only partially improved compared with embryos cultured in G1/G2. The findings of [2, 3] as well as our own later studies

[4] suggest that increased epiblast cell number correlates with an increased capacity to give rise to ESCs. Culture in G2 medium postcompaction also increased the proportion of embryos which reached the hatched blastocyst stage which was subsequently shown in [4] to be correlated with an increased capacity to give rise to primary ESC colonies. Together, these findings highlight that subsequent culture in modern culture systems can improve the efficiency of ESC derivation from embryos initially cultured in simple media.

Numerous growth factors are known to influence embryo development. However, these are not routinely included in embryo culture media. In contrast, growth factors such as leukaemia inhibitory factor (LIF) are routinely used for the isolation and maintenance of ESCs. Based on these observations, we hypothesised that addition of a growth factor to embryo culture media could further improve blastocyst development, epiblast cell number, outgrowth formation rate, and ESC derivation. In particular, we investigated whether insulin could be used to increase ESC derivation efficiency because it has previously been shown to increase ICM cell number when added to embryo culture media [5, 6].

The results from these studies showed that the inclusion of insulin in postcompaction culture medium increased the number of pluripotent cells in blastocysts. In a series of experiments we showed that insulin acted via the PI3K/GSK3 p53 pathway to shift the balance of differentiation versus pluripotency within the ICM to increase epiblast number and proportion [7]. This resulted in an increase in their capacity to give rise to outgrowths with more pluripotent cells, as well as an increase in capacity to give rise to primary ESC colonies [4]. These findings suggest that the inclusion of insulin in embryo culture medium postcompaction could be used to improve the quality of cryopreserved or fresh human embryos donated at or near compaction.

2. Impact of Culture Conditions on Epiblast Cell Number and Pluripotency

The relatively low efficiency with which hESCs can be derived has been attributed to the reduced quality of human embryos donated for this purpose [8–10]. In vitro culture of embryos is typically associated with reductions in embryo quality and viability [11]. Furthermore, human embryos for hESC derivation have often been cryopreserved for 5–10 years prior to their donation [12, 13]. As a consequence many were cultured in media now known to perturb viability, and which supported only low levels of blastocyst development. This necessitated cleavage stage transfers for the majority of IVF cycles performed.

In order to model this system, mouse embryos were cultured in relatively simple medium for the first 48 h to the 8-cell stage to approximate the culture period commonly used in human IVF experiments [1]. These experiments demonstrated that culture of embryos in simple medium retarded the development of blastocysts and significantly reduced the number of epiblast cells, which was consistent with previous studies in human and mouse [14–19]. These experiments also demonstrated that there was some capacity

to improve blastocyst development and epiblast cell number by transferring embryos to the more complex G2 medium for culture from the 8-cell stage. Despite this, initial culture in relatively simple medium had a lasting negative impact on subsequent development. Furthermore, we found that embryos cultured in simple medium were less likely to contain an epiblast and therefore lacked the capacity to generate an ESC line, irrespective of the culture medium used for the second 48 h. Additionally, assessment of outgrowths generated from these blastocysts showed that the perturbing conditions of a simple medium had lasting effects on the gene expression of the outgrowths, with altered gene expression of *Atrx* and *Nanog*.

Human embryos donated for hESC derivation are likely to have been exposed to conditions such as simple style culture medium, examples of which include HTF, Earle's, and T6, which were widely used in IVF and can still be found in use. Collectively, the findings of [1] therefore indicate that these embryos are likely to have a reduced capacity to give rise to hESCs. This in turn suggests that the characteristics of hESC lines could be affected by predonation embryo culture conditions.

As many human embryos, historically and presently, are cryopreserved at the cleavage stage [20–22] and therefore donated for ESC generation at this stage, they must be further cultured to the blastocyst stage before ESC derivation. This additional culture period represents a window where the pluripotency of embryos which have previously been exposed to perturbing culture conditions can be improved. The results of [1] demonstrated that while the quality of mouse embryos could be improved by culturing them in modern complex medium purpose designed to support embryo development from the 8-cell stage, additional interventions are necessary to fully exploit the cleavage to blastocyst culture period.

3. Insulin Stimulation of Pluripotency in Postcompaction Embryos

The inclusion of select growth factors in embryo culture media has previously been shown to be capable of improving embryo development and viability [5, 23]. However, growth factors are not routinely included in culture media commercially available for human embryo culture [24–28]. To further examine how interventions to the culture medium for the postcompaction stage embryo may affect epiblast cells and pluripotency, the growth factor insulin was added to the culture medium. Insulin has previously been shown to increase the ICM cell number of embryos [29] and was selected as the most promising candidate from a panel of growth factors previously used to improve embryo culture.

The findings of [7] demonstrated that $1.7\rho M$ insulin increased epiblast cell number without affecting ICM cell number. This resulted in a significant increase in the proportion of ICM cells which were epiblast as opposed to primitive endoderm. This novel finding suggested that insulin was acting to shift the balance of differentiation within the ICM towards more pluripotent cells, rather than acting as a general mitogenic factor and stimulating overall cell growth.

This was further highlighted by the finding that total cell number and trophectoderm cell number were also unaffected by the addition of insulin. However, as in previous studies [5, 23], there was a threshold concentration where the effect of insulin was maximal, above which further increases led to the loss of the increase in epiblast cell number. If this strategy is implemented in the human, it may be necessary to repeat these dose response experiments to establish an optimal dose in terms of epiblast cell number increases. While this work shows that insulin is able to maintain pluripotency in the ICM and direct differentiation, other growth factors may also have beneficial effects, and it is possible that a combination of growth factors may produce a synergistic effect and improve blastocyst quality and epiblast cell number.

4. Molecular Mechanism of Action of Insulin on Pluripotency in the Blastocyst

Having demonstrated that the culture of postcompaction stage embryos with insulin increases epiblast cell number, further experiments were undertaken to determine the signalling pathways behind this effect [7]. At the concentration identified as increasing epiblast cell number and proportion, insulin is known to activate the insulin receptor [29]. One of the primary second messengers of the insulin receptor is PI3K, which has previously been shown to be integral for maintaining pluripotency in ESCs [30]. Using inhibitors, it was demonstrated that PI3K activity was necessary for insulin to increase epiblast cell number (Figure 1). One target of PI3K is GSK3, which is phosphorylated by active PI3K, inactivating it. When active, GSK3 is capable of phosphorylating many second messengers which converge to reduce Nanog transcription, which is important for the retention of pluripotency [31–37]. Inhibiting GSK3, theoretically reproducing the effect of culture with insulin and active PI3K, increased epiblast cell number. Furthermore, activating GSK3 blocked insulin's ability to increase epiblast cell number without affecting the epiblast cell number of embryos cultured without insulin, replicating the effect of PI3K inhibition. These results suggest that the inactivation of GSK3 is an important component of the insulin signalling pathway in relation to increasing epiblast cell number.

The pro-apoptotic protein p53, which is also regulated by PI3K, specifically via PI3K activated ubiquitinase MDM2 [38–40], causes cell death and differentiation when active and binds to the *Nanog* promoter region to repress Nanog transcription [41]. As with GSK3, inhibition of p53 increased epiblast cell number, while activation blocked insulin-mediated epiblast increases, strongly suggesting that p53 is involved in insulin-mediated increases to epiblast cell number. Interestingly, there are multiple points of cross-reactivity between GSK3 and p53 [42–46]; however, no additional epiblast increases were found for co-inhibition of the two factors. This suggests that the potential of GSK3 inhibition to cause the accumulation of p53 [45, 46] did not have a confounding effect in these experiments. In conclusion, the results of these studies demonstrated that insulin increased the epiblast cell number via the activation of PI3K (most likely via its interaction with the insulin receptor), which subsequently inactivates the second messengers GSK3 and p53, to increase Nanog transcription and therefore promote pluripotency and the epiblast (Figure 1).

5. Expression of OCT4 and Nanog in Blastocysts

The localisation of OCT4 and Nanog in blastocysts on day 4 (early blastocysts), day 5 (predominantly expanded blastocyst), and day 6 (hatching blastocysts) was determined by immunohistochemistry [4]. Of note, it was found that at the stage of development where the literature sources suggested that OCT4 and Nanog would be restricted to the ICM and epiblast, respectively [47, 48], both were still widely expressed. A comparison of methodologies suggested that this difference is likely the result of collecting and beginning embryo culture at the zygote stage in this study, rather than at the 2-cell stage or later.

Human embryos are ubiquitously cultured from the zygote stage following in vitro fertilisation. As such, this finding suggests that future researchers who use the mouse blastocyst to model the in vitro development of human embryos should culture embryos from the zygote stage, as the discrepancy appears to produce a meaningful difference particularly with regards to epiblast development. Future work in this area should include the direct comparison of OCT4 and Nanog expression of in vitro and in vivo grown mouse embryos as well as the characterisation of Oct4 and Nanog expression in human blastocysts.

6. Effect of Insulin in Embryo Culture Medium Persists in Outgrowths

Despite the increase in Nanog positive cell number due to culture with insulin, we found that when blastocysts were plated before the transcription factor was restricted to the epiblast, outgrowths from insulin-treated embryos contained no more epiblast cells than outgrowths from control embryos. Further, our results showed that despite earlier stage blastocysts possessing more OCT4 and Nanog positive cells than later stage blastocysts, they gave rise to outgrowths with significantly fewer epiblast cells. As hESCs have been shown to be most efficiently derived from blastocyst where Oct4 has been restricted to the ICM [49], this finding supports the use of mouse blastocysts in modelling human embryo development and hESC derivation. The important finding from these experiments, however, was that when embryos were allowed to develop until Nanog was restricted to the epiblast and OCT4 was restricted to the ICM before plating, culture of embryos in insulin postcompaction resulted in the generation of outgrowths which were more likely to contain an epiblast and which contained a larger number of epiblast cells.

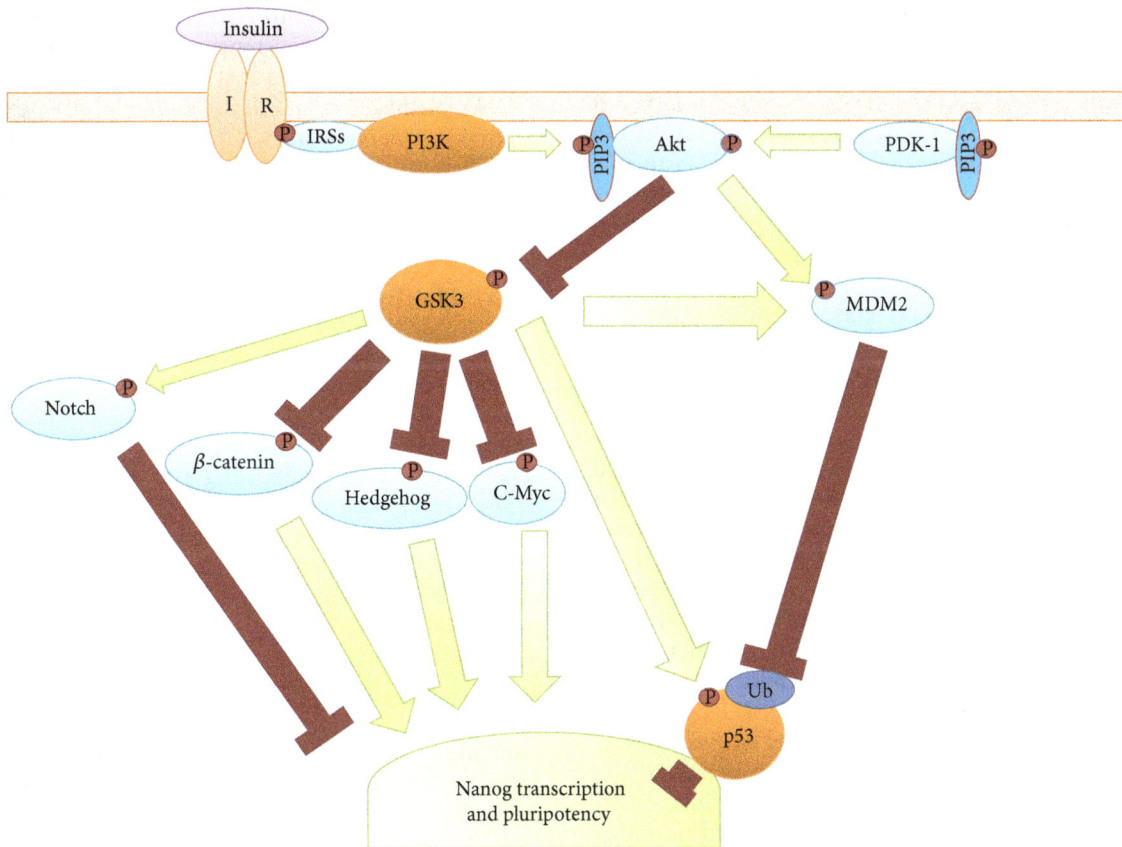

FIGURE 1: Schematic of insulin signalling and its regulation of Nanog expression and pluripotency. Green arrows indicate reactions with a stimulatory effect on their target, and red closed bars indicate reactions with a retarding effect on their target. P marks reactions where phosphorylation occurs, and Ub marks reactions where ubiquitination occurs. Insulin binds the insulin receptor (IR), a tyrosine kinase which is then able to phosphorylate the IRSs. PI3K is able to bind to the phosphorylated IRSs by its SH2 domains, resulting in activation. PI3K phosphorylates the phospholipid PIP2, producing PIP3, which can be bound by the pleckstrin homology domains of PDK-1 and Akt. Results in [7] show that activation of PI3K is necessary for insulin to increase the number of Nanog positive epiblast cells during embryo culture. When PDK-1 and Akt are colocalised to the cell membrane, PDK-1 is able to phosphorylate and activate Akt. Active Akt can phosphorylate GSK3, inactivating it. When active, GSK3 is able to phosphorylate β-catenin, Hedgehog, and c-Myc; all factors which safeguard pluripotency through interactions with other second messengers. Additionally, active GSK3 phosphorylates and protects the intracellular domain of Notch, promoting differentiation. Further, inactivation of GSK3 is necessary for insulin to increase the number of Nanog positive epiblast cells during embryo culture [7]. Akt is also able to phosphorylate and activate MDM2 which ubiquitinates the proapoptotic factor p53, causing its inactivation and removal from the nucleus, where it would bind to the *Nanog* promoter and suppresses its expression. Inactivation of p53 is necessary for insulin to increase the number of Nanog positive epiblast cells during embryo culture [7]. GSK3 and p53 are able to form a dimer, resulting in the phosphorylation of p53 and the increased activity of both factors. GSK3 is also able to phosphorylate and activate MDM2. However, despite these outcomes, the interaction of GSK3 and p53 do not have a significant effect on Nanog positive epiblast cell number during embryo culture [7].

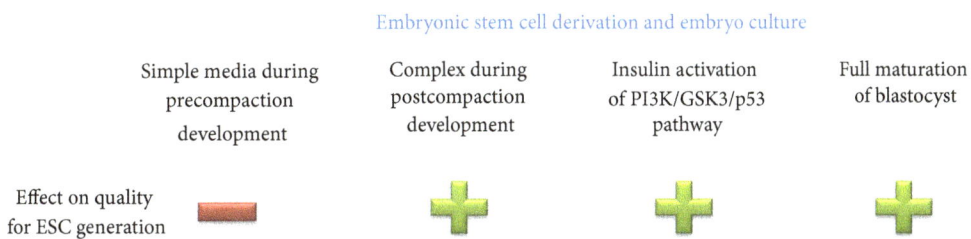

FIGURE 2: Summary of the culture effects examined in this paper and their observed effect on the retention of pluripotency towards ESC derivation.

7. Insulin in Culture Media and the Effect of ESC Colony Generation

Insulin- and control-treated blastocysts were plated on day 6, outgrown, trypsinised, and replated, and primary cell colonies with an ESC morphology were stained for OCT4 and Nanog expression to confirm pluripotency [4]. Blastocysts were twice as likely to give rise to primary ESC colonies if they were cultured with insulin for the postcompaction stage. Interestingly, examining how the inclusion of insulin interacted with this process led to the novel observation that culture of embryos with insulin increased the proportion of embryos which, at the point of plating, were at the most advanced morphological stage (hatched) and also increased the proportion of those hatched blastocysts which gave rise to ESCs. As such, hatched insulin-cultured blastocysts are more plentiful and more likely to give rise to ESCs than hatched control-cultured blastocysts. This result demonstrates that insulin improves ESC isolation through mechanisms beyond simply improving morphology, which has previously been linked to increased ESC derivation rates. It is likely that the improved capacity of insulin-cultured blastocysts with the highest morphological quality to give rise to primary ESC colonies is the result of the increased epiblast cell numbers demonstrated in both [4, 7].

Modelling of the experimental outcomes enabled conclusions to be made around the most significant characteristics that an embryo must contain to generate a primary ESC colony. The greatest predictor of a control-cultured blastocyst giving rise to a primary ESC colony was it cavitating on day 4, whereas for blastocysts cultured with insulin, the greatest predictor was being hatched on day 6. For day 4, this observation is likely the result of insulin increasing the rate of cavitation and thereby making the marker less selective. The finding on day 6 is suggestive that in the control group, hatched blastocysts, which have shown the best development, have no more epiblast cells than their more slowly developing counterparts. Both of these observations warrant further investigation.

The results of this work demonstrate that the addition of insulin to embryo culture medium from the cleavage stage to the blastocyst stage improves the efficiency with which ESCs can be generated from these embryos. As human embryos are most often donated at the cleavage stage and ESC derivation is most often attempted at the blastocyst stage, the application of this strategy has the potential to improve hESC derivation efficiency. Due to the limited availability of human embryos for ESC derivation, improving efficiency is a matter of key importance.

Future work which would be necessary to validate these findings is the expansion of mESC colonies from control- and insulin-cultured embryos to fully characterised mESC lines and the reproduction of these experiments in the human. Further, our work has shown that the effect of insulin persists beyond embryo culture through the outgrowth phase and into ESC derivation. This suggests that during embryo culture-insulin may have a permanent positive effect on cell properties and that ESC lines derived from embryos cultured with insulin may have altered characteristics. As

such, future work should include not just the characterisation of ESC lines from control and insulin-cultured embryos for pluripotency and self-renewal, but also more in depth characterisation including metabolic profile, an assessment of DNA methylation and acetylation, and gene expression, to provide a more detailed and precise picture of the quality and differentiation status of ESC lines, with a view towards investigating whether the culture of embryos with insulin results in the derivation of higher quality ESC lines.

8. Conclusion

In conclusion, the results presented in this paper show that while culture in simple medium during the cleavage stage decreases pluripotency, the inclusion of insulin in embryo culture medium from the compaction stage stimulates pluripotency supporting pathways to increase the number of epiblast cells in the fully developed blastocyst, resulting in an increased capacity to generate ESCs (Figure 2). This strategy is of particular relevance for hESC derivation where embryos are most often donated at the cleavage stage and of reduced quality.

Acknowledgments

The authors acknowledge the support of the NHMRC program grant for funding. M. Lane is a recipient of NHMRC Senior Research Fellowship.

References

[1] J. M. Campbell, M. Mitchell, M. B. Nottle, and M. Lane, "Development of a mouse model for studying the effect of embryo culture on embryonic stem cell derivation," *Stem Cells and Development*, vol. 20, no. 9, pp. 1577–1586, 2011.

[2] J. Nichols, J. Silva, M. Roode, and A. Smith, "Suppression of Erk signalling promotes ground state pluripotency in the mouse embryo," *Development*, vol. 136, no. 19, pp. 3215–3222, 2009.

[3] L. Batlle-Morera, A. Smith, and J. Nichols, "Parameters influencing derivation of embryonic stem cells from murine embryos," *Genesis*, vol. 46, no. 12, pp. 758–767, 2008.

[4] J. M. Campbell, M. Lane, I. Vassiliev, and M. Nottle, "Epiblast cell number and primary embryonic stem cell colony generation are increased by culture of cleavage stage embryos in insulin," *The Journal of Reproduction and Development*, 2012.

[5] K. Hardy and S. Spanos, "Growth factor expression and function in the human and mouse preimplantation embryo," *Journal of Endocrinology*, vol. 172, no. 2, pp. 221–236, 2002.

[6] L. Karagenc, M. Lane, and D. K. Gardner, "Granulocyte-macrophage colony-stimulating factor stimulates mouse blastocyst inner cell mass development only when media lack human serum albumin," *Reproductive BioMedicine Online*, vol. 10, no. 4, pp. 511–518, 2005.

[7] J. M. Campbell, M. Nottle, I. Vassiliev, M. Mitchell, and M. Lane, "Insulin increases epiblast cell number of in vitro cultured mouse embryos via the PI3K/GSK3/p53 pathway," *Stem Cells and Development*, vol. 21, no. 13, pp. 2430–2441, 2012.

[8] M. Mitalipova, J. Calhoun, S. Shin et al., "Human embryonic stem cell lines derived from discarded embryos," *Stem Cells*, vol. 21, no. 5, pp. 521–526, 2003.

[9] P. H. Lerou, A. Yabuuchi, H. Huo et al., "Human embryonic stem cell derivation from poor-quality embryos," *Nature Biotechnology*, vol. 26, no. 2, pp. 212–214, 2008.

[10] C. A. Cowan, I. Klimanskaya, J. McMahon et al., "Derivation of embryonic stem-cell lines from human blastocysts," *The New England Journal of Medicine*, vol. 350, no. 13, pp. 1353–1356, 2004.

[11] G. Giritharan, S. Talbi, A. Donjacour, F. Di Sebastiano, A. T. Dobson, and P. F. Rinaudo, "Effect of in vitro fertilization on gene expression and development of mouse preimplantation embryos," *Reproduction*, vol. 134, no. 1, pp. 63–72, 2007.

[12] B. J. Bankowski, A. D. Lyerly, R. R. Faden, and E. E. Wallach, "The social implications of embryo cryopreservation," *Fertility and Sterility*, vol. 84, no. 4, pp. 823–832, 2005.

[13] S. C. Klock, "Embryo disposition: the forgotten "child" of in vitro fertilization," *International Journal of Fertility and Women's Medicine*, vol. 49, no. 1, pp. 19–23, 2004.

[14] D. K. Gardner, "Dissection of culture media for embryos: the most important and less important components and characteristics," *Reproduction, Fertility and Development*, vol. 20, no. 1, pp. 9–18, 2008.

[15] M. Lane and D. K. Gardner, "Embryo culture medium: which is the best?" *Best Practice & Research. Clinical Obstetrics & Gynaecology*, vol. 21, no. 1, pp. 83–100, 2006.

[16] H. Laverge, P. De Sutler, R. Desmet, J. Van der Elst, and M. Dhont, "Prospective randomized study comparing human serum albumin with fetal cord serum as protein supplement in culture medium for in-vitro fertilization," *Human Reproduction*, vol. 12, no. 10, pp. 2263–2266, 1997.

[17] D. K. Gardner and M. Lane, "Towards a single embryo transfer," *Reproductive BioMedicine Online*, vol. 6, no. 4, pp. 470–481, 2003.

[18] D. K. Gardner and M. Lane, "Culture of viable human blastocysts in defined sequential serum-free media," *Human Reproduction*, vol. 13, no. 3, pp. 148–160, 1998.

[19] D. K. Gardner, "Blastocyst culture: toward single embryo transfers," *Human Fertility*, vol. 3, no. 4, pp. 229–237, 2000.

[20] J. Van der Elst, E. Van Den Abbeel, M. Camus, J. Smitz, P. Devroey, and A. Van Steirteghem, "Long-term evaluation of implantation of fresh and cryopreserved human embryos following ovarian stimulation with buserelin acetate-human menopausal gonadotrophin (HMG) or clomiphene citrate-HMG," *Human Reproduction*, vol. 11, no. 10, pp. 2097–2106, 1996.

[21] S. Oehninger, J. Mayer, and S. Muasher, "Impact of different clinical variables on pregnancy outcome following embryo cryopreservation," *Molecular and Cellular Endocrinology*, vol. 169, no. 1-2, pp. 73–77, 2000.

[22] Y. A. Wang, G. M. Chambers, M. Dieng, and E. A. Sullivan, "Assisted reproductive technology in Australia and New Zealand 2007," *Assisted Reproduction Technology Series*, vol. 13, pp. 1–56, 2009.

[23] B. Huppertz and A. Herrler, "Regulation of proliferation and apoptosis during development of the preimplantation embryo and the placenta," *Birth Defects Research Part C*, vol. 75, no. 4, pp. 249–261, 2005.

[24] E. Greenblatt, T. Di Beraradino, P. Chronis-Brown, D. Holt, and A. Lains, "Comparison of Global Medium and G1/G2 cleavage/blastocyst sequential media for cultre of human embryos after IVF. In:," in *European Society of Human Reproduction and Embryology (ESHRE '05)*, Copenhagan, Denmark, 2005.

[25] S. Xella, T. Marsella, D. Tagliasacchi et al., "Embryo quality and implantation rate in two different culture media: ISM1 versus Universal IVF Medium," *Fertility and Sterility*, vol. 93, no. 6, pp. 1859–1863, 2010.

[26] J. M. Baltz, "Media composition: salts and osmolality," *Methods in Molecular Biology*, vol. 912, pp. 61–80, 2012.

[27] Y. Ho, K. Wigglesworth, J. J. Eppig, and R. M. Schultz, "Preimplantation development of mouse embryos in KSOM: augmentation by amino acids and analysis of gene expression," *Molecular Reproduction and Development*, vol. 41, no. 2, pp. 232–238, 1995.

[28] M. Lane, D. K. Gardner, M. J. Hasler, and J. F. Hasler, "Use of G1.2/G2.2 media for commercial bovine embryo culture: equivalent development and pregnancy rates compared to co-culture," *Theriogenology*, vol. 60, no. 3, pp. 407–419, 2003.

[29] M. B. Harvey and P. L. Kaye, "Insulin increases the cell number of the inner cell mass and stimulates morphological development of mouse blastocysts in vitro," *Development*, vol. 110, no. 3, pp. 963–967, 1990.

[30] M. P. Storm, H. K. Bone, C. G. Beck et al., "Regulation of nanog expression by phosphoinositide 3-kinase-dependent signaling in murine embryonic stem cells," *Journal of Biological Chemistry*, vol. 282, no. 9, pp. 6265–6273, 2007.

[31] Y. Takao, T. Yokota, and H. Koide, "β-Catenin up-regulates Nanog expression through interaction with Oct-3/4 in embryonic stem cells," *Biochemical and Biophysical Research Communications*, vol. 353, no. 3, pp. 699–705, 2007.

[32] G. S. Sineva and V. A. Pospelov, "Inhibition of GSK3β enhances both adhesive and signalling activities of β-catenin in mouse embryonic stem cells," *Biology of the Cell*, vol. 102, no. 10, pp. 549–560, 2010.

[33] X. He, M. Semenov, K. Tamai, and X. Zeng, "LDL receptor-related proteins 5 and 6 in Wnt/β-catenin signaling: arrows point the way," *Development*, vol. 131, no. 8, pp. 1663–1677, 2004.

[34] A. Po, E. Ferretti, E. Miele et al., "Hedgehog controls neural stem cells through p53-independent regulation of Nanog," *The EMBO Journal*, vol. 29, no. 15, pp. 2646–2658, 2010.

[35] J. Jia, K. Amanai, G. Wang, J. Tang, B. Wang, and J. Jiang, "Shaggy/GSK3 antagonizes hedgehog signalling by regulating Cubitus interruptus," *Nature*, vol. 416, no. 6880, pp. 548–552, 2002.

[36] P. Cartwright, C. McLean, A. Sheppard, D. Rivett, K. Jones, and S. Dalton, "LIF/STAT3 controls ES cell self-renewal and pluripotency by a Myc-dependent mechanism," *Development*, vol. 132, no. 5, pp. 885–896, 2005.

[37] M. A. Gregory, Y. Qi, and S. R. Hann, "Phosphorylation by glycogen synthase kinase-3 controls c-myc proteolysis and subnuclear localization," *Journal of Biological Chemistry*, vol. 278, no. 51, pp. 51606–51612, 2003.

[38] L. D. Mayo, J. E. Dixon, D. L. Durden, N. K. Tonks, and D. B. Donner, "PTEN protects p53 from Mdm2 and sensitizes cancer cells to chemotherapy," *Journal of Biological Chemistry*, vol. 277, no. 7, pp. 5484–5489, 2002.

[39] T. M. Gottlieb, J. F. Martinez Leal, R. Seger, Y. Taya, and M. Oren, "Cross-talk between Akt, p53 and Mdm2: possible implications for the regulation of apoptosis," *Oncogene*, vol. 21, no. 8, pp. 1299–1303, 2002.

[40] Y. Ogawara, S. Kishishita, T. Obata et al., "Akt enhances Mdm2-mediated ubiquitination and degradation of p53," *Journal of Biological Chemistry*, vol. 277, no. 24, pp. 21843–21850, 2002.

[41] T. Lin, C. Chao, S. Saito et al., "p53 induces differentiation of mouse embryonic stem cells by suppressing Nanog expression," *Nature Cell Biology*, vol. 7, no. 2, pp. 165–171, 2005.

[42] G. A. Turenne and B. D. Price, "Glycogen synthase kinase3 beta phosphorylates serine 33 of p53 and activates p53's transcriptional activity," *BMC Cell Biology*, vol. 2, article 12, 2001.

[43] P. Watcharasit, G. N. Bijur, L. Song, J. Zhu, X. Chen, and R. S. Jope, "Glycogen synthase kinase-3beta (GSK3beta) binds to and promotes the actions of p53," *The Journal of biological chemistry*, vol. 278, no. 49, pp. 48872–48879, 2003.

[44] P. Watcharasit, G. N. Bijur, J. W. Zmijewski et al., "Direct, activating interaction between glycogen synthase kinase-3β and p53 after DNA damage," *Proceedings of the National Academy of Sciences of the United States of America*, vol. 99, no. 12, pp. 7951–7955, 2002.

[45] R. Kulikov, K. A. Boehme, and C. Blattner, "Glycogen synthase kinase 3-dependent phosphorylation of Mdm2 regulates p53 abundance," *Molecular and Cellular Biology*, vol. 25, no. 16, pp. 7170–7180, 2005.

[46] J. C. Ghosh and D. C. Altieri, "Activation of p53-dependent apoptosis by acute ablation of glycogen synthase kinase-3β in colorectal cancer cells," *Clinical Cancer Research*, vol. 11, no. 12, pp. 4580–4588, 2005.

[47] C. Chazaud, Y. Yamanaka, T. Pawson, and J. Rossant, "Early lineage segregation between epiblast and primitive endoderm in mouse blastocysts through the Grb2-MAPK pathway," *Developmental Cell*, vol. 10, no. 5, pp. 615–624, 2006.

[48] Y. Yamanaka, F. Lanner, and J. Rossant, "FGF signal-dependent segregation of primitive endoderm and epiblast in the mouse blastocyst," *Development*, vol. 137, no. 5, pp. 715–724, 2010.

[49] A. E. Chen, D. Egli, K. Niakan et al., "Optimal timing of inner cell mass isolation increases the efficiency of human embryonic stem cell derivation and allows generation of sibling cell lines," *Cell Stem Cell*, vol. 4, no. 2, pp. 103–106, 2009.

7

Glutamine and Alanyl-Glutamine Increase RhoA Expression and Reduce *Clostridium difficile* Toxin-A-Induced Intestinal Epithelial Cell Damage

Ana A. Q. A. Santos,[1] Manuel B. Braga-Neto,[2] Marcelo R. Oliveira,[2] Rosemeire S. Freire,[2] Eduardo B. Barros,[3] Thiago M. Santiago,[3] Luciana M. Rebelo,[3] Claudia Mermelstein,[4] Cirle A. Warren,[5] Richard L. Guerrant,[5] and Gerly A. C. Brito[1]

[1] *Department of Morphology, Faculty of Medicine, Federal University of Ceará, Delmiro de Farias, 60416-030 Fortaleza, CE, Brazil*
[2] *Department of Physiology and Pharmacology, Faculty of Medicine, Federal University of Ceará, 1127 Coronel Nunes de Melo, 60430-270 Fortaleza, CE, Brazil*
[3] *Department of Physics, Faculty of Physics, Federal University of Ceará, 922 Campus do Pici, 60455-760 Fortaleza, CE, Brazil*
[4] *Biomedical Sciences Institute, Federal University of Rio de Janeiro, 373 Avenue Carlos Chagas, 21941-902 Rio de Janeiro, RJ, Brazil*
[5] *Division of Infectious Diseases and International Health, Center for Global Health, University of Virginia, 345 Crispell Drive, Room 2709, Charlottesville, VA 22903, USA*

Correspondence should be addressed to Gerly A. C. Brito; gerlybrito@hotmail.com

Academic Editor: Reinaldo B. Oriá

Clostridium difficile is a major cause of antibiotic-associated colitis and is associated with significant morbidity and mortality. Glutamine (Gln) is a major fuel for the intestinal cell population. Alanyl-glutamine (Ala-Gln) is a dipeptide that is highly soluble and well tolerated. IEC-6 cells were used in the *in vitro* experiments. Cell morphology was evaluated by atomic force microscopy (AFM) and scanning electron microscopy (SEM). Cell proliferation was assessed by WST-1 and Ki-67 and apoptosis was assessed by TUNEL. Cytoskeleton was evaluated by immunofluorescence for RhoA and F-actin. RhoA was quantified by immunoblotting. TcdA induced cell shrinkage as observed by AFM, SEM, and fluorescent microscopy. Additionally, collapse of the F-actin cytoskeleton was demonstrated by immunofluorescence. TcdA decreased cell volume and area and increased cell height by 79%, 66.2%, and 58.9%, respectively. Following TcdA treatment, Ala-Gln and Gln supplementation, significantly increased RhoA by 65.5% and 89.7%, respectively at 24 h. Ala-Gln supplementation increased cell proliferation by 137.5% at 24 h and decreased cell apoptosis by 61.4% at 24 h following TcdA treatment. In conclusion, TcdA altered intestinal cell morphology and cytoskeleton organization, decreased cell proliferation, and increased cell apoptosis. Ala-Gln and Gln supplementation reduced intestinal epithelial cell damage and increased RhoA expression.

1. Introduction

Clostridium difficile (*C. difficile*), a gram-positive bacillus, is considered the most frequent cause of diarrhea associated with the use of antibiotics in industrialized countries and is considered a major challenge among hospitalized patients exposed to long-term antibiotic treatment resulting in increased morbidity, mortality, and length of hospitalization [1–4]. Studies from the US, Canada, and the European Union have reported increased numbers of cases of *Clostridium difficile* infection (CDI) [5, 6]. Furthermore, recent studies have demonstrated an increase in disease severity and case-fatality rates [6–10], associated with the emergence of a more virulent strain-NAP1/B1/027, that carries a binary toxin (CDT) and produces elevated quantities of A toxin (TcdA) and B toxin (TcdB) and increased numbers of spores [9, 11].

TcdA and TcdB have glucosyltransferase activity and lead to disaggregation of actin by inactivation of Rho [12, 13]. Recently, using a hamster model of infection, it has been demonstrated that either TcdA or TcdB alone, produced by isogenic mutants of *C. difficile,* may cause severe disease [14]. Additionally, similar results were found when either gene was permanently inactivated using a gene knockout system. Finally, virulence was completely attenuated when both genes were inactivated, highlighting the importance of both TcdA and TcdB [14].

TcdA induces monoglycosylation of Rho, Cdc42, and Rac,which inhibits the Rho family proteins role in the formation of actin filaments, leading to cellular restrain, loss of adhesion, and cell rounding [15–18]. Nam et al. [19] demonstrated that TcdA causes microtubule depolymerization by tubulin deacetylation through activation of HDA6, which is involved in cytokine production, alpha-tubulin deacetylation, and mucosal damage. TcdA also causes intestinal secretion, intense destruction of the mucosa, hemorrhage, and accentuated inflammation with neutrophil infiltration and production of inflammatory cytokines such as TNF-α and IL1-β [20, 21]. Additionally, TcdA induces cellular rearrangement of actin cytoskeleton into aggregates and increases secondary adhesion, by inducing Mac-1 expression in human neutrophils. Such events could be associated with the formation of pseudomembranes [22, 23].

Glutamine (Gln) is the major respiratory fuel for the intestinal epithelium, since it is a precursor for nucleotide biosynthesis and, therefore, a critical requirement for the dynamic proliferating intestinal cell population. However, glutamine has limited solubility and a tendency to hydrolyze to potentially toxic glutamate. It has been demonstrated that alanyl-glutamine (Ala-Gln) is stable, highly soluble, well tolerated, and at least as effective in driving sodium cotransport and intestinal injury repair *in vitro* [23–26] in animals [27] and in patients [28]. Glutamine supplementation influences inflammatory response, oxidative stress, apoptosis modulation, and the integrity of gut barrier [28]. Carneiro et al., 2006 [26] demonstrated that Gln and Ala-Gln significantly reduced the intestinal damage caused by TcdA in rabbit ileal loops and the amount of intestinal epithelial cell apoptosis.

In this study, we evaluated the effects of Gln or Ala-Gln supplementation on intestinal epithelial cell injury induced by TcdA.

2. Materials and Methods

2.1. Reagents, Drugs, and Toxin. Trypsin, Dulbecco's modified Eagle media (DMEM), fetal bovine serum (FBS), RPMI media, penicillin-streptomycin, sodium pyruvate, and antibiotic antimycotic solution were obtained from either Gibco BRL (Grand Island, NY, USA) or Invitrogen (Carlsbad, CA). Gln, Ala-Gln, and TcdA of *C. difficile* (c3977), tetrazolium salt WST-1 (4-[3-(4-iodophenyl)-2H-5-tetrazolio]-1-3-benzene disulfonate), bovine insulin, DAPI- and FITC-conjugated anti-mouse secondary antibodies were obtained from Sigma (St. Louis, MO, USA). Anti-RhoA monoclonal mouse primary antibody (Santa Cruz Biotechnology, CA, USA).

2.2. Cell Culture. Rat intestinal jejunal crypt cells (IEC-6, passages 7–24) were purchased from American Type Culture Collection (Rockville, MD, USA) and cultured at 37°C in a 5% CO_2 incubator. When 90–95% confluency, cells were trypsinized with 0.25% EDTA trypsin. Cells were cultivated in 75 cm^2 flasks, and media were changed twice a week. For IEC-6 cells, the maintenance cell medium was DMEM (Gibco BRL, Grand Island, NY, USA) supplemented with 5% FBS, 5 mg bovine insulin, 50 μg/mL of penicillin/streptomycin (Gibco BRL, Grand Island, NY, USA), and a final concentration of 1 mM of sodium pyruvate. The medium was changed thrice a week, according to standard culture protocols [24, 26].

2.3. Atomic Force Microscopy. In order to evaluate the effect of TcdA in IEC-6 cell morphology by atomic force microscopy (AFM), 12-well cell culture plates, with 13 mm diameter glass coverslips, were seeded with 6.25×10^4 IEC-6 cells and grown for 24 h in standard DMEM media. Then, the wells were washed and incubated for 1 h with TcdA (100 ng/mL) in standard DMEM. Since TcdA at 100 ng/mL caused severe damage on IEC-6 cell morphology, we used 10 ng/mL to evaluate the protective effect of Gln and Ala-Gln. For this, 12-well cell culture plates, with 13 mm diameter glass coverslips, were seeded with 6.25×10^4 IEC-6 cells and grown for 24 h in standard DMEM media. Then, the wells were washed and incubated for 24 h with TcdA without Gln or supplemented with 10 mM of Ala-Gln or 10 mM of Gln. Afterwards, cells were fixed to glass coverslips in 4% formaldehyde solution for 14 h. For the imaging process, the samples were air-dried for 5 min, placed on steel sample disks covered with doublesided adhesive tape and carried off to Multimode Atomic Force Microscope (Digital Instruments, Santa Barbara, CA, USA) equipped with a NanoScope IIIa controller. Scans were performed in air, and all topography images were acquired by contact mode using silicon crystal cantilevers (Veeco-probes) with a spring constant of approximately 40 N/m and tip radius of 15 nm. The AFM height data was represented as a distinct height value of the sample in a finite number of pixels (512 × 512 point scan) [29]. The clearest regions indicate the highest area, which in the control cell indicates the localization of the nucleus. All topography images were performed with a Nanoscope IIIa controller and NanoScope software (Digital Instruments, CA, USA) at room temperature. The area, height, and volume of the cells were calculated using NanoScope 5.30 R3.SR3. The volume was calculated using bearing tool, in which the calculation is performed through the volume of a set of pixels bounded by second planes [29, 30].

2.4. Scanning Electron Microscopy. Twelve-well cell culture plates, with 13 mm diameter glass coverslips, were seeded with 6.25×10^4 IEC-6 cells and grown for 24 h in standard DMEM media. Afterwards, the wells were washed and incubated for 24 h with TcdA (10 ng/mL) in DMEM without

Gln or supplemented with 10 mM of Ala-Gln or 10 mM of Gln. Cells were than fixed in 4% formaldehyde for 14 h. For imaging, the samples were fixed to samples holders with carbon adhesive tape and sputtered with a 15 nm gold layer (BALTEC MED 020 coating system) and transferred into the scanning electron microscope (TESCAN VEGA-XMU) [31, 32].

2.5. Immunofluorescence Microscopy and Digital Image Acquisition. Six-well cell culture plates were seeded with 6×10^5 IEC-6 cells and grown for 48 h in standard DMEM media (which contains Gln). Wells were washed and incubated for 24 h with TcdA (10 ng/mL) in DMEM without Gln or supplemented with 10 mM of Ala-Gln or 10 mM of Gln. IEC-6 were rinsed with PBS and fixed with 4% paraformaldehyde in PBS for 10 min at room temperature. They were then permeabilized with 0.5% Triton-X 100 in PBS for 30 min. The same solution was used for all subsequent washing steps. Cells were incubated with anti-RhoA monoclonal mouse primary antibody for 1 h at 37°C. After incubation, cells were washed for 30 min and incubated with FITC-conjugated mouse secondary antibody for 1 h at 37°C. After incubation, cells were washed for 30 min and incubated with Rhodamine-phalloidin for 30 min at 37°C. Afterwards, cells were incubated with DAPI (0.5 ug/mL) diluted in 0.9% NaCl for 5 min at 37°C and washed once with 0.9% NaCl. Cells were mounted in ProLong Gold antifade reagent (Molecular Probes) and examined with an Axiovert 100 microscope (Carl Zeiss, Germany) by using filter sets that were selective for each fluorochrome wavelength channel. Images were acquired with a C2400i integrated charge-coupled device camera (Hamamatsu Photonics, Shizuoka, Japan) and an Argus 20 image processor (Hamamatsu). Control experiments with no primary antibodies showed only faint background staining (supplementary material) [33].

2.6. Polyacrylamide Gel Electrophoresis and Immunoblotting. Six-well cell culture plates were seeded with 6×10^5 IEC-6 cells and grown for 48 h in standard DMEM media (which contains Gln). Wells were washed and incubated for 24 h with TcdA (10 ng/mL) in DMEM without Gln or supplemented with 10 mM of Ala-Gln or 10 mM of Gln. Cells were quickly washed in ice-cold PBS and 50 mL of sample buffer (4% sodium dodecyl sulphate—SDS, 20% glycerol, 0.2 M dithioethreitol, 125 mM Tris-HCl, pH 6.8) were added to the cells and boiled for 5 min. Samples were loaded in 12% SDS-polyacrylamide gels (SDS-PAGE) and transferred to PVDF membranes. Then, the PVDF membranes were incubated overnight with anti-RhoA monoclonal mouse antibody. Membranes were washed thrice with TBS-T and incubated for 1 h with peroxisome-conjugated goat anti-mouse secondary antibody. Finally, membranes washed again as described above, and the bands were visualized using the ECL plus Western Blotting Detection System (Amersham). To check sample loading, another PVDF membrane (containing the same samples in the same volume used for the other blots) was incubated with a mouse monoclonal anti-alfa-tubulin antibody (dilution 1 : 3000 in TBS-T). After three washes

in TBS-T (3 min each), the membrane was incubated with a peroxidase-conjugated goat anti-mouse antibody (dilution 1 : 7000 in TBS-T) and developed as described above. Quantification of protein bands was performed using the public domain software ImageJ (http://rsb.info.nih.gov/ij/) with data obtained from two independent experiments [33].

2.7. Proliferation Assay. IEC-6 cells were seeded in 96-well plates at the concentration of 10^5 cells per well and allowed to grow O/N until 80% of full confluence. Next day cells were washed with PBS and challenged for 24 and 48 h with TcdA at 100 ng/mL, 10 ng/mL, 1 ng/mL, and 0.1 ng/mL with or without Ala-Gln (Sigma, St. Louis, MO, USA), at 10 mM, diluted in Gln-free medium. Ala-Gln was used to evaluate cell apoptosis and proliferation since it is more stable than the Gln and similar results when compared to Gln in the morphological analysis. Cultured cells, not challenged with TcdA, served as controls. Cell proliferation reagent WST-1 (10 μL; Roche, Indianapolis, IN, USA) was added 24 and 48 hours after the treatment into 96-well plates to measure metabolic activity of viable cells. We incubated cells with WST-1 for 2 hours in controlled humidified chamber. During that time, viable cells convert WST-1 to a water-soluble formazan dye. Absorbances were measured at 450 nm using ELISA plate reader (BioTek Instruments Inc.). The absorbance directly correlates with cell number [23, 24].

2.8. TUNEL Assay. IEC-6 cells were plated in 2-well tissue culture chamber slides at a concentration of 10^5 cells per well. Cells were allowed to attach the chamber slide for 24 hours. Cells were then exposed for 24 h with TcdA 10 ng/mL and Ala-Gln 10 m M. After the treatment, cells were washed trice with 1xPBS and fixed in 4% paraformaldehyde (methanol-free) for 15 minutes, then washed trice with 1xPBS. Cells were stained with using DeadEnd Fluorometric TUNEL kit from (Promega, Madison, WI, USA). TUNEL System measures the fragmented DNA of apoptotic cells by catalytically incorporating fluorescein-12-dUTP at 3′-OH DNA ends using the enzyme TdT (terminal deoxynucleotidyl transferase). Apoptotic cells were visualized using DAPI (SouthernBiotech, Birmingham, AL, USA). Images were taken under 20X magnification using Fluorescent Olympus 1 × 71 Inverted microscope with QImaging camera, with QCapture Pro.5.1 software. We evaluated 5–8 randomly selected fields. Cells were manually counted per selected area and expressed in percentage of apoptotic cells per mm^2.

2.9. Ki67 Immunohistochemistry. Ki67 is a nuclear protein that is tightly linked to the cell cycle. It is a marker of cell proliferation. Ki67 is expressed in proliferating cells during mid-G_1 phase, increasing in level through S and G_2, and peaking in the M phase of the cell cycle [34]. IEC-6 cells were plated in 2-well tissue culture chamberslides at a concentration of 10^5 cells per well. Cells were allowed to attach to the chamberslide for 24 hours. After 24 hours, cells were washed with PBS and treated with different concentrations of TcdA (100 ng/mL, 10 ng/mL, 1 ng/mL, and 0.1 ng/mL) and 10 mM Gln/Ala-Gln

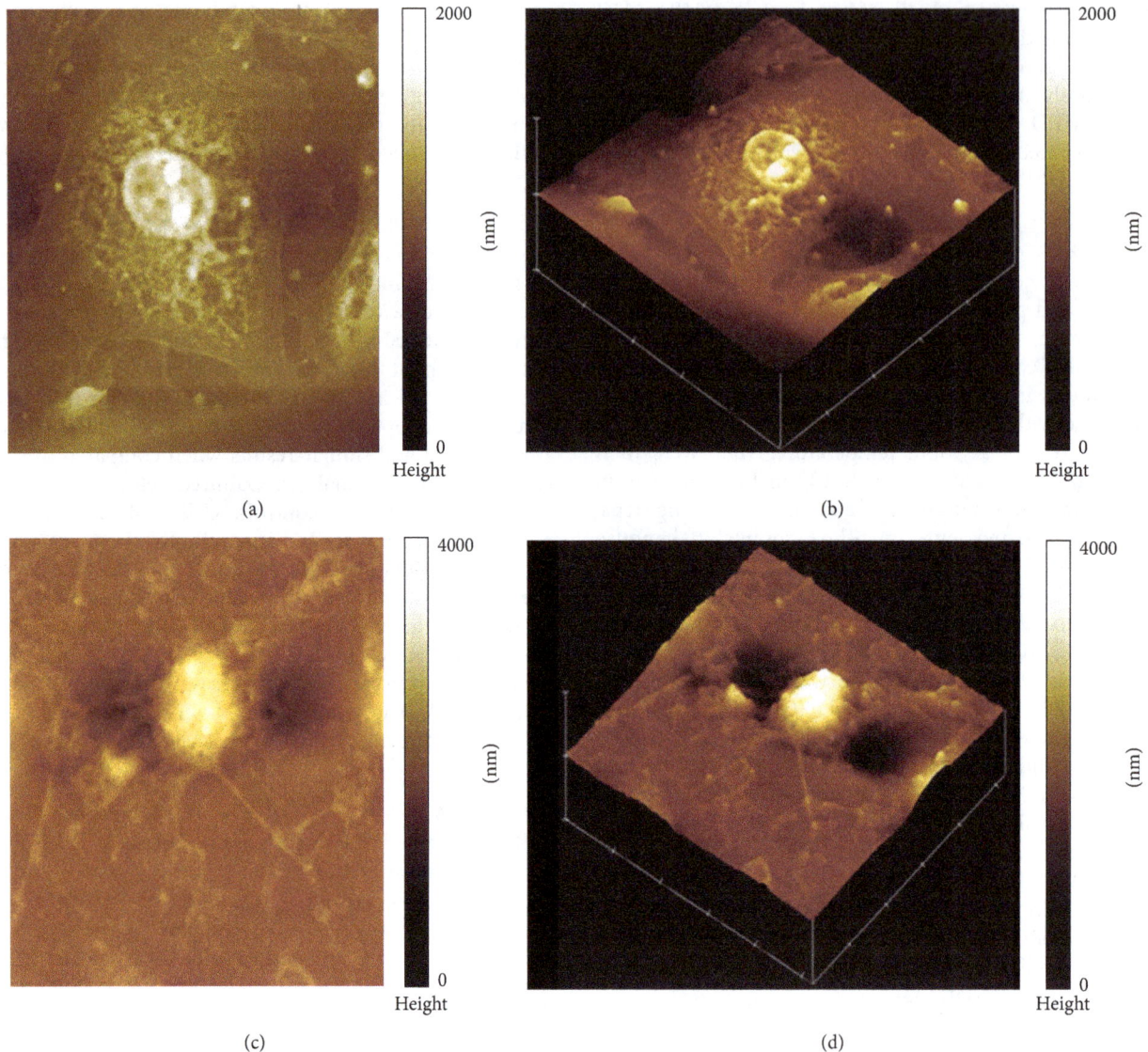

FIGURE 1: Atomic force microscopy (AFM) analysis of IEC-6 normal cell morphology with height scale (color bar): 2000 nm (a) and image of normal IEC-6 cell in 3D view (b). Effect of 1 h exposure with TcdA (100 ng/mL) on IEC-6 cell morphology with height scale (color bar): 4000 nm (c) and image of IEC-6 cell treated with TcdA (100 ng/mL) in 3D view (d). The images were obtained using AFM in contact mode (scan size $50 \times 50 \, \mu m^2$).

After the 24 hours of treatment, cells were washed trice with 1xPBS and fixed in 4% paraformaldehyde (methanol free) for 15 minutes, then washed trice with 1xPBS. IEC-6 cells were stained by using Ki67 antibody (MKI67rabbit monoclonal primary antibody) from Epitomics Inc., (Burlingame, CA, USA). Working dilution of primary antibody was 1 : 400. Secondary antibody used in procedure was anti-rabbit from DAKO, (Carpinteria, CA, USA). Staining was done in Tissue Research Core Facility at University of Virginia Medical School. Images were taken in the Core Facility using Olympus DP71 microscope and Microsuite Pathology Edition software using 20X magnifications with a 100 μm scale. We evaluated 6 to 8 randomly selected areas. Ki67-positive cells were counted manually and calculated per total number of cells, expressed as percentage of Ki67-positive cells [34].

2.10. Statistical Analyses. Results are expressed as mean ± standard error (SEM) using GraphPad Prism version 5.0 (GraphPad software, San Diego, CA, USA). Either one-way ANOVA, with Bonferroni's posttest, or unpaired Student's t-test were used to compare the differences between the experimental groups. Statistical significance was accepted at the level of $P < 0.05$.

3. Results

3.1. Effect of TcdA on IEC-6 Morphology and the Effect of Gln and Ala-Gln Treatment on Cellular Morphology and Dimensions as Evaluated through AFM. IEC-6 cells grown in normal media displayed well-preserved cytoplasm, nucleus, and nucleoli (Figures 1(a) and 1(b)). Treatment with TcdA caused

Glutamine and Alanyl-Glutamine Increase RhoA Expression and Reduce Clostridium difficile Toxin-A-Induced Intestinal Epithelial Cell Damage

65

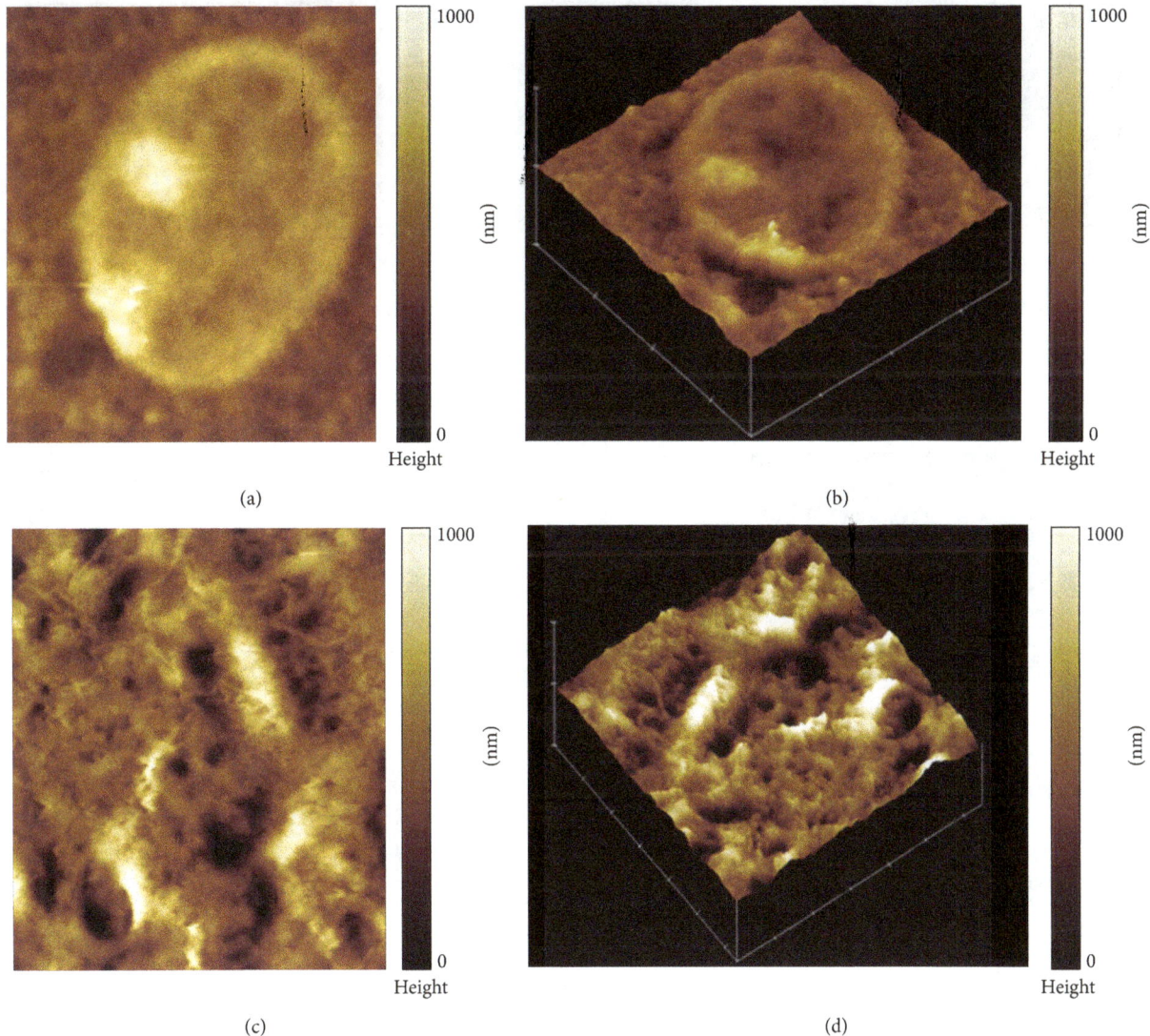

(a)

(b)

(c)

(d)

FIGURE 2: Atomic force microscopy (AFM) analysis of IEC-6 cell nucleus shape with height scale (color bar): 1000 nm (a) and image of normal IEC-6 cell nucleus in 3D view (b). Effect of 1 h exposure with TcdA (100 ng/mL) on IEC-6 cells nucleus morphology with height scale (color bar): 1000 nm (c) and image of IEC-6 cell treated with TcdA (100 ng/mL) in 3D view (d). The images were confectioned with AFM in the contact mode ($15 \times 15 \, \mu m^2$).

shrinking and compression of cytoplasmic material around the nucleus, blurring of the nuclear membrane, and condensation of nuclear elements (Figures 1(c) and 1(d)). Multiple vestigial filamentous extensions around the pyknotic cell were observed. In the presence of TcdA, the nucleus height of a representative IEC-6 cell was increased to 4000 nm (Figures 1(c) and 1(d)), compared to 2000 nm in the control group (Figures 1(a) and 1(b)). Visualization of the nucleus by AFM at higher magnification showed unchallenged IEC-6 cell nucleus to have well-defined nuclear envelope and prominent nucleoli (Figures 2(a) and 2(b)). The TcdA challenged cell had complete disruption of the nuclear envelope, condensation of chromatin, and loss of the nucleolar apparatus (Figures 2(c) and 2(d)). Measurement of cellular dimensions revealed TcdA-challenged IEC6 cells to have a 58.9% increase in cell height (Figure 3(e)), which may indicate deposition of

cytoplasmic material in the nuclear region. However, cell area and cell volume were noted to be decreased by 66.2% and 79%, respectively, compared to control (Figures 3(f) and 3(g)) ($P < 0.05$). Supplementation with 10 mM of Gln caused an increase of 46.3% and 67.6% in cell volume and area, respectively, and a reduction of 46.3% in the cell height in relation to the group treated with TcdA ($P < 0.05$, Figure 3). Supplementation with 10 mM of Ala-Gln significantly increased cell volume and area by 92.9% and 65.4%, respectively, and decreased cell height by 16.9% ($P < 0.05$ compared to TcdA-treated group).

3.2. Effect of TcdA on Rat Intestinal Epithelial Cell (IEC-6) Morphology and the Prevention of Injury by Gln and Ala-Gln through SEM.

Consistent with AFM findings, SEM also demonstrated shrinkage of IEC-6 cells in the presence of

(a)

(b)

(c)

(d)

(e)

(f)

(g)

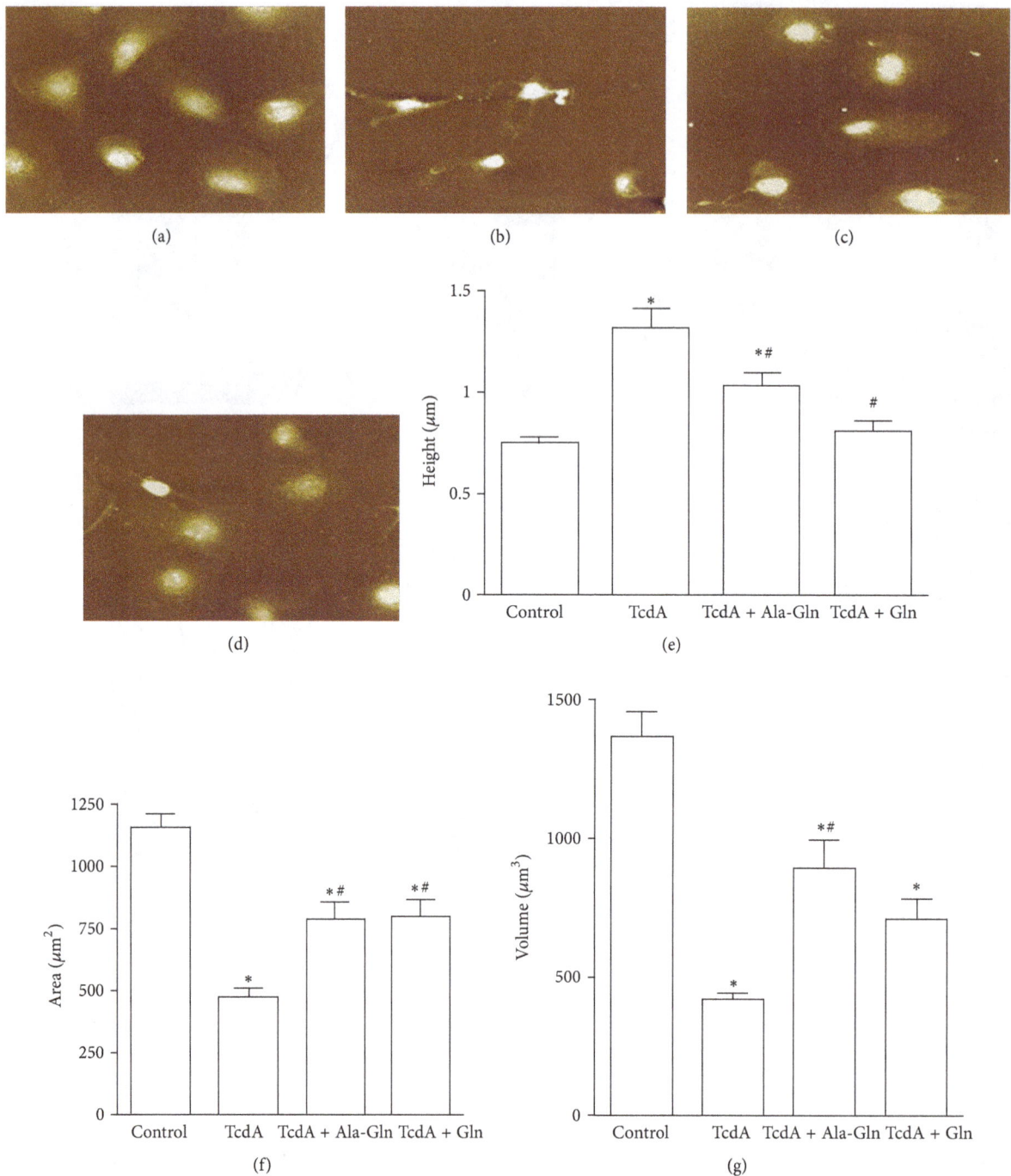

FIGURE 3: Atomic force microscopy analysis of the effects of glutamine and alanyl-glutamine supplementation on IEC-6 cell morphology following damage induced by TcdA. Effect of alanyl-glutamine (TcdA + Ala-Gln) or glutamine supplementation (TcdA + Gln), both at 10 mM, simultaneously with 24-hour exposure to TcdA at 10 ng/mL on IEC-6 cell. The images were confectioned with AFM in the contact mode with scanning of $150 \times 150 \, \mu m^2$. Control (a), TcdA (b), TcdA + Ala-Gln (c), and TcdA + Gln (d). The height (e), the area (f), and volume (g) were calculated with the nanoscope 5.30R3.SR3 software. $^*P < 0.05$ compared to control. $^\#P < 0.05$ compared to TcdA.

TcdA. Similarly, just a few thin cytoplasmic extensions or projections around the pyknotic body are left compared to the plump appearance of healthy cells in the absence of TcdA (Figure 4(b)). The cells treated with TcdA in medium contain 10 mM of Ala-Gln (Figure 4(c)) or Gln (Figure 4(d)) showed partial preservation of cell morphology displaying

a "pancake-like" shape similar to control cell incubated in medium without TcdA (Figure 4(a)).

3.3. Effect of TcdA and Ala-Gln on Cell Apoptosis. As shown in Figure 5, exposure to TcdA at 10 ng/mL for 24 h increased the percentage of TUNEL-positive cell to 15% ($P < 0.05$, com-

Glutamine and Alanyl-Glutamine Increase RhoA Expression and Reduce Clostridium difficile Toxin-A-Induced Intestinal
Epithelial Cell Damage

67

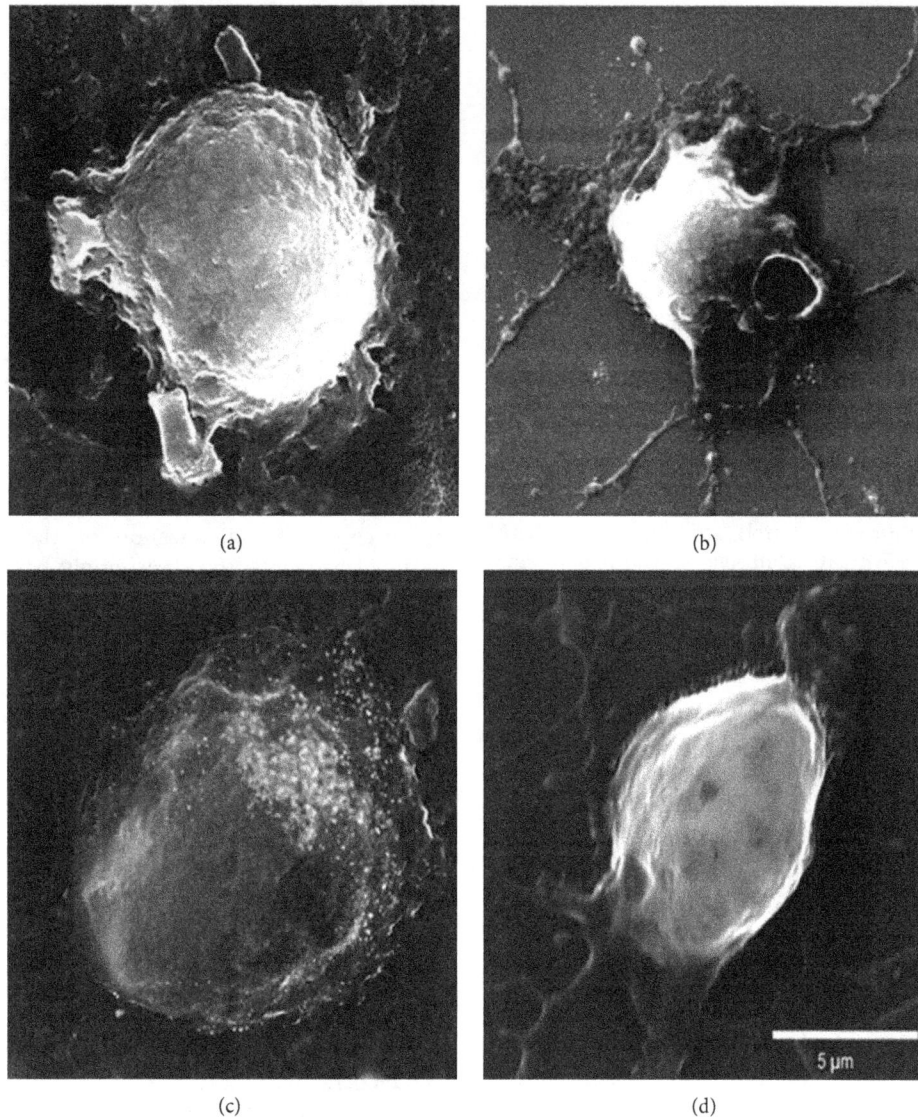

(a)

(b)

(c)

(d)

FIGURE 4: Electromicroscopy analysis of the effects of glutamine or alanyl-glutamine supplementation on IEC-6 cells following damage induced by *Clostridium difficile's* TcdA. The IEC-6 cells were divided into 4 groups: control (a), treated with TcdA at 10 ng/mL for 24 h (b), treated with TcdA at 10 ng/mL for 24 h and supplemented with 10 mM Ala-Gln (c), treated with TcdA at 10 ng/mL for 24 h and 10 mM of Gln (d). The images were confectioned with SEM, magnification 9.45X and scale bar 5 μm.

pared to control). Supplementation with Ala-Gln at 10 mM for 24 h simultaneously with TcdA exposure decreased the percentage of TUNEL-positive cell by 61.4% compared to TcdA alone ($P < 0.05$). Effect of TcdA and Ala-Gln on cell proliferation.

3.4. Indirect Evaluation of Cell Proliferation by WST Assay.
As shown in Figure 6(a), exposure to TcdA at 100 ng/mL for 24 h decreased cell proliferation by 17.0% ($P < 0.05$, compared to control). Supplementation with 10 mM of Ala-Gln prevented the antiproliferative effect of TcdA ($P < 0.05$). Similarly, as seen in Figure 6(b), after 48 h of exposure with TcdA at 100 ng/mL decreased proliferation by 20.7%, respectively ($P < 0.05$). Supplementation with Ala-Gln significantly

prevented the antiproliferative effect of TcdA at 100 ng/mL ($P < 0.05$ by one-way ANOVA).

3.5. Direct Evaluation of Cell Proliferation by Ki67 Assay.
As shown in Figure 6(c), exposure to TcdA for 24 h at 1 ng/mL, 10 ng/mL and 100 ng/mL decreased cell proliferation by 39.3%, 55.4%, and 58.4%, respectively ($P < 0.05$, compared to control group). Supplementation with Ala-Gln at 10 mM significantly prevented the antiproliferative effect of toxin A at 1, 10 ($P < 0.05$ by one-way ANOVA) and at 100 ng/mL ($P = 0.02$ by unpaired Student's t-test).

3.6. Effect of TcdA and Ala-Gln or Gln Supplementation on Distribution and Expression of f-Actin and RhoA.
As demonstrated in Figure 7(b), incubation with TcdA for

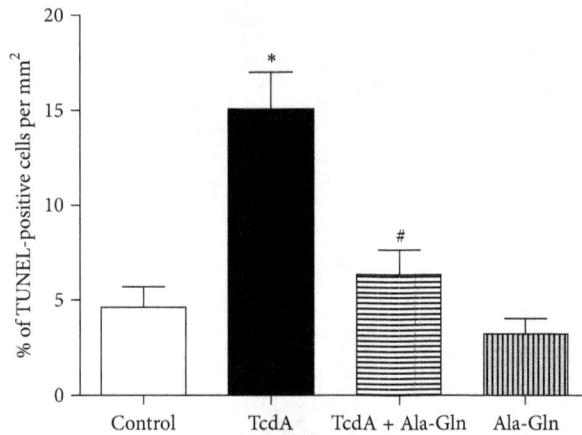

FIGURE 5: Effect of TcdA of *Clostridium difficile* and alanyl-glutamine on TUNEL cells/mm^2 percent. The IEC-6 cells were treated for 24 h medium alone (control), TcdA (10 ng/mL), TcdA + Ala-Gln (treated with TcdA 10 ng/mL and Ala-Gln 10 mM), and Ala-Gln 10 mM. $^*P < 0.05$ compared to control. $^\#P < 0.05$ compared to TcdA 100 ng/mL.

24 h caused change in F-actin distribution (red staining) causing cytoskeleton collapse around the nucleus of the IEC6 cells in the monolayer. In addition, TcdA-treated cells show decreased staining of actin filaments beyond the plasma membrane and loss of adherence. At the same time, TcdA-treated cells show RhoA (green staining) concentrated close to the nucleus overlapping the f-actin fibers as can be seen by orange staining. Additionally, as seen with DAPI staining, TcdA caused nuclear fading and nuclear pyknosis. The control cell monolayer (incubated in media without Gln) showed more homogeneous actin distribution with actin bundles attached to the plasma membrane and RhoA homogeneously distributed around the cytoplasm (Figure 7(a)). Supplementation with both Gln and Ala-Gln partially reverted the changes described in cells treated with TcdA alone (Figures 7(e) and 7(f)), showing preservation of F-actin cytoskeleton and a more organized intracellular actin network preserving actin bundles attached to the plasma membrane. The preservation of cytoskeleton was associated with increased expression of RhoA in the cytoplasm (green staining) which was confirmed by increased RhoA protein expression as detected by Western Blotting (Figures 7(g) and 7(f)). In cells not treated with TcdA, supplementation with 10 mM of Ala-Gln and Gln showed robust and organized intracellular actin network associated with increased expression of RhoA protein (Figures 7(c) and 7(d)).

4. Discussion

In this study, we used different microscopic approaches in order to demonstrate, in detail, the effect of TcdA in intestinal epithelial cell morphology and their association with cell death and proliferation and to evaluate the protective effect of Gln and Ala-Gln on TcdA-induced cell damage. High-resolution advanced microscopic techniques such as confocal laser fluorescence microscopy, atomic force microscopy, and scanning electron microscopy are capable of providing detailed morphological features and cytoskeletal information. In particular, AFM, besides its high-resolution capabilities, provides advantage over traditional microscopic techniques because it is not restricted to cell morphology, cytoskeletal elements, or organelles, but also provides quantitative information on cell volume, area, and height. We showed that TcdA, in higher concentration, caused dramatic changes in cell morphlogy as assessed by AFM associated with significant decrease in both cell volume and area and increase in cell height, as a result of shrinkage of the cell associated with aggregation of cellular material around the nucleus. In addition, TcdA caused disruption of the nuclear envelope and chromatin condensation, as early as, with 1 h of incubation. Most of these morphological changes were confirmed by SEM.

Disruption of cytoskeleton was seen by confocal microscopy. The organization of the cytoskeleton with its rapid assembly and disassembly has important role in motility, guidance, and adhesion of cells [35]. In intestinal epithelial cells, involved in maintenance of the barrier function, cytoskeletal disruption by TcdA may damage tight junctions and cause failure of focal contact formation, possibly leading to exposure of luminal pathogens to immune cells in the lamina propria [36, 37]. TcdA induces monoglycosylation of Rho, Rac, and Cdc42 at threonine 37, preventing Rho family proteins from participating in the formation of actin filaments [38]. This mechanism is believed to be the main cause for the cell morphological changes induced by TcdA, which can be a consequence of the cytoskeleton collapse as seen by confocal microscopy. Rho glycosylation by TcdA results in the disappearance of actin cables, peripheral membrane ruffling, filopodial extensions, and the disorganization of focal complexes, and ultimately resulting in complete loss of cell shape, that is, cell rounding as seen here [39, 40].

Furthermore, TcdA increased cell death as measured by TUNEL assay. The increase in TUNEL-positive cells indicate cell death, and this data, associated with the morphological feature showed here, suggest cell death by apoptosis, as reported previously [23, 25, 26], which could contribute to the disruption of intestinal epithelial barrier leading to the severe inflammatory response seen in response to *C. difficile* infection. It has been suggested that necrosis may be more important than apoptosis in *C. difficile* pathogenesis, since necrosis involves the release of cytoplasmic contents and induction of an immune response, leading to local inflammation [16]. However, it is well known that, depending on the intensity and duration of the stimulus, the damage can start with apoptosis and progress to necrosis.

Nam et al. gave new insight into the mechanisms involved in the rapid cell rounding induced by TcdA [19]. Their group demonstrated that in addition to the effect in inactivation of Rho family proteins and actin disaggregation, TcdA induces tubulin deacetylation and microtubule depolymerization through HDAC6 cytosolic tubulin deacetylase. Microtubule instability is critical to cell shape [41], cell movement [42], intracellular transport of organelles [43], and the separation

Glutamine and Alanyl-Glutamine Increase RhoA Expression and Reduce Clostridium difficile Toxin-A-Induced Intestinal Epithelial Cell Damage

69

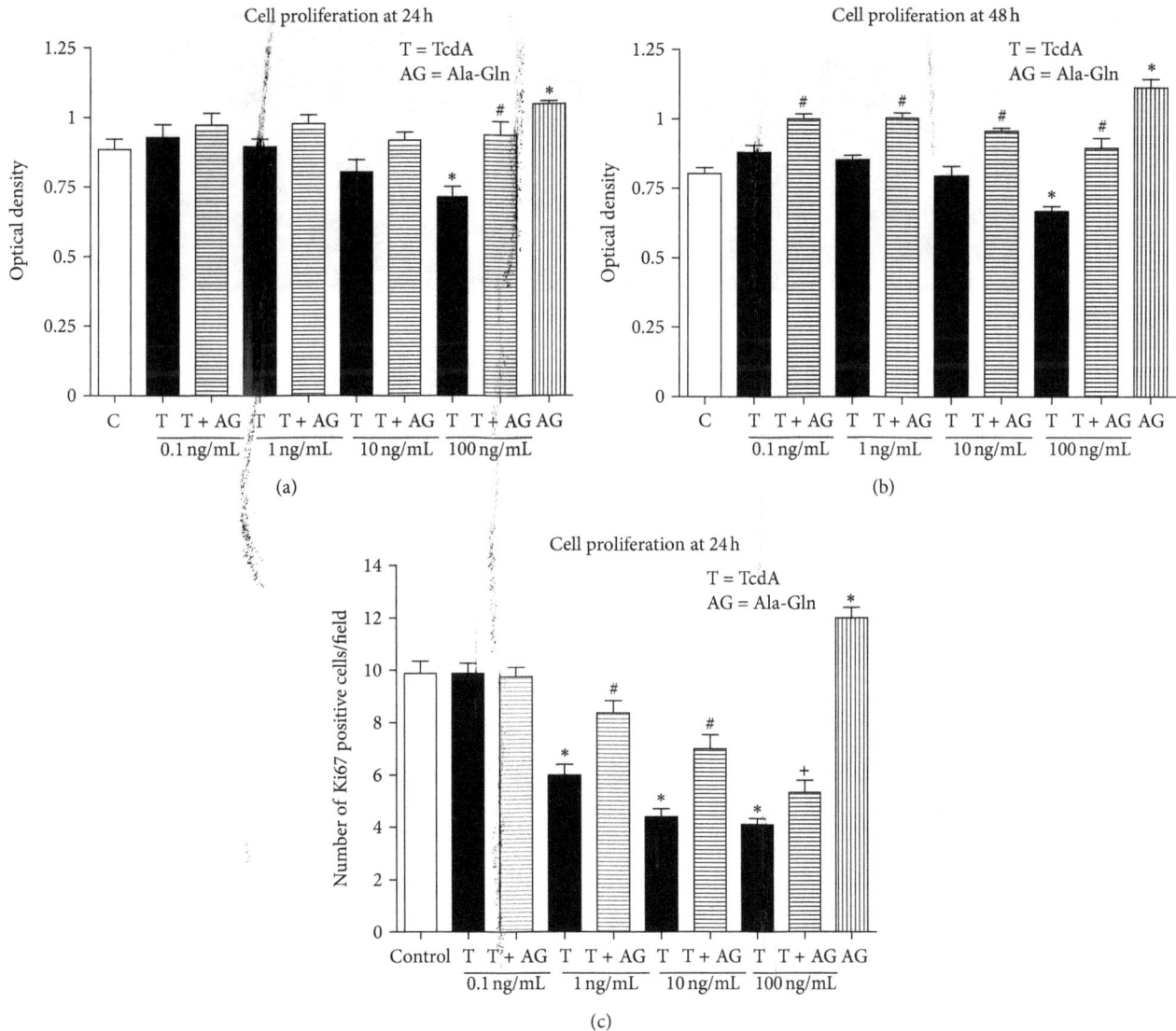

FIGURE 6: Effect of TcdA of *Clostridium difficile* and glutamine and alanyl-glutamine on cell proliferation. (a) The IEC-6 cells were exposed to TcdA (at 0.1, 1, 10, or 100 ng/mL) and supplemented or not with Ala-Gln at 10 mM for 24 h. (b) The cells were exposed to TcdA (at 0.1, 1, 10, or 100 ng/mL) and supplemented or not with Ala-Gln at 10 mM for 48 h. The WST absorbance was measured using an ELISA microplate reader at 450 nm (reference range 420–480 nm). (c) The cells treated with TcdA for 24 h, the number of cells positive for Ki67. $^*P < 0.05$ compared to control. $^\#P < 0.05$ compared to TcdA by one-way ANOVA. $^+P < 0.05$ compared to TcdA by unpaired Student's t-test.

of chromosomes during mitosis [40], which could explain the decreased proliferation induced by TcdA showed here.

Although the TcdA has well-known effect on Rho A glycosylation and inactivation, we did not observe any significant differences in RhoA concentration following TcdA exposure as compared to control. However, supplementation with Gln and Ala-Gln increased RhoA concentration even in the presence of TcdA, which could, at least in part, explain the protective effect of these micronutrients on TcdA-induced morphological changes, cell death, and inhibition of proliferation. These finding are in accordance with previous studies which showed that Gln and Ala-Gln prevented the inhibition of cell migration, apoptosis, and the initial drop in transepithelial resistance induced by TcdA [23]. These

effects may be explained by the partial preservation of the cytoskeleton as visualized here by immunofluorescence, probably as a consequence of the increase of RhoA expression. Furthermore, Gln and Ala-Gln have been reported previously to inhibit the apoptosis of T84 cells by preventing caspase 8 activation and to reduce TcdA-induced intestinal secretion and epithelial disruption [26]. Since cytoskeleton is involved in separation of chromosomes during mitosis, it is not surprising that its preservation by the micronutrients leads to increased proliferation even in the presence of TcdA. Bilban et al. showed that the main signaling pathway for cell survival is the activation of HSPs, whose action is regulated by Gln [44]. Gabai and Sherman showed, in some models of disease, that the protective effect of Gln requires

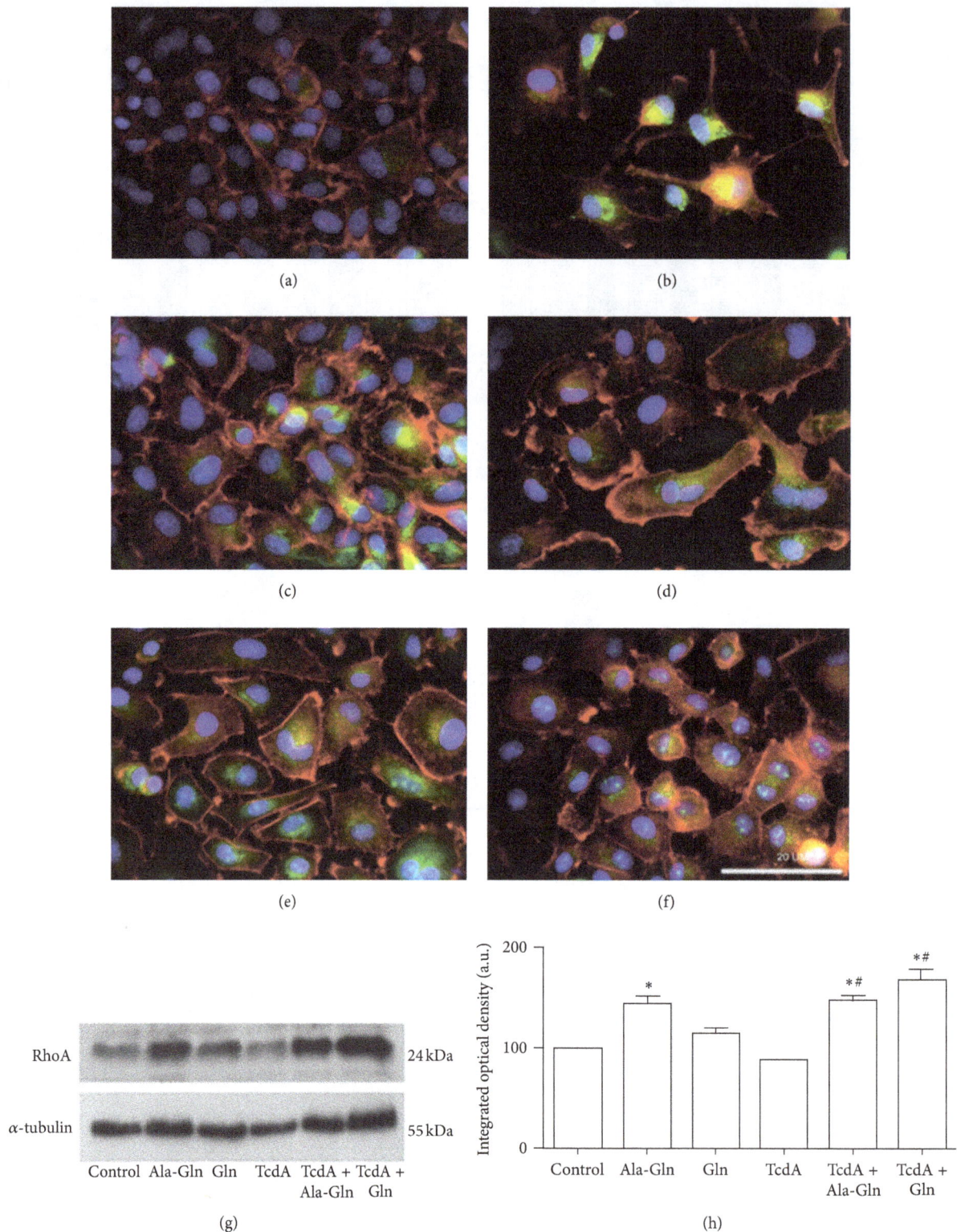

(a)

(b)

(c)

(d)

(e)

(f)

(g)

(h)

FIGURE 7: Fluorescent microscopy on distribution of f-actin fibers and RhoA and immunoblotting analysis of RhoA for the effect of TcdA of *Clostridium difficile* and the effect of micronutrients. The cells were treated with TcdA at 10 ng/mL for 24 h and supplemented or not with 10 mM of Ala-Gln or Gln and incubated with rhodamine-phalloidin, FITC-RhoA, and DAPI. The IEC-6 cells were divided into 6 groups: control (a); treated with TcdA at 10 ng/mL for 24 h (b); treated with 10 mM of alanyl-glutamine without TcdA (c); treated with 10 mM of glutamine alone without TcdA (d); treated with TcdA at 10 ng/mL for 24 h and supplemented with 10 mM alanyl-glutamine (e); treated with TcdA at 10 ng/mL for 24 h and supplemented 10 mM of glutamine (f). Immunoblotting was performed to evaluate the expression of RhoA (g and h). The quantification was done comparatively defaulting protein α-tubulin. $^{*}P < 0.05$ compared to control. $^{\#}P < 0.05$ compared to TcdA.

Glutamine and Alanyl-Glutamine Increase RhoA Expression and Reduce Clostridium difficile Toxin-A-Induced Intestinal Epithelial Cell Damage

71

induction of hemoxigenase-1 (HO-1) [45]. Our research group suggested that the pathway HO-1/carbon monoxide has a significant protective effect when there is injury caused by the TcdA,including decreased neutrophilic infiltrate in the mucosa. Studies by several authors showed that Gln and its stable derivative Ala-Gln have effective actions on cell proliferation, apoptosis, and protein synthesis [46, 47]. Additionally, we observed that supplementation with Ala-Gln or Gln alone increased RhoA protein concentration. The overproduction of RhoA may partially explain the protective effect of glutamine and alanyl-glutamine in the cytotoxicity induced by toxin A.

5. Conclusions

To the best of our knowledge, this is the first study conducted evaluating the effect of *C. difficile* toxins and Ala-Gln and Gln treatment on cell morphology using AFM. These AFM images substantiate the confocal study of the effect of TcdA on intestinal epithelial cell lines. Moreover, the aberration in intestinal epithelial cell morphology and actin filament organization mediated by TcdA may, at least, partially account for the severe intestinal mucosal disruption found in *C. difficile*-induced disease.

Conflict of Interests

The authors have no conflict of interests to declare.

Acknowledgments

This work was supported by grants from the Instituto Nacional de Ciências e Tecnologia em Biomedicina do Semiárido Brasileiro (INCT), Conselho Nacional de Pesquisa (CNPq), and Coordenação de Aperfeiçoamento de Pessoal do Ensino Superior (CAPES), Brazil. The authors gratefully acknowledge Snjezana Zaja-Milatovic for her assistance in the *in vitro* experiments and technical assistance of Conceição da Silva Martins and Maria do Socorro França Monte. Additionally, they would like to thank Carolina Pontes Soares for assistance with the immunoblots. M. B. Braga-Neto was partially supported by NIH/Fogarty International Center Global Infectious Disease Research Training (Grant no. D43 TW0006578).

References

[1] A. N. Ananthakrishnan, M. Issa, and D. G. Binion, "*Clostridium difficile* and inflammatory bowel disease," *Medical Clinics of North America*, vol. 94, no. 1, pp. 135–153, 2010.

[2] D. B. Blossom and L. C. McDonald, "The challenges posed by reemerging *Clostridium difficile* infection," *Clinical Infectious Diseases*, vol. 45, no. 2, pp. 222–227, 2007.

[3] R. Gaynes, D. Rimland, E. Killum et al., "Outbreak of *Clostridium difficile* infection in a long-term care facility: association with gatifloxacin use," *Clinical Infectious Diseases*, vol. 38, no. 5, pp. 640–645, 2004.

[4] C. A. Muto, M. Pokrywka, K. Shutt et al., "A large outbreak of *Clostridium difficile*-associated disease with an unexpected proportion of deaths and colectomies at a teaching hospital following increased fluoroquinolone use," *Infection Control and Hospital Epidemiology*, vol. 26, no. 3, pp. 273–280, 2005.

[5] L. C. McDonald, M. Owings, and D. B. Jernigan, "*Clostridium difficile* infection in patients discharged from US short-stay hospitals, 1996–2003," *Emerging Infectious Diseases*, vol. 12, no. 3, pp. 409–415, 2006.

[6] J. Pépin, N. Saheb, M. A. Coulombe et al., "Emergence of fluoroquinolones as the predominant risk factor for *Clostridium difficile*-associated diarrhea: a cohort study during an epidemic in Quebec," *Clinical Infectious Diseases*, vol. 41, no. 9, pp. 1254–1260, 2005.

[7] C. P. Kelly, "A 76-year-old man with recurrent *Clostridium difficile* associated diarrhea: review of *C difficile* infection," *Journal of the American Medical Association*, vol. 301, no. 9, pp. 954–962, 2009.

[8] V. G. Loo, L. Poirier, M. A. Miller et al., "A predominantly clonal multi-institutional outbreak of *Clostridium difficile*-associated diarrhea with high morbidity and mortality," *The New England Journal of Medicine*, vol. 353, no. 23, pp. 2442–2449, 2005.

[9] L. C. McDonald, G. E. Killgore, A. Thompson et al., "An epidemic, toxin gene-variant strain of *Clostridium difficile*," *The New England Journal of Medicine*, vol. 353, no. 23, pp. 2433–2441, 2005.

[10] J. Pépin, L. Valiquette, M. E. Alary et al., "*Clostridium difficile*-associated diarrhea in a region of Quebec from 1991 to 2003: a changing pattern of disease severity," *Canadian Medical Association Journal*, vol. 171, no. 5, pp. 466–472, 2004.

[11] J. Freeman, M. P. Bauer, S. D. Baines et al., "The changing epidemiology of *Clostridium difficile* infections," *Clinical Microbiology Reviews*, vol. 23, no. 3, pp. 529–549, 2010.

[12] T. McGuire, P. Dobesh, D. Klepser, M. Rupp, and K. Olsen, "Clinically important interaction between statin drugs and *Clostridium difficile* toxin?" *Medical Hypotheses*, vol. 73, no. 6, pp. 1045–1047, 2009.

[13] I. Castagliuolo, C. P. Kelly, B. S. Qiu, S. T. Nikulasson, J. Thomas LaMont, and C. Pothoulakis, "IL-11 inhibits *Clostridium difficile* toxin A enterotoxicity in rat ileum," *American Journal of Physiology*, vol. 273, no. 2, pp. G333–G341, 1997.

[14] M. Riegler, R. Sedivy, C. Pothoulakis et al., "*Clostridium difficile* toxin B is more potent than toxin A in damaging human colonic epithelium in vitro," *Journal of Clinical Investigation*, vol. 95, no. 5, pp. 2004–2011, 1995.

[15] D. Drudy, S. Fanning, and L. Kyne, "Toxin A-negative, toxin B-positive *Clostridium difficile*," *International Journal of Infectious Diseases*, vol. 11, no. 1, pp. 5–10, 2007.

[16] H. Genth, S. C. Dreger, J. Huelsenbeck, and I. Just, "*Clostridium difficile* toxins: more than mere inhibitors of Rho proteins," *The International Journal of Biochemistry and Cell Biology*, vol. 40, no. 4, pp. 592–597, 2008.

[17] H. Kim, S. H. Rhee, C. Pothoulakis, and J. T. LaMont, "*Clostridium difficile* toxin A binds colonocyte Src causing dephosphorylation of focal adhesion kinase and paxillin," *Experimental Cell Research*, vol. 315, no. 19, pp. 3336–3344, 2009.

[18] M. Sauerborn, P. Leukel, and C. von Eichel-Streiber, "The C-terminal ligand-binding domain of *Clostridium difficile* toxin A (TcdA) abrogates TcdA-specific binding to cells arid prevents mouse lethality," *FEMS Microbiology Letters*, vol. 155, no. 1, pp. 45–54, 1997.

[19] H. J. Nam, J. K. Kang, S. K. Kim et al., "*Clostridium difficile* toxin A decreases acetylation of tubulin, leading to microtubule

depolymerization through activation of histone deacetylase 6, and this mediates acute inflammation," *Journal of Biological Chemistry*, vol. 285, no. 43, pp. 32888–32896, 2010.

[20] G. A. C. Brito, M. H. L. P. Souza, A. A. Melo-Filho et al., "Role of pertussis toxin A subunit in neutrophil migration and vascular permeability," *Infection and Immunity*, vol. 65, no. 3, pp. 1114–1118, 1997.

[21] K. Weiss, "Toxin-binding treatment for *Clostridium difficile*: a review including reports of studies with tolevamer," *International Journal of Antimicrobial Agents*, vol. 33, no. 1, pp. 4–7, 2009.

[22] G. A. C. Brito, J. Fujji, B. A. Carneiro-Filho, A. A. M. Lima, T. Obrig, and R. L. Guerrant, "Mechanism of *Clostridium difficile* toxin A-induced apoptosis in T84 cells," *The Journal of Infectious Diseases*, vol. 186, no. 10, pp. 1438–1447, 2002.

[23] G. A. C. Brito, B. Carneiro-Filho, R. B. Oriá, R. V. Destura, A. A. M. Lima, and R. L. Guerrant, "*Clostridium difficile* toxin A induces intestinal epithelial cell apoptosis and damage: role of Gln and Ala-Gln in toxin A effects," *Digestive Diseases and Sciences*, vol. 50, no. 7, pp. 1271–1278, 2005.

[24] M. B. Braga-Neto, C. A. Warren, R. B. Oriá et al., "Alanylglutamine and glutamine supplementation improves 5-fluorouracil-induced intestinal epithelium damage in vitro," *Digestive Diseases and Sciences*, vol. 53, no. 10, pp. 2687–2696, 2008.

[25] G. A. C. Brito, G. W. Sullivan, W. P. Ciesla Jr., H. T. Carper, G. L. Mandell, and R. L. Guerrant, "*Clostridium difficile* toxin A alters in vitro-adherent neutrophil morphology and function," *The Journal of Infectious Diseases*, vol. 185, no. 9, pp. 1297–1306, 2002.

[26] B. A. Carneiro, J. Fujii, G. A. C. Brito et al., "Caspase and bid involvement in *Clostridium difficile* toxin A-induced apoptosis and modulation of toxin A effects by glutamine and alanylglutamine in vivo and in vitro," *Infection and Immunity*, vol. 74, no. 1, pp. 81–87, 2006.

[27] B. A. Carneiro-Filho, R. B. Oriá, K. Wood Rea et al., "Alanylglutamine hastens morphologic recovery from 5-fluorouracil-induced mucositis in mice," *Nutrition*, vol. 20, no. 10, pp. 934–941, 2004.

[28] G. P. Oliveira, C. M. Dias, P. Pelosi, and P. R. M. Rocco, "Understanding the mechanisms of glutamine action in critically ill patients," *Anais da Academia Brasileira de Ciencias*, vol. 82, no. 2, pp. 417–430, 2010.

[29] K. D. Jandt, "Developments and perspectives of scanning probe microscopy (SPM) on organic materials systems," *Materials Science and Engineering R*, vol. 21, no. 5-6, pp. 221–295, 1998.

[30] C. S. Quirino, G. O. Leite, L. M. Rebelo et al., "Healing potential of Pequi (*Caryocar coriaceum* Wittm.) fruit pulp oil," *Phytochemistry Letters*, vol. 2, no. 4, pp. 179–183, 2009.

[31] C. Heneweer, M. Schmidt, H. W. Denker, and M. Thie, "Molecular mechanisms in uterine epithelium during trophoblast binding: the role of small GTPase RhoA in human uterine Ishikawa cells," *Journal of Experimental and Clinical Assisted Reproduction*, vol. 2, no. 4, pp. 1–11, 2005.

[32] C. Schwan, B. Stecher, T. Tzivelekidis et al., "*Clostridium difficile* toxin CDT induces formation of microtubule-based protrusions and increases adherence of bacteria," *PLoS Pathogens*, vol. 5, no. 10, Article ID e1000626, 2009.

[33] A. C. B. Possidonio, M. L. Senna, D. M. Portilho et al., "α-cyclodextrin enhances myoblast fusion and muscle differentiation by the release of IL-4," *Cytokine*, vol. 55, no. 2, pp. 280–287, 2011.

[34] J. Gerdes, H. Lemke, and H. Baisch, "Cell cycle analysis of a cell proliferation associated human nuclear antigen defined by the monoclonal antibody Ki-67," *Journal of Immunology*, vol. 133, no. 4, pp. 1710–1715, 1984.

[35] D. Docheva, D. Padula, C. Popov, W. Mutschler, H. Clausen-Schaumann, and M. Schieker, "Researching into the cellular shape, volume and elasticity of mesenchymal stem cells, osteoblasts and osteosarcoma cells by atomic force microscopy: stem cells," *Journal of Cellular and Molecular Medicine*, vol. 12, no. 2, pp. 537–552, 2008.

[36] H. Kim, E. Kokkotou, X. Na et al., "*Clostridium difficile* toxin A-induced colonocyte apoptosis involves p53-dependent p21(WAF1/CIP1) induction via p38 mitogen-activated protein kinase," *Gastroenterology*, vol. 129, no. 6, pp. 1875–1888, 2005.

[37] H. Kim, S. H. Rhee, C. Pothoulakis, and J. T. LaMont, "Inflammation and apoptosis in *Clostridium difficile* enteritis is mediated by PGE2 up-regulation of fas ligand," *Gastroenterology*, vol. 133, no. 3, pp. 875–886, 2007.

[38] I. Just, M. Wilm, J. Selzer et al., "The enterotoxin from *Clostridium difficile* (ToxA) monoglucosylates the Rho proteins," *Journal of Biological Chemistry*, vol. 270, no. 23, pp. 13932–13936, 1995.

[39] I. Castagliuolo, M. Riegler, A. Pasha et al., "Neurokinin-1 (NK-1) receptor is required in *Clostridium difficile*-induced enteritis," *Journal of Clinical Investigation*, vol. 101, no. 8, pp. 1547–1550, 1998.

[40] N. Fernandez, Q. Chang, D. W. Buster, D. J. Sharp, and A. Ma, "A model for the regulatory network controlling the dynamics of kinetochore microtubule plus-ends and poleward flux in metaphase," *Proceedings of the National Academy of Sciences of the United States of America*, vol. 106, no. 19, pp. 7846–7851, 2009.

[41] J. S. Popova and M. M. Rasenick, "Gβγ mediates the interplay between tubulin dimmers and microtubules in the modulation of Gq signaling," *Journal of Biological Chemistry*, vol. 278, no. 36, pp. 34299–34308, 2003.

[42] J. Gao, L. Huo, X. Sun et al., "The tumor suppressor CYLD regulates microtubule dynamics and plays a role in cell migration," *Journal of Biological Chemistry*, vol. 283, no. 14, pp. 8802–8809, 2008.

[43] A. D. Bicek, E. Tüzel, A. Demtchouk et al., "Anterograde microtubule transport drives microtubule bending in LLC-PK1 epithelial cells," *Molecular Biology of the Cell*, vol. 20, no. 12, pp. 2943–2953, 2009.

[44] M. Bilban, A. Haschemi, B. Wegiel, B. Y. Chin, O. Wagner, and L. E. Otterbein, "Heme oxygenase and carbon monoxide initiate homeostatic signaling," *Journal of Molecular Medicine*, vol. 86, no. 3, pp. 267–279, 2008.

[45] V. L. Gabai and M. Y. Sherman, "Invited review: interplay between molecular chaperones and signaling pathways in survival of heat shock," *Journal of Applied Physiology*, vol. 92, no. 4, pp. 1743–1748, 2002.

[46] K. D. Singleton, V. E. Beckey, and P. E. Wischmeyer, "Glutamine prevents activations of NF-κB and stress kinase pathways, attenuates inflammatory cytokine release, and prevents

Glutamine and Alanyl-Glutamine Increase RhoA Expression and Reduce Clostridium difficile Toxin-A-Induced Intestinal
Epithelial Cell Damage

73

acute respiratory distress syndrome (ARDS) following sepsis,"
Shock, vol. 24, no. 6, pp. 583–589, 2005.

[47] K. D. Singleton and P. E. Wischmeyer, "Glutamine's protection against sepsis and lung injury is dependent on heat shock protein 70 expression," *American Journal of Physiology*, vol. 292, no. 5, pp. R1839–R1845, 2007.

8

LRP-1: A Checkpoint for the Extracellular Matrix Proteolysis

Nicolas Etique, Laurie Verzeaux, Stéphane Dedieu, and Hervé Emonard

CNRS FRE 3481 MEDyC (Matrice Extracellulaire et Dynamique Cellulaire), Laboratoire SiRMa (Signalisation et Récepteurs Matriciels), Université de Reims Champagne-Ardenne (URCA), Moulin de la Housse, Bât. 18, Chemin des Rouliers, BP 1039, 51687 Reims Cedex 2, France

Correspondence should be addressed to Hervé Emonard; herve.emonard@univ-reims.fr

Academic Editor: Davide Vigetti

Low-density lipoprotein receptor-related protein-(LRP-1) is a large endocytic receptor that binds more than 35 ligands and exhibits signaling properties. Proteinases capable of degrading extracellular matrix (ECM), called matrix proteinases in this paper, are mainly serine proteinases: the activators of plasminogen into plasmin, tissue-type (tPA) and urokinase-type (uPA) plasminogen activators, and the members of the matrix metalloproteinase (MMP) family. LRP-1 is responsible for clearing matrix proteinases, complexed or not with inhibitors. This paper attempts to summarize some aspects on the cellular and molecular bases of endocytic and signaling functions of LRP-1 that modulate extra- and pericellular levels of matrix proteinases.

1. Introduction

Extracellular matrix (ECM) remodeling occurs in both physiological and pathological situations [1]. Tissue homeostasis depends on a strict equilibrium between synthesis and degradation of ECM macromolecules. In contrast, fibrotic pathologies are classically related to a defect or an increased ECM breakdown, while an excessive proteolytic degradation is the hallmark of inflammatory processes or tumor invasion. Numerous proteolytic enzymes are able to degrade ECM macromolecules, including the serine proteinases tissue-type plasminogen activator (tPA) and urokinase-type plasminogen activator (uPA) [2] and the members of the matrix metalloproteinase (MMP) family [3].

A series of specific or nonspecific inhibitors controls the activities of these powerful catalytic enzymes. Thus, the pan-protease inhibitor α2-macroglobulin (α2M) binds to and inhibits active members of the four classes of proteolytic enzymes [4]. More specifically, the serine proteinase inhibitors (serpins) and the plasminogen activator inhibitors (PAI) 1 and 2 block the activity of tPA and uPA [5]. Tissue inhibitors of metalloproteinases (TIMPs) inhibit the activity of MMPs [6] and also of adamalysins (a disintegrin and metalloproteinases, ADAMs) [7]. Besides this level of control, receptor-mediated endocytosis is an emergent and efficient biological mechanism to regulate extra- or pericellular levels of proteolytic enzymes by internalizing them for catabolism in lysosomes [8]. This paper briefly describes the main molecules involved in these events and reviews the different roles of low-density lipoprotein (LDL) receptor-related protein-(LRP-1) in controlling extracellular matrix remodeling.

2. Plasminogen Activators and Their Inhibitors

Urokinase-type plasminogen activator (uPA) and tissue-type PA (tPA) are serine proteinases that catalyze the conversion of the zymogen plasminogen to the active serine proteinase plasmin [9]. Plasmin degrades numerous ECM macromolecules including laminin, fibronectin, and proteoglycans, triggers the activation of pro-MMPs, and activates or releases growth factors from ECM including latent-transforming growth factor β and vascular endothelial growth factor. Both α2M and the serpin α2-antiplasmin inhibit its activities [10]. Pro-uPA is synthesized as a one-chain molecule that is cleaved at a single peptide bond (K158-I159 in human uPA) by various proteases including plasmin to give active two-chain uPA of 55 kDa. Human tPA was first purified as a single-chain form of approximately 70 kDa. A limited attack of the R275-I276 bond by plasmin generates a two-chain tPA. The plasminogen activation activity of single-chain tPA is 10- to 50-fold lower than that of the two-chain form [2]. The PA inhibitors PAI-1 and PAI-2 efficiently inhibited tPA and uPA catalytic activities [11].

The binding of uPA to its cell-surface receptor (uPAR) increases the affinity of uPAR for vitronectin and integrins, thus promoting cell adhesion, [12]. Interestingly enough by disrupting these interactions, PAI-1 detaches cells not only from vitronectin but also from fibronectin and collagen matrices [13]. This deadhesive property exhibited by PAI-1 could explain, at least partly, why paradoxically PAI-1 appears to be essential for cancer cell invasion and angiogenesis [14].

3. Matrix Metalloproteinases and Their Inhibitors

MMPs are the major matrix-degrading proteases due to the wide variety of their substrates and their role in numerous physiopathological processes [15, 16]. They belong to a large family of zinc-dependent endopeptidases. In humans, MMPs are represented by 23 members divided into two groups based on their localization (secreted or membrane-bound) or in five groups based on their domain organization and their substrate preference (collagenases, gelatinases, stromelysins, matrilysins, and membrane-type) [3, 17]. The general structure of MMPs consists in three domains that are common to almost all MMPs: the prodomain of about 80 amino acids, the catalytic metalloproteinase domain of about 170 amino acids, and the hemopexin domain of about 200 amino acids (except for MMP-7, -26, and -23). MMPs are secreted as a proenzyme, an enzymatically inactive state that results from the interaction between the "cysteine switch" motif in the prodomain and the zinc ion of the catalytic site [18]. The activation of these zymogens is an important regulatory step of MMP activity and occurs after the disruption of this interaction [15]. This process requires the proteolytic removal of the pro-domain by intracellular convertases such as furin or by extracellular proteinases (MMPs, plasmin, ..., etc.). A chemical perturbation of the cysteine-zinc interaction by SH reagents, by chaotropic agents (*in vitro*), or by antioxidant has been shown as sufficient to activate proMMPs [18].

After their activation, MMPs are regulated by two major types of endogenous inhibitors: α2M and TIMPs [18]. α2M is a plasma glycoprotein produced in the liver. Four nearly identical, disulfide-bonded domains of 180 kDa compose this 772 kDa protein. Inhibition mechanism involves the presentation of a cleavable region that, once proteolytically cleaved, induces a conformational change entrapping the proteinase that becomes covalently anchored by transacylation. Such a molecular complex is rapidly cleared by LRP-1-mediated endocytosis [19].

TIMPs are 184–194 amino acid proteins that have been described to form 1 : 1 stoichiometric complexes with active MMPs leading to the inhibition of their proteolytic activity. Four TIMPs (TIMP-1, -2, -3, and -4) have been identified in humans, inhibiting all MMPs tested so far, except TIMP-1 that

was reported as being a poor inhibitor for MT1-MMP, MT3-MMP, MT5-MMP, and MMP-19 [6, 18, 19]. All structurally characterized inhibitory TIMP-metalloproteinase complexes are closely similar. Within the metalloproteinase active site, the catalytic zinc atom is chelated by the N-terminal amino group and the carbonyl group of cysteine 1 [20]. TIMPs are also able to interact with proMMPs: TIMP-2, TIMP-3 or TIMP-4 with proMMP-2 and TIMP-1 or TIMP-3, with proMMP-9 [20]. These complexes are stabilized by interaction between the TIMP C-terminal domain and hemopexin domain of the zymogen. Since these interactions do not involve the N-terminal domain of the TIMP, such molecular complexes are capable of interacting with a second MMP molecule. Except for the role of proMMP-2-TIMP-2 in the MT1-MMP-mediated activation of proMMP-2 [21], their functional significance remains unclear [20].

More recently, TIMPs have been reported to induce various biological processes (cell survival, differentiation, epithelial-mesenchymal transition, ..., etc.) independently from their MMP-inhibitory activity [22, 23]. These effects involved an interaction with specific cell-surface receptors leading to signaling pathway activation. For example, TIMP-1 promotes cell survival in erythroleukemic cells after binding with a CD44/proMMP-9 complex receptor [24, 25] and in breast and lung epithelial cells after interacting with a CD63/integrin-β1 complex receptor [26, 27].

4. Low-Density Lipoprotein Receptor-Related Protein-1

4.1. General Features. LRP-1 is the first member of a receptor family related to the LDL receptor [28]. The receptor for α2M-proteinase complexes [29] and CD91, which interacts with heat-shock proteins at the surface of antigen-presenting cells [30], corresponds to LRP-1. It is synthesized as a single-chain molecule processed by furin in the trans-Golgi compartment into a 515 kDa α-chain and an 85 kDa β-chain which remain non-covalently associated at the cell surface [8]. The extracellular α-chain contains four basic amino acid residue-rich domains that interact with a number of ligands including proteins involved in lipoprotein metabolism, ECM proteins, growth factors, proteinases, and proteinase-inhibitor complexes. The transmembrane β-chain contains a cytoplasmic tail of 100 amino-acid residues including two NPxY motifs, necessary to trigger endocytosis and capable of interacting with many adaptors and signaling proteins.

The endocytic clearance of various ligands and signaling properties confer a main role to LRP-1 in a variety of pathophysiological processes including lipid metabolism, neurodegenerative diseases, blood-brain-barrier integrity, atherosclerosis, and cancer [8]. The importance of LRP-1 is confirmed by the lethality of mice carrying LRP-1 gene deletion at an early stage of embryonic development [31].

4.2. Endocytic Function. The LRP-1-mediated endocytic internalization of active proteinases linked to the pan-proteinase inhibitor α2M represents a general process to eliminate the excess of active proteinases from cellular

TABLE 1: Main matrix proteinases and specific inhibitors known to bind to LRP-1.

Serine proteinases, serpins, and serine proteinase/serpin complexes		
tPA		tPA, uPA/PAI-1
	PAI-1	
(pro)uPA		uPA/PAI-2
MMPs, TIMPs, and MMP/TIMP complexes		
(pro)MMP-2/TSP-1, -2	TIMP-1	
		(pro)MMP-2/TIMP-2
(pro)MMP-9	TIMP-2	(pro)MMP-9/TIMP-1
(pro)MMP-13	TIMP-3	
Other matrix proteinases		
Heparanase precursor		
Procathepsin-D		
ADAMTS-5		

environment [19, 32]. Here, we review additional LRP-1-mediated endocytosis that occurs independently from α2M to regulate extracellular proteinase activities (Table 1).

4.2.1. Serine Proteinases and Inhibitors. The binding of tPA to cell surface has first been described through PAI-1-dependent [33] and PAI-1-independent [34] receptors. These receptors have been rapidly identified as being LRP-1 [35, 36]. Orth and colleagues [37] confirmed that tPA, under its free form or complexed to PAI-1, binds to LRP-1 to be intracellularly degraded. Also, LRP-1 was shown to mediate the internalization of uPA associated to PAI-1 [31, 38] and PAI-2 [39]. Pro-uPA binds to purified LRP-1 with affinity 15 to 20 fold, weaker than that of the uPA/PAI-1 complex [40]. In contrast, PAI-1 was described to interact with LRP-1 with high affinity when associated with proteinases [41]. These data strongly suggest that the binding of proteinase to PAI-1 could modify PAI-1 conformation, revealing a cryptic high-affinity binding site for LRP-1.

Besides its ability to link to LRP-1 to be internalized [40], pro-uPA is activated upon binding to uPAR [42]. Interestingly, both uPAR endocytosis and uPA catabolism are dependent on PAI-1 [43]. These important data support the role of LRP-1 in promoting the cell-surface PA activity by facilitating the clearance of uPA/PAI-1-occupied uPAR and the regeneration of unoccupied uPAR at the cell surface [44, 45]. Such a process requires a direct binding between uPAR and LRP-1 [46]. This cycle of binding uPA/PAI-1 to uPAR followed by association with LRP-1, internalization, and intracellular dissociation and recycling of unoccupied uPAR and free LRP-1 to the cell surface can explain, at least in part, the promigratory effect of PAI-1 observed in invasive cells [47].

4.2.2. Matrix Metalloproteases and Inhibitors. In addition to its effect on uPA and tPA activities, LRP-1 also regulates extracellular levels of MMP-2, -9, and -13 [48]. As for uPA and tPA, the endocytosis of MMP-2 and MMP-13 involves preliminary binding to adjacent receptors. Indeed, when bound to thrombospondin-2 (TSP-2), proMMP-2 first associates with an unknown cell-surface heparin-sulfate proteoglycan before

interacting with LRP-1 [49]. When complexed with its specific inhibitor TIMP-2, proMMP-2 first binds to an unidentified coreceptor before being internalized by LRP-1 [50]. Likewise, the endocytic clearance of MMP-13 by LRP-1 requires a two-step process, involving a first binding to a 170 kDa co-receptor [51]. The internalization of MMP-9 by LRP-1 requires a more simple mechanism. Thus, proMMP-9/TIMP-1 directly interacts with LRP-1, leading to its endocytic uptake and degradation by lysosomal proteases [52]. The analysis of this interaction reveals that the hemopexin domain of MMP-9 contains a binding site for LRP-1 [53].

Although Hahn-Dantona et al. failed to demonstrate a direct interaction between TIMP-1 and LRP-1 in their *in vitro* study [52], our unpublished data reveal that LRP-1 could bind and endocytose TIMP-1 in neurons, in an MMP-independent way. Furthermore, noncomplexed TIMP-2 [50] and TIMP-3 [54, 55] also bind directly to LRP-1 to be internalized.

4.2.3. Other Matrix Proteinases. Heparanase-1 is secreted as an inactive heparanase precursor. Once activated, this endoglycosidase degrades heparan sulfate and consequently alters the stability of ECM [56]. The group of Guido David [57] has clearly identified LRP-1 as one of the receptors able to mediate the uptake of secreted heparanase precursor and its intracellular trafficking to the site of activation process. Recently, ADAM with thrombospondin motifs 5 (ADAMTS-5), a major aggrecan-degrading enzyme in cartilage, has been shown to be endocytosed by LRP-1 [58].

The aspartic proteinase cathepsin-D (cath-D) is capable of degrading ECM in an acidic microenvironment [59]. Recently, Liaudet-Coopman and colleagues [60, 61] identified pro-cath-D as the first ligand of the extracellular domain of LRP-1 β-chain.

4.3. Signaling Function. Additionally, LRP-1 acts in signaling pathways [8, 28, 62]. We recently demonstrated that the abrogation of LRP-1 expression inhibited migration and invasive capacities of thyroid carcinoma cells despite a strong stimulation of pericellular MMP-2 and uPA proteolytic activities [63]. We identified ERK and JNK as the main molecular relays by which LRP-1 regulates focal adhesion disassembly of malignant cells to support invasion [64].

A stimulating study reveals that LRP-1-mediated endocytosis of tPA and tPA/PAI-1 complex is accompanied by a decrease in tPA mRNA transcription [65], suggesting that a secreted protein could regulate its own biosynthesis. Furthermore, the binding of tPA to LRP-1 triggers intracellular signal transduction to induce the expression of another matrix proteinase, MMP-9, both in microvascular endothelial cells [66] and fibroblasts [67]. Probably more surprising, the binding of proteinase inhibitors to LRP-1 also induces MMP-9 expression, as demonstrated for the serpin nexin-1 in a mammary tumor model [68], and activate α2M in macrophage-derived cell lines [69]. Recently, the knockdown of LRP-1 expression in human glioblastoma U87 cells revealed that LRP-1 promoted cell migration and invasion by inducing the expression of MMP-2 and MMP-9 [70].

Altogether, these data indicate a close link between MMP-9 and LRP-1: from one of its ligands to a product of LRP-1-induced expression. This suggests important functions for MMP-9 in normal and pathophysiological conditions.

4.4. Regulation of LRP-1 Cell-Surface Level and Endocytic Activity by Shedding.

Most membrane proteins, including type I and type II transmembrane proteins, are subjected to a shedding process, that is, the proteolytic cleavage of their extracellular part or ectodomain [71]. LRP-1 also constitutes a membrane target for numerous proteinases. The LRP-1 ectodomain consists in the entire extracellular α-chain (515 kDa) noncovalently associated to the extracellular part (55 kDa) of the transmembrane β-chain [72].

The product of LRP-1 shedding, the soluble LRP-1 (sLRP-1) α-chain, was first detected in human plasma and serum [73]. A metalloproteinase, cleaving LRP-1 at the membrane-proximal region of the β-chain, was described in human BeWo choriocarcinoma cells [72]. Since this work was completed, different metalloproteinases have been identified, mainly among the ADAM family. Thus, ADAM-10 and ADAM-17 are associated to LRP-1 shedding in human brain [74]. We recently showed that ADAM-12 exhibited sheddase activity towards LRP-1 in human HT1080 fibrosarcoma cells [75]. Additionally, we reported that MT-MMP, first described to degrade LRP-1 in small fragments [76], was able to generate sLRP-1 in medium conditioned by HT1080 cells in culture [75]. Besides these metalloproteinases, tPA and BACE-1 were described to mediate shedding of LRP-1 [77, 78]. It has been reported that, during cerebral ischemia, tPA induces the shedding of LRP-1 from perivascular astrocytes followed by the development of cerebral edema [79]. These authors demonstrated that the interaction between tPA and LRP-1 in perivascular astrocytes induced Akt phosphorylation, leading to an increase of permeability in the blood-brain barrier.

Soluble LRP-1, which is composed of the entire extracellular α-chain and noncovalently associated extracellular part of the transmembrane β-chain [72], retains ligand-binding capacity and acts as a decoy receptor [80]. Thus, Quinn and colleagues first reported that the addition of sLRP-1 to cultured rat hepatocytes resulted in an inhibition of tPA clearance [73]. Immunoprecipitation assays confirmed that tPA interacted with LRP-1 [72]. Also, LRP-1 shedding from human lung fibroblasts impairs endocytosis of MMP-2 and -9 [81]. We similarly reported that the inhibition of LRP-1 shedding increased MMP-2 and -9 activities, in cultures of human endometrial explants [82] and fibrosarcoma cells [75]. Finally, we recently demonstrated that TIMP-3 bound to sLRP-1, which is resistant to endocytosis, retained its inhibitory activity against metalloproteinases [55].

5. Conclusion

Understanding the precise role of LRP-1 in the regulation of ECM breakdown remains an exciting challenge, as it appears to be a multifunctional "Swiss knife." Thus, besides the endocytosis of proteinases, LRP-1 mediates the clearance of their own inhibitors [8]. Moreover, LRP-1 acts as a membrane receptor that transduces intracellular signals to induce the MMP-9 expression [66–69]. Finally, the proteolytic cleavage of LRP-1 at the cell surface solubilizes the LRP-1 ectodomain, which conserves ligand-binding capacities. Such a property could allow matrix proteinases—but also inhibitors—to increase their extracellular half-life time by escaping from endocytic clearance mediated by membrane-LRP 1.

Despite a strong stimulation of pericellular MMP-2 and uPA proteolytic activities, carcinoma cell invasion decreased by LRP-1 silencing [63]. This result clearly indicates that, depending on parameters yet to be elucidated, the signaling function of LRP-1 can counteract or override its endocytic function.

Another paradox is represented by a proteinase, tPA, for instance, which can either be endocytosed by LRP-1 or solubilize the ectodomain of LRP-1. At the molecular level, interactions between tPA and LRP-1 will be different according to the event: binding to domains 2 and 4 of the LRP-1 α-chain for endocytic pathway or cleaving at a single site of both α- and β-chains at the vicinity of the cell surface for shedding LRP-1 ectodomain. Which is determinant the enzyme or the cell?

On the whole, these data strongly suggest that LRP-1 does not act alone but with membrane partners, which vary according to numerous parameters including cell origin, ECM composition, pathological conditions, and so forth. In this way, we recently demonstrated [83] that LRP-1 forms complex with the hyaluronan receptor CD44, which may bind proMMP-9 [24]. The identification of these partners could represent a key to the understanding of the LRP-1 roles in ECM remodeling.

Acknowledgments

The authors thank Claude-Annie Turlier for editorial assistance. They are supported by the Université de Reims Champagne-Ardenne and the Centre National de la Recherche Scientifique (CNRS), France. They acknowledge support from the Agence Nationale pour la Recherche (ANR-08-MNPS-042-02, TIMPAD project). Laurie Verzeaux and Stéphane Dedieu were recipients of grants from the Région Champagne-Ardenne.

References

[1] P. Lu, V. M. Weaver, and Z. Werb, "The extracellular matrix: a dynamic niche in cancer progression," *Journal of Cell Biology*, vol. 196, no. 4, pp. 395–406, 2012.

[2] P. A. Andreasen, R. Egelund, and H. H. Petersen, "The plasminogen activation system in tumor growth, invasion, and metastasis," *Cellular and Molecular Life Sciences*, vol. 57, no. 1, pp. 25–40, 2000.

[3] G. Murphy and H. Nagase, "Progress in matrix metalloproteinase research," *Molecular Aspects of Medicine*, vol. 29, no. 5, pp. 290–308, 2008.

[4] W. Borth, "α2-Macroglobulin, a multifunctional binding protein with targeting characteristics," *The FASEB Journal*, vol. 6, no. 15, pp. 3345–3353, 1992.

[5] B. R. Binder, G. Christ, F. Gruber et al., "Plasminogen activator inhibitor 1: physiological and pathophysiological roles," *News in Physiological Sciences*, vol. 17, no. 2, pp. 56–61, 2002.

[6] G. Murphy, "Tissue inhibitors of metalloproteinases," *Genome Biology*, vol. 12, no. 11, article 233, 2011.

[7] G. Murphy, "The ADAMs: signalling scissors in the tumour microenvironment," *Nature Reviews Cancer*, vol. 8, no. 12, pp. 929–941, 2008.

[8] A. P. Lillis, L. B. Van Duyn, J. E. Murphy-Ullrich, and D. K. Strickland, "LDL receptor-related protein 1: unique tissue-specific functions revealed by selective gene knockout studies," *Physiological Reviews*, vol. 88, no. 3, pp. 887–918, 2008.

[9] K. Danø, P. A. Andreasen, J. Grøndahl-Hansen, P. Kristensen, L. S. Nielsen, and L. Skriver, "Plasminogen activators, tissue degradation, and cancer," *Advances in Cancer Research*, vol. 44, pp. 139–266, 1985.

[10] E. I. Deryugina and J. P. Quigley, "Cell surface remodeling by plasmin: a new function for an old enzyme," *Journal of Biomedicine and Biotechnology*, vol. 2012, Article ID 564259, 21 pages, 2012.

[11] V. Ellis, T.-C. Wun, N. Behrendt, E. Ronne, and K. Dano, "Inhibition of receptor-bound urokinase by plasminogen-activator inhibitors," *The Journal of Biological Chemistry*, vol. 265, no. 17, pp. 9904–9908, 1990.

[12] S. M. Kanse, C. Kost, O. G. Wilhelm, P. A. Andreasen, and K. T. Preissner, "The urokinase receptor is a major vitronectin-binding protein on endothelial cells," *Experimental Cell Research*, vol. 224, no. 2, pp. 344–353, 1996.

[13] R.-P. Czekay, K. Aertgeerts, S. A. Curriden, and D. J. Loskutoff, "Plasminogen activator inhibitor-1 detaches cells from extracellular matrices by inactivating integrins," *Journal of Cell Biology*, vol. 160, no. 5, pp. 781–791, 2003.

[14] K. Bajou, A. Noël, R. D. Gerard et al., "Absence of host plasminogen activator inhibitor 1 prevents cancer invasion and vascularization," *Nature Medicine*, vol. 4, no. 8, pp. 923–928, 1998.

[15] K. Kessenbrock, V. Plaks, and Z. Werb, "Matrix metalloproteinases: regulators of the tumor microenvironment," *Cell*, vol. 141, no. 1, pp. 52–67, 2010.

[16] H. Hua, M. Li, T. Luo, Y. Yin, and Y. Jiang, "Matrix metalloproteinases in tumorigenesis: an evolving paradigm," *Cellular and Molecular Life Sciences*, vol. 68, no. 23, pp. 3853–3868, 2011.

[17] D. Bourboulia and W. G. Stetler-Stevenson, "Matrix metalloproteinases (MMPs) and tissue inhibitors of metalloproteinases (TIMPs): positive and negative regulators in tumor cell adhesion," *Seminars in Cancer Biology*, vol. 20, no. 3, pp. 161–168, 2010.

[18] H. Nagase, R. Visse, and G. Murphy, "Structure and function of matrix metalloproteinases and TIMPs," *Cardiovascular Research*, vol. 69, no. 3, pp. 562–573, 2006.

[19] A. H. Baker, D. R. Edwards, and G. Murphy, "Metalloproteinase inhibitors: biological actions and therapeutic opportunities," *Journal of Cell Science*, vol. 115, no. 19, pp. 3719–3727, 2002.

[20] K. Brew and H. Nagase, "The tissue inhibitors of metalloproteinases (TIMPs): an ancient family with structural and functional diversity," *Biochimica et Biophysica Acta*, vol. 1803, no. 1, pp. 55–71, 2010.

[21] Y. Nishida, H. Miyamori, E. W. Thompson, T. Takino, Y. Endo, and H. Sato, "Activation of matrix metalloproteinase-2 (MMP-2) by membrane type 1 matrix metalloproteinase through an artificial receptor for ProMMP-2 generates active MMP-2," *Cancer Research*, vol. 68, no. 21, pp. 9096–9104, 2008.

[22] R. Chirco, X.-W. Liu, K.-K. Jung, and H.-R. C. Kim, "Novel functions of TIMPs in cell signaling," *Cancer and Metastasis Reviews*, vol. 25, no. 1, pp. 99–113, 2006.

[23] W. G. Stetler-Stevenson, "Tissue inhibitors of metalloproteinases in cell signaling: metalloproteinase-independent biological activities," *Science Signaling*, vol. 1, no. 27, article re6, 2008.

[24] E. Lambert, L. Bridoux, J. Devy et al., "TIMP-1 binding to proMMP-9/CD44 complex localized at the cell surface promotes erythroid cell survival," *International Journal of Biochemistry and Cell Biology*, vol. 41, no. 5, pp. 1102–1115, 2009.

[25] L. Bridoux, N. Etique, E. Lambert et al., "A crucial role for Lyn in TIMP-1 erythroid cell survival signalling pathway," *FEBS Letters*, vol. 587, no. 10, pp. 1524–1528, 2013.

[26] K.-K. Jung, X.-W. Liu, R. Chirco, R. Fridman, and H.-R. C. Kim, "Identification of CD63 as a tissue inhibitor of metalloproteinase-1 interacting cell surface protein," *The EMBO Journal*, vol. 25, no. 17, pp. 3934–3942, 2006.

[27] Y. Xia, N. Yeddula, M. Leblanc et al., "Reduced cell proliferation by IKK2 depletion in a mouse lung-cancer model," *Nature Cell Biology*, vol. 14, no. 3, pp. 257–265, 2012.

[28] J. Herz and D. K. Strickland, "LRP: a multifunctional scavenger and signaling receptor," *The Journal of Clinical Investigation*, vol. 108, no. 6, pp. 779–784, 2001.

[29] D. K. Strickland, J. D. Ashcom, S. Williams, W. H. Burgess, M. Migliorini, and W. Scott Argraves, "Sequence identity between the α2-macroglobulin receptor and low density lipoprotein receptor-related protein suggests that this molecule is a multifunctional receptor," *The Journal of Biological Chemistry*, vol. 265, no. 29, pp. 17401–17404, 1990.

[30] J. Stebbing, P. Savage, S. Patterson, and B. Gazzard, "All for CD91 and CD91 for all," *Journal of Antimicrobial Chemotherapy*, vol. 53, no. 1, pp. 1–3, 2004.

[31] J. Herz, D. E. Clouthier, and R. E. Hammer, "LDL receptor-related protein internalizes and degrades uPA-PAI-1 complexes and is essential for embryo implantation," *Cell*, vol. 71, no. 3, pp. 411–421, 1992.

[32] T. Kristensen, S. K. Moestrup, J. Gliemann, L. Bendtsen, O. Sand, and L. Sottrup-Jensen, "Evidence that the newly cloned low-density-lipoprotein receptor related protein (LRP) is the α2-macroglobulin receptor," *FEBS Letters*, vol. 276, no. 1-2, pp. 151–155, 1990.

[33] P. A. Morton, D. A. Owensby, B. E. Sobel, and A. L. Schwartz, "Catabolism of tissue-type plasminogen activator by the human hepatoma cell line Hep G2. Modulation by plasminogen activator inhibitor type 1," *The Journal of Biological Chemistry*, vol. 264, no. 13, pp. 7228–7235, 1989.

[34] G. Bu, P. A. Morton, and A. L. Schwartz, "Identification and partial characterization by chemical cross-linking of a binding protein for tissue-type plasminogen activator (t-PA) on rat hepatoma cells: a plasminogen activator inhibitor type 1-independent t-PA receptor," *The Journal of Biological Chemistry*, vol. 267, no. 22, pp. 15595–15602, 1992.

[35] G. Bu, S. Williams, D. K. Strickland, and A. L. Schwartz, "Low density lipoprotein receptor-related protein/α2-macroglobulin receptor is an hepatic receptor for tissue-type plasminogen activator," *Proceedings of the National Academy of Sciences of the United States of America*, vol. 89, no. 16, pp. 7427–7431, 1992.

[36] K. Orth, E. L. Madison, M.-J. Gething, J. F. Sambrook, and J. Herz, "Complexes of tissue-type plasminogen activator and its serpin inhibitor plasminogen-activator inhibitor type 1 are

internalized by means of the low density lipoprotein receptor-related protein/α2-macroglobulin receptor," *Proceedings of the National Academy of Sciences of the United States of America*, vol. 89, no. 16, pp. 7422–7426, 1992.

[37] K. Orth, T. Willnow, J. Herz, M. J. Gething, and J. Sambrook, "Low density lipoprotein receptor-related protein is necessary for the internalization of both tissue-type plasminogen activator-inhibitor complexes and free tissue-type plasminogen activator," *The Journal of Biological Chemistry*, vol. 269, no. 33, pp. 21117–21122, 1994.

[38] A. Nykjaer, C. M. Petersen, B. Moller et al., "Purified α2-macroglobulin receptor/LDL receptor-related protein binds urokinase plasminogen activator inhibitor type-1 complex. Evidence that the α2-macroglobulin receptor mediates cellular degradation of urokinase receptor-bound complexes," *The Journal of Biological Chemistry*, vol. 267, no. 21, pp. 14543–14546, 1992.

[39] D. Croucher, D. N. Saunders, and M. Ranson, "The urokinase/PAI-2 complex: a new high affinity ligand for the endocytosis receptor low density lipoprotein receptor-related protein," *The Journal of Biological Chemistry*, vol. 281, no. 15, pp. 10206–10213, 2006.

[40] M. Z. Kounnas, J. Henkin, W. S. Argraves, and D. K. Strickland, "Low density lipoprotein receptor-related protein/α2-macroglobulin receptor mediates cellular uptake of prourokinase," *The Journal of Biological Chemistry*, vol. 268, no. 29, pp. 21862–21867, 1993.

[41] S. Stefansson, S. Muhammad, X.-F. Cheng, F. D. Battey, D. K. Strickland, and D. A. Lawrence, "Plasminogen activator inhibitor-1 contains a cryptic high affinity binding site for the low density lipoprotein receptor-related protein," *The Journal of Biological Chemistry*, vol. 273, no. 11, pp. 6358–6366, 1998.

[42] M. V. Cubellis, P. Andreasen, P. Ragno, M. Mayer, K. Dano, and F. Blasi, "Accessibility of receptor-bound urokinase to type-1 plasminogen activator inhibitor," *Proceedings of the National Academy of Sciences of the United States of America*, vol. 86, no. 13, pp. 4828–4832, 1989.

[43] M. V. Cubellis, T.-C. Wun, and F. Blasi, "Receptor-mediated internalization and degradation of urokinase is caused by its specific inhibitor PAI-1," *The EMBO Journal*, vol. 9, no. 4, pp. 1079–1085, 1990.

[44] A. Nykjaer, M. Conese, E. I. Christensen et al., "Recycling of the urokinase receptor upon internalization of the uPA:serpin complexes," *The EMBO Journal*, vol. 16, no. 10, pp. 2610–2620, 1997.

[45] J.-C. Zhang, R. Sakthivel, D. Kniss, C. H. Graham, D. K. Strickland, and K. R. McCrae, "The low density lipoprotein receptor-related protein/α2-macroglobulin receptor regulates cell surface plasminogen activator activity on human trophoblast cells," *The Journal of Biological Chemistry*, vol. 273, no. 48, pp. 32273–32280, 1998.

[46] R.-P. Czekay, T. A. Kuemmel, R. A. Orlando, and M. G. Farquhar, "Direct binding of occupied urokinase receptor (uPAR) to LDL receptor-related protein is required for endocytosis of uPAR and regulation of cell surface urokinase activity," *Molecular Biology of the Cell*, vol. 12, no. 5, pp. 1467–1479, 2001.

[47] B. Chazaud, R. Ricoux, C. Christov, A. Plonquet, R. K. Gherardi, and G. Barlovatz-Meimon, "Promigratory effect of plasminogen activator inhibitor-1 on invasive breast cancer cell populations," *American Journal of Pathology*, vol. 160, no. 1, pp. 237–246, 2002.

[48] H. Emonard, G. Bellon, P. de Diesbach, M. Mettlen, W. Hornebeck, and P. J. Courtoy, "Regulation of matrix metalloproteinase (MMP) activity by the low-density lipoprotein receptor-related protein (LRP). A new function for an 'old friend'," *Biochimie*, vol. 87, no. 3-4, pp. 369–376, 2005.

[49] Z. Yang, D. K. Strickland, and P. Bornstein, "Extracellular matrix metalloproteinase 2 levels are regulated by the low density lipoprotein-related scavenger receptor and thrombospondin 2," *The Journal of Biological Chemistry*, vol. 276, no. 11, pp. 8403–8408, 2001.

[50] H. Emonard, G. Bellon, L. Troeberg et al., "Low density lipoprotein receptor-related protein mediates endocytic clearance of pro-MMP-2-TIMP-2 complex through a thrombospondin-independent mechanism," *The Journal of Biological Chemistry*, vol. 279, no. 52, pp. 54944–54951, 2004.

[51] O. Y. Barmina, H. W. Walling, G. J. Fiacco et al., "Collagenase-3 binds to a specific receptor and requires the low density lipoprotein receptor-related protein for internalization," *The Journal of Biological Chemistry*, vol. 274, no. 42, pp. 30087–30093, 1999.

[52] E. Hahn-Dantona, J. F. Ruiz, P. Bornstein, and D. K. Strickland, "The low density lipoprotein receptor-related protein modulates levels of matrix metalloproteinase 9 (MMP-9) by mediating its cellular catabolism," *The Journal of Biological Chemistry*, vol. 276, no. 18, pp. 15498–15503, 2001.

[53] P. E. Van den Steen, I. Van Aelst, V. Hvidberg et al., "The hemopexin and O-glycosylated domains tune gelatinase B/MMP-9 bioavailability via inhibition and binding to cargo receptors," *The Journal of Biological Chemistry*, vol. 281, no. 27, pp. 18626–18637, 2006.

[54] L. Troeberg, K. Fushimi, R. Khokha, H. Emonard, P. Ghosh, and H. Nagase, "Calcium pentosan polysulfate is a multifaceted exosite inhibitor of aggrecanases," *The FASEB Journal*, vol. 22, no. 10, pp. 3515–3524, 2008.

[55] S. D. Scilabra, L. Troeberg, K. Yamamoto et al., "Differential regulation of extracellular tissue inhibitor of metalloproteinases-3 levels by cell membrane-bound and shed low density lipoprotein receptor-related protein 1," *The Journal of Biological Chemistry*, vol. 288, no. 1, pp. 332–342, 2013.

[56] U. Barash, V. Cohen-Kaplan, I. Dowek, R. D. Sanderson, N. Ilan, and I. Vlodavsky, "Proteoglycans in health and disease: new concepts for heparanase function in tumor progression and metastasis," *FEBS Journal*, vol. 277, no. 19, pp. 3890–3903, 2010.

[57] V. Vreys, N. Delande, Z. Zhang et al., "Cellular uptake of mammalian heparanase precursor involves low density lipoprotein receptor-related proteins, mannose 6-phosphate receptors, and heparan sulfate proteoglycans," *The Journal of Biological Chemistry*, vol. 280, no. 39, pp. 33141–33148, 2005.

[58] K. Yamamoto, L. Troeberg, S. D. Scilabra et al., "LRP-1-mediated endocytosis regulates extracellular activity of ADAMTS-5 in articular cartilage," *The FASEB Journal*, vol. 27, no. 2, pp. 511–521, 2013.

[59] P. Briozzo, M. Morisset, F. Capony, C. Rougeot, and H. Rochefort, "In vitro degradation of extracellular matrix with M(r) 52,000 cathepsin D secreted by breast cancer cells," *Cancer Research*, vol. 48, no. 13, pp. 3688–3692, 1988.

[60] M. Beaujouin, C. Prébois, D. Derocq et al., "Pro-cathepsin D interacts with the extracellular domain of the β chain of LRP1 and promotes LRP1-dependent fibroblast outgrowth," *Journal of Cell Science*, vol. 123, no. 19, pp. 3336–3346, 2010.

[61] D. Derocq, C. Prébois, M. Beaujouin et al., "Cathepsin D is partly endocytosed by the LRP1 receptor and inhibits LRP1-regulated intramembrane proteolysis," *Oncogene*, vol. 31, no. 26, pp. 3202–3212, 2012.

[62] P. May, E. Woldt, R. L. Matz, and P. Boucher, "The LDL receptor-related protein (LRP) family: an old family of proteins with new physiological functions," *Annals of Medicine*, vol. 39, no. 3, pp. 219–228, 2007.

[63] S. Dedieu, B. Langlois, J. Devy et al., "LRP-1 silencing prevents malignant cell invasion despite increased pericellular proteolytic activities," *Molecular and Cellular Biology*, vol. 28, no. 9, pp. 2980–2995, 2008.

[64] B. Langlois, G. Perrot, C. Schneider et al., "LRP-1 promotes cancer cell invasion by supporting ERK and inhibiting JNK signaling pathways," *PLoS ONE*, vol. 5, no. 7, Article ID e11584, 2010.

[65] M. M. Hardy, J. Feder, R. A. Wolfe, and G. Bu, "Low density lipoprotein receptor-related protein modulates the expression of tissue-type plasminogen activator in human colon fibroblasts," *The Journal of Biological Chemistry*, vol. 272, no. 10, pp. 6812–6817, 1997.

[66] X. Wang, S.-R. Lee, K. Arai et al., "Lipoprotein receptor-mediated induction of matrix metalloproteinase by tissue plasminogen activator," *Nature Medicine*, vol. 9, no. 10, pp. 1313–1317, 2003.

[67] K. Hu, J. Yang, S. Tanaka, S. L. Gonias, W. M. Mars, and Y. Liu, "Tissue-type plasminogen activator acts as a cytokine that triggers intracellular signal transduction and induces matrix metalloproteinase-9 gene expression," *The Journal of Biological Chemistry*, vol. 281, no. 4, pp. 2120–2127, 2006.

[68] B. Fayard, F. Bianchi, J. Dey et al., "The serine protease inhibitor protease nexin-1 controls mammary cancer metastasis through LRP-1-mediated MMP-9 expression," *Cancer Research*, vol. 69, no. 14, pp. 5690–5698, 2009.

[69] L. C. Cáceres, G. R. Bonacci, M. C. Sánchez, and G. A. Chiabrando, "Activated α2 macroglobulin induces matrix metalloproteinase 9 expression by low-density lipoprotein receptor-related protein 1 through MAPK-ERK1/2 and NF-κB activation in macrophage-derived cell lines," *Journal of Cellular Biochemistry*, vol. 111, no. 3, pp. 607–617, 2010.

[70] H. Song, Y. Li, J. Lee, A. L. Schwartz, and G. Bu, "Low-density lipoprotein receptor-related protein 1 promotes cancer cell migration and invasion by inducing the expression of matrix metalloproteinases 2 and 9," *Cancer Research*, vol. 69, no. 3, pp. 879–886, 2009.

[71] M. Hartmann, A. Herrlich, and P. Herrlich, "Who decides when to cleave an ectodomain?" *Trends in Biochemical Sciences*, vol. 38, no. 3, pp. 111–120, 2013.

[72] K. A. Quinn, V. J. Pye, Y.-P. Dai, C. N. Chesterman, and D. A. Owensby, "Characterization of the soluble form of the low density lipoprotein receptor-related protein (LRP)," *Experimental Cell Research*, vol. 251, no. 2, pp. 433–441, 1999.

[73] K. A. Quinn, P. G. Grimsley, Y.-P. Dai, M. Tapner, C. N. Chesterman, and D. A. Owensby, "Soluble low density lipoprotein receptor-related protein (LRP) circulates in human plasma," *The Journal of Biological Chemistry*, vol. 272, no. 38, pp. 23946–23951, 1997.

[74] Q. Liu, J. Zhang, H. Tran et al., "LRP1 shedding in human brain: roles of ADAM10 and ADAM17," *Molecular Neurodegeneration*, vol. 4, no. 1, article 17, 2009.

[75] C. Selvais, L. D'Auria, D. Tyteca et al., "Cell cholesterol modulates metalloproteinase-dependent shedding of low-density lipoprotein receptor-related protein-1 (LRP-1) and clearance function," *The FASEB Journal*, vol. 25, no. 8, pp. 2770–2781, 2011.

[76] D. V. Rozanov, E. Hahn-Dantona, D. K. Strickland, and A. Y. Strongin, "The low density lipoprotein receptor-related protein LRP is regulated by membrane type-1 matrix metalloproteinase (MT1-MMP) proteolysis in malignant cells," *The Journal of Biological Chemistry*, vol. 279, no. 6, pp. 4260–4268, 2004.

[77] R. Polavarapu, M. C. Gongora, H. Yi et al., "Tissue-type plasminogen activator-mediated shedding of astrocytic low-density lipoprotein receptor-related protein increases the permeability of the neurovascular unit," *Blood*, vol. 109, no. 8, pp. 3270–3278, 2007.

[78] C. A. F. von Arnim, A. Kinoshita, I. D. Peltan et al., "The low density lipoprotein receptor-related protein (LRP) is a novel β-secretase (BACE1) substrate," *The Journal of Biological Chemistry*, vol. 280, no. 18, pp. 17777–17785, 2005.

[79] J. An, C. Zhang, R. Polavarapu, X. Zhang, X. Zhang, and M. Yepes, "Tissue-type plasminogen activator and the low-density lipoprotein receptor related protein induce Akt phosphorylation in the ischemic brain," *Blood*, vol. 112, no. 7, pp. 2787–2794, 2008.

[80] P. G. Grimsley, K. A. Quinn, and D. A. Owensby, "Soluble low-density lipoprotein receptor-related protein," *Trends in Cardiovascular Medicine*, vol. 8, no. 8, pp. 363–368, 1998.

[81] M. Wygrecka, J. Wilhelm, E. Jablonska et al., "Shedding of low-density lipoprotein receptor-related protein-1 in acute respiratory distress syndrome," *American Journal of Respiratory and Critical Care Medicine*, vol. 184, no. 4, pp. 438–448, 2011.

[82] C. Selvais, H. P. Gaide Chevronnay, P. Lemoine et al., "Metalloproteinase-dependent shedding of low-density lipoprotein receptor-related protein-1 ectodomain decreases endocytic clearance of endometrial matrix metalloproteinase-2 and -9 at menstruation," *Endocrinology*, vol. 150, no. 8, pp. 3792–3799, 2009.

[83] G. Perrot, B. Langlois, J. Devy et al., "LRP-1:CD44, identification of a new cell surface complex regulating tumor cell adhesion," *Molecular and Cellular Biology*, vol. 32, no. 16, pp. 3293–3307, 2012.

9

Expression and Function of NUMB in Odontogenesis

Haitao Li,[1] Amsaveni Ramachandran,[2] Qi Gao,[2] Sriram Ravindran,[2] Yiqiang Song,[2] Carla Evans,[3] and Anne George[2]

[1] *Department of Orthodontics, College of Dental Medicine, Nova Southeastern University, 3200 S. University Drive, Fort Lauderdale, FL 33328, USA*
[2] *Brodie Tooth Development Genetics & Regenerative Medicine Research Laboratory, Department of Oral Biology (M/C 690), College of Dentistry, University of Illinois at Chicago, Chicago IL 60612, USA*
[3] *Department of Orthodontics, College of Dentistry, University of Illinois at Chicago, 801 S. Paulina Street, Chicago IL 60612, USA*

Correspondence should be addressed to Anne George; anneg@uic.edu

Academic Editor: Avina Paranjpe

NUMB is a multifunctional protein implicated to function in self-renewal and differentiation of progenitors in several tissues. To characterize the transcripts and to analyze the expression pattern of NUMB in odontogenesis, we isolated 2 full-length clones for NUMB from mouse dental pulp mRNA. One novel sequence contained 200 bp insertion in the phosphotyrosine binding domain (PTB). Confocal microscopy analysis showed strong NUMB expression in human dental pulp stem cells (hDPSC) and preameloblasts. Western blot analysis indicated that NUMB isoforms were differentially expressed in various dental tissues. Immunohistochemical analysis showed that in postnatal mouse tooth germs, NUMB was differentially expressed in the preameloblasts, odontoblasts, cervical loop region, and in the dental pulp stem cells during development. Interestingly, overexpression of NUMB in HAT-7, a preameloblast cell line, had dramatic antagonizing effects on the protein expression level of activated Notch 1. Further analysis of the Notch signaling pathway showed that NUMB significantly downregulates sonic hedgehog (Shh) expression in preameloblasts. Therefore, we propose that NUMB maintains ameloblast progenitor phenotype at the cervical loop by downregulating the activated Notch1 protein and thereby inhibiting the mRNA expression of Shh.

1. Introduction

NUMB protein was initially identified in *Drosophila*. Its name comes from the fact that the gene mutation causes flies to lose their sensory neurons and become "numb." NUMB is demonstrated to play roles in lineage commitment by both gain- and loss-of-functions approaches. Its function has been attributed to asymmetric distribution in daughter cells as well as interaction with Notch signaling components. After its identification, NUMB has been extensively studied in the development of sensory organs, and subsequently in cancer research [1]. We recently identified NUMB as one of the genes that are differentially expressed in rat immortalized odontoblast cells (T4-4) [2].

The *Drosophila* NUMB is a membrane-associated protein in the sensory organ precursor cells, and its mutation is lethal resulting from the abnormality of the peripheral nervous systems [3]. Spana et al. identified that the unequal distribution of NUMB in the descendents controls the cell fate of the neurons projecting towards two directions: daughter cells remain as stem cells from maintaining high level NUMB expression, and daughter cells differentiate into neurons with the loss of NUMB expression [4]. NUMB knockout mice die around embryonic day 11.5. They exhibit severe defects in cranial neural tube closure and precocious neuron production, indicating that NUMB promotes progenitor cell fates [5]. Mammalian NUMB is not only expressed in the embryonic tissues but also in most adult tissues, suggesting more complex functions other than neurogenesis [6]. To date, four mammalian NUMB isoforms were identified from alternatively spliced transcripts in neuronal lineage cells [7]. The two members of PRRL isoforms (have long proline rich region) were expressed in early mouse embryonic ages (E7–E10) and were undetectable at E13 when the cells are in rapid expansion stage. The two members of the PRRS isoforms (have short proline rich region) were detected in all developmental stages and adult brains [7, 8]. This differential expression pattern is associated with the function of NUMB in maintaining progenitor cells in the early phase of cortical neurogenesis [9] verses its function in self-renewal and differentiation in late stages of neuron development [10]. Recently, three novel isoforms of NUMB were identified in the human extravillous trophoblast [11].

In rodent incisors, the cervical loop situated at the posterior end of the epithelium consists of progenitor cells which are capable of continuously amplifying and differentiating into amelobasts and support the lifelong growth of the enamel on the labial side [12]. These progenitor cells in the labial cervical loop give rise to the transit-amplifying cells which will further differentiate into preameloblasts and eventually into well-differentiated enamel forming ameloblasts [13]. Notch and sonic hedgehog (Shh) signaling pathway have been shown to be important in regulating this transformation [14].

Seidel et al. demonstrated that Shh produced by the transit amplifying/preameloblasts signals to the stem cells in the cervical loop for continuous ameloblast regeneration. After injecting a hedgehog pathway inhibitor in the mouse mandible, new enamel formation on the labial surface of the mouse incisors was completely blocked. However, the stem cell survival was not endangered due to the fact that normal amelobasts as well as the enamel were able to form after the removal of the hedgehog signal inhibitor [15]. These results clearly revealed the existence of a positive feedback signal loop between the differentiating cells and the stem cells. Shh produced by the differentiating amelobasts signals to the stem cells located in the labial cervical loop to continuously provide progenies in order to support the differentiation demand. In addition, Shh is also important in regulating the preameloblasts to elongate, polarize, and deposit enamel matrix [16].

Notch 1, a transmembrane protein, has been shown to be important in specifying dental cell-type identity [14]. Notch signaling regulates the maintenance, fate decision, and proliferation of progenitor cells in multiple tissues such as bone marrow and various neuronal tissues [17]. Notch 1 expressing cells in the cervical loop respond to FGF signaling secreted by the adjacent mesenchyme during mitogenesis and cell fate decision [18]. Inhibition of Notch signaling leads to impaired growth of the mouse incisor cervical loop with reduced proliferation and increased apoptosis. In addition, the inhibition also reduced the population of the ameloblast precursors [19].

NUMB inhibits Notch signaling through interaction with the Notch intracellular domain and promotes its ubiquitination and degradation [20]. NUMB was also shown to control the intracellular trafficking of the Notch to suppress its function [21]. Studies also indicated that the expression of NUMB correlates with the suppression of hedgehog signaling [22]. NUMB is also involved in the TP53-activated pathway [23], endocytosis [24], and the determination of cell polarity [25]. These studies suggest that there is a finely tuned regulatory network centered by NUMB in determination of cell fates [26].

Even though the expression of NUMB has been reported in several tissues, there are no reports thus far during tooth development. In this study, we identified several NUMB isoforms that are differently expressed in dental tissues. They were expressed in the cervical loop, immature ameloblasts, odontoblasts, and in specific dental pulp cells. Overexpression of NUMB in preameloblast cells inhibited both the activated Notch 1 protein and Shh expression. Therefore, we propose that NUMB regulates ameloblasts differentiation by modulating the expression levels of Notch 1 protein and Shh expressions.

2. Materials and Methods

2.1. RT-PCR. Dental pulp tissues were isolated from postnatal day 5 mouse tooth germs. Single cell suspension was derived by digesting the dental pulp tissues at $37°C$ for 30 min under constant agitation in a digestion cocktail containing (0.05% trypsin and 0.1% collagenase P). Dental pulp cells were cultured in αMEM with addition of 10% fetal bovine serum and 50 U/mL of penicillin and streptomycin for 7 days. RNA was extracted from the cultures using TRIzol reagent (Invitrogen, CA, USA) according to the manufacturer's instructions. cDNA library was generated using SuperScript First-Strand Synthesis System (Invitrogen, CA, USA). Primers were designed to amplify the longest NUMB open reading frame using sequence derived from Ensembl Genome Browser database with the additions of the endonuclease sites on the forward and reverse primers. The primer sequences are Forward: $5'$-SalI+ATGAACAAACTACGGC-$3'$ and Reverse: $5'$-SacI+TAAAGTTCTATTTCAAAT-$3'$. PCR products were amplified using high-fidelity Taq with $65°C$ annealing temperature and 2 minutes extension time.

2.2. Immunohistochemistry (IHC). E13.5, E16.5, E18.5 mouse embryos, postnatal day 3, day 5, and day 7 mouse heads were fixed in 10% neutral buffered formalin for 3 days at $4°C$ and processed for paraffin section. Serial sections were used for immunohistochemical analysis for NUMB (abcam ab14140) and activated Notch 1 (abcam ab8925) using the Vectastain ABC Kit (Vector Labs, CA, USA) following the manufacturer's instructions. Sections were imaged using Zeiss microscope.

2.3. Immunocytochemistry (ICC). Human dental pulp stem cells (DPSCs), T4-4 preodontoblasts, and HAT-7 cells were plated on sterilized cover slips placed in 6-well culture dishes. Cells were fixed in 10% neutral buffered formalin at $4°C$ for 2 hours. Cells were permeabilized with 0.5% Triton in PBS for 30 minutes, then blocked in 5% BSA for 1 hour at room temperature, and incubated with primary antibody NUMB (abcam ab14140) and activated Notch 1 (abcam ab8925) overnight at $4°C$. TRITC-conjugated goat anti-rabbit secondary antibody and FITC-conjugated goat anti-rabbit secondary antibody were utilized for the visualization of the proteins of interest. Cells were counterstained

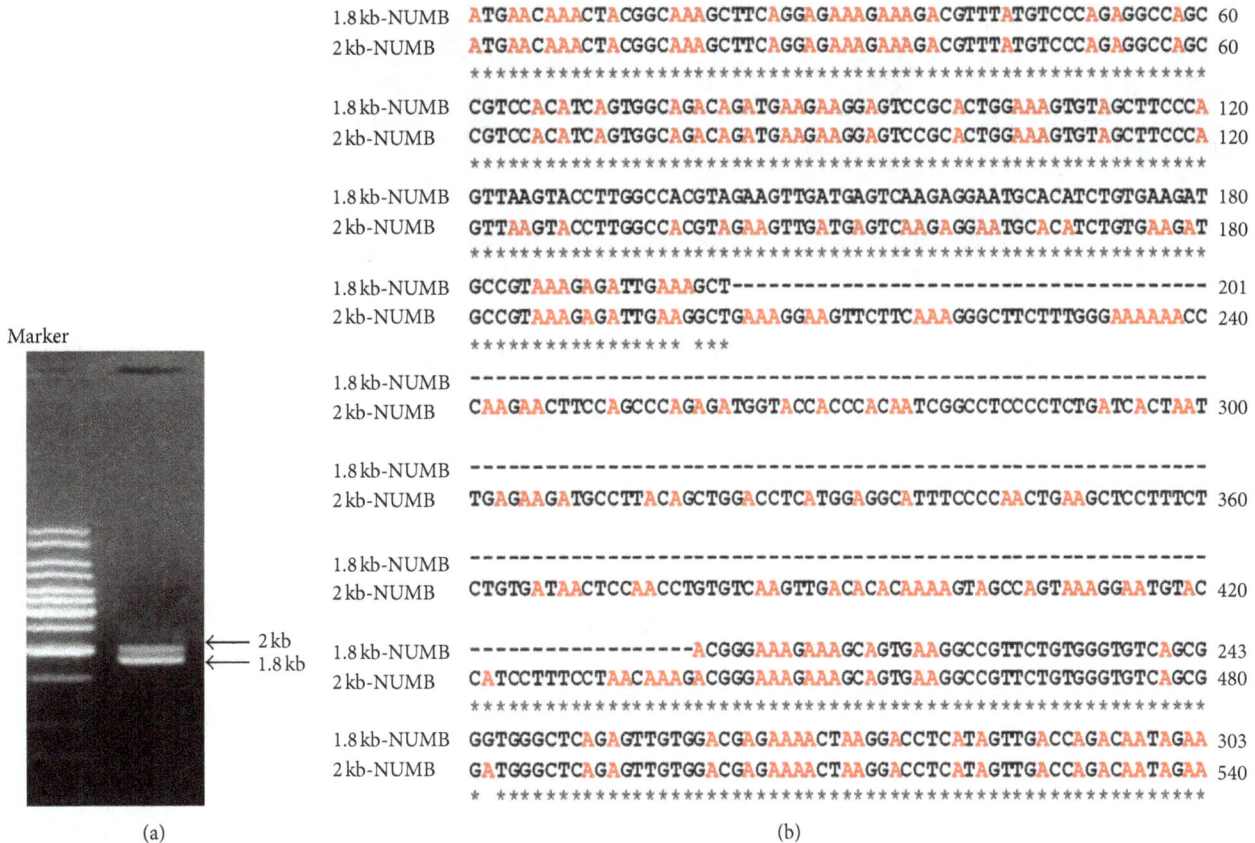

Marker

2 kb
1.8 kb

(a)

1.8 kb-NUMB	ATGAACAAACTACGGCAAAGCTTCAGGAGAAAGAAAGACGTTTATGTCCCAGAGGCCAGC	60
2 kb-NUMB	ATGAACAAACTACGGCAAAGCTTCAGGAGAAAGAAAGACGTTTATGTCCCAGAGGCCAGC	60
	**	
1.8 kb-NUMB	CGTCCACATCAGTGGCAGACAGATGAAGAAGGAGTCCGCACTGGAAAGTGTAGCTTCCCA	120
2 kb-NUMB	CGTCCACATCAGTGGCAGACAGATGAAGAAGGAGTCCGCACTGGAAAGTGTAGCTTCCCA	120
	**	
1.8 kb-NUMB	GTTAAGTACCTTGGCCACGTAGAAGTTGATGAGTCAAGAGGAATGCACATCTGTGAAGAT	180
2 kb-NUMB	GTTAAGTACCTTGGCCACGTAGAAGTTGATGAGTCAAGAGGAATGCACATCTGTGAAGAT	180
	**	
1.8 kb-NUMB	GCCGTAAAGAGATTGAAAGCT---------------------------------------	201
2 kb-NUMB	GCCGTAAAGAGATTGAAGGCTGAAAGGAAGTTCTTCAAAGGGCTTCTTTGGGAAAAAACC	240
	***************** ***	
1.8 kb-NUMB	--	
2 kb-NUMB	CAAGAACTTCCAGCCCAGAGATGGTACCACCCACAATCGGCCTCCCCTCTGATCACTAAT	300
1.8 kb-NUMB	--	
2 kb-NUMB	TGAGAAGATGCCTTACAGCTGGACCTCATGGAGGCATTTCCCCAACTGAAGCTCCTTTCT	360
1.8 kb-NUMB	--	
2 kb-NUMB	CTGTGATAACTCCAACCTGTGTCAAGTTGACACACAAAAGTAGCCAGTAAAGGAATGTAC	420
1.8 kb-NUMB	-----------------ACGGGAAAGAAAGCAGTGAAGGCCGTTCTGTGGGTGTCAGCG	243
2 kb-NUMB	CATCCTTTCCTAACAAAGACGGGAAAGAAAGCAGTGAAGGCCGTTCTGTGGGTGTCAGCG	480
	*************************************	
1.8 kb-NUMB	GGTGGGCTCAGAGTTGTGGACGAGAAAACTAAGGACCTCATAGTTGACCAGACAATAGAA	303
2 kb-NUMB	GATGGGCTCAGAGTTGTGGACGAGAAAACTAAGGACCTCATAGTTGACCAGACAATAGAA	540
	* ***	

(b)

FIGURE 1: (a) RT-PCR analysis of NUMB isoforms in day 7 dental pulp cells. RT-PCR analysis of mRNA extracted from day 7 dental pulp culture shows two PCR bands corresponding to 2 kb and 1.8 kb obtained using primers designed to amplify the longest NUMB reading frame. (b) The alignment of the sequence obtained from 1.8 kb and the 2 kb PCR fragments. The 2 kb PCR fragment contains 236 bp insertion in the phosphotyrosine binding domain (PTB) 201 bp downstream of the initiation start site.

with nuclei-specific fluorescent stain, DAPI (Vectashield Mounting Medium, CA, USA), and imaged using confocal microscopy with corresponding fluorescent channel.

2.4. Generation of Plasma Membrane Patches. Plasma membrane patches from HAT-7 cells were prepared as described previously [27, 28]. The patches were fixed in 4% paraformaldehyde and immunostained with anti-NUMB antibody.

2.5. Total Protein Lysate. Total protein lysate was prepared using RIPA buffer (50 mM Tris, 150 mM NaCl, 0.1% SDS, 0.5% Na. Deoxycholate, and 1% NP40) in the presence of proteinase inhibitor cocktail (Sigma, MO, USA). For protein extraction from cell lines, the cells were cultured in 100 mm petri dish and washed with ice-cold PBS at the time of harvest. Cells were lysed with 700 μL of RIPA buffer, scraped off from the culture dish, vortexed for 10 seconds, and incubated on ice for 10 minutes. The supernatant was collected after centrifugation at 14 k rpm at 4°C. For protein extraction from whole tooth buds, day 4 postnatal mouse tooth buds were dissected under the microscope. The tooth buds were frozen with liquid nitrogen, ground in a mortar and pestle with the addition of the RIPA buffer supplemented with proteinase inhibitor cocktail, and processed as above.

2.6. Western Blot Analysis. Total protein extracted from cell cultures and embryonic tooth germs was quantified using Bradford protein assay with an MBA 2000 spectrometer (Perkin Elmer, MA, USA). A total of 35–40 μg of protein were loaded on the 10% SDS polyacrylamide gel for separation. Well-separated proteins were transferred onto a nitrocellulose membrane at 4°C. The nitrocellulose membrane was then blocked with 5% nonfat milk in PBS for 1 hour and then incubated with primary antibody in 3% BSA/PBS overnight (NUMB: abcam ab14140; 1/1500 dilution; activated Notch 1: abcam; ab8925 1/1600; tubulin 1/3000) under constant agitation at 4°C. The blot was incubated with the secondary antibody (goat antirabbit conjugated with HRP 1/3000) at room temperature for 1 hour. Chemiluminescent Western Blot Substrate (Thermo Scientific Pierce ECL, IL, USA) was utilized for developing, and the blot was exposed on CL-XPosure Film (Thermo Scientific, IL, USA).

2.7. Culturing HAT-7 Cells. HAT-7 cells [29] were maintained in DMEM/F-12 medium with addition of 10% FBS and 50 U/mL of penicillin and streptomycin. To overexpress NUMB, HAT-7 cells were plated in 6-well culture dish 24 hour in advance with 90% confluence in antibiotic-free medium. Plasmids including pSR-GFP/Neo (gifts from Wang et al. [30]); GFP-NUMB (NUMB isoform 4) (gifts from

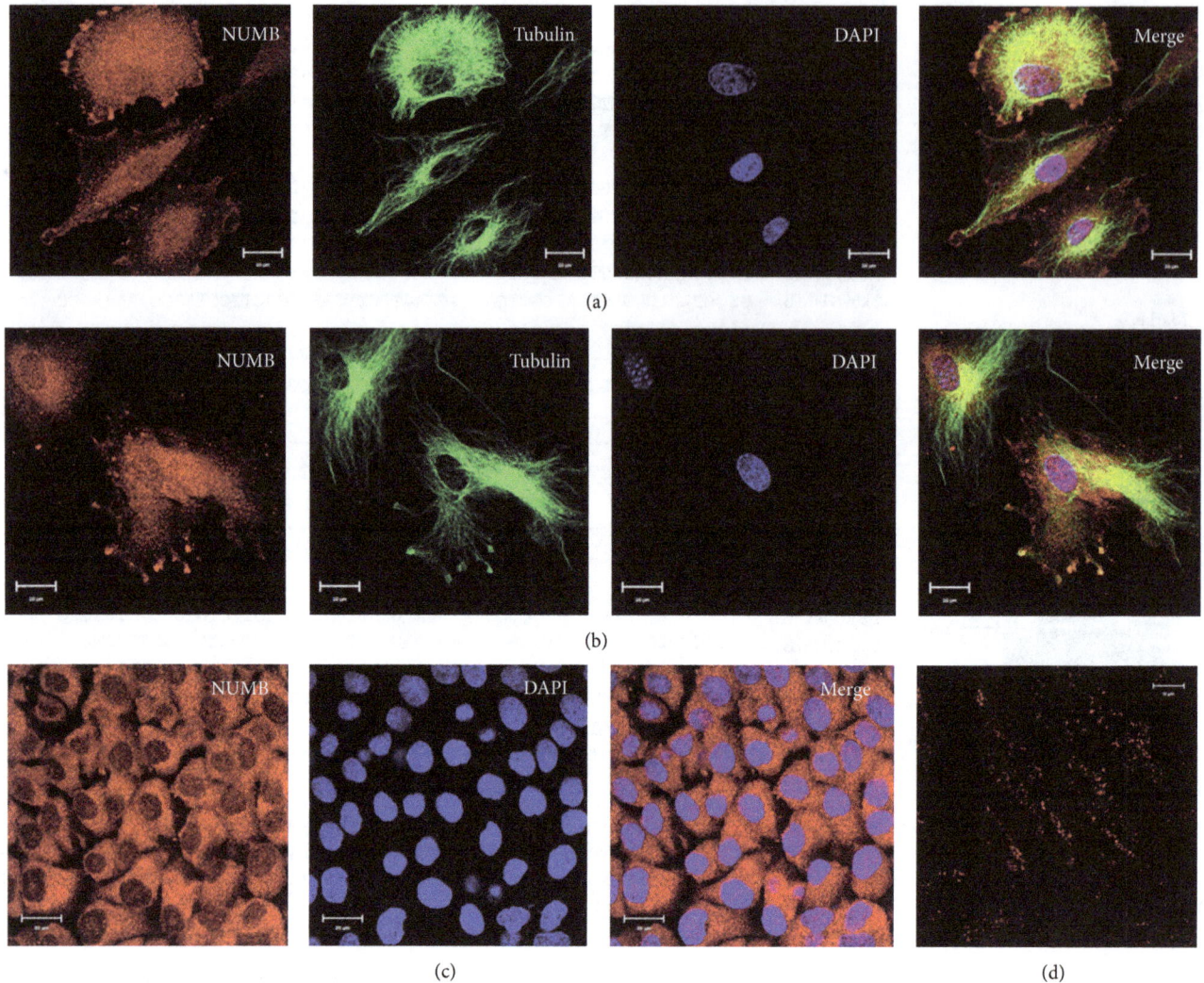

FIGURE 2: Localization of NUMB in various odontogenic cells: immunocytochemistry staining of NUMB expression in T4-4 odontoblast cells (a), human dental pulp stem cell (b), and preameloblast cells (c). Membrane patches isolated from preameloblast HAT-7 cells stained using NUMB antibody (d).

Nishimura and Kaibuchi [31]) were reconstituted in serum-free and antibiotics-free DMEM/F-12 medium for transfection by Lipofectamine 2000 (Invitrogen). Transfection medium was replaced with fresh culture medium in 24 hours. Cells were allowed to recover for 48 hours before being placed in selection medium. The selection medium contains 1 to 100 dilution of the G418 neomycin sulfate (Sigma) stock solution (100 mg/mL) in DMEM/F-12 medium. The G418 concentration was determined from the killing curve where 50% of the cells were killed in 48 hours. Cells were cultured in selection medium for 6 weeks before any experiments were carried out in order to completely remove the nontransfected or transiently transfected cells. Cells were maintained in the selection medium hereafter. For the PCR-based Notch 1 superarray signaling pathway analysis, transiently transfected cells for 48 hours were utilized.

2.8. Functional Analysis of NUMB in Notch Signaling Using Pathway-Specific PCR-Based Superarray. HAT-7 cells were transiently transfected with NUMB-GFP expression vector and control Neomycin-GFP expression vector. RNA was extracted using Qiagen RNA easy kit following the manufacturer's protocol. cDNA was generated using RT First-Strand cDNA Synthesis Kit (C-03 Qiagen, CA, USA) from 4 μg of RNA. Genomic DNA was eliminated by on column DNase digestion during RNA extraction and before reverse transcription. RT^2 Profiler PCR Array System was used for analyzing the mouse Notch 1 signaling pathway (PAMM-059C).

3. Results

3.1. Expression of NUMB Transcripts in Dental Tissue. To demonstrate the presence of NUMB transcripts in dental pulp cells, we performed RT-PCR using cDNA generated from day 7 primary dental pulp cultures. Primers were designed to amplify the longest NUMB open reading frame. Two specific PCR fragments were amplified (Figure 1(a)). Sequence

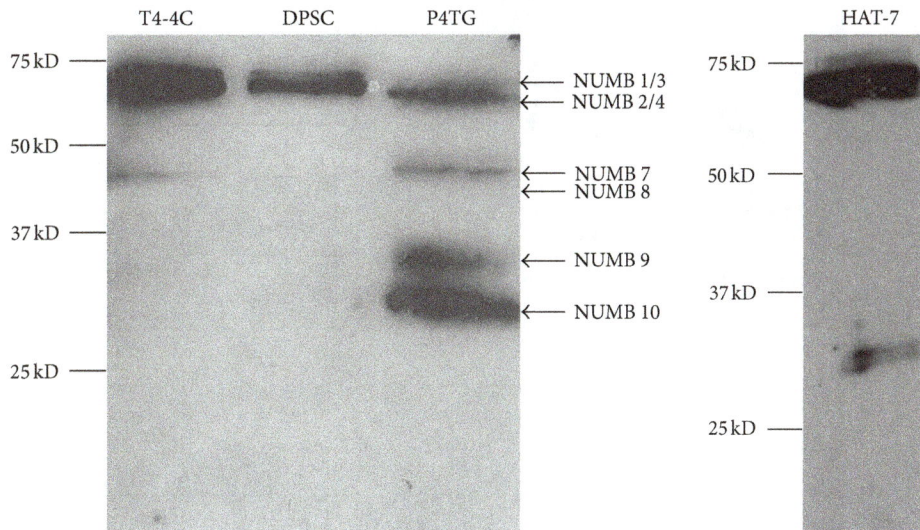

FIGURE 3: Western blot analysis of NUMB isoforms in odontogenic cells: total protein lysates were obtained from odontoblasts (T4-4 cells); dental pulp stem cells (DPCs); preameloblasts (HAT-7); tooth germs isolated from postnatal day 4 molars (P4TG). Notice that NUMB isoforms are differentially expressed in these tissues.

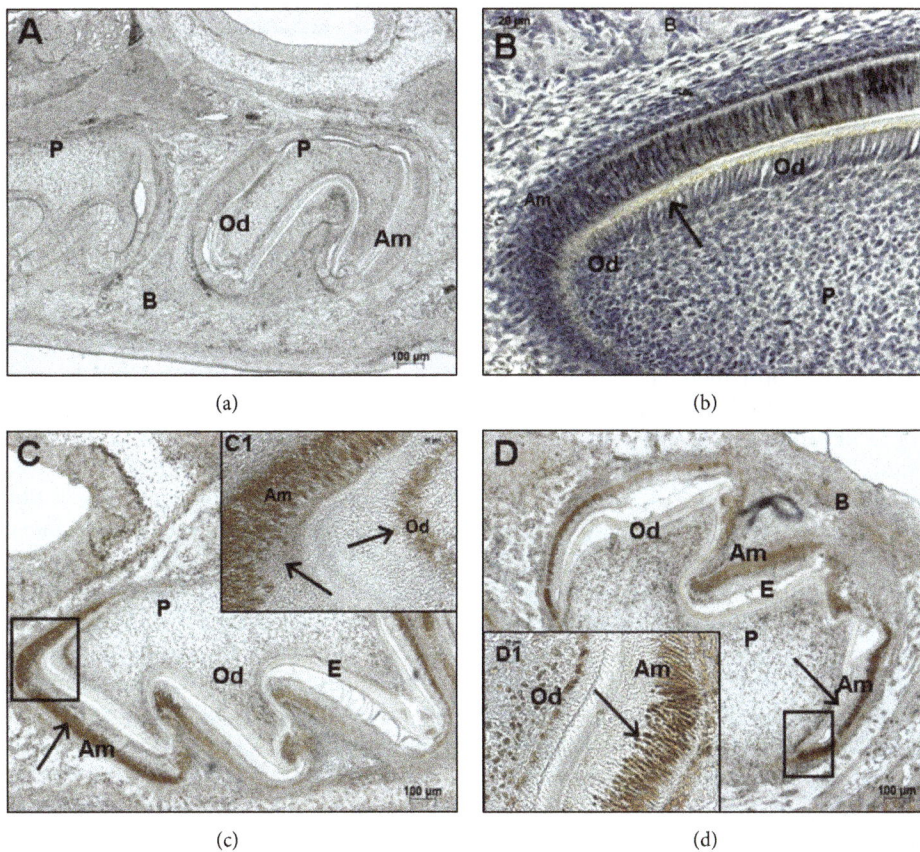

FIGURE 4: Localization of NUMB in odontogenic tissues at postnatal day 3: localization of NUMB in P3 incisors (B) and molars (C and D). NUMB can be detected in the ameloblasts (Am); odontoblasts (Od); ameloblast progenitors (cervical loop); in the dental pulp cells (P) in close vicinity to the odontoblasts. (D1) shows the immature ameloblasts, immature odontoblasts, and dental pulp cells from incisor and molar, respectively. (a) is the negative control. Scale bar represents 100 μm for (a), (c), and (d) and 20 μm for C1 and D1, respectively.

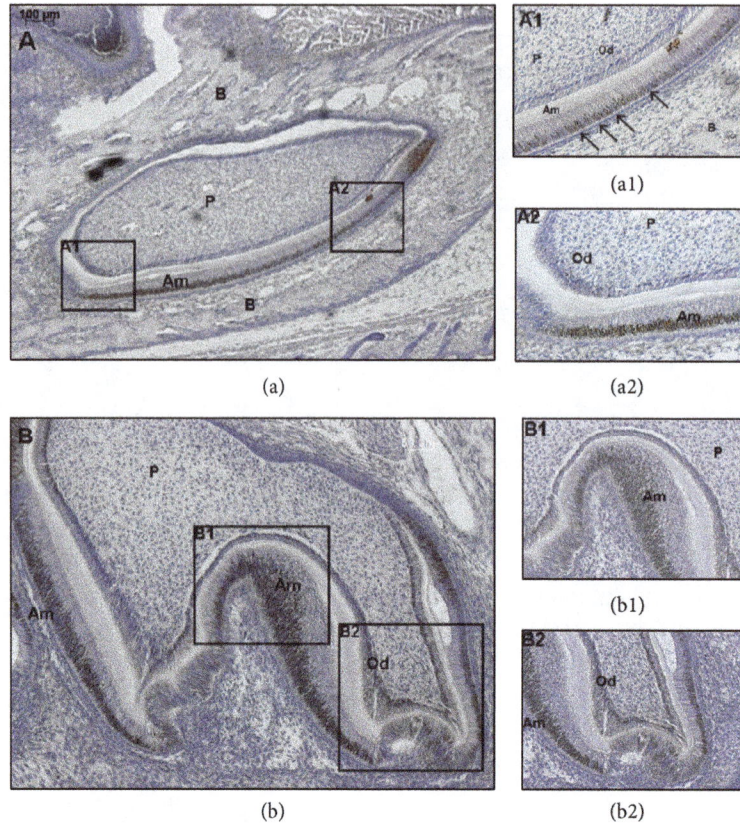

FIGURE 5: Localization of NUMB in odontogenic tissues at postnatal day 5: NUMB expression can be detected in the ameloblasts and odontoblasts. Notice that in (a1), NUMB is positively stained in some ameloblasts but negatively stained in the adjacent cells (indicated by arrows). Scale bar represents 100 μm for (a); 50 μm for (b); 20 μm for (a1), (a2), and (b1), (b2), respectively.

analysis indicated that the 1.8 kb PCR fragment is consistent with the mRNA sequence of NUMB isoform 4. The 2 kb PCR fragment contains 236 bp insertion in the phosphotyrosine binding domain (PTB) 201 bp downstream of the initiation codon (Figure 1(b)). However, protein prediction analysis from the 2 kb DNA fragment did not lead to a longer protein product but a protein of 100 amino acids that contained an early termination codon.

3.2. Expression of NUMB in Dental Cell Lines.

To investigate the subcellular localization of NUMB, immunocytochemical analysis was performed on human dental pulp stem cells (DPSCs), T4-4 preodontoblasts, and HAT-7 preameloblast cells (Figure 2). NUMB was clearly expressed in these 3 cell types. Specifically, NUMB was localized on the cell membrane, within the cytoplasm, and in the nucleus in only the DPSCs. In order to confirm the localization of NUMB on the cell membrane, plasma membrane patches from HAT-7 cells were used in immunostaining. The positively stained cell membrane patches by NUMB antibody confirmed the presence of NUMB on the cell membrane of HAT-7 cells (Figure 2(d)).

In order to identify the NUMB isoforms that are present in the dental tissue, Western blot analysis was performed using total protein lysates prepared from postnatal day 4 tooth germ and from the dental cell lines (Figure 3). All

the high molecular weight NUMB proteins ~72 KD and ~66 KDa were detected in all the dental tissues (proteins that have 1 KD difference were unable to be separated). To our surprise, multiple low molecular weight proteins located between 25 KD and 50 KD were also detected in the tooth germ as well as in HAT-7 cells. Since there are no published reports on the proteolytic cleavage of NUMB, we speculated that these could be due to NUMB isoforms expressed in dental tissues that have not yet been identified. While this work was in progress, two new isoforms, namely, NUMB 5 and NUMB 6 were published [32]. These proteins were described to be rare and were transiently expressed in cancer cells. The novel isoforms identified by Karaczyn et al. [32] appeared to have higher molecular weight than the proteins we identified. For the sake of simplicity, in Figure 3, we have described the 37–50 KD proteins as NUMB 7 and NUMB 8 and the 25–37 KD proteins as NUMB 9 and NUMB10. This needs to be further validated by mass spectrometry or protein sequencing analysis and determine if it is similar to the NUMB 7 and 8 isoform reported by Haider et al. in human placenta [11]. The ameloblast cell line, HAT-7 cells, expresses high levels of NUMB 1/2 and NUMB 3/4, as well as NUMB 9/10 but is deficient in the expression of NUMB 7/8. The dental pulp cells show very low levels of low molecular weight NUMB isoform expression. Among the high molecular weight NUMB isoforms, the NUMB 2/4

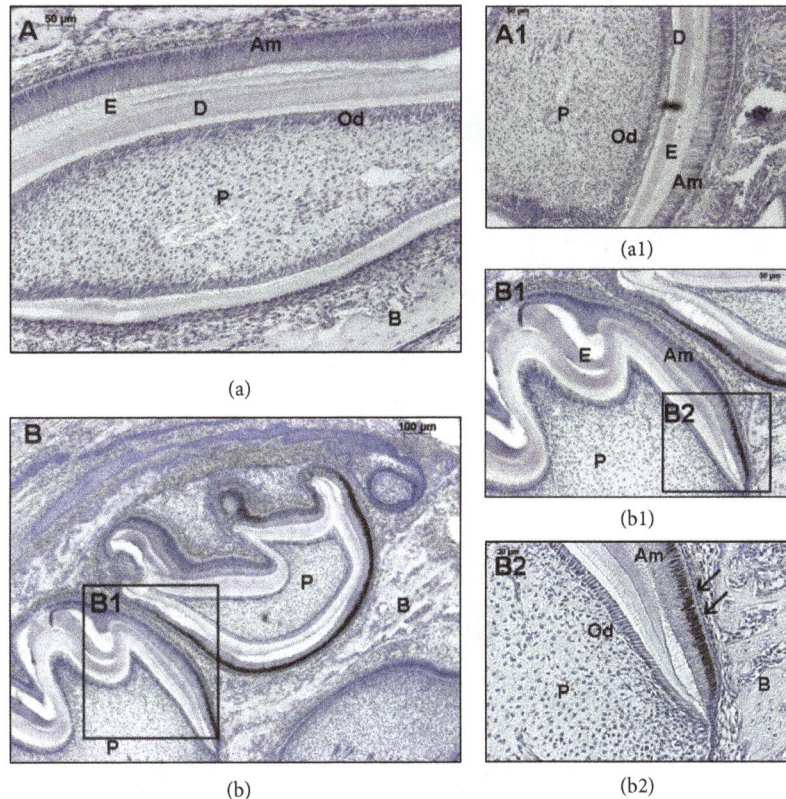

FIGURE 6: Localization of NUMB in odontogenic tissues at postnatal day 7: NUMB expression is more defined in the immature ameloblasts, and very low expression levels can be detected in the immature odontoblasts. Scale bar represents 50 μm for (a), 100 μm for (b), and 50 and 20 μm for (a1), (b1), and (b2), respectively.

appeared to be the major isoforms in the dental tissue. These data suggest that there is differential expression of NUMB isoforms in dental tissues.

3.3. Expression of NUMB in Developing Tooth Germs.

NUMB expression patterns were analyzed in the developing tooth germs from embryonic stage 13.5 to postnatal day 7 by immunohistochemistry. At E13.5, E16.5, and E18.5, there was no detectable NUMB expression in the developing tooth germs (data not shown). However, during postnatal tooth development, there was strong NUMB expression. Mouse incisors have enamel and dentin present on the labial surface, and only dentin on the lingual surface [33]. In P3 mouse incisor, NUMB protein was strongly expressed in the ameloblasts on the labial surface, in the cervical loop, in the odontoblasts at both labial and lingual surfaces, and in the dental pulp cells in close vicinity to the odontoblasts. In the P3 molar, NUMB expression can be detected in the ameloblasts, immature odontoblasts, and the dental pulp cells adjacent to the odontoblasts (Figure 4). In the P5 incisor, NUMB is localized in the ameloblasts and immature odontoblasts, and the expression is less intense compared to the P3 dental tissues. Also, there was no expression of NUMB in the dental pulp cells (Figure 5). In P7 dental tissues, the NUMB expression is more specifically limited to the stem cells of the stratum intermedium (Figure 6). Overall, in the postnatal dental tissues, NUMB is expressed in the ameloblast progenitors in the cervical loop of the incisor, immature ameloblasts and odontoblasts, and in the dental pulp cells in the vicinity of the odontoblasts. Note that some ameloblasts in P5 and P7 express NUMB, but their neighboring cells do not, indicating that NUMB is asymmetrically distributed in the daughter cells.

3.4. NUMB Downregulates Activated Notch 1.

The *in vivo* temporal and spatial expression pattern of NUMB in the developing tooth germ suggested that NUMB may play a role in regulating ameloblast differentiation. Our study shows that NUMB has a defined expression pattern in the ameloblast lineage cells, more specifically in the ameloblast progenitors and preamelobasts. Several publications have reported that Notch signaling is critical in ameloblast differentiation, and NUMB is involved in regulating Notch1 proteolysis. Therefore, we investigated the relationship of NUMB and Notch 1 in preameloblasts.

We first established that NUMB and Notch 1 colocalized in the same cell types. Activated Notch 1 was found to be strongly expressed in ameloblasts, odontoblasts, and dental pulp cells (see Supplementary Data 1 available online at http://dx.doi.org/10.1155/2013/182965). In order to study the functional relationship between NUMB and activated Notch 1, we generated NUMB overexpressing HAT-7 cells (Figure 7(a)). The activated Notch 1 expression pattern was evaluated by immunocytochemical analysis. Confocal

(a)

(b)

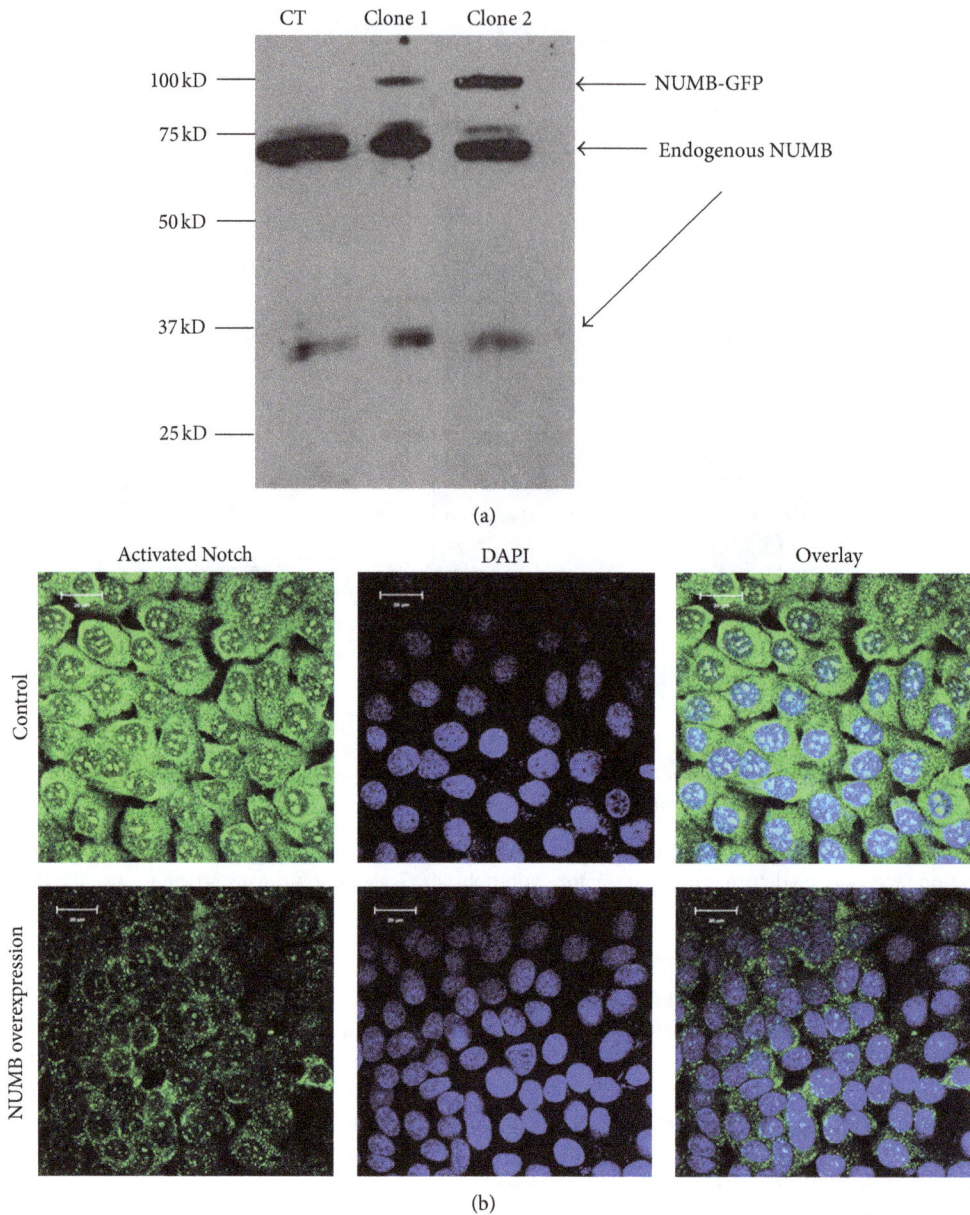

FIGURE 7: Overexpression of NUMB-GFP in HAT-7 cells: (a) Western blot analysis to detect overexpression of NUMB-GFP in HAT-7 cells. Total proteins were isolated from 2 clones, namely, clone 1 and 2. (b) Immunohistochemical analysis detected that Notch1 protein expression is dramatically reduced in the NUMB overexpressing cell line (stable cell line clone 2 was used). The green fluorescence represents activated Notch1 protein.

microscopy clearly indicated that the NUMB overexpressing HAT-7 cells had dramatically reduced expression of activated Notch 1 represented by the expression of green fluorescence (Figure 7(b)).

Next, we investigated the key players in the Notch 1 signaling pathway using a PCR-based array. The RT^2 Profiler PCR Array profiles the expression of 84 genes involved in Notch signaling. Results in Figure 9 show that NUMB mRNA is 7.5 times upregulated in the NUMB overexpressing cells compared to the control vector transfected cells. Majority of the genes in the Notch 1 signaling pathway were downregulated. Noteworthy among them were Shh, Fos, Fzd2, and PParg with P values < 0.05 (Figure 8).

4. Discussion

While we were isolating the NUMB isoforms expressed in dental pulp cells using primers designed to amplify the longest reading frame, we repeatedly obtained two PCR fragments. The 1.8 kb fragment is consistent with published NUMB isoform 4 cDNA sequence, while the 2 kb PCR fragment contained a 237 bp insertion in the PTB domain. The predicted protein sequence indicated an early termination codon which allows only 103 amino acids being translated. The NUMB antibody utilized in this study was raised against the NUMB C-terminal polypeptide (the synthetic peptide derived from residues 600 to the C-terminus of human NUMB). Therefore, isoforms that end upstream of this

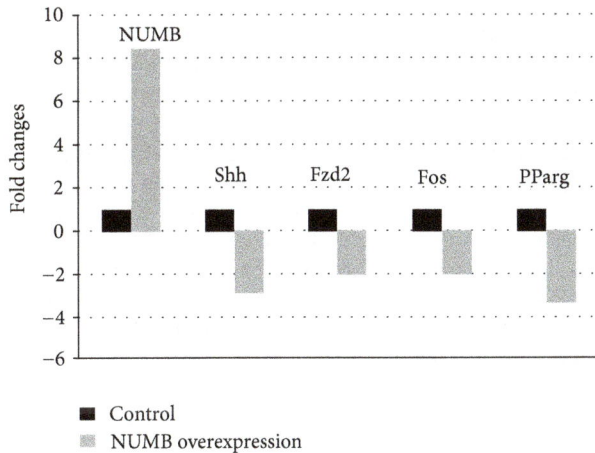

FIGURE 8: Notch signaling pathway PCR array. Genes from the Notch1 signaling pathway that were affected by NUMB overexpression.

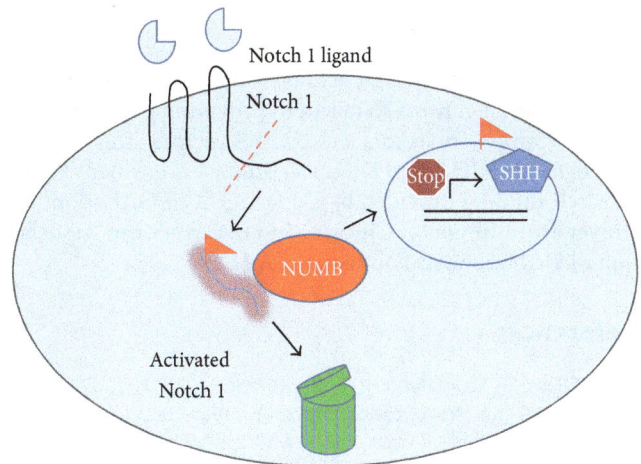

FIGURE 9: Hypothetical model. NUMB regulates stem cell properties at two levels. NUMB downregulates the expression of the activated Notch 1 protein level in the progenitor cells at the cervical loop and thereby inhibits the Shh expression in the adjacent preameloblasts. Thus, NUMB influences the expression of downstream genes such as Notch1 and Shh and thus performs critical role in ameloblast differentiation.

polypeptide were not detected. The total protein lysate from the whole tooth germ contained all the detectable NUMB isoforms. The high molecular weight NUMB isoform which is the major published isoform is expressed in all dental tissues. 40 KDa isoform was detected in the odontoblasts, and the low molecular weight NUMB isoform (30 KDa) was predominantly expressed in the ameloblasts. While this paper was in preparation, 2 new NUMB isoforms, namely, NUMB 5 and 6 were reported and identified in the human cancer cells [32]. These two new isoforms are reported to be above 50 KD with 10 KD difference. They were characterized as being expressed rarely and transiently in these cancer cells. Recently, NUMB 7, 8, and 9 isoforms were identified by RT-PCR cloning in the human extravillous trophoblast [11]. In this study, four possible NUMB isoforms were detected in dental tissues. Further analysis is required to confirm the protein sequences. The differentially expressed NUMB isoforms in various dental tissues and possibly at different stages of tooth development suggest that NUMB might play specific functional role during tooth development.

Subcellular localization of NUMB in various odontogenic cell lines showed its localization in the cytoplasm, nucleus, and the cell membrane. Nuclear NUMB is particularly interesting, and its functional role in the nucleus has yet to be determined. Immunohistochemical analysis showed prominent NUMB expression during early tooth development. No NUMB expression was detected during embryonic development of the tooth. Strong NUMB expression was detected in the early postnatal development and was mainly localized in immature ameloblasts, preamelobasts, cervical loop, immature odontoblasts, and dental pulp cells.

As NUMB is regulated by Notch 1, we therefore examined the effect of overexpression of NUMB on Notch expression in preameloblasts. Results from this study showed that overexpression of NUMB dramatically reduced the expression of activated Notch 1 protein expression in HAT-7 cells. Further, we examined if the reduced expression of Notch1 had an effect on downstream signaling events. Results from this study showed that with NUMB overexpression (7.5-fold increase), several genes in the Notch signaling pathway were downregulated. Most notable was the downregulation of sonic hedgehog.

The preameloblast HAT-7 cells used in this study were derived from the rat incisor cervical loop and its proximity [29]. This region contains stem cell population as well as immature preameloblasts. Based on our findings, we suggest that NUMB can have multiple effects on ameloblast differentiation by exerting its effect on both the stem cells and the preameloblasts. In the stem cells at the cervical loop, NUMB is highly expressed, and this downregulates Notch 1 which in turn inhibits Shh mRNA expression. Thus, the expression of NUMB is responsible for the maintenance of the progenitor cells in the cervical loop. During development, NUMB regulates the differentiation of preameloblasts into ameloblasts. In day 3, NUMB is expressed in the preameloblasts, but by day 7 there is less NUMB expression and thereby relieving its influence on suppressing the expression of Shh. This differential expression of NUMB and Shh in preameloblasts might be necessary for its differentiation and polarization. This observation corroborates well with published reports that demonstrate the critical role that NUMB plays in regulating cell polarization and differentiation [23–25].

Based on our findings, we propose a hypothetical model as shown in Figure 9. NUMB regulates the putative stem cell number and property by inhibiting Notch 1 signaling and thereby downregulating the expression of Shh. At the same time, NUMB can influence the differentiation of preameloblasts to ameloblasts. The signaling mechanism involved in this differentiation cascade will be pursued in the future.

Conflict of Interests

The authors have no conflict of interests with any of the companies/manufactories whose products were utilized in this work.

Acknowledgments

The authors would like to acknowledge the help from Dr. Shawn Li and Dr. Kozo Kaibuchi in providing them with the NUMB expression vectors and control vectors. This work is supported by NIH Grant DE 11657, Brodie Endowment Fund, research funding provided by Department of Orthodontics, University of Illinois at Chicago, and the American Association of Orthodontists Foundation (AAOF).

References

[1] S. Pece, S. Confalonieri, P. R. Romano, and P. P. Di Fiore, "NUMB-ing down cancer by more than just a NOTCH," *Biochimica et Biophysica Acta*, vol. 1815, no. 1, pp. 26–43, 2011.

[2] J. Hao, K. Narayanan, T. Muni, A. Ramachandran, and A. George, "Dentin matrix protein 4, a novel secretory calcium-binding protein that modulates odontoblast differentiation," *The Journal of Biological Chemistry*, vol. 282, no. 21, pp. 15357–15365, 2007.

[3] T. Uemura, S. Shepherd, L. Ackerman, L. Y. Jan, and Y. N. Jan, "Numb, a gene required in determination of cell fate during sensory organ formation in Drosophila embryos," *Cell*, vol. 58, no. 2, pp. 349–360, 1989.

[4] E. P. Spana, C. Kopczynski, C. S. Goodman, and C. Q. Doe, "Asymmetric localization of numb autonomously determines sibling neuron identity in the Drosophila CNS," *Development*, vol. 121, no. 11, pp. 3489–3494, 1995.

[5] W. Zhong, M. Jiang, M. D. Schonemann et al., "Mouse numb is an essential gene involved in cortical neurogenesis," *Proceedings of the National Academy of Sciences of the United States of America*, vol. 97, no. 12, pp. 6844–6849, 2000.

[6] A. Gulino, L. Di Marcotullio, and I. Screpanti, "The multiple functions of Numb," *Experimental Cell Research*, vol. 316, no. 6, pp. 900–906, 2010.

[7] S. E. Dho, M. B. French, S. A. Woods, and C. J. McGlade, "Characterization of four mammalian numb protein isoforms. Identification of cytoplasmic and membrane-associated variants of the phosphotyrosine binding domain," *The Journal of Biological Chemistry*, vol. 274, no. 46, pp. 33097–33104, 1999.

[8] J. M. Verdi, A. Bashirullah, D. E. Goldhawk et al., "Distinct human NUMB isoforms regulate differentiation vs. proliferation in the neuronal lineage," *Proceedings of the National Academy of Sciences of the United States of America*, vol. 96, no. 18, pp. 10472–10476, 1999.

[9] P. H. Petersen, K. Zou, S. Krauss, and W. Zhong, "Continuing role for mouse Numb and Numbl in maintaining progenitor cells during cortical neurogenesis," *Nature Neuroscience*, vol. 7, no. 8, pp. 803–811, 2004.

[10] Y. Zhou, J. B. Atkins, S. B. Rompani et al., "The mammalian Golgi regulates numb signaling in asymmetric cell division by releasing ACBD3 during mitosis," *Cell*, vol. 129, no. 1, pp. 163–178, 2007.

[11] M. Haider, Q. Qiu, M. Bani-Yaghoub, B. K. Tsang, and A. Gruslin, "Characterization and role of NUMB in the human extravillous trophoblast," *Placenta*, vol. 32, no. 6, pp. 441–449, 2011.

[12] T. A. Mitsiadis, A. S. Tucker, C. De Bari, M. T. Cobourne, and D. P. Rice, "A regulatory relationship between Tbx1 and FGF signaling during tooth morphogenesis and ameloblast lineage determination," *Developmental Biology*, vol. 320, no. 1, pp. 39–48, 2008.

[13] C. E. Smith and H. Warshawsky, "Cellular renewal in the enamel organ and the odontoblast layer of the rat incisor as followed by radioautography using 3H thymidine," *Anatomical Record*, vol. 183, no. 4, pp. 523–561, 1975.

[14] T. A. Mitsiadis, E. Hirsinger, U. Lendahl, and C. Goridis, "Delta-Notch signaling in odontogenesis: correlation with cytodifferentiation and evidence for feedback regulation," *Developmental Biology*, vol. 204, no. 2, pp. 420–431, 1998.

[15] K. Seidel, C. P. Ahn, D. Lyons et al., "Hedgehog signaling regulates the generation of ameloblast progenitors in the continuously growing mouse incisor," *Development*, vol. 137, no. 22, pp. 3753–3761, 2010.

[16] H. R. Dassule, P. Lewis, M. Bei, R. Maas, and A. P. McMahon, "Sonic hedgehog regulates growth and morphogenesis of the tooth," *Development*, vol. 127, no. 22, pp. 4775–4785, 2000.

[17] J. J. Breunig, J. Silbereis, F. M. Vaccarino, N. Šestan, and P. Rakic, "Notch regulates cell fate and dendrite morphology of newborn neurons in the postnatal dentate gyrus," *Proceedings of the National Academy of Sciences of the United States of America*, vol. 104, no. 51, pp. 20558–20563, 2007.

[18] H. Harada, P. Kettunen, H. Jung, T. Mustonen, Y. A. Wang, and I. Thesleff, "Localization of putative stem cells in dental epithelium and their association with Notch and FGF signaling," *Journal of Cell Biology*, vol. 147, no. 1, pp. 105–120, 1999.

[19] S. Felszeghy, M. Suomalainen, and I. Thesleff, "Notch signalling is required for the survival of epithelial stem cells in the continuously growing mouse incisor," *Differentiation*, vol. 80, no. 4-5, pp. 241–248, 2010.

[20] M. A. McGill and C. J. McGlade, "Mammalian Numb proteins promote Notch1 receptor ubiquitination and degradation of the Notch1 intracellular domain," *The Journal of Biological Chemistry*, vol. 278, no. 25, pp. 23196–23203, 2003.

[21] M. A. McGill, S. E. Dho, G. Weinmaster, and C. J. McGlade, "Numb regulates post-endocytic trafficking and degradation of notch1," *The Journal of Biological Chemistry*, vol. 284, no. 39, pp. 26427–26438, 2009.

[22] L. Di Marcotullio, E. Ferretti, A. Greco et al., "Numb is a suppressor of Hedgehog signalling and targets Gli1 for Itch-dependent ubiquitination," *Nature Cell Biology*, vol. 8, no. 12, pp. 1415–1423, 2006.

[23] T. Ito, H. Y. Kwon, B. Zimdahl et al., "Regulation of myeloid leukaemia by the cell-fate determinant Musashi," *Nature*, vol. 466, no. 7307, pp. 765–768, 2010.

[24] D. Berdnik, T. Török, M. González-Gaitán, and J. A. Knoblich, "The endocytic protein α-adaptin is required for numb-mediated asymmetric cell division in Drosophila," *Developmental Cell*, vol. 3, no. 2, pp. 221–231, 2002.

[25] S. X. Atwood and K. E. Prehoda, "aPKC phosphorylates Miranda to polarize fate determinants during neuroblast asymmetric cell division," *Current Biology*, vol. 19, no. 9, pp. 723–729, 2009.

[26] G. Chapman, L. Liu, C. Sahlgren, C. Dahlqvist, and U. Lendahl, "High levels of Notch signaling down-regulate Numb and Numblike," *Journal of Cell Biology*, vol. 175, no. 4, pp. 535–540, 2006.

[27] R. Stan, W. Gregory Roberts, D. Predescu et al., "Immunoisolation and partial characterization of endothelial plasmalemmal vesicles (caveolae)," *Molecular Biology of the Cell*, vol. 8, no. 4, pp. 595–605, 1997.

[28] S. Ravindran, Q. Gao, A. Ramachandran, S. Blond, S. A. Predescu, and A. George, "Stress chaperone GRP-78 functions

in mineralized matrix formation," *The Journal of Biological Chemistry*, vol. 286, no. 11, pp. 8729–8739, 2011.

[29] S. Kawano, T. Morotomi, T. Toyono et al., "Establishment of dental epithelial cell line (HAT-7) and the cell differentiation dependent on Notch signaling pathway," *Connective Tissue Research*, vol. 43, no. 2-3, pp. 409–412, 2002.

[30] Z. Wang, S. Sandiford, C. Wu, and S. S. Li, "Numb regulates cell-cell adhesion and polarity in response to tyrosine kinase signalling," *EMBO Journal*, vol. 28, no. 16, pp. 2360–2373, 2009.

[31] T. Nishimura and K. Kaibuchi, "Numb controls integrin endo-cytosis for directional cell migration with aPKC and PAR-3," *Developmental Cell*, vol. 13, no. 1, pp. 15–28, 2007.

[32] A. Karaczyn, M. Bani-Yaghoub, R. Tremblay et al., "Two novel human NUMB isoforms provide a potential link between development and cancer," *Neural Development*, vol. 5, no. 1, article 31, 2010.

[33] T. A. Mitsiadis and D. Graf, "Cell fate determination during tooth development and regeneration," *Birth Defects Research C*, vol. 87, no. 3, pp. 199–211, 2009.

10

Urine Bikunin as a Marker of Renal Impairment in Fabry's Disease

Antonio Junior Lepedda,[1] Laura Fancellu,[2] Elisabetta Zinellu,[1] Pierina De Muro,[1] Gabriele Nieddu,[1] Giovanni Andrea Deiana,[2] Piera Canu,[2] Daniela Concolino,[3] Simona Sestito,[3] Marilena Formato,[1] and Gianpietro Sechi[2]

[1] Dipartimento di Scienze Biomediche, University of Sassari, Via Muroni 25, 07100 Sassari, Italy
[2] Dipartimento di Medicina Clinica e Sperimentale, University of Sassari, Viale San Pietro 10, 07100 Sassari, Italy
[3] Unità Operativa di Pediatria Universitaria, Azienda Ospedaliera "Pugliese-Ciaccio", Viale Pio X, 88100 Catanzaro, Italy

Correspondence should be addressed to Antonio Junior Lepedda; ajlepedda@uniss.it and Gianpietro Sechi; gpsechi@uniss.it

Academic Editor: Achilleas D. Theocharis

Fabry's disease is a rare lysosomal storage disorder caused by the deficiency of α-galactosidase A that leads to the accumulation of neutral glycosphingolipids in many organs including kidney, heart, and brain. Since end-stage renal disease represents a major complication of this pathology, the aim of the present work was to evaluate if urinary proteoglycan/glycosaminoglycan excretion could represent a useful marker for monitoring kidney function in these patients at high risk. Quali-quantitative and structural analyses were conducted on plasma and urine from 24 Fabry's patients and 43 control subjects. Patients were sorted for presence and degree of renal impairment (proteinuria/renal damage). Results showed that levels of urine bikunin, also known as urinary trypsin inhibitor (UTI), are significantly higher in patients with renal impairment than in controls. In this respect, no differences were evidenced in plasma chondroitin sulfate isomers level/structure indicating a likely direct kidney involvement. Noteworthy, urine bikunin levels are higher in patients since early symptoms of renal impairment occur (proteinuria). Overall, our findings suggest that urine bikunin level, as well as proteinuria, could represent a useful parameter for monitoring renal function in those patients that do not present any symptoms of renal insufficiency.

1. Introduction

Fabry's disease (FD) is a panethnic, X-linked lysosomal storage disorder due to deficiency of α-galactosidase A [1]. This lysosomal enzyme normally breaks down neutral glycosphingolipids, particularly globotriaosylceramide (Gb3), catalyzing the hydrolytic cleavage of the terminal molecule of galactose. The consequent accumulation of these glycosphingolipids in many cell types and tissues results in several clinical signs and symptoms [1]. The prevalence of Fabry's disease has been estimated to range from 1 in 117,000 to up to 1 in 40,000, but it might be much higher since it is likely that many patients are not identified, because of either the nonspecificity of clinical features or the scarce suspicion of the clinician for the disease [1]. The α-galactosidase A gene (GLA-gene) is located on the long arm of chromosome X in position Xq22, and it has recently been sequenced [1]. More than 400 mutations have been identified so far. Depending on the type of mutation there may be different clinical forms of the disease. In particular, GLA-gene mutations resulting in a total absence of α-galactosidase A activity usually lead to a more severe form of FD [1]. Disease manifestations usually start in childhood, with intermittent acroparesthesias, sometimes associated with episodic fever, hypohidrosis, gastrointestinal symptoms, typical vascular skin lesions named angiokeratomas, and corneal opacities [2]. After the 3rd decade of life the progression of the pathology frequently leads to renal damage [3], cardiac manifestations, and high propensity to develop brain ischemic stroke, resulting in decreased life expectancy [4]. In particular, end-stage renal disease, with proteinuria and progressive renal failure, is a major cause of morbidity and mortality in FD. Renal damage seems mainly to be caused by diffuse deposition of glycosphingolipids in glomeruli, tubular system, and vasculature. In routine clinical practice, general proteinuria and microalbuminuria are considered the best biomarkers of FD nephropathy [3], although, recently, many new markers, including Gb3 or specific proteins such as N-acetyl-β-D-glucosaminidase and cystatin C, have been suggested to improve decision making [5]. However, the efficacy of all of the biomarkers currently in use for Fabry's nephropathy remains uncertain.

Several studies evidenced variations in plasma/urine glycosaminoglycans in physiological and pathological conditions. Glycosaminoglycans (GAGs) are linear and complex polysaccharides, composed of a variable number of repeating disaccharide units, each containing a hexuronic acid glycosidically linked to a hexosamine residue. GAGs have been found in many tissues and in biological fluids such as blood, plasma and urine [6]. Plasma GAGs represent components of intact proteoglycans (PGs) mainly of hepatic and endothelial origin, secreted into blood as well as products of tissue PG degradation. Chondroitin sulfate (CS), the main GAG type in plasma, is derived from both the cell surfaces and the extracellular matrix. A portion of CS is covalently bound to a protein core to form bikunin that is principally synthesized and secreted by the liver [7]. CS chains are short, consisting of 12–18 disaccharides units, and present a charge density of about 30–40% as previously reported [8, 9]. Most of circulating bikunin is present as the light chain of Inter-α-Inhibitor (IαI) family molecules [10]. Bikunin is a serine protease inhibitor and it occurs in plasma as well as in many tissues [10]. It is also excreted in urine and referred to as urinary trypsin inhibitor (UTI). In urine, GAGs consist mainly of heparan sulfate (HS), CS, and, in negligible quantity, dermatan sulfate [6]. In the general population, bikunin has been reported to occur at higher levels in various pathological conditions exhibiting chronic inflammation, including cancer [11], chronic glomerulonephritis [12, 13], kidney transplantation [14], type 1 diabetes [15], and systemic lupus erythematosus [16]. Moreover, we reported variations of urine bikunin levels during the physiological menstrual cycle in fertile women [17]. It has also been suggested that bikunin may be a useful marker for renal damage [18], liver disease [19], and brain contusion [20], suggesting a potential application in patients with FD. Since kidney damage is a frequent and severe complication of FD, we thought of evaluating plasma and urine GAGs levels in this pathology to assess if their urinary excretion could represent a useful early marker of kidney impairment in patients with Fabry's disease.

2. Methods

2.1. Samples.
Analyses were conducted on fasting blood-plasma and first-morning urine samples from 24 Fabry's patients of both sexes, aged from 20 to 61 years, and 43 age- and gender-matched healthy controls. All FD patients were diagnosed by identifying the mutation in the GLA-gene and showed reduced activity of α-galactosidase A enzyme in plasma (Table 1). Among patients, 13 did not present evidence of chronic renal damage (NRD); the other 11, instead,

presented, for more than 3 months, pathological abnormalities indicative of chronic renal damage (RD), including abnormalities in blood or urine tests (i.e., serum creatinine; proteinuria/microalbuminuria; and abnormalities in urine sediment) or imaging studies (e.g., renal ultrasound). More in detail, 5 RD patients presented with only proteinuria, whereas 6 presented with overt renal damage. In both FD patients and controls the serum levels of three classic markers of inflammation such as erythrocyte sedimentation rate (ESR) (by means of stopped flow capillary microspectrophotometry), C-reactive protein (CRP) (by means of chemiluminescence assay), and α1- and α2-globulins (by means of capillary electrophoresis) were evaluated. Informed consent was obtained before enrolment. Institutional review board approval was obtained. The study was conducted in accordance with the ethical principles of the current Declaration of Helsinki.

2.2. Plasma CS Isomers Analysis.
GAGs isolation was performed by a microanalytic preparative method, as previously described [8]. Briefly, 500 μL of plasma samples was subjected to proteolytic treatment with papain. Plasma CS isomers were isolated by anion exchange chromatography (DEAE-Sephacel) and precipitated with 5 volumes of ethanol at $-20°C$ for 24 h. Subsequently, they were subjected to depolymerization by using chondroitin ABC lyase (0.1U per 100 μg hexuronic acid), and the unsaturated disaccharides were derivatized with 12.5 mmol/L 2-aminoacridone (AMAC).

Separation of CS-derived unsaturated disaccharides (ΔDi) was performed by fluorophore-assisted carbohydrate electrophoresis (FACE). Images were acquired by means of Gel Doc XR system (Bio-Rad) and analyzed by using Quantity One v4.6.3 software (Bio-Rad). A calibration curve was set up by using home-made CS isomers obtained from a pool of plasma samples, assayed for hexuronate content [21], and subjected to disaccharides analysis [8]. CS levels were reported as μg of hexuronic acid per mL of plasma (μg_{UA}/mL), and CS charge density was evaluated as ratio between 4-sulfated Δ-disaccharides (ΔDi-4S) and total unsaturated disaccharides (ΔDi-4S + ΔDi-nonS).

2.3. Urine GAGs/Bikunin Analysis.
First-morning urine samples (about 50 mL) were collected and, after centrifugation at 3000 g for 15 min at 4°C, the sediment of broken cells or tissues and other solid materials was discarded. Urine GAGs/bikunin containing fraction was obtained by anion exchange chromatography (DEAE-Sephacel resin) as previously described [15]. Briefly, clarified urines were applied to a column packed with about 6 mL of resin, previously equilibrated with a buffer containing 0.02 M Tris-HCl, 0.15 M NaCl (pH 8.6). After exhaustive washing, urinary GAGs/bikunin were eluted with a buffer containing 0.02 M Tris-HCl, 2 M LiCl (pH 8.6), and assayed for hexuronate content by the method of Bitter and Muir, using glucuronic acid as a standard [21]. Hexuronate levels were normalized for the urinary creatinine concentration, formerly determined by the Jaffè method (Sentinel CH, Milan, Italy). Urinary GAGs/bikunin composition was determined by discontinuous electrophoresis on cellulose

TABLE 1: Clinical and genetic features of Fabry's patients.

Patient	Age	Gender	Renal involvement	GLA mutations	α-galactosidase A (nmol/mL/h)*	ERT
1	59	F	RD	Cys172Tyr	1.70	Y
2	26	F	RD	Cys172Tyr	4.68	Y
3	30	M	NRD	Cys172Tyr	3.70	Y
4	38	M	NRD	Cys172Tyr	5.30	Y
5	33	M	proteinuria	Cys172Tyr	2.80	Y
6	27	M	RD	Cys172Tyr	3.40	Y
7	46	F	RD	846_847delTC	3.50	Y
8	24	F	proteinuria	846_847delTC	3.50	Y
9	20	F	proteinuria	846_847delTC	3.30	Y
10	58	F	RD	846_847delTC	3.90	Y
11	36	M	proteinuria	846_847delTC	0.48	Y
12	36	F	NRD	Arg112His	2.80	N
13	61	F	NRD	Gln57Arg	11.10	N
14	62	F	NRD	Gln57Arg	14.80	N
15	23	F	NRD	Gln57Arg	10.40	N
16	58	F	NRD	Gln57Arg	9.80	Y
17	53	F	NRD	Asp313Tyr	15.40	Y
18	52	F	NRD	Arg227Gln	8.10	N
19	31	M	NRD	Arg227Gln	2.50	Y
20	42	M	RD	Arg227Gln	0.10	Y
21	32	M	proteinuria	Arg227Gln	0.70	Y
22	25	M	NRD	Arg227Gln	2.50	Y
23	19	F	NRD	Arg227Gln	6.60	Y
24	43	F	NRD	IVS3+G>A	9.70	Y

NRD: no-renal disease; RD: renal disease; ERT: enzyme replacement therapy; Y: under ERT therapy; N: no ERT therapy.
*At the time of diagnosis.

acetate plates [12–17], according to Cappelletti et al. [22]. Analytes were resolved by three electrophoretic steps in 0.25 M barium acetate running buffer, pH 5.0. Titan III-H cellulose acetate plate (6.0 × 7.5 cm, Helena BioSciences) was first soaked in distilled H_2O for about 1.5 cm and immediately blotted between filter papers. Then, the opposite end was soaked in 0.1 M barium acetate buffer, pH 5.0, for 5.5 cm, leaving a narrow band (2–4 mm large), apparently dry, where 5 μg as uronic acid of each sample was loaded. Electrophoresis was carried out at 5 mA for about 6 minutes followed by incubation of the plate in 0.1 M barium acetate, pH 5.0, for 2 minutes. The second electrophoretic step was carried out at 15 mA for 14 minutes. Subsequently, the plate was soaked in 0.1 M barium acetate buffer, pH 5.0, containing 15% (v/v) ethanol for 2 minutes. A third electrophoretic step was carried out at 12 mA for 17 minutes. Finally, electrophoretic profiles were detected following 0.1% (w/v) Alcian Blue staining. Images were acquired by means of GS-800 calibrated densitometer (Bio-Rad) and analyzed by using Quantity One v4.6.3 software (Bio-Rad). GAGs were expressed as relative percentages.

The GAGs/bikunin fractions were assessed for both urine bikunin protein content and presence of urine bikunin fragments by performing SDS-PAGE followed by highly sensitive Coomassie Brilliant Blue G-250 staining (limit of detection: 20 ng of protein) [23] on untreated samples and after chondroitin ABC lyase digestion. The latter was performed in a buffer containing 0.1 M ammonium acetate (pH 8.0) using 0.1 U of chondroitin ABC lyase (Sigma Aldrich) per 100 μg of hexuronate at 37°C overnight. Samples were added with 4X SDS-buffer containing 250 mM Tris (pH 6.8), 8% SDS (w/v), 8% DTT (w/v), 40% glycerol (v/v), and 0.0008% bromophenol blue (w/v) and boiled for 5 minutes before electrophoresis. Urine bikunin was resolved by Tris-glycine SDS-PAGE in 1 mm thick 15% T, 3% C running gel, using a MiniProtean II cell vertical slab gel electrophoresis apparatus (Bio-Rad). Electrophoresis was carried out at 50 V for 15 minutes and subsequently at 150 V until the bromophenol dye front reached the lower limit of the gel. Then, gels were fixed in 30% ethanol (v/v), 2% phosphoric acid (v/v) for 1 h, washed twice in 2% phosphoric acid (v/v) for 10 minutes, equilibrated in 18% ethanol (v/v), 2% phosphoric acid (v/v), and 15% ammonium sulfate (w/v) for 30 minutes, and stained in the same solution containing 0.02% Coomassie Brilliant Blue G-250 (w/v) for 48 h. Gel images were acquired by using GS-800 calibrated densitometer (Bio-Rad) at 63 μm resolution.

2.4. Statistical Analysis. Student's t-test for unpaired samples was used to compare plasma and urinary GAGs levels between Fabry's patients and control subjects, using

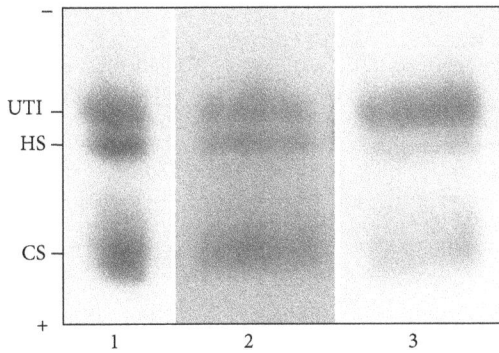

FIGURE 1: Representative cellulose acetate electrophoretic profiles of urine glycosaminoglycans/UTI from control subjects (lane 2) and Fabry's patients with renal disease (lane 3). Lane 1: mixture of standard GAGs/UTI (UTI: urinary trypsin inhibitor/urine bikunin; HS: heparan sulfate; CS: chondroitin sulfate).

the software package Sigma Stat 3 (Systat Software). Pearson's correlation analysis was performed to evaluate the association between plasma CS and UTI levels and between UTI and serum creatinine levels. Significance was set at $P < 0.05$.

3. Results

Quali-quantitative GAGs analyses were conducted in plasma and urine from 24 Fabry's patients and 43 control subjects. Patients were sorted in RD patients, either with only proteinuria ($n = 5$) or with overt renal damage ($n = 6$), and NRD patients, with no evidence of chronic renal damage ($n = 13$). Qualitative analysis by electrophoresis on cellulose acetate (Figure 1) followed by image analysis allowed evaluating percentages of each purified urinary glycosaminoglycan/proteoglycan, namely, urine bikunin (UTI), HS, and CS. Quali-quantitative results with regard to urine are reported in Table 2 and Figure 2. Hexuronate content analysis evidenced higher urinary GAGs levels (+48%) in patients group in respect to controls ($P = 0.03$). After sorting the group of patients as mentioned previously, it was evident that the major differences found had to be ascribed to the occurrence of renal damage since NRD patients showed urinary GAGs levels not significantly different from controls ($P = 0.99$). The patients group showed quite different electrophoretic profiles with respect to controls (Table 2, Figure 2). After integration, quali-quantitative results evidenced levels of urine bikunin 2.8 times higher in the patients group in respect to controls ($P = 0.005$). Notably, this difference was significant for RD patients only, who showed about 3.8 times higher levels of urine bikunin as compared with control subjects ($P = 0.0001$). To evaluate if urine bikunin levels in RD patients were associated with the degree of renal impairment, the results for RD patients were analyzed after sorting the group in RD patients with only proteinuria (early renal impairment) and RD patients with overt renal damage (Table 3, Figure 3). In this respect, no major differences were evidenced either in GAGs/bikunin levels or in their distribution among the two subgroups, indicating that the increase of bikunin excretion in RD patients is likely an early biochemical event that occurs

at the onset of renal impairment. In order to verify if urine bikunin was in its intact form, we performed SDS-PAGE analysis on GAGs/bikunin fractions as a whole and after chondroitin sulfate removal by chondroitin ABC lyase treatment. In this respect, no significant bikunin fragmentation was evidenced in urine samples from either patients or controls (Figure 4). Furthermore, the SDS-PAGE analysis allowed confirming the higher urine bikunin levels in RD patients.

To rule out the possibility that higher urine bikunin levels in patients could be ascribed, at least partly, to higher levels of plasma bikunin, we assayed plasma CS isomers by FACE analysis evidencing no differences in either plasma CS levels or charge density between patients and control subjects (Table 4, Figures 5 and 6). Furthermore, no correlation between plasma CS and urine bikunin levels (Figure 7) or between serum creatinine and urine bikunin levels was found suggesting a direct kidney involvement in the higher UTI excretion of Fabry's patients. In absence of overt infection, eleven FD patients out of twenty-four (45.8%) presented with at least one marker of inflammation altered in serum (ESR or CRP, or α-1 and α-2 globulins). These markers were altered in only 10% of controls.

4. Discussion

Fabry's disease is a multisystemic disorder in which progressive renal impairment, along with cardiac and central nervous system involvement, plays a major role in lowering life quality and expectancy [3, 4]. In a retrospective study on Fabry's patients with renal involvement, Branton et al. showed that 50% of patients had proteinuria by 35 years of age and 100% by 52 years of age [24]. Moreover, 50% of male patients presented with chronic renal insufficiency (CRI) by 42 years of age. These authors evidenced that after the development of CRI, there was a rapid decline in glomerular filtration rate leading to end-stage renal disease within 4.1 years. The enzyme replacement therapy (ERT) seems to represent a valid tool to partly counteract the natural progression of Fabry's disease in combination with renoprotective treatments, such as ACE inhibitors, which are known to be effective in slowing disease progression in other chronic proteinuric kidney diseases [25, 26]. It has also been shown that ERT may be effective in preserving normal renal function in children [27]. Nevertheless, in FD, the diagnosis of an early renal dysfunction is likely of primary importance to aid the clinician in decision making, in designing therapeutic interventions, and in following the natural disease progression or the effects of specific treatments [28].

This paper is the first report to point out that urine bikunin levels are significantly higher in FD patients with renal impairment compared to healthy controls. This finding suggests that the amount of this proteoglycan in urine, as well as proteinuria, could represent an early biomarker of renal impairment in Fabry's patients, useful in monitoring renal functionality also in those patients without overt renal damage.

Interestingly, in our study also several FD patients treated with ERT showed elevated levels of bikunin in urine.

TABLE 2: Urine GAGs/UTI levels and distribution in Fabry's patients and healthy control subjects.

	All patients (n = 24)	RD Patients (n = 11)	NRD Patients (n = 13)	Controls (n = 43)	All patients versus Controls (P)	RD Patients versus NRD patients (P)	RD Patients versus Controls (P)	NRD Patients versus Controls (P)
Uronic acid (μg_{UA}/mg Cr)	3.83 ± 1.69	4.61 ± 1.63	2.60 ± 0.88	2.59 ± 1.59	**0.03**	**0.009**	**0.004**	0.99
UTI (%)	52.8 ± 20.9	63.2 ± 18.3	36.6 ± 12.9	28.5 ± 17.6	**0.0009**	**0.004**	**<0.0001**	0.29
HS (%)	20.1 ± 9.1	17.4 ± 10.0	24.4 ± 5.6	30.3 ± 9.3	**0.003**	0.113	**0.002**	0.14
CS (%)	27.0 ± 14.4	19.4 ± 9.0	39.0 ± 13.2	41.2 ± 11.9	**0.004**	**0.002**	**<0.0001**	0.70
UTI (μg_{UA}/mg Cr)	2.29 ± 1.79	3.15 ± 1.79	0.93 ± 0.44	0.82 ± 0.81	**0.005**	**0.006**	**0.0001**	0.74
HS (μg_{UA}/mg Cr)	0.66 ± 0.26	0.67 ± 0.28	0.63 ± 0.24	0.72 ± 0.37	0.57	0.77	0.71	0.58
CS (μg_{UA}/mg Cr)	0.89 ± 0.36	0.79 ± 0.23	1.03 ± 0.50	1.04 ± 0.65	0.38	0.18	0.23	0.97

GAGs/PGs levels are reported as μg of uronic acid (UA) per mg of creatinine.
UTI, HS and CS levels are calculated from total UA content and relative percentages of each GAG.
RD: renal disease (proteinuria/renal damage); NRD: no-renal disease.
Significant differences are reported in bold ($P < 0.05$).

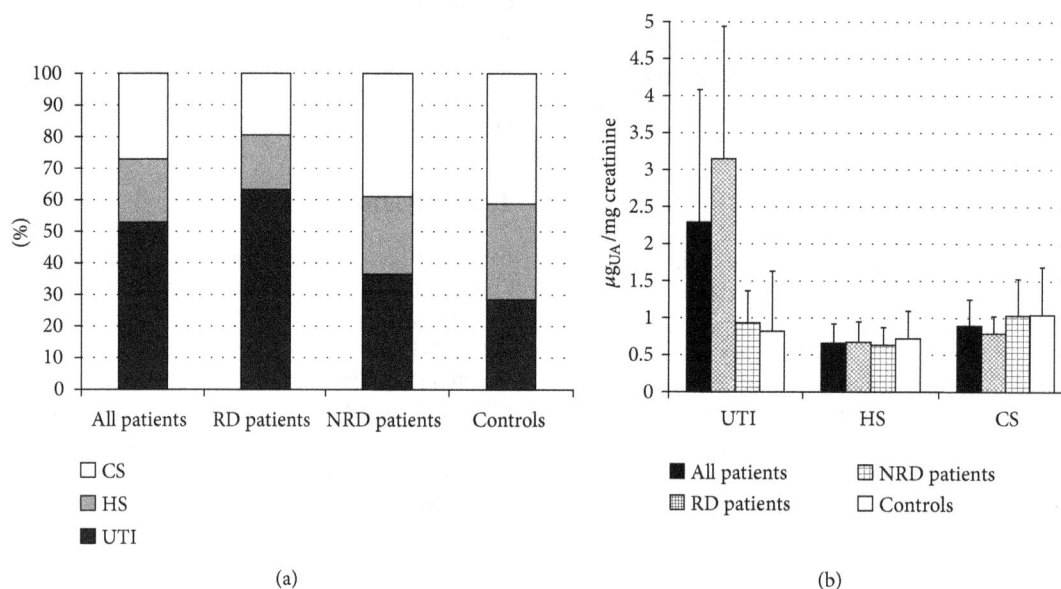

(a)   (b)

FIGURE 2: Diagrams reporting percentages (a) and levels (b) of urinary trypsin inhibitor (UTI), heparan sulfate (HS), and chondroitin sulfate (CS) in the totality of patients, patients with renal disease (RD), patients without renal disease (NRD), and controls. UA: uronic acid.

Although in this study we did not plan a formal design aimed to evaluate the urinary bikunin levels at baseline and after a period of ERT, this finding may be supportive of several clinical data indicating that ERT may change significantly the natural progression of the disease if started before the establishment of irreversible organ lesion [25, 29].

Plasma and urine levels of bikunin are related to its anti-inflammatory activity [18], and several studies have associated high plasma and/or urine levels of this proteoglycan with various diseases exhibiting chronic inflammation [11–16]. Notably, several clinical and laboratory findings, as the occurrence of episodic and unexplained fever in some patients with FD, and/or the increased serum levels of ESR, CRP, and α1- and α2-globulin (observed also in 45.8% of our patients with FD) indicate the likely occurrence of both systemic and local inflammation in this pathology [30, 31]. The activation of the inflammatory biochemical pathways in FD, as it may occur in other lysosomal storage disorders (LSDs), is probably related to secondary inappropriate activation of the immune system, in response to storage, resulting in chronic inflammation [32, 33]. It is noteworthy that in LSDs, the occurrence of systemic inflammation contributes to pathogenesis, predates the onset of clinical signs, and may determine the appearance in plasma and/or urine of secondary metabolites which may act as biomarkers that could be useful in following disease progression and may become a target for adjunctive therapy [32, 33]. Importantly, in FD, the ERT is frequently only partially effective in many patients, either due to a late start of this therapy or because of the secondary activation of biochemical mechanisms other than glycosphingolipids storage that also contribute to the pathogenesis of FD. A better understanding of the secondary biochemical pathways involved in FD pathogenesis, as the ones involved in chronic inflammation, may foster the discovery of new therapeutic

TABLE 3: Urinary glycosaminoglycans/UTI levels and distribution in Fabry's patients presenting with either only proteinuria or renal damage.

	RD Patients with proteinuria ($n = 5$)	RD Patients with renal damage ($n = 6$)	(P)
Uronic acid (μg_{UA}/mg Cr)	4.08 ± 1.40	4.91 ± 1.77	0.44
UTI (%)	52.7 ± 13.6	69.2 ± 18.8	0.16
HS (%)	23.8 ± 8.9	13.8 ± 9.2	0.11
CS (%)	23.5 ± 4.8	17.0 ± 10.3	0.27
UTI (μg_{UA}/mg Cr)	2.27 ± 1.19	3.64 ± 1.96	0.24
HS (μg_{UA}/mg Cr)	0.88 ± 0.07	0.55 ± 0.29	0.05
CS (μg_{UA}/mg Cr)	0.92 ± 0.21	0.72 ± 0.22	0.18

GAGs/PGs levels are reported as μg of uronic acid (UA) per mg of creatinine.
UTI, HS and CS levels are calculated from total UA content and relative percentages of each GAG.

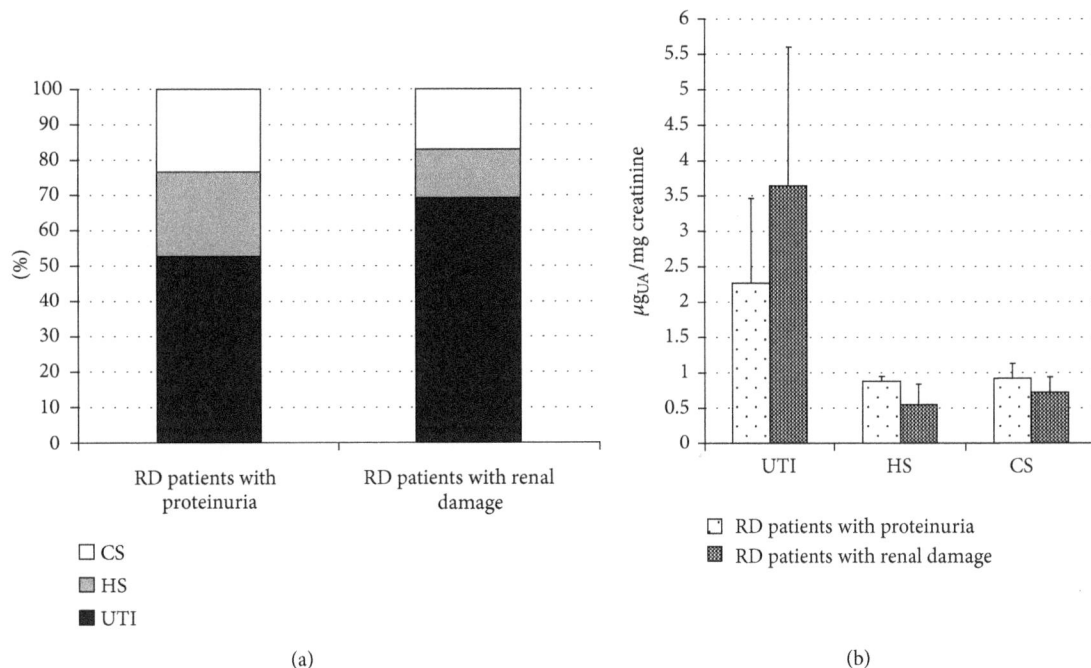

FIGURE 3: Diagrams reporting percentages (a) and levels (b) of urinary trypsin inhibitor (UTI), heparan sulfate (HS), and chondroitin sulfate (CS) in RD patients with proteinuria and RD patients with renal damage. UA: uronic acid.

TABLE 4: Plasma CS isomers levels and structure in Fabry's patients and healthy control subjects obtained by FACE analysis of fluorophore-labeled unsaturated disaccharides.

	Patients ($n = 24$)	Controls ($n = 43$)	Patients versus Controls (P)
CS isomers (μg_{UA}/mL plasma)	6.07 ± 2.96	5.61 ± 2.99	0.66
*CS charge density (%)	41.9 ± 5.9	42.8 ± 8.6	0.72

*CS charge density was evaluated as ratio between ΔDi-4S and the sum of ΔDi-nonS and ΔDi-4S.

approaches, adjunctive to ERT, potentially of benefit in this pathology.

Since in our Fabry's patients no correlation was found between plasma CS and urine bikunin levels and no differences were evidenced in plasma CS level/structure between patients and controls, the origin of higher levels of bikunin in urine may imply a direct kidney involvement. Recent RT-PCR analysis studies documenting that several human organs, including the kidney, express bikunin fit this possibility [34]. Moreover, no significant correlation was found between urine bikunin levels and serum creatinine; thus the only impairment of renal function in FD patients seems insufficient to explain higher urine bikunin levels. Nevertheless, the origin of urine bikunin levels and the mechanisms by which urine levels are elevated in our Fabry's patients remains unclear and need to be further evaluated.

5. Conclusions

Our data indicate that urine bikunin levels may be an early biomarker of renal impairment in patients with FD. Moreover, higher urine levels of this secondary metabolite in FD patients suggest the secondary activation, in response to glycosphingolipids storage, of biochemical pathways related to inflammation. Further studies are necessary to test our

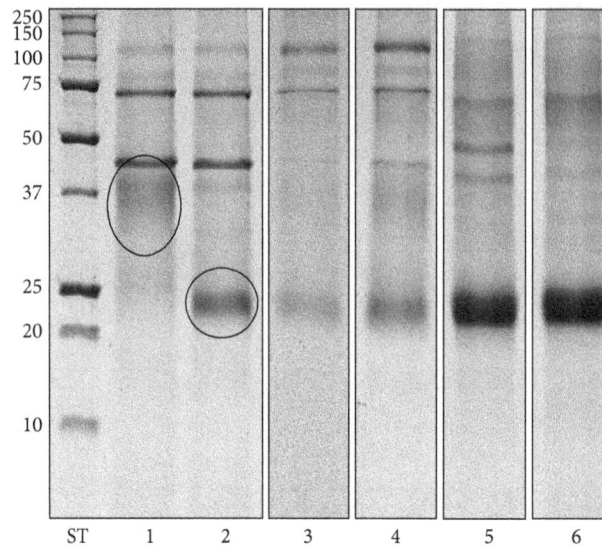

FIGURE 4: Representative SDS PAGE profiles of nontreated (lane 1) and Chase ABC-treated GAGs-containing fractions from controls (lane 2). Chase ABC-treated GAGs-containing fractions from controls (lane 3), NRD patients (lane 4), RD patients with proteinuria (lane 5), and RD patients with renal damage (lane 6). In lanes 3, 4, 5, and 6, aliquots corresponding to 2 mg of creatinine were loaded. Ovals indicate intact UTI (lane 1) and UTI depleted of chondroitin sulfate chains (lane 2). ST: molecular weight standards (kDa).

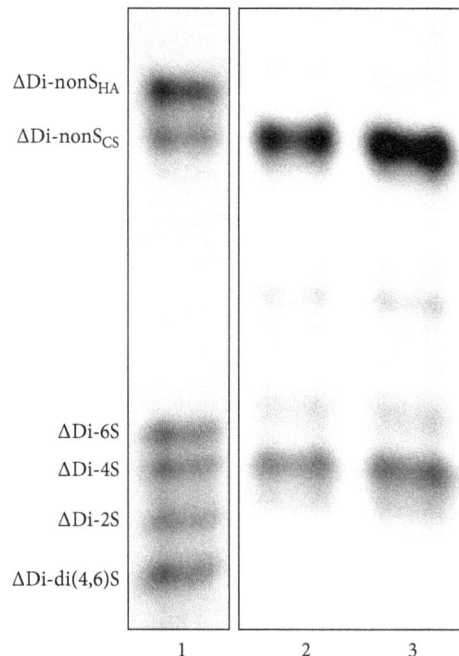

FIGURE 5: Representative FACE profiles of fluorophore-labeled unsaturated disaccharides (ΔDi) obtained from plasma CS isomers of both controls (lane 2) and Fabry's patients (lane 3). Lane 1: mixture of commercial standard ΔDi (ΔDi-nonS$_{HA}$, 2-acetamido-2-deoxy-3-O-(4-deoxy-α-L-threo-hex-4-enepyranosyluronic acid)-4-D-glucose; ΔDi-nonS$_{CS}$, 2-acetamido-2-deoxy-3-0-(4-deoxy-α-L-threo-hex-4-enepyranosyluronic acid)-4-D-galactose; ΔDi-6S, 2-acetamido-2-deoxy-3-0-(4-deoxy-α-L-threo-hex-4-enepyranosyluronic acid)-6-O-sulpho-D-galactose; ΔDi-4S, 2-acetamido-2-deoxy-3-0-(4-deoxy-α-L-threo-hex-4-enepyranosyluronic acid)-4-O-sulpho-D-galactose; ΔDi-mono2S, 2-acetamido-2-deoxy-3-O-(4-deoxy-2-O-sulpho-α-L-threo-hex-4-enepyranosyluronic acid)-D-galactose; ΔDi-di(4,6)S, 2-acetamido-2-deoxy-3-O-(4-deoxy-α-L-threo-hex-4-enepyranosyluronic acid)-4,6-O-sulpho-D-galactose).

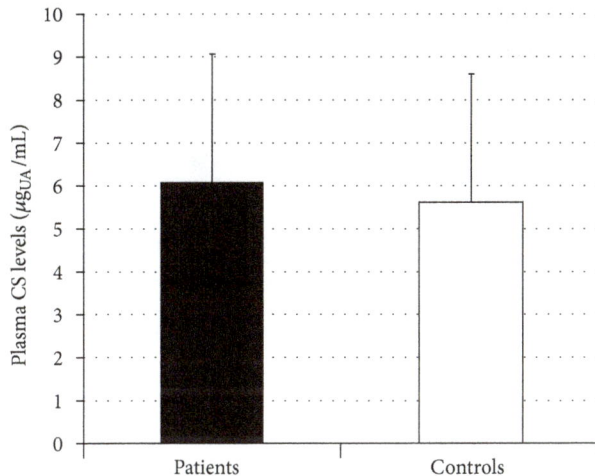

FIGURE 6: Diagram showing levels of plasma chondroitin sulfate isomers in Fabry's patients and controls. UA: uronic acid.

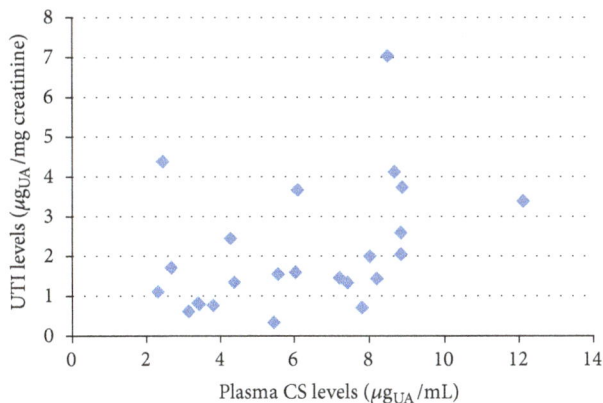

FIGURE 7: Scatter plot showing levels of urine bikunin (UTI), as μg of uronic acid (UA) per mg of creatinine, in relation to plasma CS isomers levels, as μg of UA per mL of plasma, in Fabry's patients.

findings in a larger cohort of FD patients and to investigate any existing correlation between urine bikunin levels, progression of FD, and changes in response to ERT.

Conflict of Interests

The authors declare that they have no conflict of interests.

Acknowledgments

This study was supported by Regione Autonoma della Sardegna (P.O.R. Sardegna F.S.E. 2007/2013, Asse IV Capitale Umano, Obiettivo operativo 1.3-Linea di attività 1.3.1) and a free research grant from Shire Italy-Human Genetic Therapies Unit.

References

[1] Y. A. Zarate and R. J. Hopkin, "Fabry's disease," *The Lancet*, vol. 372, no. 9647, pp. 1427–1435, 2008.

[2] R. J. Desnick and R. O. Brady, "Fabry disease in childhood," *Journal of Pediatrics*, vol. 144, no. 5, Supplement, pp. S20–S26, 2004.

[3] F. Breunig and C. Wanner, "Update on Fabry disease: kidney involvement, renal progression and enzyme replacement therapy," *Journal of Nephrology*, vol. 21, no. 1, pp. 32–37, 2008.

[4] R. Schiffmann, "Fabry disease," *Pharmacology and Therapeutics*, vol. 122, no. 1, pp. 65–77, 2009.

[5] M. Torralba-Cabeza, S. Olivera, D. A. Hughes, G. M. Pastores, R. N. Mateo, and J. Pérez-Calvo, "Cystatin C and NT-proBNP as prognostic biomarkers in Fabry disease," *Molecular Genetics and Metabolism*, vol. 104, no. 3, pp. 301–307, 2011.

[6] J. D. Esko, K. Kimata, and U. Lindahl, "Proteoglycans and sulfated glycosaminoglycans," in *Essential of Glycobiology*, A. Varki, R. D. Cummings, J. D. Esko et al., Eds., Chapter 16, Cold Spring Harbor Laboratory Press, Cold Spring Harbor, NY, USA, 2nd edition, 2009.

[7] M. Ly, F. E. Leach, T. N. Laremore, T. Toida, I. J. Amster, and R. J. Linhardt, "The proteoglycan bikunin has a defined sequence," *Nature Chemical Biology*, vol. 7, no. 11, pp. 827–833, 2011.

[8] A. Zinellu, S. Pisanu, E. Zinellu et al., "A novel LIF-CE method for the separation of hyalurnan- and chondroitin sulfate-derived disaccharides: application to structural and quantitative analyses of human plasma low- and high-charged chondroitin sulfate isomers," *Electrophoresis*, vol. 28, no. 14, pp. 2439–2447, 2007.

[9] E. Zinellu, A. J. Lepedda, A. Cigliano et al., "Association between human plasma chondroitin sulfate isomers and carotid atherosclerotic plaques," *Biochemistry Research International*, vol. 2012, Article ID 281284, 6 pages, 2012.

[10] L. Zhuo, V. C. Hascall, and K. Kimata, "Inter-α-trypsin inhibitor, a covalent protein-glycosaminoglycan- protein complex," *Journal of Biological Chemistry*, vol. 279, no. 37, pp. 38079–38082, 2004.

[11] H. Matsuzaki, H. Kobayashi, T. Yagyu et al., "Plasma bikunin as a favorable prognostic factor in ovarian cancer," *Journal of Clinical Oncology*, vol. 23, no. 7, pp. 1463–1472, 2005.

[12] P. De Muro, R. Faedda, A. Satta et al., "Urinary glycosaminoglycan composition in chronic glomerulonephritis," *Journal of Nephrology*, vol. 18, no. 2, pp. 154–160, 2005.

[13] P. De Muro, R. Faedda, A. E. Satta et al., "Quali-quantitative analysis of urinary glycosaminoglycans for monitoring glomerular inflammatory activity," *Scandinavian Journal of Urology and Nephrology*, vol. 41, no. 3, pp. 230–236, 2007.

[14] P. De Muro, R. Faedda, A. Masala et al., "Kidney post-transplant monitoring of urinary glycosaminoglycans/proteoglycans and monokine induced by IFN-γ (MIG)," *Clinical and Experimental Medicine*, vol. 13, no. 1, pp. 59–65, 2013.

[15] P. De Muro, P. Fresu, G. Tonolo et al., "A longitudinal evaluation of urinary glycosaminoglycan excretion in normoalbuminuric type 1 diabetic patients," *Clinical Chemistry and Laboratory Medicine*, vol. 44, no. 5, pp. 561–567, 2006.

[16] P. De Muro, R. Faedda, M. Formato et al., "Urinary glycosaminoglycans in patients with systemic lupus erythematosus," *Clinical and Experimental Rheumatology*, vol. 19, no. 2, pp. 125–130, 2001.

[17] P. De Muro, G. Capobianco, M. Formato et al., "Glycosaminoglycan and transforming growth factor β1 changes in human plasma and urine during the menstrual cycle, in vitro fertilization treatment, and pregnancy," *Fertility and Sterility*, vol. 92, no. 1, pp. 320–327, 2009.

[18] M. J. Pugia, R. Valdes Jr., and S. A. Jortani, "Bikunin (urinary trypsin inhibitor): structure, biological relevance, and measurement," *Advances in Clinical Chemistry*, vol. 44, pp. 223–245, 2007.

[19] S. De Lin, R. Endo, H. Kuroda et al., "Plasma and urine levels of urinary trypsira inhibitor in patients with chronic liver diseases and hepatocellular carcinoma," *Journal of Gastroenterology and Hepatology*, vol. 19, no. 3, pp. 327–332, 2004.

[20] K. Sakai, H. Okudera, and K. Hongo, "Significant elevation of urinary trypsin inhibitor in patients with brain contusion—a preliminary report," *Journal of Clinical Neuroscience*, vol. 10, no. 6, pp. 677–679, 2003.

[21] T. Bitter and H. M. Muir, "A modified uronic acid carbazole reaction," *Analytical Biochemistry*, vol. 4, no. 4, pp. 330–334, 1962.

[22] R. Cappelletti, M. Del Rosso, and V. P. Chiarugi, "A new electrophoretic method for the complete separation of all known animal glycosaminoglycans in a monodimensional run," *Analytical Biochemistry*, vol. 99, no. 2, pp. 311–315, 1979.

[23] A. Zinellu, A. Lepedda Jr., S. Sotgia et al., "Albumin-bound low molecular weight thiols analysis in plasma and carotid plaques by CE," *Journal of Separation Science*, vol. 33, no. 1, pp. 126–131, 2010.

[24] M. H. Branton, R. Schiffmann, S. G. Sabnis et al., "Natural history of fabry renal disease: influence of α-galactosidase a activity and genetic mutations on clinical course," *Medicine*, vol. 81, no. 2, pp. 122–138, 2002.

[25] A. Mehta, M. Beck, P. Elliott et al., "Enzyme replacement therapy with agalsidase alfa in patients with Fabry's disease: an analysis of registry data," *The Lancet*, vol. 374, no. 9706, pp. 1986–1996, 2009.

[26] D. G. Warnock and M. L. West, "Diagnosis and management of kidney involvement in fabry disease," *Advances in Chronic Kidney Disease*, vol. 13, no. 2, pp. 138–147, 2006.

[27] M. Ries, J. T. R. Clarke, C. Whybra et al., "Enzyme-replacement therapy with agalsidase alfa in children with Fabry disease," *Pediatrics*, vol. 118, no. 3, pp. 924–932, 2006.

[28] R. Schiffmann, S. Waldek, A. Benigni, and C. Auray-Blais, "Biomarkers of fabry disease nephropathy," *Clinical Journal of the American Society of Nephrology*, vol. 5, no. 2, pp. 360–364, 2010.

[29] E. Ortu, L. Fancellu, G. Sau et al., "Primary motor cortex hyperexcitability in Fabry's disease," *Clinical Neurophysiology*, 2013.

[30] A. C. Vedder, É. Biró, J. M. F. G. Aerts, R. Nieuwland, G. Sturk, and C. E. M. Hollak, "Plasma markers of coagulation and endothelial activation in Fabry disease: impact of renal impairment," *Nephrology Dialysis Transplantation*, vol. 24, no. 10, pp. 3074–3081, 2009.

[31] Y. Kikumoto, Y. Kai, H. Morinaga et al., "Fabry disease exhibiting recurrent stroke and persistent inflammation," *Internal Medicine*, vol. 49, no. 20, pp. 2247–2252, 2010.

[32] A. H. Futerman and G. Van Meer, "The cell biology of lysosomal storage disorders," *Nature Reviews Molecular Cell Biology*, vol. 5, no. 7, pp. 554–565, 2004.

[33] E. B. Vitner, F. M. Platt, and A. H. Futerman, "Common and uncommon pathogenic cascades in lysosomal storage diseases," *Journal of Biological Chemistry*, vol. 285, no. 27, pp. 20423–20427, 2010.

[34] H. Itoh, M. Tomita, T. Kobayashi, H. Uchino, H. Maruyama, and Y. Nawa, "Expression of inter-α-trypsin inhibitor light chain (bikunin) in human pancreas," *Journal of Biochemistry*, vol. 120, no. 2, pp. 271–275, 1996.

11

Expression of Syndecan-4 and Correlation with Metastatic Potential in Testicular Germ Cell Tumours

Vassiliki T. Labropoulou,[1,2] **Spyros S. Skandalis,**[3] **Panagiota Ravazoula,**[4] **Petros Perimenis,**[5] **Nikos K. Karamanos,**[3] **Haralabos P. Kalofonos,**[2] **and Achilleas D. Theocharis**[3]

[1] *Hematology Division, Department of Internal Medicine, University Hospital of Patras, 26500 Patras, Greece*
[2] *Clinical Oncology Laboratory, Division of Oncology, Department of Medicine, University Hospital of Patras, 26500 Patras, Greece*
[3] *Laboratory of Biochemistry, Department of Chemistry, University of Patras, 26500 Patras, Greece*
[4] *Department of Pathology, University Hospital of Patras, 26500 Patras, Greece*
[5] *Division of Urology, Department of Medicine, University Hospital of Patras, 26500 Patras, Greece*

Correspondence should be addressed to Achilleas D. Theocharis; atheoch@upatras.gr

Academic Editor: Martin Götte

Although syndecan-4 is implicated in cancer progression, there is no information for its role in testicular germ cell tumours (TGCTs). Thus, we examined the expression of syndecan-4 in patients with TGCTs and its correlation with the clinicopathological findings. Immunohistochemical staining in 71 tissue specimens and mRNA analysis revealed significant overexpression of syndecan-4 in TGCTs. In seminomas, high percentage of tumour cells exhibited membranous and/or cytoplasmic staining for syndecan-4 in all cases. Stromal staining for syndecan-4 was found in seminomas and it was associated with nodal metastasis ($P = 0.04$), vascular/lymphatic invasion ($P = 0.01$), and disease stage ($P = 0.04$). Reduced tumour cell associated staining for syndecan-4 was observed in nonseminomatous germ cell tumours (NSGCTs) compared to seminomas. This loss of syndecan-4 was associated with nodal metastasis ($P = 0.01$), vascular/lymphatic invasion ($P = 0.01$), and disease stage ($P = 0.01$). Stromal staining for syndecan-4 in NSGCTs did not correlate with any of the clinicopathological variables. The stromal expression of syndecan-4 in TGCTs was correlated with microvessel density ($P = 0.03$). Our results indicate that syndecan-4 is differentially expressed in seminomas and NSGCTs and might be a useful marker. Stromal staining in seminomas and reduced levels of syndecan-4 in tumour cells in NSGCTs are related to metastatic potential, whereas stromal staining in TGCTs is associated with neovascularization.

1. Introduction

Testicular germ cell tumour (TGCT), although relatively rare, is the most common malignancy in men between 15 and 35 years old age group with increasing incidence in the past decades [1, 2]. TGCTs have become one of the most curable solid neoplasms, due to the advantage of diagnostic and therapeutic methods, but still the prognosis of highly advanced cases with bulky metastatic lesions is generally poor. Histologically, the TGCTs can be classified as seminomas germ cell tumours, which originate from undifferentiated germ cells, and nonseminomatous germ cell tumours (NSGCTs), which are arise from undifferentiated (embryonal carcinoma) and differentiated multipotent cells [3]. NSGCTs

are generally more aggressive and the histological classification to seminoma or NSGCTs is the most important criterion for the selection of the treatment strategy. In patients with clinical stage I NSGCTs other biological markers apart from the percentage of embryonal carcinoma and the presence of vascular invasion, which are reliable prognostic indicators to identify patients at high risk for occult retroperitoneal disease, have not yet been shown to be of prognostic significance [4]. It has been shown that the presence of vascular invasion is associated with gain of a region at 17q12 and more specifically with the expression of inflammatory cytokine CCL2 in NSGCTs of stage I [5]. We demonstrated recently that the aggressiveness of testicular germ cell tumour cell lines is associated with increased expression of matrix metalloproteinases (MMPs) and reduced expression of tissue inhibitors of matrix metalloproteinases (TIMPs) [6]. Hence it is important to evaluate novel markers for the development and prognosis of TGCTs.

Several studies have already focused on the role of proteoglycans in human tumours [7–11]. Accumulation of versican, an extracellular matrix proteoglycan, has been shown to correlate to the metastatic potential of testicular tumours [12]. Syndecans are integral membrane proteoglycans that are implicated in cell-cell recognition and cell-matrix interactions [11, 13]. Syndecans have a short cytoplasmic domain, one transmembrane, and one extracellular domain. The latter bearing heparan sulphate and or chondroitin sulphate glycosaminoglycan chains are capable of binding various growth factors and matrix molecules [13]. Syndecan-1 is the most thoroughly investigated member of the syndecan family and downregulation of cell membrane syndecan-1 is regarded as initial step towards malignant transformation in various malignancies [11, 13]. Although various studies have focused on the role of other syndecans in malignancies, little is known about the role of syndecan-4 in tumour development. Syndecan-4 mediates breast cancer cell adhesion and spreading [14] but also binds proangiogenic growth factors and cytokines and modulates growth factor/growth factor receptor interactions regulating angiogenic processes [15, 16]. Syndecan-4 potentiates Wnt5a signaling and enhances invasion and metastasis of melanoma cells [17]. The cell surface levels of syndecan-4 are reduced by Wnt5a signaling that promotes its ubiquitination and degradation thus regulating cell adhesion and migration [18]. Syndecan-4 interacts with chemokines through HS chains and promotes tumour cell migration and invasion [19, 20] but also regulates the invasion of K-ras mutant cells in collagen lattice together with integrin $\alpha2\beta1$ and MT1-MMP [21]. Taken into account the proved role of syndecans in malignancies and the structure/function similarities among syndecans we aimed to study the expression profile of syndecan-4 in TGCTs as well as its association with the metastatic potential of these tumours.

2. Material and Methods

2.1. Cell Lines and Cultures.

The human seminoma cell line JKT-1 was a gift from Patrick Fenichel (University of Nice-Sophia-Antipolis, Faculty of Medicine, Nice, France) [22]. JKT-1 cells were cultured up to 38 passages to avoid the drift of these cells. The molecular signature of JKT-1 cells used in our study was described previously concerning the expression of seminoma markers (placenta alkaline phosphatase, NANOG, OCT3/4, AP2γ, and HIWI) [6]. Early passages of JKT-1 cells used in our study express a signature of markers which is still near from the one expressed by seminoma cells allowing their use as a model to study seminomas. Human embryonal carcinoma cell line NTERA-2 and teratocarcinoma cell line NCCIT were purchased from American Type Culture Collection (ATCC, Manassas, VA, USA). The NTERA-2 and JKT-1 cell lines were cultured in DMEM supplemented with 10% fetal calf serum. The NCCIT cell line was cultured in RPMI 1640 supplemented with 10% fetal calf serum. All culture media contained 100 UI/mL penicillin and 100 UI/mL streptomycin. The cell lines were cultured in a humidified atmosphere containing 5% CO_2 at 37°C.

2.2. Patients and Tissue Samples.

Primary tumours were obtained at surgery from nine patients with TGCTs (five with seminoma and four with NSGCTs). Six control healthy testicular tissues were taken from autopsies. All tissue samples were frozen immediately and subjected to RNA extraction.

A retrospective study was performed including 71 patients with TGCTs who had undergone orchiectomy in our hospital. Patients were further treated according to their stage, the histological type, and specific predictive and prognostic factors. Patients with stage I seminoma were treated with 2 cycles of adjuvant chemotherapy based on carboplatin, while patients with stage I NSGCTs were treated with 2–4 cycles of chemotherapy based on bleomycin, etoposide, and carboplatin. Patients with stage II disease were treated with 4 cycles of adjuvant chemotherapy based on bleomycin, etoposide, and carboplatin, while in patients with stage III disease ifosfamide was added in the therapeutic pattern. RPLND was selected for the treatment of patients with NSGCTs with identified residual disease after completion of adjuvant chemotherapy. Tissue samples were selected from the archives of the Pathology Department of the University Hospital of Patras. None of the patients had received prior chemotherapy or irradiation. The median age of the patients at the time of surgery was 30 years, with a range of 17–78 years. All experiments were performed after obtaining informed consent according to the institutional guidelines. Tumour staging and histopathologic findings were assessed according to the American Joint Committee on Cancer. Clinicopathological characteristics of the patients are summarized in Table 1. After an initial review of all available hematoxylin-eosin stained slides of surgical specimens, serial sections from a representative paraffin block of each case were immunostained. The study was performed in accordance with the precepts established by the Helsinki Declaration, approved by Ethic Committee of Patras University Hospital and patients were enrolled after giving written consent. All data were analyzed anonymously.

2.3. Immunohistochemistry.

Syndecan-4 expression was examined immunohistochemically by using the D-16 goat polyclonal antibody (sc 9499, Santa Cruz, USA) and avidin-biotin-peroxidase complex (Dako Co., Copenhagen, Denmark). Tissue samples were fixed in 10% buffered formalin and embedded in paraffin. Serial 5 μm sections were taken and deparaffinized with xylene and dehydrated with 98% ethanol. Antigen retrieval was performed in a microwave oven in 10 mM citric acid buffer (pH 6.0). Endogenous peroxidase activity was quenched with 3% hydrogen peroxide for 5 min at room temperature. Nonspecific protein binding of the antibodies was blocked by incubation with 3% normal swine serum in PBS for 20 min at room temperature. Slides were incubated with anti-syndecan-4 polyclonal antibody

TABLE 1: Clinicopathological characteristics of the 71 patients with TGCTs.

Variable	n	%
Histological type		
Seminoma	33	46.5
Median age: 35 years		
Nonseminoma	38	53.5
Median age: 26 years		
Embryonal carcinoma	8	11.3
Teratoma	5	7.0
Mixed type	25	35.2
Tumour size (T)		
T_1	26	36.6
T_2	41	57.7
T_3	4	5.6
Vascular-lymphatic invasion		
Negative	32	45.1
Positive	39	54.9
Nodal status (N)		
N_0	36	50.7
N_1	9	12.7
N_2	22	31.0
N_3	4	5.6
Distant metastases (M)		
M_0	63	88.7
M_1	7	9.9
M_2	1	1.4
Stage		
I	36	50.7
II	27	38.0
III	8	11.3

diluted 1 : 150 in PBS containing 1% normal swine serum for 1 hr at room temperature. Obtained antigen-antibody complexes were visualized by 30 min incubation at room temperature, using biotinylated rabbit anti-mouse antibody diluted 1 : 200 and the avidin-biotin-peroxidase technique (Dako Co., Copenhagen, Denmark). The staining was developed with 3,3-diaminobenzidine (DAB)/hydrogen peroxide for 5 min at room temperature and slides were counterstained with hematoxylin. A positive tissue control and a negative reagent control (without primary antibody) were run in parallel. The level of syndecan-4 immunoreactivity in epithelial and stromal cells was expressed by scoring the percentage of syndecan-4 positive cells into three groups: high staining >30% of the cells stained; low staining 10–30% of the cells stained; and negative staining <10% of the cells stained. Syndecan-4 immunoreactivity in the tumour stroma was scored as follows: 0, no staining; 1+, moderate; 2+, strong staining. The level of stromal components immunostaining was graded by scoring the percentage of positivity into two groups: negative (<10% of stromal cells and negative staining of the stroma) and positive (>10% of stromal cells or/and moderate or strong staining of the stroma). Three independent researchers randomly evaluated the specimens using this method.

Endothelial cells in tumour tissues were stained immunohistochemically as described previously [12]. After examination of the slides, six random fields at high magnification (×250) were chosen to be evaluated for the number of microvessels in each slide. The number of microvessels in each section represents the mean of the six independent measurements. The evaluation was performed by three independent investigators blinded to the clinicopathological characteristics and syndecan-4 expression in the corresponding tissues.

2.4. Reverse Transcriptase-Polymerase Chain Reaction (RT-PCR) Analysis. Total RNA was isolated from cell cultures and tissues using the NucleoSpin RNA/Protein extraction kit from Macherey-Nagel (MN GmbH & Co., Germany) following DNase treatment to remove DNA contaminations according to the manufacturers' instructions. Total RNA (1 μg) was reverse transcribed using the PrimeScript 1st strand cDNA synthesis kit (Takara Inc.) using random 6 mers primers provided according to standard protocol suggested. For PCR, primers for syndecan-4 "CTCCTAGAAGGCCGATACTTCT and GGACCTCCGTTCTCTCAAAGAT" and the reference gene glyceraldehyde-3-phosphate dehydrogenase (GAPDH) "ACATCATCCCTGCCTCTACTGG and AGTGGGTGTCGCTGTTGAAGTC" were used.

PCR was performed for 35 cycles (initial denaturation at 94°C for 2 min, denaturation at 94°C for 1 min, annealing at 60°C for 1 min, and extension at 72°C for 1 min in each cycle and final extension at 72°C for 10 min) using 50 ng of template according to DyNAzyme II kit (Finnzymes, Finland). PCR products for syndecan-4 and GAPDH were separated by 2% agarose gel electrophoresis and visualized by ethidium bromide staining. The amounts of PCR products were determined by measuring the fluorescence of the bands using UNIDocMV program (UVI Tech). Relative fluorescence for syndecan-4 was obtained by dividing the fluorescence value for syndecan-4 by that of GAPDH.

2.5. Statistical Analysis. Data were analyzed using GraphPad Prism (Version 3.0 GraphPad Software Inc., San Diego, CA, USA). Statistical analyses were performed using the Fisher's exact tests to evaluate the associations between clinicopathologic variables and syndecan-4 expression. All tests were two tailed and statistical significance was set at $P < 0.05$. To estimate statistical significance of the differences in RT-PCR analyses as well as of microvessel number with stromal expression of syndecan-4, a two-tailed Student's t-test was used.

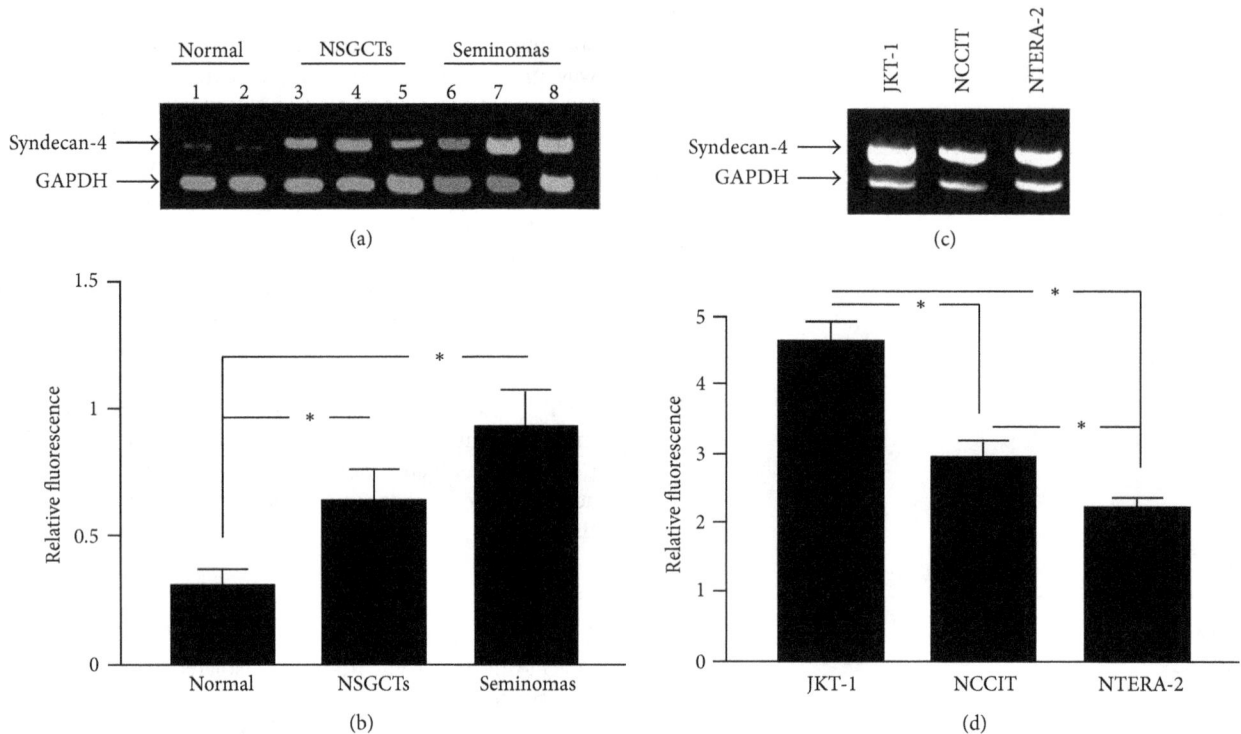

FIGURE 1: Expression of syndecan-4 in testicular germ cell tumours and cell lines. (a) Indicative RT-PCR analyses of syndecan-4 compared to reference gene GAPDH in two control normal testicular tissues (lanes 1 and 2), three NSGCTs (lanes 3, 4, and 5), and in three seminomas (lanes 6, 7 and 8). (b) Semiquantitative analysis of syndecan-4 expression in normal testicular tissues ($n = 6$), NSGCTs ($n = 4$), and seminomas ($n = 5$). (c) RT-PCR analyses of syndecan-4 compared to GAPDH in testicular GCT cell lines. (d) Semiquantitative analysis of syndecan-4 in TGCT cell lines. The data are presented as the median ± SE and analysed using two-tailed Student's t-test ($^*P < 0.05$).

3. Results

3.1. RT-PCR Analysis for Syndecan-4 Expression in GCTs.
A limited number of tissue samples obtained from patients with TGCTs as well as normal testicular tissues were analyzed for the expression level of syndecan-4 by RT-PCR. As shown in Figures 1(a) and 1(b) low expression for syndecan-4 was found in normal testicular tissues (relative fluorescence median ± SE, 0.30 ± 0.06). Statistically significant increase in the expression for syndecan-4 was detected in both NSGCTs (relative fluorescence median ± SE, 0.64 ± 0.11) and seminomas (relative fluorescence median ± SE, 0.79 ± 0.14) (Figures 1(a) and 1(b)), suggesting a higher expression of syndecan-4 by tumour cells or activated stromal cells. To evaluate the expression levels of syndecan-4 in TGCT cell lines, we performed RT-PCR analysis in three tumour cell lines (Figures 1(c) and 1(d)). Syndecan-4 is highly expressed in seminoma cell line JKT-1 (relative fluorescence median ± SE, 4.64 ± 0.26), whereas statistically significant lower expression was detected in teratocarcinoma cell line NCCIT (relative fluorescence median ± SE, 2.90 ± 0.31) and embryonal carcinoma cell line NTERA-2 (relative fluorescence median ± SE, 2.2 ± 0.19).

3.2. Histological Overview of the Patients.
A retrospective study for the expression of syndecan-4 in testicular TGCTs was performed in a population of 71 patients. The histological review (Table 1) of the primary tumours revealed 33 patients (46.5%) with seminoma and 38 patients (53.5%) with NSGCTs. The median age at the time of surgery was 35 years (range 21–78 year) for the patients with seminoma and 26 years (range 17–65) for the patients with NSGCTs. Patients with NSGCTs were divided into three groups: 8 (11.3%) with embryonal carcinoma, 5 (7.0%) with teratoma, and 25 (35.2%) with mixed type TGCTs. Twenty-six of the patients were of T_1 stage, whereas 41 and 4 patients were of T_2 and T_3 stage, respectively. Among the 71 patients with TGCTs, 39 patients (54.9%) were positive for vascular and/or lymphatic invasion and in 35 patients (49.3%) nodal spread of the disease was observed. Only 8 patients (11.3%) were positive for distant metastases (M_1 and M_2). Finally, the categorization of the patients showed that 36 patients (50.7%) were of stage I, whereas 27 (38.0%) and 8 (11.3%) patients were of stage II and stage III, respectively.

3.3. Immunohistochemical Expression of Syndecan-4 in GCTs and Correlation with Clinicopathological Variables.
To evaluate the expression of syndecan-4 by tumour and stromal cells, we performed immunohistochemistry in tissue sections. In normal tests ($n = 4$) weak staining for syndecan-4 was observed in the normal seminiferous tubules showing a cytoplasmic as well as membranous localization with the prominent staining to be detected in the basal cells. No staining was

TABLE 2: Syndecan-4 expression in 71 patients with testicular tumours.

Histological type	Syndecan-4 positive tumour cells			Syndecan-4 stromal staining	
	<10%	10–30%	>30%	Negative	Positive
Seminoma	0	1	32	18	15
NSGCTs	2	15	21	16	22

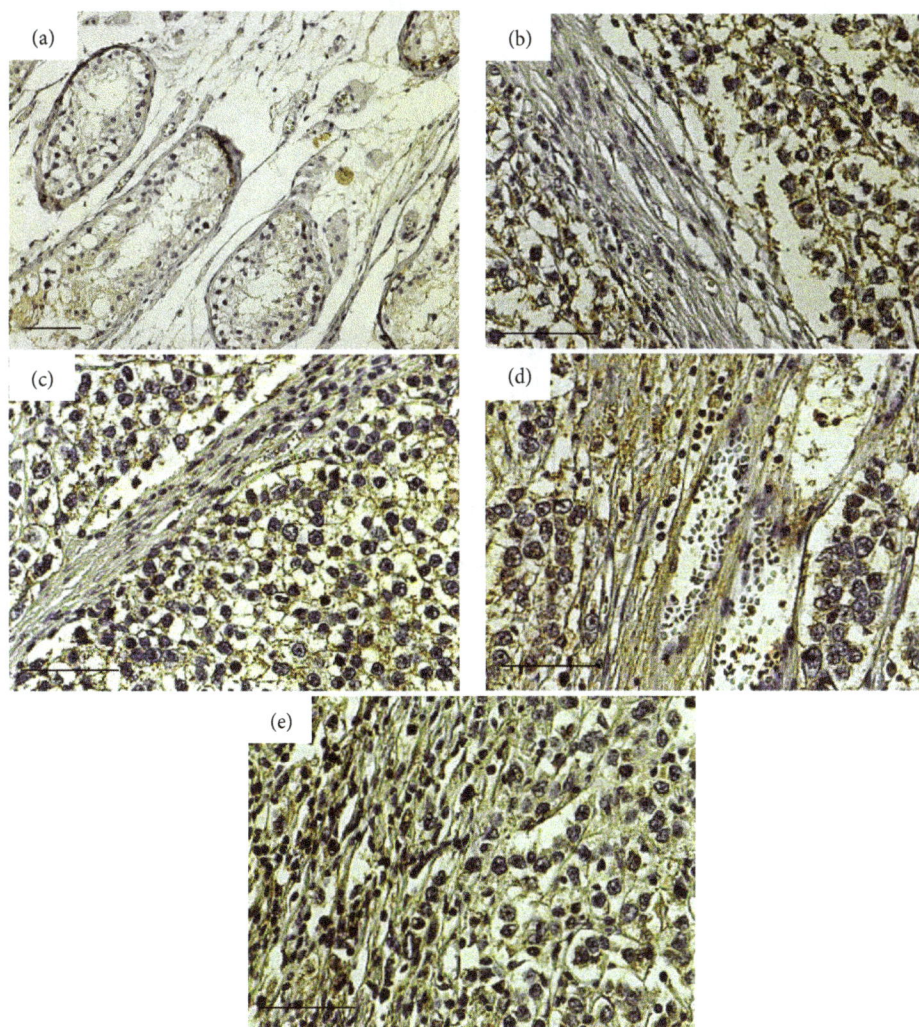

FIGURE 2: Syndecan-4 is highly expressed in seminomas. Weak staining for syndecan-4 in the seminiferous tubules in normal testicular tissue (a). Tumour cell associated staining for syndecan-4 in stage I seminomas ((b) and (c)) and tumour cell associated and stromal staining for syndecan-4 in stage II seminomas ((d) and (e)). Scale bar denotes 50 μm.

observed in the interstitial connective tissue in the interlobular septa surrounding the seminiferous tubules (Figure 2). Syndecan-4 expression in seminoma (Figures 2(b)–2(e)) and NSGCTs (Figures 3(a)–3(d)) was observed in tumour cells, stromal components, or both. High percentage (>30%) of tumour cells positive for syndecan staining was found in 32/33 (97.0%) patients with seminoma and in 21/38 (55.2%) patients with NSGCTs (Table 2). Stromal syndecan-4 staining was observed in 15/33 (45.5%) patients with seminoma and in 22/38 (57.9%) patients with NSGCTs (Table 2). Syndecan-4 was present in both cell membrane and cytoplasm of tumour cells in seminoma (Figures 2(b)–2(e)) and NSGCTs (Figures 3(a)–3(d)). Stromal syndecan-4 staining was seen both in stromal cells and in collagen tissue mainly in seminomas of advanced stage (Figures 2(d) and 2(e)), and in NSGCTs independently of disease stage (Figure 3). Since high staining for syndecan-4 was observed in tumour cells in all patients with seminoma (Figures 2(b)–2(e)), no correlation with the various clinicopathological variables was demonstrated. In contrast, the stromal expression of syndecan-4 in patients with seminoma was associated with the nodal status ($P = 0.04$), vascular/lymphatic invasion ($P = 0.01$), and stage of

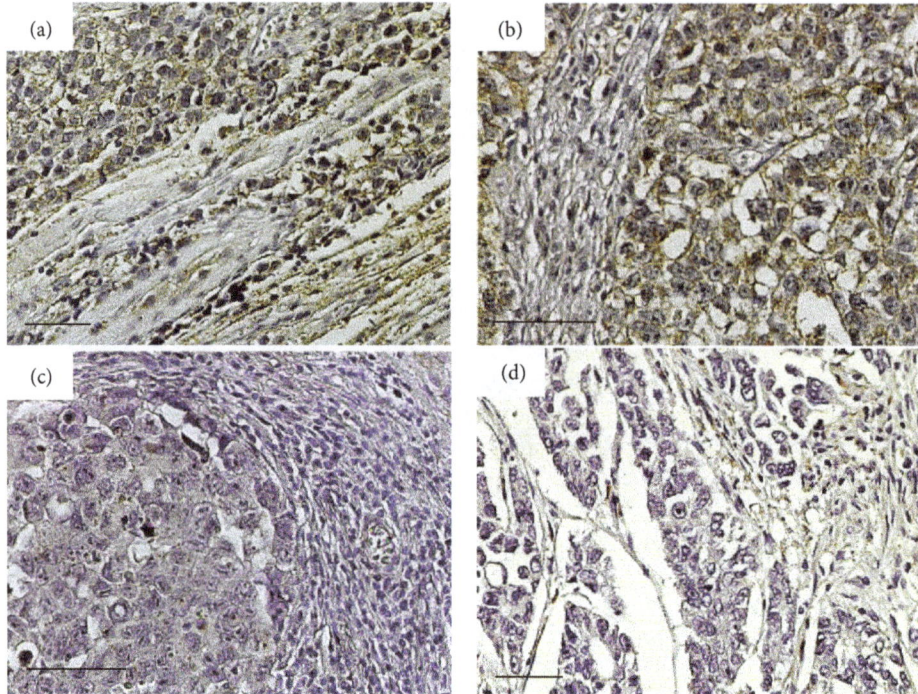

FIGURE 3: Loss of syndecan-4 staining in aggressive NSGCTs. Tumour cell associated staining and variable stromal staining for syndecan-4 in stage I teratoma/seminoma (a) and stage I embryonal/seminoma (b). Variable stromal staining and loss of tumour cell associated immunoreactivity for syndecan-4 in stage II embryonal/yolk sac (c) and embryonal stage III tumours (d). Scale bar denotes 50 μm.

TABLE 3: The association between syndecan-4 stromal staining and the clinicopathologic variables of 33 patients with seminoma.

Variable	Negative	Positive	Statistics
Tumour size (T)			
T_1	9	6	$P = 0.73$
$T_2 + T_3$	9	9	
Nodal status (N)			
N_0	14	6	$P = 0.04$
$N_1 + N_2$	4	9	
Vascular-lymphatic invasion			
Negative	15	6	$P = 0.01$
Positive	3	9	
Disease stage			
I	14	6	$P = 0.04$
II	4	9	

disease ($P = 0.04$) (Figures 2(b)–2(e) and Table 3). Stromal syndecan-4 staining was not related to any of the clinico-pathologic variable in NSGCTs (Table 4). In NSGCTs less tumour cell associated staining for syndecan-4 was observed in patients with advanced disease stage (Figures 3(c) and 3(d)). In NSGCTs 17/38 patients showed low syndecan-4 expression in contrast to seminoma where 22/23 patients exhibited high syndecan-4 staining (Table 2). This loss of syndecan-4 by tumour cells in NSGCTs was associated with nodal metastasis ($P = 0.01$), vascular and lymphatic invasion ($P = 0.01$), and disease stage ($P = 0.01$) (Figure 3 and

Table 4). A clear trend for correlation of lower tumour cells associated staining for syndecan-4 with tumour size and distant metastases was observed as well, but no statistical significance was reached (Table 4).

3.4. Correlation between the Stromal Expression of Syndecan-4 and Microvessel Density. The expression of syndecan-4 in the tumour stroma was correlated with the microvessel density in TGCTs. Figure 4 shows that increased staining for syndecan-4 in the tumour stroma was significantly associated with increased microvessel numbers in TGCTs, suggesting an implication in neovascularization.

4. Discussion

Syndecans are directly implicated in cancer progression [11, 13]. The aim of the present study was to investigate the expression of syndecan-4 in seminomatous and NSGCTs and to examine all possible associations with the malignant behavior of these tumours. In both seminomatous TGCTs and NSGCTs, significantly increased expression for syndecan-4 was detected in tumour cells. Previously, syndecan-4 has been reported to correlate significantly with high histological grade and negative estrogen receptor status [23], suggesting it to be a marker of poor prognosis in breast cancer. Another study failed to confirm this but instead found syndecan-4 expression to be independent of histological tumour grade and histological tumour type [24]. In our previous study, we demonstrated that estradiol does not affect the levels of syndecan-4 in breast cancer cells through ERα signaling

TABLE 4: The association between syndecan-4 stromal and tumour cells staining and the clinicopathologic variables of 38 patients with NSGCTs.

Variable	Stromal staining		Statistics	Syndecan-4 positive tumour cells		Statistics
	Negative	Positive		$\leq 30\%$	>30%	
Tumour size (T)						
T_1	6	4	$P = 0.27$	2	8	$P = 0.14$
$T_2 + T_3$	10	18		15	13	
Nodal status (N)						
N_0	8	8	$P = 0.51$	3	13	$P = 0.01$
$N_1 + N_2 + N_3$	8	14		14	8	
Distant metastases (M)						
M_0	14	16	$P = 0.43$	11	19	$P = 0.11$
$M_1 + M_2$	2	6		6	2	
Vascular-lymphatic invasion						
Negative	6	5	$P = 0.47$	1	10	$P = 0.01$
Positive	10	17		16	11	
Disease stage						
I	8	8	$P = 0.51$	3	13	$P = 0.01$
II + III	8	14		14	8	

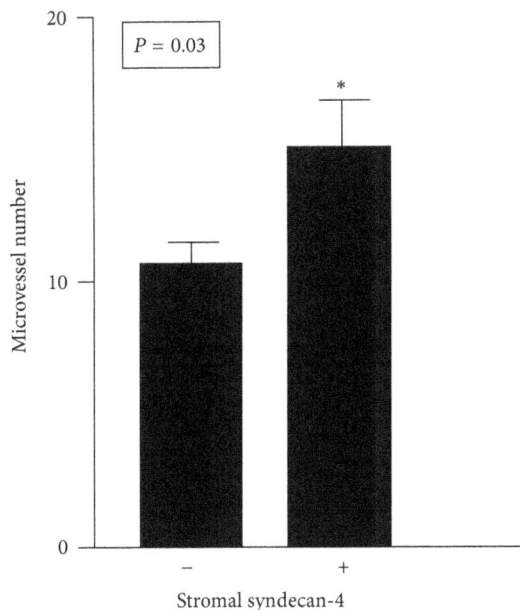

FIGURE 4: Stromal syndecan-4 promotes angiogenesis. Correlation between stromal syndecan-4 expression and microvessel number in TGCTs. Two-tailed P value was obtained by Student's t-test.

although the levels of syndecan-2 are regulated by hormonal treatment [25]. The effects of syndecans in tumour progression may be dependent on organ and tumour type. In this study, the overexpression of syndecan-4 in tumour cells may facilitate the transmission of growth signals in these cells since syndecans are important coreceptors for various growth factors. It has been shown that soluble and membrane-bound forms of syndecan-1 play different roles at different stages of breast cancer progression. The release of soluble syndecan-1 from cell membrane by proteolytic degradation marks a switch from a proliferative to an invasive phenotype in cancer cells [26].

The reduction of syndecans by cancer cell surface is associated with reduced levels of E-cadherin and induction of epithelial to mesenchymal transition (EMT) [26–28]. EMT results in the conversion of malignant epithelial cells into cells with a mesenchymal phenotype and clinically more aggressive tumours. Decreased expression of epithelial syndecan-1 has been reported to be associated with dedifferentiating cancer cells or increasing metastatic potential and to correlate with a poor prognosis in head and neck, gastric, colorectal, laryngeal, cholangiocarcinoma, malignant mesothelioma, hepatocellular, and non-small-cell lung tumours [28–36].

Syndecan-4 is a focal adhesion component in a range of cell types, adherent to several different matrix molecules [37, 38], activating protein kinase C-alpha (PKCa), focal adhesion kinase (FAK), and small GTPase Rho to promote cell adhesion and migration [39–46]. FGF-2 treatment of melanoma cells resulted in the reduction in syndecan-4 expression and downregulation of FAK Y397-phosphorylation thus decreasing cell attachment on FN and promoting their migration [47]. Syndecan-4 overexpressing cells form larger and denser focal adhesions, correlated to stronger attachment and decreased cell migration [48], whereas lack of syndecan-4 engagement promotes amotile fibroblast phenotype where FAK and Rho signaling are downregulated and filopodia are extended [49]. These results suggest that a directed homeostasis in syndecan-4 levels supports optimal migration. Our study revealed that seminomatous TGCTs are characterized by much higher staining of syndecan-4 in tumour cells compared to NSGCTs. The lower staining of syndecan-4 in tumour cells is significantly correlate, with nodal metastasis, vascular and lymphatic invasion, and disease stage in NSGCTs. Identical results were obtained by analysis of syndecan-4 expression in TGCT cell lines. Less aggressive seminoma cells JKT-1 [6] exhibited higher expression levels of syndecan-4 more than aggressive NSGCT cell lines such as embryonal carcinoma cell line NTERA-2 and teratocarcinoma cell line NCCIT. Although increased levels of syndecan-4 in tumour cells may promote cell growth, the imbalanced upregulation of syndecan-4 in seminomas may be related to the lower metastatic potential of these cells, which is a general characteristic of this type of testicular tumours. The lower expression of syndecan-4 in NSGCTs compared to seminomas but still higher than that found in the corresponding normal cells is significantly correlated to the metastatic potential of these tumours. These results strengthen the current opinion that the balanced expression of syndecans by tumour cells regulates their spreading.

Both seminomas and NSGCTs have shown stromal staining for syndecan-4. The presence of syndecan-4 in the tumour stroma was associated with nodal metastasis, vascular and lymphatic invasion, and disease stage only in seminomas. Such stroma immunoreactivity was also reported for syndecan-1 in reactive stromal cells [30, 50, 51]. Since many epithelial mitogens, including FGFs, hepatocyte growth factor (HGF), and heparin-binding epidermal growth factor (HB-EGF), bind to glycosaminoglycan chains of syndecans, it is speculated that syndecans store several growth factors within the tumour stroma and the accumulation of syndecans may contribute to the extensive angiogenesis and stromal proliferation. The expression of syndecans by stromal fibroblasts may create a favorable microenvironment for accelerated tumour cell growth by storing and presenting growth factors to the carcinoma cells. Furthermore, experimental and clinical data have shown that the expression of syndecan-1 by the stromal fibroblasts promotes breast carcinoma growth *in vivo* and stimulates tumour angiogenesis [52, 53]. Our study demonstrates for first time the stromal distribution of syndecan-4 in malignancies. Syndecan-4 present not only in extracellular matrix but also in stromal cells may play a tumour promoting role in TGCTs. Syndecan-4 stromal staining is significantly associated with neovascularization in TGCTs and the metastatic potential only in seminomas and may be involved in the proliferation of reactive stroma, the promotion of angiogenesis, and the formation of chemotactic gradient of growth factors within tumour stroma. In contrast, the stromal expression of syndecan-4 in NSGCTs, which are more aggressive in general, is not important for tumour cells dissemination and this may be only regulated by lower expression of syndecan-4 in tumour cells that directly affects their migratory ability.

5. Conclusions

In conclusion, seminomas and NSGCTs are two different categories of testicular tumours with different expression profiles for syndecan-4. Loss of syndecan-4 overexpression on the surface of tumour cells in NSGCTs is correlated with aggressiveness in contrast to less aggressive seminomas where

syndecan-4 is highly expressed constantly. Furthermore, stromal expression of syndecan-4 promotes angiogenesis in TGCTs and metastatic potential only in seminomas. Our data suggest that syndecan-4 represents a biological marker in patients with TGCTs and further studies can be performed in order to determine the clinical utility of syndecan-4 expression in predicting occult lymph node disease in patients with stage I NSGCTs. The identification of reliable prognostic risk factors for those patients remains one of the most challenging issues in assigning patients to the best therapeutic options according to their individual risk profiles for metastasis.

References

[1] X. Peng, X. Zeng, S. Peng, D. Deng, and J. Zhang, "The association risk of male subfertility and testicular cancer: a systematic review," *PloS ONE*, vol. 4, no. 5, Article ID e5591, 2009.

[2] D. J. Vidrine, J. E. H. M. Hoekstra-Weebers, H. J. Hoekstra, M. A. Tuinman, S. Marani, and E. R. Gritz, "The effects of testicular cancer treatment on health-related quality of life," *Urology*, vol. 75, no. 3, pp. 636–641, 2010.

[3] A. Diéz-Torre, U. Silván, O. De Wever, E. Bruyneel, M. Mareel, and J. Aréchaga, "Germinal tumor invasion and the role of the testicular stroma," *International Journal of Developmental Biology*, vol. 48, no. 5-6, pp. 545–557, 2004.

[4] A. Heidenreich, I. A. Sesterhenn, F. K. Mostofi, and J. W. Moul, "Prognostic risk factors that identify patients with clinical stage I nonseminomatous germ cell tumors at low risk and high risk for metastasis," *Cancer*, vol. 83, no. 5, pp. 1002–1011, 1998.

[5] D. C. Gilbert, I. Chandler, B. Summersgill et al., "Genomic gain and over expression of CCL2 correlate with vascular invasion in stage I non-seminomatous testicular germ-cell tumours," *International Journal of Andrology*, vol. 34, no. 4, pp. e114–e121, 2011.

[6] E. Milia-Argeiti, E. Huet, V. T. Labropoulou et al., "Imbalance of MMP-2 and MMP-9 expression versus TIMP-1 and TIMP-2 reflects increased invasiveness of human testicular germ cell tumours," *International Journal of Andrology*, vol. 35, no. 6, pp. 835–844, 2012.

[7] A. D. Theocharis, D. H. Vynios, N. Papageorgakopoulou, S. S. Skandalis, and D. A. Theocharis, "Altered content composition and structure of glycosaminoglycans and proteoglycans in gastric carcinoma," *International Journal of Biochemistry and Cell Biology*, vol. 35, no. 3, pp. 376–390, 2003.

[8] S. S. Skandalis, V. T. Labropoulou, P. Ravazoula et al., "Versican but not decorin accumulation is related to malignancy in mammographically detected high density and malignant-appearing microcalcifications in non-palpable breast carcinomas," *BMC Cancer*, vol. 11, article 314, 2011.

[9] A. D. Theocharis, "Human colon adenocarcinoma is associated with specific post-translational modifications of versican and decorin," *Biochimica et Biophysica Acta*, vol. 1588, no. 2, pp. 165–172, 2002.

[10] J. Tímár, K. Lapis, J. Dudás, A. Sebestyén, L. Kopper, and I. Kovalszky, "Proteoglycans and tumor progression: janus-faced molecules with contradictory functions in cancer," *Seminars in Cancer Biology*, vol. 12, no. 3, pp. 173–186, 2002.

[11] A. D. Theocharis, S. S. Skandalis, G. N. Tzanakakis, and N. K. Karamanos, "Proteoglycans in health and disease: novel roles

for proteoglycans in malignancy and their pharmacological targeting," *FEBS Journal*, vol. 277, no. 19, pp. 3904–3923, 2010.

[12] V. T. Labropoulou, A. D. Theocharis, P. Ravazoula et al., "Versican but not decorin accumulation is related to metastatic potential and neovascularization in testicular germ cell tumours," *Histopathology*, vol. 49, no. 6, pp. 582–593, 2006.

[13] J. R. Couchman, "Transmembrane signaling proteoglycans," *Annual Review of Cell and Developmental Biology*, vol. 26, pp. 89–114, 2010.

[14] D. M. Beauvais and A. C. Rapraeger, "Syndecan-1-mediated cell spreading requires signaling by $\alpha_v\beta_3$ integrins in human breast carcinoma cells," *Experimental Cell Research*, vol. 286, no. 2, pp. 219–232, 2003.

[15] S. Clasper, S. Vekemans, M. Fiore et al., "Inducible expression of the cell surface heparan sulfate proteoglycan syndecan-2 (fibroglycan) on human activated macrophages can regulate fibroblast growth factor action," *Journal of Biological Chemistry*, vol. 274, no. 34, pp. 24113–24123, 1999.

[16] C. Mundhenke, K. Meyer, S. Drew, and A. Friedl, "Heparan sulfate proteoglycans as regulators of fibroblast growth factor-2 receptor binding in breast carcinomas," *American Journal of Pathology*, vol. 160, no. 1, pp. 185–194, 2002.

[17] M. P. O'Connell, J. L. Fiori, E. K. Kershner et al., "Heparan sulfate proteoglycan modulation of Wnt5A signal transduction in metastatic melanoma cells," *Journal of Biological Chemistry*, vol. 284, no. 42, pp. 28704–28712, 2009.

[18] L. Carvallo, R. Muñoz, F. Bustos et al., "Non-canonical Wnt signaling induces ubiquitination and degradation of Syndecan4," *Journal of Biological Chemistry*, vol. 285, no. 38, pp. 29546–29555, 2010.

[19] S. Brule, V. Friand, A. Sutton, F. Baleux, L. Gattegno, and N. Charnaux, "Glycosaminoglycans and syndecan-4 are involved in SDF-1/CXCL12-mediated invasion of human epitheloid carcinoma HeLa cells," *Biochimica et Biophysica Acta*, vol. 1790, no. 12, pp. 1643–1650, 2009.

[20] F. Charni, V. Friand, O. Haddad et al., "Syndecan-1 and syndecan-4 are involved in RANTES/CCL5-induced migration and invasion of human hepatoma cells," *Biochimica et Biophysica Acta*, vol. 1790, no. 10, pp. 1314–1326, 2009.

[21] K. Vuoriluoto, G. Högnäs, P. Meller, K. Lehti, and J. Ivaska, "Syndecan-1 and -4 differentially regulate oncogenic K-ras dependent cell invasion into collagen through $\alpha_2\beta_1$ integrin and MT1-MMP," *Matrix Biology*, vol. 30, no. 3, pp. 207–217, 2011.

[22] A. Bouskine, A. Vega, M. Nebout, M. Benahmed, and P. Fénichel, "Expression of embryonic stem cell markers in cultured JKT-1, a cell line derived from a human seminoma," *International Journal of Andrology*, vol. 33, no. 1, pp. 54–63, 2010.

[23] F. Baba, K. Swartz, R. Van Buren et al., "Syndecan-1 and syndecan-4 are overexpressed in an estrogen receptor-negative, highly proliferative breast carcinoma subtype," *Breast Cancer Research and Treatment*, vol. 98, no. 1, pp. 91–98, 2006.

[24] M. E. Lendorf, T. Manon-Jensen, P. Kronqvist, H. A. B. Multhaupt, and J. R. Couchman, "Syndecan-1 and syndecan-4 are independent indicators in breast carcinoma," *Journal of Histochemistry and Cytochemistry*, vol. 59, no. 6, pp. 615–629, 2011.

[25] O. C. Kousidou, A. Berdiaki, D. Kletsas et al., "Estradiol-estrogen receptor: a key interplay of the expression of syndecan-2 and metalloproteinase-9 in breast cancer cells," *Molecular Oncology*, vol. 2, no. 3, pp. 223–232, 2008.

[26] V. Nikolova, C.-Y. Koo, S. A. Ibrahim et al., "Differential roles for membrane-bound and soluble syndecan-1 (CD138) in breast cancer progression," *Carcinogenesis*, vol. 30, no. 3, pp. 397–407, 2009.

[27] M. Kato, S. Saunders, H. Nguyen, and M. Bernfield, "Loss of cell surface syndecan-1 causes epithelia to transform into anchorage-independent mesenchyme-like cells," *Molecular Biology of the Cell*, vol. 6, no. 5, pp. 559–576, 1995.

[28] S. Leppa, K. Vleminckx, F. Van Roy, and M. Jalkanen, "Syndecan-1 expression in mammary epithelial tumor cells is E-cadherin-dependent," *Journal of Cell Science*, vol. 109, no. 6, pp. 1393–1403, 1996.

[29] A. Anttonen, M. Kajanti, P. Heikkilä, M. Jalkanen, and H. Joensuu, "Syndecan-1 expression has prognostic significance in head and neck carcinoma," *British Journal of Cancer*, vol. 79, no. 3-4, pp. 558–564, 1999.

[30] J. P. Wiksten, J. Lundin, S. Nordling et al., "Epithelial and stromal syndecan-1 expression as predictor of outcome in patients with gastric cancer," *International Journal of Cancer*, vol. 95, no. 1, pp. 1–6, 2001.

[31] M. Fujiya, J. Watari, T. Ashida et al., "Reduced expression of syndecan-1 affects metastatic potential and clinical outcome in patients with colorectal cancer," *Japanese Journal of Cancer Research*, vol. 92, no. 10, pp. 1074–1081, 2001.

[32] J. Klatka, "Syndecan-1 expression in laryngeal cancer," *European Archives of Oto-Rhino-Laryngology*, vol. 259, no. 3, pp. 115–118, 2002.

[33] K. Harada, S. Masuda, M. Hirano, and Y. Nakanuma, "Reduced expression of syndecan-1 correlates with histologic dedifferentiation, lymph node metastasis, and poor prognosis in intrahepatic cholangiocarcinoma," *Human Pathology*, vol. 34, no. 9, pp. 857–863, 2003.

[34] S. Kumar-Singh, W. Jacobs, K. Dhaene et al., "Syndecan-1 expression in malignant mesothelioma: correlation with cell differentiation, WT1 expression, and clinical outcome," *Journal of Pathology*, vol. 186, no. 3, pp. 300–305, 1998.

[35] A. Matsumoto, M. Ono, Y. Fujimoto, R. L. Gallo, M. Bernfield, and Y. Kohgo, "Reduced expression of syndecan-1 in human hepatocellular carcinoma with high metastatic potential," *International Journal of Cancer*, vol. 74, no. 5, pp. 482–491, 1997.

[36] L. Shah, K. L. Walter, A. C. Borczuk et al., "Expression of syndecan-1 and expression of epidermal growth factor receptor are associated with survival in patients with nonsmall cell lung carcinoma," *Cancer*, vol. 101, no. 7, pp. 1632–1638, 2004.

[37] A. Woods and J. R. Couchman, "Syndecan 4 heparan sulfate proteoglycan is a selectively enriched and widespread focal adhesion component," *Molecular Biology of the Cell*, vol. 5, no. 2, pp. 183–192, 1994.

[38] A. Woods and J. R. Couchman, "Syndecan-4 and focal adhesion function," *Current Opinion in Cell Biology*, vol. 13, no. 5, pp. 578–583, 2001.

[39] A. C. Rapraeger, "Syndecan-regulated receptor signaling," *Journal of Cell Biology*, vol. 149, no. 5, pp. 995–997, 2000.

[40] A. Woods, R. L. Longley, S. Tumova, and J. R. Couchman, "Syndecan-4 binding to the high affinity heparin-binding domain of fibronectin drives focal adhesion formation in fibroblasts," *Archives of Biochemistry and Biophysics*, vol. 374, no. 1, pp. 66–72, 2000.

[41] S. A. Wilcox-Adelman, F. Denhez, and P. F. Goetinck, "Syndecan-4 modulates focal adhesion kinase phosphorylation," *Journal of Biological Chemistry*, vol. 277, no. 36, pp. 32970–32977, 2002.

[42] L. A. Cary, J. F. Chang, and J.-L. Guan, "Stimulation of cell migration by overexpression of focal adhesion kinase and its association with Src and Fyn," *Journal of Cell Science*, vol. 109, no. 7, pp. 1787–1794, 1996.

[43] S. Saoncella, F. Echtermeyer, F. Denhez et al., "Syndecan-4 signals cooperatively with integrins in a Rho-dependent manner in the assembly of focal adhesions and actin stress fibers," *Proceedings of the National Academy of Sciences of the United States of America*, vol. 96, no. 6, pp. 2805–2810, 1999.

[44] S.-T. Lim, R. L. Longley, J. R. Couchman, and A. Woods, "Direct binding of syndecan-4 cytoplasmic domain to the catalytic domain of protein kinase Cα (PKCα) increases focal adhesion localization of PKCα," *Journal of Biological Chemistry*, vol. 278, no. 16, pp. 13795–13802, 2003.

[45] E.-S. Oh, A. Woods, and J. R. Couchman, "Syndecan-4 proteoglycan regulates the distribution and activity of protein kinase C," *Journal of Biological Chemistry*, vol. 272, no. 13, pp. 8133–8136, 1997.

[46] A. Horowitz, E. Tkachenko, and M. Simons, "Fibroblast growth factor-specific modulation of cellular response by syndecan-4," *Journal of Cell Biology*, vol. 157, no. 4, pp. 715–725, 2002.

[47] G. Chalkiadaki, D. Nikitovic, A. Berdiaki et al., "Fibroblast growth factor-2 modulates melanoma adhesion and migration through a syndecan-4-dependent mechanism," *International Journal of Biochemistry and Cell Biology*, vol. 41, no. 6, pp. 1323–1331, 2009.

[48] R. L. Longley, A. Woods, A. Fleetwood, G. J. Cowling, J. T. Gallagher, and J. R. Couchman, "Control of morphology, cytoskeleton and migration by syndecan-4," *Journal of Cell Science*, vol. 112, no. 20, pp. 3421–3431, 1999.

[49] K. S. Midwood, Y. Mao, H. C. Hsia, L. V. Valenick, and J. E. Schwarzbauer, "Modulation of cell-fibronectin matrix interactions during tissue repair," *Journal of Investigative Dermatology Symposium Proceedings*, vol. 11, no. 1, pp. 73–78, 2006.

[50] M. J. Stanley, M. W. Stanley, R. D. Sanderson, and R. Zera, "Syndecan-1 expression is induced in the stroma of infiltrating breast carcinoma," *American Journal of Clinical Pathology*, vol. 112, no. 3, pp. 377–383, 1999.

[51] D. Mennerich, A. Vogel, I. Klaman et al., "Shift of syndecan-1 expression from epithelial to stromal cells during progression of solid tumours," *European Journal of Cancer*, vol. 40, no. 9, pp. 1373–1382, 2004.

[52] T. Maeda, C. M. Alexander, and A. Friedl, "Induction of syndecan-1 expression in stromal fibroblasts promotes proliferation of human breast cancer cells," *Cancer Research*, vol. 64, no. 2, pp. 612–621, 2004.

[53] T. Maeda, J. Desouky, and A. Friedl, "Syndecan-1 expression by stromal fibroblasts promotes breast carcinoma growth *in vivo* and stimulates tumor angiogenesis," *Oncogene*, vol. 25, no. 9, pp. 1408–1412, 2006.

12

Extracellular Matrix Degradation and Tissue Remodeling in Periprosthetic Loosening and Osteolysis: Focus on Matrix Metalloproteinases, Their Endogenous Tissue Inhibitors, and the Proteasome

Spyros A. Syggelos,[1] **Alexios J. Aletras,**[2] **Ioanna Smirlaki,**[2] **and Spyros S. Skandalis**[2]

[1] *Department of Anatomy, Histology, Embryology, Medical School, University of Patras, 26500 Patras, Greece*
[2] *Laboratory of Biochemistry, Department of Chemistry, University of Patras, 26500 Patras, Greece*

Correspondence should be addressed to Spyros S. Skandalis; skandalis@upatras.gr

Academic Editor: Achilleas D. Theocharis

The leading complication of total joint replacement is periprosthetic osteolysis, which often results in aseptic loosening of the implant, leading to revision surgery. Extracellular matrix degradation and connective tissue remodeling around implants have been considered as major biological events in the periprosthetic loosening. Critical mediators of wear particle-induced inflammatory osteolysis released by periprosthetic synovial cells (mainly macrophages) are inflammatory cytokines, chemokines, and proteolytic enzymes, mainly matrix metalloproteinases (MMPs). Numerous studies reveal a strong interdependence of MMP expression and activity with the molecular mechanisms that control the composition and turnover of periprosthetic matrices. MMPs can either actively modulate or be modulated by the molecular mechanisms that determine the debris-induced remodeling of the periprosthetic microenvironment. In the present study, the molecular mechanisms that control the composition, turnover, and activity of matrix macromolecules within the periprosthetic microenvironment exposed to wear debris are summarized and presented. Special emphasis is given to MMPs and their endogenous tissue inhibitors (TIMPs), as well as to the proteasome pathway, which appears to be an elegant molecular regulator of specific matrix macromolecules (including specific MMPs and TIMPs). Furthermore, strong rationale for potential clinical applications of the described molecular mechanisms to the treatment of periprosthetic loosening and osteolysis is provided.

1. Pathobiology of Periprosthetic Loosening Process

The total hip or knee replacement is an operation whereby the damaged cartilage and the subchondral sclerotic bone of the hip or knee joint are surgically replaced with artificial materials. The continuous improvement of the materials and the surgical techniques have given comfort to patients suffering from painful diseases of the joints, such as primary osteoarthritis and secondary ones caused by rheumatoid arthritis, posttraumatic conditions, congenital dysplasia or dislocation, and aseptic necrosis of the femoral head. After the improvement in prophylaxis against infection, aseptic loosening of endoprostheses represents the predominant complication of this operation, which usually occurs during the second decade, after the primary arthroplasty. Although many reports have been published on the pathogenesis of periprosthetic loosening, the precise biological mechanisms responsible for this process have not yet been completely elucidated.

Wear-generated particular debris at the interface between implant components is associated with chronic inflammation and osteolysis, limits the lifespan of the implants, and is the main cause of initiating this destructive process. However, many other factors, such as cyclic loading or micromotion of the implants and hydrostatic fluid pressure, have also been implicated revealing the high heterogeneity in the histology of the tissue around the prosthesis [1]. Evidence in support

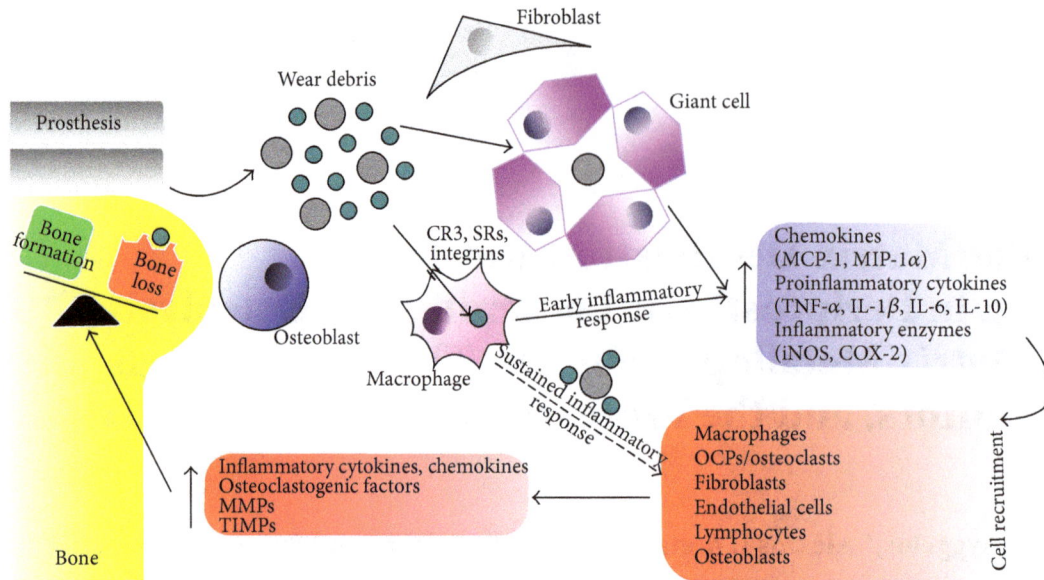

FIGURE 1: Schematic representation of periprosthetic loosening and osteolysis. Implant-derived wear debris induces an early inflammatory response from the resident or infiltrating macrophages in the periprosthetic tissue. Small particles are phagocytosed, whereas the larger induce fusion of macrophages and giant cell formation. Activated macrophages release proinflammatory cytokines, chemokines, and enzymes that recruit multiple cell types within periprosthetic tissue, which are further activated by the particles resulting in sustained inflammation, increased secretion of cytokines/chemokines/osteoclastogenic factors/MMPs/TIMPs, and osteolysis.

of the central role of wear debris in periprosthetic loosening and osteolysis includes the observations that osteolysis is correlated with higher wear rates [2] and that vast numbers of wear particles are found associated with the periprosthetic interfacial membrane removed during revision surgery [3–5]. Furthermore, experimental systems have demonstrated that particulate debris can induce osteolysis in a variety of animal models [6–12] and inflammatory responses in cultured macrophages [8, 13–17]. Wear debris may include particles from all the various components of the prosthesis (such as polyethylene, metal, and ceramic) as well as bone cement [18]. Since cellular responses are highly dependent upon the composition, size, and shape of particles, the type of prosthesis and bearing surface used may have a significant impact on the potential for development of osteolysis [19].

The release of implant-derived particles induces a cellular host response, which initially is taking place in the pseudocapsular tissue (PCT). This membranous tissue is formed postoperatively around the artificial joint and practically replaces the normal joint capsular tissue, which is usually removed during the primary joint replacement procedure. The most important and active cells in this tissue are macrophages and fibroblasts, which after their interaction with the wear debris produce most of the soluble chemical factors and mediators, which are going to be analyzed below. Additionally these soluble factors migrate through the joint fluid (pseudosynovial fluid, PSSF) in the layer between the implant and the bone (interface), where they continue their action, mainly affecting the bony tissue. Finally the fibrous interface tissue (IFT), between the prosthesis and the bone, is formed and this leads to failure of the implant, which becomes loose. The communication

of the interface layer with the space of the initial foreign body reaction is described as effective joint space, may result an early micromotion of the implant, and could be related to the surgical technique [20]. The interface tissue is heavily infiltrated with several different cell types, mainly macrophages, lymphocytes, fibroblasts, endothelial cells, and osteoclast precursors (OCPs)/osteoclasts. Beside enhanced and chronic inflammatory reactions in the periprosthetic region, the cellular recruitment to this region is promoted by induced chemokine expression [21–25]. Macrophages activation by phagocytosis of the wear debris particles, which are impervious to enzymatic degradation, has been shown to be the principle pathophysiologic mechanism in particle-induced periprosthetic osteolysis. Activated macrophages secrete proinflammatory and osteoclastogenic cytokines as well as proteolytic enzymes exacerbating the inflammatory response leading to activation of a periprosthetic osteolytic cascade (Figure 1). It is known that particles smaller than 8–10 μm are phagocytosed by macrophages, while bigger particles induce giant cell reaction and are associated with such cells [26]. However, it has been reported that contact between wear particles and macrophages without phagocytosis is also important for the signal transduction of cytokines and activation of macrophages [14].

The identity of the macrophage surface receptors responsible for recognition of the particles and the full repertoire of signaling cascades initiated or modified by particle binding remain poorly understood although macrophages are the best-characterized cellular target for particle action. One important consideration in determining the involvement of specific cell surface receptors is the extent to which different particles become opsonized with host serum proteins prior

Extracellular Matrix Degradation and Tissue Remodeling in Periprosthetic Loosening and Osteolysis: Focus on Matrix
Metalloproteinases, Their Endogenous Tissue Inhibitors, and the Proteasome

113

to phagocytosis. There is evidence that polyethylene activates complement [27], and this would argue in favor of a role for complement receptors, such as complement receptor 3 (CR3), in particle uptake. Indeed, CR3 expressing phagocytes have been detected in granulomatous lesions associated with hip replacement [28]. An involvement of CR3 in particle action is also supported by observations that antibodies against CR3 reduce particle uptake [14] and that activation of this receptor mimics several aspects of downstream signaling by particles in that MAP kinases [29] and the transcription factors nuclear factor κB (NF-κB) and activator protein-1 (AP-1) are activated [14, 30], and production of proinflammatory cytokines [14] and chemokines [30] is elevated. By contrast, research on alveolar macrophage response to environmental particulate matter has implicated scavenger receptors (SRs), such as scavenger receptor A (macrophage receptor with collagenous structure; MARCO), in opsonin-independent uptake of titanium particles [31], suggesting that different particles may use different surface receptors. Accordingly, Rakshit and coworkers suggested the involvement of opsonization, complement, and integrin receptors, including CR3 and fibronectin receptors, in polymethylmethacrylate action, and an involvement of scavenger receptors (scavenger receptor A) in macrophages responses to titanium [32]. This would provide an intriguing explanation of the abilities of different types of wear debris to elicit particle type-specific responses in cultured macrophages. The concept that opsonization may differentially regulate uptake of different compositions of wear debris is also supported by observations that the spectra of adherent human serum proteins demonstrate a level of particle specificity [33].

Other cell types that are abundant within the periprosthetic tissue are fibroblasts and osteoclasts. Frequently, a proliferation of periprosthetic fibroblasts, which constitute a major source of proinflammatory and osteoclastogenic mediators [34–37], is accompanied by tissue hypervascularization. Periprosthetic fibroblasts exposed to wear and/or proinflammatory mediators are a major source of the receptor activator for nuclear factor κB ligand (RANKL) required to drive osteoclastogenesis in patients with osteolysis (discussed below). Particles can also induce production in cultured fibroblasts of proinflammatory mediators, collagenases, and stromelysins [36, 37], which contribute to the development of osteolysis and chemokines, which promote the recruitment of increased numbers of osteoclast precursors to periprosthetic tissues. The final cellular consequence of particle action is an excess of osteoclast activity, which results in uncontrolled bone erosion. Osteoclasts, which are the unique cell type capable to resorb bone, are derived from circulating hematopoietic cells of the monocyte/macrophage lineage. Therefore, wear particles might increase osteoclast activity either by generation of functional osteoclasts from osteoclast precursor cells within the periprosthetic space or recruitment of osteoclast precursor cells from the blood or both [19]. However, it is not only an increased osteoclastic bone resorption due to particle exposure that can disrupt the balance in the bone remodeling process, but also a reduced bone formation caused by a direct negative impact of particles on osteoblasts [38]. As shown by Lochner and coworkers, wear particles

can alter the metabolism of human primary osteoblasts [39]. In particular, metallic particles in the wear debris of cemented hip endoprostheses can compromise the vitality and activity of bone cells and bone matrix. In consequence, this may lead to a reduction of implant integration strength. Osteoblasts are rather responsible for bone formation but can indirectly participate in bone degeneration by changing cell viability and expression of specific chemokines as well as directly by the secretion of preosteolytic mediators and specific proteinases.

Collectively, the extensive body of research on *in vitro* cellular responses to wear debris suggests that while an inflammatory response by macrophages is central to the development of periprosthetic osteolysis, the detailed nature of this response will vary based upon several parameters, including prosthetic type, patterns of wear, cellular crosstalk, host factors, and cell-associated/extracellular molecular effectors.

2. Matrix Metalloproteinases, Their Endogenous Tissue Inhibitors, and Cytokines/Chemokines in the Periprosthetic Extracellular Matrix

Extracellular matrix (ECM) degradation and connective tissue remodeling around implants have been considered as major biological events in the periprosthetic loosening. Critical mediators of wear particle-induced inflammatory osteolysis released by periprosthetic synovial cells (mainly macrophages) are inflammatory cytokines (such as tumor necrosis factor-α (TNF-α), interleukin- (IL-) 1β, IL-6, and IL-10), chemokines (monocyte chemoattractant protein-1 (MCP-1), macrophage inflammatory protein-1alpha (MIP-1α)), inflammatory enzymes (inducible nitric oxide synthase (iNOS), cyclooxygenase-2 (COX-2)), and proteolytic enzymes, mainly matrix metalloproteinases (MMPs).

TNF-α, IL-1β, and IL-6 are known to be important molecules involved in the foreign body reaction process, and their upregulation is considered to be a marker of inflammation. They are well recognized as key proinflammatory cytokines that provoke cellular proliferation, stimulate osteoclast formation, and increase bone resorption around prostheses [40–44]. In particular, TNF-α has a catabolic effect on bone. It can upregulate bone resorption in cultured mouse calvaria by a prostaglandin-independent mechanism and stimulates osteoblasts to produce osteoresorptive factors such as IL-6 and prostaglandin E_2 (PGE$_2$) [45, 46]. High levels of TNF-α have been detected in periprosthetic tissues of loose endoprostheses with focal osteolysis [47]. It has also been shown to exhibit a synergistic effect with titanium particles, when added in osteoblast culture [48]. IL-1β induces differentiation and proliferation of osteoclasts as well as the production of MMPs and PGE$_2$ from fibroblasts and synovial cells [49, 50]. It also reduces the osteocalcin production by the osteoblasts [51]. According to Jiranek and coworkers, IL-1β might play a significant role in the formation of IFT, because of its stimulatory activity on fibroblasts [52]. Kusano and coworkers have shown that IL-1β augments bone resorption in mouse calvaria culture *in vitro*, by inducing MMP-2, MMP-3, MMP-9, and MMP-13 production [53]. IL-6 is strongly implicated in bone catabolism. It is produced by the osteoblasts and induces bone resorption [54]. It also stimulates the formation of osteoclast-like cells in long-term human marrow cultures [55]. In periprosthetic tissues from loose orthopaedic implants with osteolysis, IL-6 levels are much higher than in tissues from loose implants without bone loss [47]. The role of prostaglandins in mediating pseudomembrane-associated bone resorption remains questionable. It is proposed, from *in vitro* studies, that prostaglandins play an important role in bone resorption [56]. Periprosthetic tissue, cultured in the presence of indomethacin, showed less bone resorptive capacity. Other investigators have shown that conditioned media from predialysed periprosthetic tissue cultures maintained their ability to cause bone resorption, indicating that the prostaglandins, removed by dialysis, had no effect whatsoever upon bone resorption [57]. Therefore, prostaglandins may be implicated in the loosening process through complex mechanisms involving interactions with MMPs and cytokines. On the other hand, IL-10 is synthesized by activated immune cells, in particular monocytes/macrophages, and has profound anti-inflammatory and immunoregulatory effects. This anti-inflammatory cytokine diminishes the expression of inflammatory mediators, inhibits antigen presentation, and induces expression of endogenous TNF-α inhibitors (soluble TNF receptors) to suppress the effects of proinflammatory cytokines in periprosthetic tissues [47, 58].

Chemokines play pivotal roles in the recruitment of inflammatory and immune cells subsequent to the development of periprosthetic inflammation following wear particle generation. MCP-1 and MIP-1α are two chemokines involved in this adverse process by recruiting monocytes/macrophages and lymphocytes to the site around prostheses and play important roles in periprosthetic osteolysis [59–61]. Previous studies suggest that high levels of inflammatory enzymes, such as iNOS and COX-2, are also present in the tissues around prostheses and therefore may account for periprosthetic bone resorption [62]. Macrophages are the major inflammatory cells accounting for this response. iNOS is closely involved in regulating inflammatory responses and COX-2 is induced by many cytokines, such as TNF-α and IL-1β, and the overexpression of these two enzymes plays a key role in chronic inflammatory diseases [63]. Furthermore, iNOS and COX-2, as well as TNF-α and IL-6, are inductive regulators of osteoclastogenesis [64].

A key role in periprosthetic ECM remodeling and destruction belongs to MMPs because of their ability to degrade in concert most extracellular matrix components, such as collagens, gelatin, elastin, laminin, fibronectin, or proteoglycan core proteins. MMPs contain four well-defined domains: a signal peptide, a propeptide with a conserved cysteine residue, a catalytic domain with a Zn-binding site, and a hemopexin-like domain at the COOH-terminal region, and they are frequently subgrouped based on substrate specificities and sequence characteristics. There are six main families of MMPs: collagenases (MMP-1, MMP-8, and MMP-13), gelatinases (MMP-2 and MMP-9), stromelysins (MMP-3, MMP-10, and MMP-11), matrilysins (MMP-7 and MMP-26), membrane-type MMPs (MT-MMPs: MMP-14, -15, -16, -17, -24, and -25), and other MMPs, which are not categorized in any of the previous groups (MMP-12, -19, -20, -21, -23, -27, and -28). The expression of MMPs is under tight control at the transcription level and their proteolytic activity is regulated posttranslationally in several ways [65]. MMPs are synthesized as zymogens, which are then activated extracellularly, with the exception of MMP-11 (stromelysin 3), MT-MMPs, MMP-21, MMP-23, and MMP-28. Although pro-MMPs can be activated *in vitro* by various proteolytic and nonproteolytic means, the *in vivo* activation mechanisms have not yet been completely clarified. Further, the proteolytic activity of MMPs is regulated by specific tissue inhibitors of metalloproteinases (TIMPs). Four TIMPs have been identified (named TIMP-1 to -4), which form high-affinity 1:1 noncovalent complexes with all active MMPs, thereby inhibiting their action. TIMPs inhibit all MMPs tested so far, but TIMP-1 is a poor inhibitor for MT3-MMP, MT5-MMP, and MMP-19. TIMP-3 has been shown to inhibit members of the ADAM (a disintegrin and metalloproteinase) family (ADAM-10, -12, and -17) and ADAMTSs (ADAM with thrombospondin motifs) (ADAMTS-1, -4, and -5). TIMP-1 inhibits ADAM-10. While TIMP-1-null mice and TIMP-2-null mice do not exhibit obvious abnormalities, TIMP-3 ablation in mice causes lung emphysema-like alveolar damage [66] and faster apoptosis of mammary epithelial cells after weaning [67], indicating that TIMP-3 is a major regulator of metalloproteinase activities *in vivo*. However, the functions of TIMPs go beyond the inhibition of MMPs and are also partakers in the activation and coactivation of others [68]. The balance between the levels of activated MMPs and free TIMPs determines in part the net MMP activity. In addition to regulating the MMPs, TIMPs have also been shown to have angiogenic and growth factor-like activities [69].

Numerous studies have demonstrated that specific MMPs and TIMPs are expressed in periprosthetic tissues and are critically involved in the bone resorption and subsequent implant failure (Tables 1 and 2). In a study conducted by Takei and coworkers, the mRNA expression patterns of 16 different types of MMPs in synovium-like interface tissues between bone and prosthesis of loose artificial hip joints were analyzed to evaluate which MMPs were present at the mRNA level and possibly contributed to periprosthetic loosening [70]. It was shown that periprosthetic tissues were characterized by highly elevated expression of MMP-1, MMP-9, MMP-10, MMP-12, and MMP-13; moderate expression of MMP-2, MMP-7, MMP-8, MMP-11, MT1-MMP (MMP-14), MT2-MMP (MMP-15), MT3-MMP (MMP-16), MT4-MMP (MMP-17), and MMP-19; lower expression of MMP-3; and little significance of MMP-20. Quantitative analysis of mRNA expression of their endogenous inhibitors (TIMPs) in periprosthetic tissues showed a significant upregulation of TIMP-1, -2, and -3 mRNA expressions in contrast to the decreased levels of TIMP-4 [71]. On protein level, strong immunoreactivity was observed for the extracellular matrix metalloproteinase inducer (EMMPRIN/CD147) in the lining-like layers, sublining area, and vascular endothelium of synovium-like interface tissue around loosened prostheses.

Extracellular Matrix Degradation and Tissue Remodeling in Periprosthetic Loosening and Osteolysis: Focus on Matrix
Metalloproteinases, Their Endogenous Tissue Inhibitors, and the Proteasome

115

TABLE 1: Matrix metalloproteinases (MMPs) in periprosthetic microenvironment (expression and/or activity: ↑ with bold data: high; ↑ without bold data: moderate).

MMPs		Substrates	Expression and/or activity in periprosthetic microenvironment [References]
Collagenases			
Contain hemopexin domain and peptide linking with catalytic domain	**MMP-1** (interstitial collagenase; collagenase 1) **MMP-8** (neutrophil collagenase; collagenase 2) **MMP-13** (collagenase 3)	Collagen type I, III, V, VII, VIII, X, gelatin, IL-1β, MMP-2, -9, fibronectin	↑**MMP-1** [39, 58, 70, 72, 80, 82, 90] ↑MMP-8 [70] ↑**MMP-13** [70, 82, 89]
Gelatinases			
High substrate specificity to native collagen and gelatin	**MMP-2** (gelatinase A; 72 kDa metalloproteinase) **MMP-9** (gelatinase B; 92 kDa metalloproteinase)	Collagen type IV, V, VII, X, proteoglycans, gelatin, elastin, laminin	↑**MMP-2** [58, 70, 73, 74, 76, 77, 80, 90] ↑**MMP-9** [58, 70, 76, 77, 80, 90, 91]
Stromelysins			
Metalloproteinases of stroma	**MMP-3** (stromelysin 1) **MMP-10** (stromelysin 2) **MMP-11** (stromelysin 3)	Proteoglycans, fibronectin, laminin, elastin, gelatin, plasminogen, vitronectin, fibrinogen, fibrin, collagen type III, IV, V, antithrombin III, MMP-1, -2, -8, -9, -13	↑MMP-3 [58, 70, 74, 80] ↑**MMP-10** [70] ↑MMP-11 [70]
Matrilysins			
The smallest among MMPs, lack of hemopexin domain	**MMP-7** (matrilysin, metalloendopeptidase) **MMP-26** (matrilysin-2, endometase)	Collagen type IV, proteoglycans, glycoproteins, gelatin	↑MMP-7 [70]
Membrane-type MMPs			
(A) Transmembrane-type MMPs	**MMP-14** (MT1-MMP) **MMP-15** (MT2-MMP) **MMP-16** (MT3-MMP) **MMP-24** (MT5-MMP)	Collagen type I, II, III, gelatin, elastin, laminin, fibronectin, fibrin, proteoglycans, proMMP-2, proMMP-13	↑**MMP-14** [70, 73] ↑MMP-15 [70] ↑MMP-16 [70]
(B) GPI-anchored MMPs	**MMP-17** (MT4-MMP) **MMP-25** (MT6-MMP)		↑MMP-17 [70]
Other MMPs			
MMPs that are not categorized in any of the previous groups	**MMP-12** (macrophage metalloelastase) **MMP-19** **MMP-20** (enamelysin) **MMP-21, MMP-23** **MMP-27, MMP-28**		↑**MMP-12** [70] ↑MMP-19 [70]

Moreover, double immunofluorescence labeling revealed EMMPRIN/MMP-1 double-positive cells in lining-like areas and the sublining area of interface tissue. These data indicated that EMMPRIN expression was upregulated in interface tissues, and that locally accumulated EMMPRIN may modulate MMP-1 expression [72].

In another study, Nawrocki and coworkers used immunohistochemistry (IHC) to identify the cells responsible for the synthesis of MMPs in the periprosthetic microenvironment [73]. MMP-2 (gelatinase A) and its activator MT1-MMP were strongly detected in macrophages and multinucleated giant cells in contact with polyethylene wear debris. Similar results have been also obtained by other IHC studies on MMP-2 in

this pathological process [58, 74, 75]. Indeed, these studies reported the expression of MMP-2, as well as those of other MMPs, such as MMP-9 and MMP-1 and, in a more restricted pattern, MMP-3, in macrophages, fibroblasts, and endothelial cells. The strong expression of MMP-2 and its activator MT1-MMP in phagocytic cells of periprosthetic samples suggests their contribution to aseptic loosening of prosthetic components. These data are supported by the observation that high levels of gelatinolytic activities were also previously detected in the same type of lesion [76–79]. Of particular interest was the colocalization of MMP-2, MT1-MMP, and TIMP-2 in the same cells [73]. The strong expression of TIMP-2 in interface tissue around implants was also reported by Ishiguro and

TABLE 2: Tissue inhibitors of metalloproteinases (TIMPs) in periprosthetic microenvironment (expression and/or activity: ↑ with bold data: high; ↓ without bold data: low).

TIMPs	Preferred MMP/ADAM/ADAMTS	Expression and/or activity in periprosthetic microenvironment [References]
TIMP-1	Most MMPs, ADAM-10 (inhibition). MT3-MMP, MT5-MMP, MMP-19 (weak inhibition)	↑**TIMP-1** [71, 74, 76, 78, 80, 82]
TIMP-2	Most MMPs (inhibition). MMP-2 (activation)	↑**TIMP-2** [71, 73, 76, 80, 82]
TIMP-3	Most MMPs, ADAM-10, -12, -17, and ADAMTS-1, -4, -5 (inhibition). MMP-2, MT3-MMP (activation)	↑**TIMP-3** [71]
TIMP-4	Most MMPs (inhibition)	↓TIMP-4 [71]

coworkers [80]. These data support the concept of Strongin and coworkers, who postulated that proMMP-2 activation could be mediated by a trimolecular stoichiometric complex involving MMP-2, TIMP-2, and MT1-MMP [81]. More specifically, these authors demonstrated that the activated form of MT1-MMP acts as a cell surface TIMP-2 receptor. The MT1-MMP/TIMP-2 complex may in turn serve as a receptor for proMMP-2, leading to its processing into the active enzyme. Interestingly, the detection of a soluble type of MT1-MMP (~56 kDa) in synovial and pseudosynovial fluid of patients with rheumatoid arthritis, osteoarthritis, and loose arthroplasty endoprostheses has been previously reported, without clarifying the origin of this type or its activation state. It was proposed that this form was probably processed proteolytically from the transmembrane type of MT1-MMP [75]. A protein band of ~56 kDa was also detected in periprosthetic tissues extracts and pseudosynovial fluids from loose arthroplasty endoprostheses that was ascribed to a soluble form of MT1-MMP [82].

The contribution of different members of the MMP family in gelatinolytic and collagenolytic potential was evaluated by gelatin zymography, and the degradation of synthetic dinitrophenyl-Pro-Gln-Gly-Ile-Ala-Gly-Gln-D-Arg (DNP-S) together with reverse phase high performance liquid chromatography, respectively [82]. Activated species of both MMP-1 and MMP-13 were identified in most periprosthetic tissues, which could be responsible for the detected DNP-S-degrading activity, while the gelatinases MMP-2 and MMP-9 did not contribute in this potential, since they mainly existed in complex with TIMP-2 and TIMP-1, respectively. These data indicated that MMP-1 and MMP-13 may play a key role in the degradation of periprosthetic ECM, since they degrade native type-I and type-III collagens. Moreover, they may directly contribute to bone resorption, by removing the osteoid layer from calcified bone, facilitating the osteoclastic bone resorption [83–85]. Accordingly, it has been previously reported that periprosthetic tissue extracts exhibited high TIMP-free collagenolytic activity although TIMP-1 and

TIMP-2 have been detected in periprosthetic tissues [78, 79]. TIMPs produced by pseudosynoviocytes may be released into synovial fluid to limit MMP proteolysis, but their localization far from local degradation sites leads to the hypothesis of a disruption of the MMP-TIMP balance in favor of MMPs surrounding wear particles.

Immunohistochemical study of the plasminogen activation system, which is closely associated with MMP activities, disclosed localization in periprosthetic tissues of urokinase plasminogen activator (uPA), uPA-receptor (uPAR), and tissue type plasminogen activator (tPA) in macrophages with phagocytosed metal, polyethylene, cement particles, or accompanying pieces of necrotic bone [86]. Plasminogen activator inhibitor-1 (PAI-1) staining was present in the neighboring areas that stained for uPA or tPA, but PAI-1 staining was also found overlapping and outside these areas. These findings suggest a role for the uPA/uPAR and PAI-1 in activation and focalization of extracellular matrix degradation in periprosthetic tissues. The expression of the plasminogen activation system by macrophages containing phagocytosed material suggests undegradable microdebris as a possible initiating and perpetuating stimulus for a proteolytic activation cascade, which may contribute to loosening of the prosthesis. In contrast to most ECM-degrading proteases, uPA has restricted substrate specificity. Although uPA best-documented proteolytic action is the conversion of inactive plasminogen to active plasmin, it has been also reported that it is able to activate the cell surface MT1-MMP proenzyme [87]. Like uPA, plasmin is also a serine protease but, in contrast to uPA, has broad substrate specificity. Apart from native collagen, plasmin can degrade most proteins present in the ECM. It can also activate the precursor forms of a number of MMPs, such as MMP-3, MMP-9, MMP-12, and MMP-13 [88]. Elevated protein levels of MMP-13, together with uPA and PAI-1 in periprosthetic pseudocapsular and interface tissues were also reported by Diehl and coworkers [89]. However, no significant correlation between the protein expression of these factors and years from arthroplasty to revision or to type of fixation (cemented versus cementless) was observed.

It should be noted that the physical characteristics of wear particles (size, shape, and sintering temperature) as well as their amounts in the periprosthetic tissues can modify the toxicity of the biomaterials and the production of cytokines, MMPs, and TIMPs by various cell types. For example, macrophages seemed to release MMPs (MMP-1, -2, and -9) in proportion to the amount of particulate debris at the prosthetic interface [90]. Laquerriere and coworkers demonstrated that sintering temperature (that modify crystal size and surface area) had little effect on MMPs and TIMPs production. Nonphagocytable particles induced more MMP-9, although phagocytable particles induced more IL-1β release. The shape of the particles was the most important factor since needle-shaped particles induced the most significant upregulated expression of MMPs (mostly MMP-9) and IL-1β [91]. In another study, human osteoblasts were incubated with particles experimentally generated in the interface between hip stems with rough and smooth surface finishings as well as different material compositions [39].

Extracellular Matrix Degradation and Tissue Remodeling in Periprosthetic Loosening and Osteolysis: Focus on Matrix
Metalloproteinases, Their Endogenous Tissue Inhibitors, and the Proteasome

117

The results revealed distinct effects on the cytokine release of human osteoblasts towards particulate debris. Thereby, human osteoblasts released increased levels of IL-6 and IL-8 after treatment with metallic wear particles. The expression of VEGF was slightly induced by all particle entities at lower concentrations. Apoptotic rates were enhanced for osteoblasts exposed to all the tested particles. Furthermore, the *de novo* synthesis of type 1 collagen was reduced and the expression of MMP-1 was considerably increased. Therefore, by the secretion of degrading effectors, osteoblasts may actively contribute to matrix weakening.

3. Molecular Mechanisms Controlling the Periprosthetic Microenvironment: Implication of MMPs/TIMPs and an Emerging Role for Proteasome

A large body of studies reveals a strong interdependence of MMP expression and activity with the molecular mechanisms that control the composition and turnover of periprosthetic ECMs. MMPs can either actively modulate or be modulated by the molecular mechanisms that determine the debris-induced remodeling of the periprosthetic microenvironment (summarized in Figure 2). One likely mechanism whereby particulate debris may induce osteoclast generation and activation is an indirect one, mediated through the actions of proinflammatory mediators that can act on osteoclast precursors and, most importantly, modulate the RANKL/osteoprotegerin (OPG) ratio through actions on cells within the periprosthetic tissue. RANKL is a type II homotrimeric transmembrane protein, which is normally expressed on osteoblastic cell membrane but is also expressed by fibroblasts and activated T cells [92]. Binding of RANKL to RANK on preosteoclasts (OCPs) activates NF-κB and Jun N-terminal kinases (JNKs) pathways to induce cell differentiation [93]. NF-κB is likely the most notable transcription factor implicated in wear debris action. This protein complex, long known as a key regulator of inflammatory gene expression, is also emerging as an important player during osteoclastogenesis. Supporting evidence for a role of NF-κB in periprosthetic osteolysis comes from observations that deficiency of NF-κB in mice protects against titanium-induced calvarial osteolysis [94], and that inhibition of NF-κB blocks wear debris induction of osteoclastogenesis *in vitro* [95, 96]. Osteoblasts also secrete OPG, a soluble decoy receptor for RANKL, which strongly binds to RANKL and effectively inhibits its activity on preosteoclasts differentiation and maturation [97]. OPG is a glycoprotein possessing 4 cysteine-rich domains at its Nterminus by which it binds to RANKL, whereas its C-terminus contains 22 homologous death domains of unknown function and a heparin binding domain by which the glycoprotein interacts with matrix macromolecules, such as glycosaminoglycans and proteoglycans. Importantly, any imbalance in the RANKL/OPG ratio impairs normal bone remodeling and evidence suggests a role of RANKL/OPG ratio in wear debris-induced osteolysis. In particular, it has been shown that RANKL blockade with OPG [98, 99] or RANK:Fc (RANKL antagonist consisting of the extracellular region of RANK fused to the Fc portion of human IgG1), or by using mice genetically deficient in RANK prevents wear debris-induced osteolysis in the murine calvarial model [100]. Moreover, wear debris can increase the RANKL/OPG ratio in murine calvarial tissues [101], and several reports have identified elevated RANKL expression in IFTs [102–105]. However, the fact that several different cell types within the periprosthetic tissue are capable of RANKL expression, including osteoblasts, fibroblasts, T lymphocytes, and also macrophages and giant cells [102–106], makes the cellular basis for elevated RANKL expression very complicated.

Several MMPs are overexpressed and correlated with osteoclast differentiation, maturation and activation by interfering with the RANK/RANKL/OPG system in inflammation and cancer [107]. MMP-9 is likely to play an important role in the recruitment of osteoclasts at inflammatory and metastatic sites since the use of chemical inhibitors or anti-sense oligonucleotides against MMP-9 abrogated the recruitment of osteoclasts [108]. Franco and coworkers showed that doxycycline (Dox), which can suppress the enzyme activity of MMP-9 [109], as well as MMP-9 inhibitor (MMP-9 inhibitor I), downregulated the expression of RANKL-induced osteoclast maturation genes in conjunction with the suppression of RANKL-induced osteoclastogenesis [110]. These findings indicated that MMP-9 induced by RANKL plays a role as an upstream effector of osteoclast gene expression, and, as such, it may also be a regulator of osteoclastogenesis. Previous studies reported that MMP inhibitor (RP59794) [111] or MMP-9 gene knockout [112] reduced osteoclast migration, which results in reduction of the resorption process in the growth plate and, as a consequence, attenuated development of bone marrow cavity. However, since the latter studies treated the aspect of osteoclast migration, but not differentiation, the study by Franco and coworkers is the first to report the involvement of MMP-9 activity in RANKL-induced osteoclastogenesis. In another report, MMP-7 could solubilize RANKL in mouse models of prostate and breast cancer promoting osteoclast activation and osteolysis [113]. The limiting step in RANKL-dependent osteoclastogenesis is the contact of RANKL-expressing osteoblasts with RANK on the cell surface of osteoclasts. This limitation is prohibited by proteolytic cleavage of RANKL from the cell surface through the action of MMP-7 and cathepsin G. Importantly, it has been shown in tumor-induced bone disease that soluble RANKL retains its activity and is liberated at the tumor-bone interface promoting osteoclastogenesis without the necessity of direct interaction of osteoblast with osteoclasts [113–115].

Possible accumulation of cell membrane and matrix proteoglycans at the inflammatory periprosthetic ECM may also modulate the RANK/RANKL/OPG system through both MMP-independent and -dependent manners. For example, it has been shown that myeloma cells decrease OPG availability by internalizing it through binding to glycosaminoglycan side chains of surface syndecan-1 and degradation to lysosomes, thereby regulating its inhibitory effect on RANKL [116]. Moreover, shed syndecan-1 secreted by myeloma cells may also bind OPG [117] and block its inhibitory activity to RANKL triggering further osteoclast differentiation and activation. Syndecan ectodomain shedding is an important regulatory mechanism, because it rapidly changes cell surface receptor dynamics and generates soluble ectodomains that can function as paracrine or autocrine effectors or competitive inhibitors. Strong evidence indicates the involvement of several MMPs in syndecan cleavage in vitro and in vivo [118]. Matrilysin (MMP-7) cleaves syndecan-1 [119], gelatinases MMP-2 and MMP-9 can cleave syndecans-1, -2, and -4 [120, 121], whereas the membrane-associated metalloproteinases MT1-MMP and MT3-MMP are known to cleave syndecan-1 [122]. Taken together, these data suggest a critical role of certain members of MMPs in interfering with the RANK/RANKL signaling axis by directly and/or indirectly regulating OPG and RANKL availability, thereby modulating osteoclast generation and activation within the periprosthetic tissue (Figure 2).

An important observation in several studies was that specific gene responses were induced in different cell types of the periprosthetic microenvironment by an initial and early particulate biomaterial-cell interaction. The differential gene expression indicated that particle-cell interactions activated specific signaling events and transcription factors. Vermes and coworkers have found that particles rapidly activated protein tyrosine phosphorylation and induced the nuclear transcription factor NF-κB in osteoblasts [123]. The rapid kinetics of the activation suggested that the particles elicited signals before the phagocytosis process. Importantly, inhibition of NF-κB function by either tyrosine kinase inhibitors or antioxidants reversed the suppressive effect of titanium particles on procollagen a1[I] gene expression suggesting a functional relationship in osteoblasts between tyrosine phosphorylation, NF-κB activation, and collagen gene expression. Thus, particle-cell interactions before their phagocytosis appear to initiate an intracellular tyrosine phosphorylation cascade that targets the nuclear activation of the inducible transcription factor NF-κB.

A role for protein tyrosine kinases (PTKs) in regulation of the activation of MMPs/TIMPs in ion-induced activation of macrophages was suggested by Luo and coworkers [124]. In particular, cobalt (Co) and chromium (Cr) ions, two corrosion products found in the periprosthetic environment of metal-on-metal prostheses, were shown to upregulate MMP-1, TIMP-1, and cytokines (such as TNF-α) in cultures of human U937 macrophages. The inhibitory effect of genistein suggested the implication of PTKs in the induction of MMP-1 and TIMP-1 expressions by Co^{2+} and Cr^{3+} ions in macrophages, the most important cellular target of wear debris. Genistein, a soy isoflavonoid that is a natural broad spectrum PTK (such as EGFR, PDGFR, IGFR, Src) inhibitor, has been shown to regulate the transcription of several MMPs and their endogenous inhibitors (TIMPs) by breast cancer cells [125, 126]. Moreover, previous in vitro studies demonstrated that genistein downregulates the expression of vascular endothelial growth factor (VEGF), which is a major signaling protein that contributes to angiogenesis [127]. VEGF is produced by multiple cell types, including macrophages and osteoblasts [128, 129]. It exerts its biological activity by binding to two TK receptors, VEGF receptor-1 (VEGFR-1; Flt-1) and VEGFR-2 (Flk-1/KDR) [130]. VEGF

is actively involved in the process of inflammation, osteoclastogenesis, and bone resorption [131–133] and probably plays an important role in wear debris-induced inflammatory osteolysis since the periprosthetic tissues at the bone-implant interface show a high degree of vascularization [134]. Notably, Luo and coworkers showed a more potent effect of herbimycin A, Src kinase-specific inhibitor, on the expression of MMP-1 and TIMP-1 compared to genistein [124], providing strong evidence for a critical role of Src kinases in modulating the expression levels of MMP-1 and TIMP-1 in macrophages in the presence of Co^{2+} and Cr^{3+} ions (Figure 2). Other ions released from hip prostheses, such as titanium [135] and nickel [136], have been shown to stimulate TNF-a in a manner similar to Co and Cr, suggesting that other ions may also modulate tyrosine kinase activity probably affecting the amounts and activities of MMPs/TIMPs in periprosthetic ECMs.

Moreover, the adhesion of macrophages to phosphorylcholine-polymer coated surfaces stimulated the expression of MMP-1 and TIMP-1, suggesting that cell adhesion induced a remodeling of the macrophage ECM. The inhibition of the expression of these genes by genistein and herbimycin A suggested that PTKs were also implicated in this remodeling [124]. Interestingly, this kind of materials-stimulated expression of genes implicated in ECM remodeling was also observed in fibroblasts induced by three-dimensional collagen [137, 138]. In these studies, alpha1beta1 and alpha2beta1 integrins mediated the signals inducing downregulation of collagen gene expression and upregulation of MMP-1, respectively. Therefore, the potential impact of macrophage surface integrins-evoked signals on the periprosthetic microenvironment should be further investigated to better understand the cellular effects of particles liberated from the articular surface of prostheses (Figure 2).

One major part of the organism's first line of defense against infection is a family of pattern recognition receptors (PRRs) called the Toll-like receptors (TLRs). TLRs are transmembrane proteins found in various cells and recognize infectious and endogenous threats, so-called danger signals, which evoke inflammation and assist adaptive immune reactions. It has been suggested that TLRs play a role in periprosthetic tissues and arthritic synovium. Tamaki and coworkers found that peri-implant tissues were well equipped with TLRs and, in aseptic loosening, monocytes/macrophages were the main TLR-expressing cells [139]. This could lead to production of inflammatory cytokines and MMPs after phagocytosis of wear debris derived from an implant. A major conclusion of the study was that inflammatory cells in both aseptic and septic tissues were equipped with TLRs, providing them with responsiveness to both endogenous and exogenous TLR ligands. In this line, the high expression of TLRs in the periprosthetic tissues could be potentially important, as they can reflect occurrence of subclinical biofilms on the prosthetic surfaces. Activation of TLRs has been suggested to modulate the expression levels of certain MMPs but not TIMPs. In a recent study, Lisboa and coworkers showed that activation of TLR-2 and TLR-4, two TLR members expressed by a variety of human cells that participate in the recognition

Extracellular Matrix Degradation and Tissue Remodeling in Periprosthetic Loosening and Osteolysis: Focus on Matrix Metalloproteinases, Their Endogenous Tissue Inhibitors, and the Proteasome

119

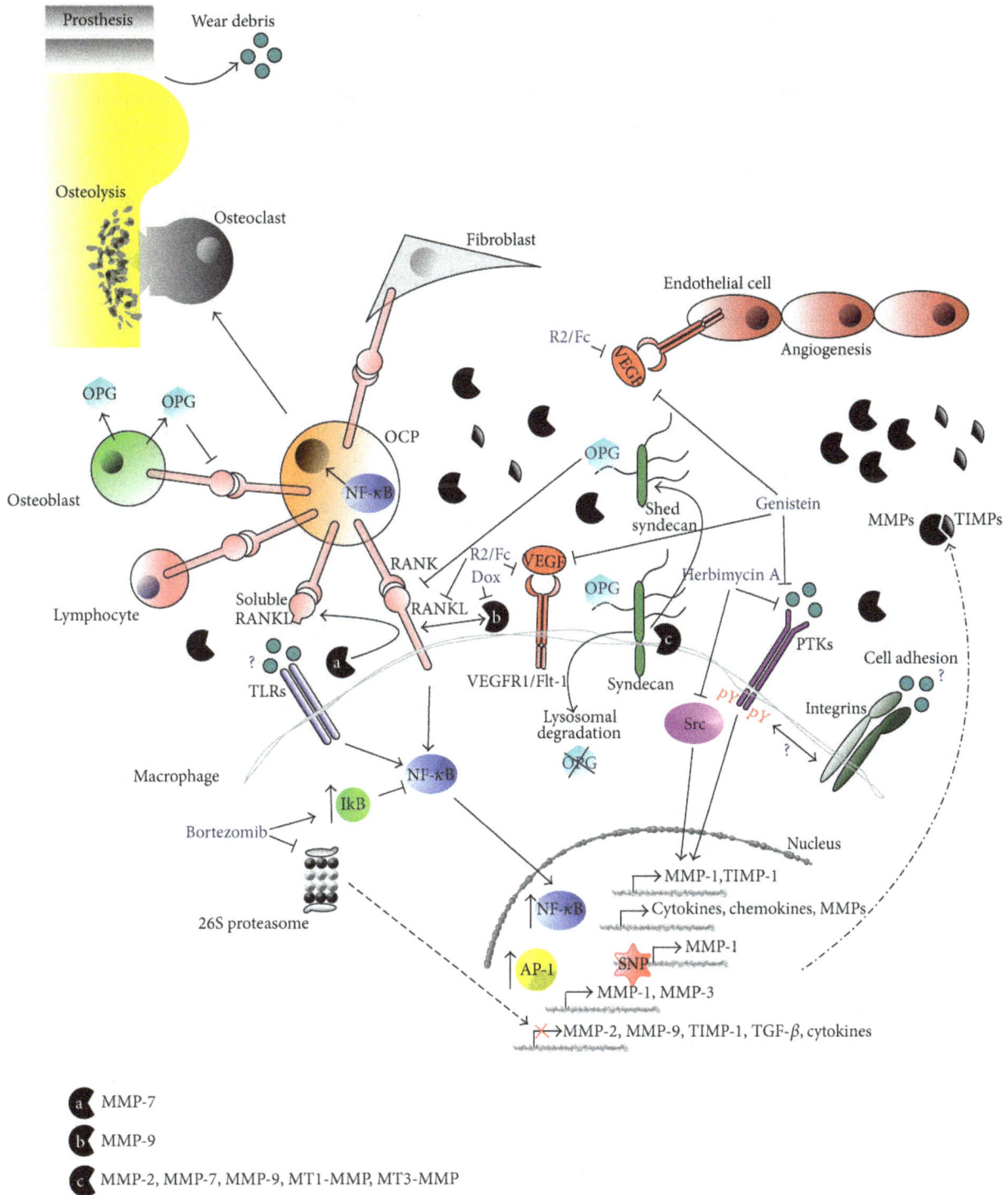

FIGURE 2: Hypothetical model of the molecular mechanisms that control periprosthetic microenvironment and potential molecular targeting with regard to the expression and activity of MMPs/TIMPs to prevent osteolysis (*see text for details*).

of bacterial lipoproteins and lipopolysaccharides (LPS) [140], induced an increase in the secretion of MMPs-1, -3, and -10 by cultured periodontal fibroblasts, and this was mediated via the p38, JNK1/2, and NF-κB pathways [141]. It is likely that there is a broad variation in the response of cells to TLR ligands that is dependent on the type of stimulus in the periprosthetic microenvironment. Therefore, the possibility of potentiation of MMPs activation concomitant with TLR activation in periprosthetic tissues needs to be further investigated (Figure 2).

Genetic variation may determine individual responses in terms of susceptibility to osteolysis and recovery. Expression levels of the MMPs at both the mRNA and protein levels can be affected by the introduction or loss of transcription binding sites by single nucleotide polymorphisms (SNPs). SNPs are the most common sequence variation in the human genome and can affect coding sequences, splicing, or transcription regulation. In a case control, it was shown that a single-nucleotide polymorphism (SNP) of MMP-1 was highly associated with total hip replacement aseptic failure [142]. This SNP existed within a promoter region of the gene and as such may have a direct effect on the amount of gene expression (Figure 2). However, the mechanisms of MMP gene regulation are still not fully delineated, and it is likely that many more functionally important elements in their promoter regions are yet to be identified [143–145]. Moreover, investigation of SNPs in the TIMP genes would be a necessary complement for any study of MMP SNPs, given the evidence that the MMP-to-TIMP ratio plays a role in defining overall MMP activity.

In another line of research, Ortiz-Lazareno and coworkers found that the proteasome inhibitor MG-132 significantly diminished proinflammatory cytokines (TNF-α, IL-1β, IL-6) release by U937 macrophages, whereas, induced a decrease in the membrane receptors TNF-R1 and IL-1R1 and an increase in the soluble receptors sTNF-R1 and sIL-1R1. However, MG-132 increased the IL-6R and decreased sIL-6R [146]. In another report, Mao and coworkers investigated the effects of Ti particles and the specific proteasome inhibitor bortezomib on the secretory profile of inflammatory cytokines, chemokines, and inflammatory enzymes in a murine macrophage cell line [147]. It was shown that Ti particles increased the production of TNF-α, IL-1β, IL-6, IL-10, MCP-1, MIP-1α, iNOS, and COX-2 in this cell line, while bortezomib inhibited the expression of all factors, except IL-10, in a time-dependent manner. Bortezomib, a potent, reversible and selective inhibitor of the chymotryptic activity of the proteasome, prevents the degradation of IκB proteins, which mask the nuclear localization sequence of NF-κB, therefore inhibiting the translocation of NF-κB into the nucleus and further inhibiting the transcription and secretion of inflammatory mediators. It is known that NF-κB regulates the transcription of a variety of genes of inflammatory cytokines (TNF-α, IL-1β, IL-2, IL-6, and GM-CSF (granulocytemacrophage colony-stimulating factor)), chemokines (IL-8, MCP-1, and MIP-1α), and inflammatory enzymes (iNOS and COX-2) [148]. Therefore, bortezomib may inhibit the proinflammatory factors mentioned earlier through inhibition of NF-κB activity. Moreover, proteasome inhibition with bortezomib alters the binding of other transcription factors to the promoter region of several molecular effectors, thus modulating their expression levels [149]. The anti-inflammatory cytokine IL-10 induced by bortezomib inhibits TNF-α gene expression via inhibiting NF-κB activity or by directly inhibiting TNF-α itself [150, 151]. It should be noted that bortezomib (as well as other proteasome inhibitors) exhibits beneficial effect on bone metabolism as it inhibits osteoclastic function and promotes osteoblastic activity by inhibiting NF-κB activation induced by the RANK-RANKL signaling axis, which is the master regulator of differentiation and activation of osteoclasts [152–155].

The inhibitory effects of bortezomib and MG-132 on the secretion of inflammatory cytokines and their receptors by macrophages suggested the potential involvement of the proteasome pathway in periprosthetic loosening and osteolysis process. The proteasome is a major cellular protease complex that functions as the main driver of intracellular degradation of a wide variety of cellular proteins implicated in several physiological and pathological cellular functions [156]. Interestingly, the proteasome pathway controls via transcriptional and posttranslational mechanisms the concentration and turnover of several ECM macromolecules (including proteoglycans/glycosaminoglycans, MMPs/TIMPs, and collagens) [157]. Importantly, the proteasome provides a link between the regulation of extracellular proteolytic events to intracellular proteolysis by modulating MMP/TIMP expression and activity. In particular, it has been shown that proteasome blockade by proteasome inhibitors resulted in a marked modification of gene expression and activity of MMPs (upregulation of MMP-1, -3 and downregulation of MMP-2, -9) and TIMPs (downregulation of TIMP-1). Moreover, proteasome inhibition regulated also the synthesis and activity of other ECM constituents, such as TGF-β (downregulation), decorin (upregulation), and collagen type-I and type-IV (downregulation) [157]. Therefore, since matrix remodeling and degradation can be tightly regulated by proteasome activities, its modulation may be considered as a novel strategy to control the properties of periprosthetic ECMs as has been recently suggested for tumor microenvironment (Figure 2).

4. Potential Therapeutic Perspectives

Much progress has recently been made in understanding the molecular and cellular mechanisms whereby prosthetic wear debris can ultimately cause aseptic loosening and osteolysis. However, the complex nature of the interactions between wear particles and periprosthetic cells as well as the multiple intracellular signaling pathways activated by such interactions results in the reality that development of therapeutic approaches to the treatment of periprosthetic osteolysis is long overdue. In the following lines, we will try to address strong rationale for potential clinical applications of the described molecular mechanisms for periprosthetic loosening and osteolysis treatment.

The finding that proteasome inhibitors (i.e., bortezomib, MG-132) altered the macrophage secretory profile of inflammatory cytokines, chemokines, and inflammatory enzymes [146, 147], which play a key-role in the inflammatory response of periprosthetic tissue to wear debris, reveals a critical role for proteasome in the development of periprosthetic loosening and osteolysis. In a recent review, we have highlighted the novel approach of targeting the proteasome as a mechanism to control the synthesis and bioactivity of ECM effectors in tumors, since the proteasome appears to be an elegant molecular regulator of specific matrix macromolecules [157]. From the data described in the present review, proteasome signaling pathway emerges as a promising target to selectively regulate the synthesis and activity of inflammatory factors (such as TNF-α, IL-1β, IL-6, IL-8), their membrane receptors, and matrix degrading effectors (such as specific MMPs) in the periprosthetic microenvironment. To this aim, a promising agent is bortezomib, which exhibits multiple functions by interfering also with other intracellular signaling pathways such as the RANK-RANKL system thereby regulating new bone formation by both inducing osteoblastic function and inhibiting osteoclastogenesis. However, the elevated production of reactive oxygen species (ROS) by activated macrophages and osteoclasts in the presence of wear particles [158] should be considered in this context, since proteasome inhibitors have been also shown to induce ROS [159], which further attenuate the proteasomal system activation [160]. This proteasomal inhibition would potentially result in the accumulation of phosphorylated c-Jun and activation of AP-1 that ultimately induce MMP-1 and MMP-3 expression levels [160]. Therefore, proteasome inhibitors may have a synergistic effect with wear particles on ROS production and strongly induce the expression of specific MMPs within the periprosthetic microenvironment although their inhibitory effect on inflammatory cytokines and their receptors as well as other MMPs has been documented. Moreover, MG-132 has been found to significantly downregulate TIMP-1 expression in organ interface tissue cultures and primary IFT fibroblast cultures (Aletras and coworkers, unpublished data), which is in line with the findings of Fineschi and coworkers in dermal fibroblasts [161], revealing the complex and questionable role of proteasome in regulating distinct molecular effectors that would potentially be beneficial for periprosthetic osteolysis treatment. Therefore, the efficacy of proteasome inhibitors (such as bortezomib) to prevent periprosthetic loosening and osteolysis caused by implant-derived particles is an emerging concept and needs to be further investigated.

Taking under consideration these data, it should be investigated whether an alternative strategy associated with proteasome activation would be more beneficial in the treatment of periprosthetic loosening. Several activators of the proteasome, such as isoflavonoids, should be tested in order to reverse the effects on the expression levels of specific MMPs and TIMPs described previously, as a result of the reduced proteasome activity in IFT. Proteasome activation might be further induced by combined treatment with activators of nuclear factor erythroid 2-related factor 2 (Nrf2), such as sulforaphane [162]. Notably, Nrf2 upregulates the transcription of multiple antioxidant enzymes providing an effective means of reducing elevated ROS levels in IFT.

The observation that genistein and herbimycin A strongly attenuated the expression of MMP-1 and TIMP-1 by macrophages implied that tyrosine kinases play also an essential role in the signaling pathways regulating the remodeling of macrophage ECM in the periprosthetic microenvironment [124]. Therefore, PTKs (e.g., Src kinases) may serve as an additional target for selective inhibition of periprosthetic osteolysis. Importantly, proteasome is implicated in this process since it has been reported that herbimycin A targets the degradation of tyrosine kinases by the 20S proteasome [163]. Moreover, apart from its inhibitory action on PTKs, genistein was found to downregulate the expression of VEGF, a major

angiogenic factor in periprosthetic microenvironment. The interactive network of the VEGF/Flt-1 and RANKL/RANK pathways may play important roles in the initiation, progression, and resolution of aseptic loosening. In a study by Ren and coworkers, it was shown that VEGF may be actively involved in the regulation of RANK/RANKL gene expression, and that it exerted a regulatory effect on the development of particle-induced inflammatory osteoclastogenesis through its unique Flt-1, rather than Flk-1, receptor located on monocyte/macrophage cell lineages [164]. In particular, they found that treatment with R2/Fc (a VEGF neutralizing antibody) but not SU5416 (an Flk-1 receptor inhibitor) resulted in the inhibition of polyethylene particle-enhanced VEGF/Flt-1 signaling and inflammatory osteolysis by trapping VEGF in the periprosthetic milieu (Figure 2). Taken together, these findings provide the biological rationale for a combined VEGF/Flt-1- and RANKL/RANK-targeted treatment strategy, especially in the early stages of wear debris-induced inflammatory response. The fact that the RANK/RANKL/OPG system is of crucial importance for the development of periprosthetic osteolysis together with the finding that Dox inhibits RANKL-induced osteoclastogenesis by its inhibitory effect on MMP-9 enzyme activity [110] provides a reasonable rationale for a pharmaceutical advantage of tetracycline antibiotics against periprosthetic osteolysis. It should be noted that this class of antibiotics, including Dox, has been effectively utilized for the treatment of bone resorptive diseases because of their activity to suppress osteoclastogenesis induced by RANKL.

Given that excessive osteoclast activity represents the cellular endpoint of osteolysis, it is not surprising that the bisphosphonate class of osteoclast inhibitors have come in for much discussion as possible therapeutic agents for this disease. Again, however, despite promising results in animal models, there is no clinical evidence supporting the effectiveness of these drugs in the treatment of osteolysis patients. Alendronate inhibits wear debris-induced osteolysis in the rat loaded tibial implant model of osteolysis [165] and in a similar canine model [9] and is also effective in preventing osteolysis in the murine calvarial model [6]. A single dose of zoledronic acid administered directly after surgery also suppressed particle-induced osteolysis in mouse calvaria [166]. Bisphosphonates inhibit osteoclast formation by blocking the mevalonate pathway of isoprenoid biosynthesis. Their potential effect in periprosthetic osteolysis should be also considered with regard to their ability to inhibit the enzymatic activity of various MMPs. Certain bisphosphonates showed beneficial effects as a result of altering the expression pattern of MMPs/TIMPs by inhibiting and increasing the gene and protein expression of several MMPs and TIMPs, respectively, in breast cancer cells. In particular, it has been shown that zoledronic acid suppressed the expression of metalloproteinases MMP-2, -9, the membrane type MT1- and MT2-MMP, whereas it increased the expression of their endogenous tissue inhibitors [167].

Though not extensively studied, other mechanisms that should be further investigated with regard to their contribution to the remodeling of periprosthetic ECM include SNPs of certain MMP/TIMP genes as well as the involvement of TLRs in periprosthetic inflammation. An SNP of MMP1 gene was highly associated with total hip replacement aseptic failure [142]. It should be noted that MMPs do not possess only degrading functions but they also play protective and anti-inflammatory roles. Therefore, the association that exists with a particular polymorphic form of MMP-1 does not necessarily show that particular form is associated with increased MMP-1 activity; in fact, the opposite may be true. The possibility that SNP markers may serve as predictors of implant survival and aid in pharmacogenomic prevention of total joint replacement failure should be further investigated. A more comprehensive analysis of MMP and TIMP SNPs is thus required, and given the coverage by existing genome-wide association study (GWAS) platforms, a candidate gene approach is justified. Regarding TLRs, strong evidence indicated that macrophages, which are the most important cellular targets of wear debris, are the main TLR-expressing cells in periprosthetic microenvironment. The increased secretion of MMPs by combined TLR activation may be an important factor that should also be considered during treatment of periprosthetic loosening and osteolysis.

Extended information is available regarding the action of several nonsteroidal anti-inflammatory drugs (NSAIDs) upon significant for the loosening process effector molecules, which though originates from *in vitro* studies with articular chondrocytes and synovial or dental pulp fibroblasts [168–173]. However, little information is available in the literature about their possible role in retarding the periprosthetic loosening and bone resorption process. To this aim, we tested the effect of four widely used NSAIDS (i.e., aceclofenac, piroxicam, tenoxicam, and indomethacin) on cytokine, MMP, TIMP, and prostanoid production by IFT from patients with aseptic loosening of total arthroplasty [174]. The results showed that all the tested drugs exerted uniformly an inhibitory effect on IL-6 and TNF-α, both known to directly cause osteoclastic bone resorption, independently of PGE2 [175–177]. Moreover, all of them modified specific MMPs (MMP-1, MMP-2, MMP-3, and MMP-9) expression and activity, although these drugs did not have a statistically clear effect on MMPs, which might reflect individual responses in terms of susceptibility to osteolysis. However, NSAIDs had a profound stimulatory effect on TIMP-1 production. Interestingly, paracetamol, which was used as a neutral drug, significantly decreased the synthesis of TNF-α and gelatinases (MMP-2 and MMP-9). Considering these observations, NSAIDs could reduce the ability of periprosthetic membrane to cause bone resorption, which is in line with previous reports that have shown that piroxicam, which exhibited about the same effects as the other tested NSAIDs, significantly decreased the IFT-induced resorptive process [178]. Consequently, *in vivo* long-term clinical trials may shed light on the possibility of a beneficial effect of specific NSAIDs on the loosening process.

5. Concluding Remarks

The elucidation and understanding of the cellular and molecular mechanisms that control the composition, turnover, and activity of matrix macromolecules within the periprosthetic microenvironment exposed to wear debris is highly important for the development of novel therapeutic approaches to the treatment of periprosthetic loosening and osteolysis. One ultimate target would be to disrupt the vicious cycle between the inflammatory response to wear debris particles induced by the secreted proinflammatory and osteoclastogenic cytokines and the periprosthetic osteolytic cascade governed by the uncontrolled action of MMPs. Considering the multicomplex biological mechanisms underlying the particle-induced periprosthetic loosening and osteolysis described in the present review, it may be crucial to develop and use combinations of conventional therapeutic agents as well as new approaches targeting specific extracellular, cell surface, and intracellular molecular effectors and apply them in clinical practice.

Acknowledgment

The authors wish to thank Dr. Nikos Karamanos for valuable discussions and suggestions to improve the quality of this paper.

References

[1] M. Sundfeldt, L. V. Carlsson, C. B. Johansson, P. Thomsen, and C. Gretzer, "Aseptic loosening, not only a question of wear: a review of different theories," *Acta Orthopaedica*, vol. 77, no. 2, pp. 177–197, 2006.

[2] J. H. Dumbleton, M. T. Manley, and A. A. Edidin, "A literature review of the association between wear rate and osteolysis in total hip arthroplasty," *Journal of Arthroplasty*, vol. 17, no. 5, pp. 649–661, 2002.

[3] T. P. Schmalzried, M. Jasty, and W. H. Harris, "Periprosthetic bone loss in total hip arthroplasty. Polyethylene wear debris and the concept of the effective joint space," *Journal of Bone and Joint Surgery A*, vol. 74, no. 6, pp. 849–863, 1992.

[4] K. Hirakawa, T. W. Bauer, B. N. Stulberg et al., "Comparison and quantitation of wear debris of failed total hip and total knee arthroplasty," *Journal of Biomedical Materials Research*, vol. 31, pp. 257–263, 1996.

[5] K. J. Margevicius, T. W. Bauer, J. T. McMahon, S. A. Brown, and K. Merritt, "Isolation and characterization of debris in membranes around total joint prostheses," *Journal of Bone and Joint Surgery A*, vol. 76, no. 11, pp. 1664–1675, 1994.

[6] E. M. Schwarz, E. B. Benz, A. P. Lu et al., "Quantitative small-animal surrogate to evaluate drug efficacy in preventing wear debris-induced osteolysis," *Journal of Orthopaedic Research*, vol. 18, no. 6, pp. 849–855, 2000.

[7] P. J. Millett, M. J. Allen, and M. P. G. Bostrom, "Effects of alendronate on particle-induced osteolysis in a rat model," *Journal of Bone and Joint Surgery A*, vol. 84, no. 2, pp. 236–249, 2002.

[8] K. D. Merkel, J. M. Erdmann, K. P. McHugh, Y. Abu-Amer, F. P. Ross, and S. L. Teitelbaum, "Tumor necrosis factor-α mediates orthopedic implant osteolysis," *American Journal of Pathology*, vol. 154, no. 1, pp. 203–210, 1999.

[9] A. S. Shanbhag, C. T. Hasselman, and H. E. Rubash, "The John Charnley Award. Inhibition of wear debris mediated osteolysis in a canine total hip arthroplasty model," *Clinical Orthopaedics and Related Research*, no. 344, pp. 33–43, 1997.

Extracellular Matrix Degradation and Tissue Remodeling in Periprosthetic Loosening and Osteolysis: Focus on Matrix
Metalloproteinases, Their Endogenous Tissue Inhibitors, and the Proteasome

123

[10] P. H. Wooley, R. Morren, J. Andary et al., "Inflammatory responses to orthopaedic biomaterials in the murine air pouch," *Biomaterials*, vol. 23, no. 2, pp. 517–526, 2002.

[11] B. A. Warme, N. J. Epstein, M. C. D. Trindade et al., "Proinflammatory mediator expression in a novel murine model of titanium-particle-induced intramedullary inflammation," *Journal of Biomedical Materials Research B*, vol. 71, no. 2, pp. 360–366, 2004.

[12] S.-Y. Yang, B. Wu, L. Mayton et al., "Protective effects of IL-1Ra or vIL-10 gene transfer on a murine model of wear debris-induced osteolysis," *Gene Therapy*, vol. 11, no. 5, pp. 483–491, 2004.

[13] T. A. Blaine, R. N. Rosier, J. E. Puzas et al., "Increased levels of tumor necrosis factor-α and interleukin-6 protein and messenger RNA in human peripheral blood monocytes due to titanium particles," *Journal of Bone and Joint Surgery A*, vol. 78, no. 8, pp. 1181–1192, 1996.

[14] Y. Nakashima, D.-H. Sun, M. C. D. Trindade et al., "Signaling pathways for tumor necrosis factor-α and interleukin-6 expression in human macrophages exposed to titanium-alloy particulate debris *in vitro*," *Journal of Bone and Joint Surgery A*, vol. 81, no. 5, pp. 603–615, 1999.

[15] J. A. Wimhurst, R. A. Brooks, and N. Rushton, "Inflammatory responses of human primary macrophages to particulate bone cements *in vitro*," *Journal of Bone and Joint Surgery B*, vol. 83, no. 2, pp. 278–282, 2001.

[16] E. Ingham, T. R. Green, M. H. Stone, R. Kowalski, N. Watkins, and J. Fisher, "Production of TNF-α and bone resorbing activity by macrophages in response to different types of bone cement particles," *Biomaterials*, vol. 21, no. 10, pp. 1005–1013, 2000.

[17] W. J. Maloney, R. E. James, and R. L. Smith, "Human macrophage response to retrieved titanium alloy particles *in vitro*," *Clinical Orthopaedics and Related Research*, no. 322, pp. 268–278, 1996.

[18] K. J. Saleh, I. Thongtrangan, and E. M. Schwarz, "Osteolysis: medical and surgical approaches," *Clinical Orthopaedics and Related Research*, no. 427, pp. 138–147, 2004.

[19] P. E. Purdue, P. Koulouvaris, B. J. Nestor, and T. P. Sculco, "The central role of wear debris in periprosthetic osteolysis," *HSS Journal*, vol. 2, no. 2, pp. 102–113, 2006.

[20] T. P. Schmalzried, M. Jasty, and W. H. Harris, "Periprosthetic bone loss in total hip arthroplasty. Polyethylene wear debris and the concept of the effective joint space," *Journal of Bone and Joint Surgery A*, vol. 74, no. 6, pp. 849–863, 1992.

[21] T. T. Glant, J. J. Jacobs, G. Molnar, A. S. Shanbhag, M. Valyon, and J. O. Galante, "Bone resorption activity of particulate-stimulated macrophages," *Journal of Bone and Mineral Research*, vol. 8, no. 9, pp. 1071–1079, 1993.

[22] M. Goppelt-Struebe and M. Stroebel, "Synergistic induction of monocyte chemoattractant protein-1 (MCP-1) by platelet-derived growth factor and interleukin-1," *FEBS Letters*, vol. 374, no. 3, pp. 375–378, 1995.

[23] S. M. Horowitz and J. B. Gonzales, "Inflammatory response to implant particulates in a macrophage/osteoblast coculture model," *Calcified Tissue International*, vol. 59, no. 5, pp. 392–396, 1996.

[24] N. Ishiguro, T. Kojima, T. Ito et al., "Macrophage activation and migration in interface tissue around loosening total hip arthroplasty components," *Journal of Biomedical Materials Research*, vol. 35, pp. 399–406, 1997.

[25] Y. Kadoya, P. A. Revell, N. Al-Saffar, A. Kobayashi, G. Scott, and M. A. R. Freeman, "Bone formation and bone resorption in failed total joint arthroplasties: histomorphometric analysis with histochemical and immunohistochemical technique," *Journal of Orthopaedic Research*, vol. 14, no. 3, pp. 473–482, 1996.

[26] A. Pizzoferrato, S. Stea, A. Sudanese et al., "Morphometric and microanalytical analyses of alumina wear particles in hip prostheses," *Biomaterials*, vol. 14, no. 8, pp. 583–587, 1993.

[27] D. H. DeHeer, J. A. Engels, A. S. DeVries et al., "In situ complement activation by polyethylene wear debris," *Journal of Biomedical Materials Research*, vol. 54, pp. 12–19, 2001.

[28] S. Santavirta, V. Hoikka, A. Eskola, Y. T. Konttinen, T. Paavilainen, and K. Tallroth, "Aggressive granulomatous lesions in cementless total hip arthroplasty," *Journal of Bone and Joint Surgery B*, vol. 72, no. 6, pp. 980–984, 1990.

[29] R. Rezzonico, R. Chicheportiche, V. Imbert, and J.-M. Dayer, "Engagement of CD11b and CD11c β2 integrin by antibodies or soluble CD23 induces IL-1β production on primary human monocytes through mitogen- activated protein kinase-dependent pathways," *Blood*, vol. 95, no. 12, pp. 3868–3877, 2000.

[30] R. Rezzonico, V. Imbert, R. Chicheportiche, and J.-M. Dayer, "Ligation of CD11b and CD11c β2 integrins by antibodies or soluble CD23 induces macrophage inflammatory protein 1α (MIP-1α) and MIP-1β production in primary human monocytes through a pathway dependent on nuclear factor-κB," *Blood*, vol. 97, no. 10, pp. 2932–2940, 2001.

[31] A. Palecanda, J. Paulauskis, E. Al-Mutairi et al., "Role of the scavenger receptor MARCO in alveolar macrophage binding of unopsonized environmental particles," *The Journal of Experimental Medicine*, vol. 189, no. 9, pp. 1497–1506, 1999.

[32] D. S. Rakshit, J. T. E. Lim, K. Ly et al., "Involvement of complement receptor 3 (CR3) and scavenger receptor in macrophage responses to wear debris," *Journal of Orthopaedic Research*, vol. 24, no. 11, pp. 2036–2044, 2006.

[33] D.-H. Sun, M. C. D. Trindade, Y. Nakashima et al., "Human serum opsonization of orthopedic biomaterial particles: protein-binding and monocyte/macrophage activation *in vitro*," *Journal of Biomedical Materials Research A*, vol. 65, no. 2, pp. 290–298, 2003.

[34] M. Horiki, T. Nakase, A. Myoui et al., "Localization of RANKL in osteolytic tissue around a loosened joint prosthesis," *Journal of Bone and Mineral Metabolism*, vol. 22, no. 4, pp. 346–351, 2004.

[35] J. M. W. Quinn, N. J. Horwood, J. Elliott, M. T. Gillespie, and T. J. Martin, "Fibroblastic stromal cells express receptor activator of NF-κB ligand and support osteoclast differentiation," *Journal of Bone and Mineral Research*, vol. 15, no. 8, pp. 1459–1466, 2000.

[36] J. Yao, T. T. Glant, M. W. Lark et al., "The potential role of fibroblasts in periprosthetic osteolysis: fibroblast response to titanium particles," *Journal of Bone and Mineral Research*, vol. 10, no. 9, pp. 1417–1427, 1995.

[37] M. Manlapaz, W. J. Maloney, and R. L. Smith, "*In vitro* activation of human fibroblasts by retrieved titanium alloy wear debris," *Journal of Orthopaedic Research*, vol. 14, no. 3, pp. 465–472, 1996.

[38] C. Vermes, R. Chandrasekaran, J. J. Jacobs, J. O. Galante, K. A. Roebuck, and T. T. Glant, "The effects of particulate wear debris, cytokines, and growth factors on the functions of MG-63 osteoblasts," *Journal of Bone and Joint Surgery A*, vol. 83, no. 2, pp. 201–211, 2001.

[39] K. Lochner, A. Fritsche, A. Jonitz et al., "The potential role of human osteoblasts for periprosthetic osteolysis following exposure to wear particles," *International Journal of Molecular Medicine*, vol. 28, no. 6, pp. 1055–1063, 2011.

[40] A. Sabokbar, O. Kudo, and N. A. Athanasou, "Two distinct cellular mechanisms of osteoclast formation and bone resorption in periprosthetic osteolysis," *Journal of Orthopaedic Research*, vol. 21, no. 1, pp. 73–80, 2003.

[41] C.-T. Wang, Y.-T. Lin, B.-L. Chiang, S.-S. Lee, and S.-M. Hou, "Over-expression of receptor activator of nuclear factor-κB ligand (RANKL), inflammatory cytokines, and chemokines in periprosthetic osteolysis of loosened total hip arthroplasty," *Biomaterials*, vol. 31, no. 1, pp. 77–82, 2010.

[42] B. Möller and P. M. Villiger, "Inhibition of IL-1, IL-6, and TNF-α in immune-mediated inflammatory diseases," *Springer Seminars in Immunopathology*, vol. 27, no. 4, pp. 391–408, 2006.

[43] R. Horai, S. Saijo, H. Tanioka et al., "Development of chronic inflammatory arthropathy resembling rheumatoid arthritis in interleukin I receptor antagonist-deficient mice," *The Journal of Experimental Medicine*, vol. 191, no. 2, pp. 313–320, 2000.

[44] M. Caicedo, J. J. Jacobs, and N. J. Hallab, "Inflammatory bone loss in joint replacements: the mechanisms," *Journal Musculoskeletal Medicine*, vol. 27, pp. 209–216, 2010.

[45] U. H. Lerner and A. Ohlin, "Tumor necrosis factors α and β can stimulate bone resorption in cultured mouse calvariae by a prostaglandin-independent mechanism," *Journal of Bone and Mineral Research*, vol. 8, no. 2, pp. 147–155, 1993.

[46] S. M. Horowitz and M. A. Purdon, "Mediator interactions in macrophage/particulate bone resorption," *Journal of Biomedical Materials Research*, vol. 29, no. 4, pp. 477–484, 1995.

[47] J. Chiba, H. E. Rubash, K. J. Kim, and Y. Iwaki, "The characterization of cytokines in the interface tissue obtained from failed cementless total hip arthroplasty with and without femoral osteolysis," *Clinical Orthopaedics and Related Research*, no. 300, pp. 304–312, 1994.

[48] H. Takei, D. Pioletti, S. Y. Kwon, and P. Sung, "Combined effect of titanium particles and TNF-a on the production of IL-6 by osteoblast-like cells," *Journal of Biomedical Materials Research*, vol. 52, pp. 382–387, 2000.

[49] M. Gowen, D. D. Wood, E. J. Ihrie, M. K. McGuire, and R. G. Russell, "An interleukin 1 like factor stimulates bone resorption *in vitro*," *Nature*, vol. 306, no. 5941, pp. 378–380, 1983.

[50] T. Akatsu, N. Takahashi, N. Udagawa et al., "Role of prostaglandins in interleukin-1-induced bone resorption in mice *in vitro*," *Journal of Bone and Mineral Research*, vol. 6, no. 2, pp. 183–190, 1991.

[51] S. B. Goodman, R. C. Chin, S. S. Chiou, D. J. Schurman, S. T. Woolson, and M. P. Masada, "A clinical-pathologic-biochemical study of the membrane surrounding loosened and nonloosened total hip arthroplasties," *Clinical Orthopaedics and Related Research*, no. 244, pp. 182–187, 1989.

[52] W. A. Jiranek, M. Machado, M. Jasty et al., "Production of cytokines around loosened cemented acetabular components: analysis with immunohistochemical techniques and in situ hybridization," *Journal of Bone and Joint Surgery A*, vol. 75, no. 6, pp. 863–879, 1993.

[53] K. Kusano, C. Miyaura, M. Inada et al., "Regulation of matrix metalloproteinases (MMP-2,-3,-9, and -13) by interleukin-1 and interleukin-6 in mouse calvaria: association of MMP induction with bone resorption," *Endocrinology*, vol. 139, no. 3, pp. 1338–1345, 1998.

[54] Y. Ishimi, C. Miyaura, C. H. Jin et al., "IL-6 is produced by osteoblasts and induces bone resorption," *Journal of Immunology*, vol. 145, no. 10, pp. 3297–3303, 1990.

[55] N. Kurihara, D. Bertolini, T. Suda, Y. Akiyama, and G. D. Roodman, "IL-6 stimulates osteoclast-like multinucleated cell formation in long term human marrow cultures by inducing IL-1 release," *Journal of Immunology*, vol. 144, no. 11, pp. 4226–4230, 1990.

[56] S. R. Goldring, M. Jasty, M. S. Roelke, C. M. Rourke, F. R. Bringhurst, and W. H. Harris, "Formation of a synovial-like membrane at the bone-cement interface. Its role in bone resorption and implant loosening after total hip replacement," *Arthritis and Rheumatism*, vol. 29, no. 7, pp. 836–842, 1986.

[57] A. M. Appel, W. G. Sowder, S. W. Siverhus, C. N. Hopson, and J. H. Herman, "Prosthesis-associated pseudomembrane-induced bone resorption," *British Journal of Rheumatology*, vol. 29, no. 1, pp. 32–36, 1990.

[58] M. Takagi, Y. T. Konttinen, S. Santavirta et al., "Extracellular matrix metalloproteinases around loose total hip prostheses," *Acta Orthopaedica Scandinavica*, vol. 65, no. 3, pp. 281–286, 1994.

[59] E. A. Fritz, T. T. Glant, C. Vermes, J. J. Jacobs, and K. A. Roebuck, "Chemokine gene activation in human bone marrow-derived osteoblasts following exposure to particulate wear debris," *Journal of Biomedical Materials Research A*, vol. 77, no. 1, pp. 192–201, 2006.

[60] Z. Huang, T. Ma, P.-G. Ren, R. L. Smith, and S. B. Goodman, "Effects of orthopedic polymer particles on chemotaxis of macrophages and mesenchymal stem cells," *Journal of Biomedical Materials Research A*, vol. 94, no. 4, pp. 1264–1269, 2010.

[61] S. B. Goodman and T. Ma, "Cellular chemotaxis induced by wear particles from joint replacements," *Biomaterials*, vol. 31, no. 19, pp. 5045–5050, 2010.

[62] M. Hukkanen, S. A. Corbett, J. Batten et al., "Aseptic loosening of total hip replacement. Macrophage expression of inducible nitric oxide synthase and cyclo-oxygenase-2, together with peroxynitrite formation, as a possible mechanism for early prosthesis failure," *Journal of Bone and Joint Surgery B*, vol. 79, no. 3, pp. 467–474, 1997.

[63] S. G. Harris, J. Padilla, L. Koumas, D. Ray, and R. P. Phipps, "Prostaglandins as modulators of immunity," *Trends in Immunology*, vol. 23, no. 3, pp. 144–150, 2002.

[64] K. T. Steeve, P. Marc, T. Sandrine, H. Dominique, and F. Yannick, "IL-6, RANKL, TNF-alpha/IL-1: interrelations in bone resorption pathophysiology," *Cytokine and Growth Factor Reviews*, vol. 15, no. 1, pp. 49–60, 2004.

[65] H. Nagase, R. Visse, and G. Murphy, "Structure and function of matrix metalloproteinases and TIMPs," *Cardiovascular Research*, vol. 69, no. 3, pp. 562–573, 2006.

[66] K. J. Leco, P. Waterhouse, O. H. Sanchez et al., "Spontaneous air space enlargement in the lungs of mice lacking tissue inhibitor of metalloproteinases-3 (TIMP-3)," *The Journal of Clinical Investigation*, vol. 108, pp. 817–829, 2001.

[67] J. E. Fata, K. J. Leco, E. B. Voura et al., "Accelerated apoptosis in the Timp-3-deficient mammary gland," *The Journal of Clinical Investigation*, vol. 108, no. 6, pp. 831–841, 2001.

[68] J. L. English, Z. Kassiri, I. Koskivirta et al., "Individual Timp deficiencies differentially impact pro-MMP-2 activation," *The Journal of Biological Chemistry*, vol. 281, no. 15, pp. 10337–10346, 2006.

[69] J. M. Ray and W. G. Stetler-Stevenson, "The role of matrix metalloproteases and their inhibitors in tumour invasion, metastasis and angiogenesis," *European Respiratory Journal*, vol. 7, no. 11, pp. 2062–2072, 1994.

Extracellular Matrix Degradation and Tissue Remodeling in Periprosthetic Loosening and Osteolysis: Focus on Matrix
Metalloproteinases, Their Endogenous Tissue Inhibitors, and the Proteasome

125

[70] I. Takei, M. Takagi, S. Santavirta et al., "Messenger ribonucleic acid expression of 16 matrix metalloproteinases in bone-implant interface tissues of loose artificial hip joints," *Journal of Biomedical Materials Research*, vol. 52, pp. 613–620, 2000.

[71] K. Sasaki, M. Takagi, J. Mandelin et al., "Quantitative analysis of mRNA expression of TIMPs in the periprosthetic interface tissue of loose hips by real-time PCR system," *Journal of Biomedical Materials Research*, vol. 58, no. 6, pp. 605–612, 2001.

[72] T.-F. Li, S. Santavirta, I. Virtanen, M. Könönen, M. Takagi, and Y. T. Konttinen, "Increased expression of EMMPRIN in the tissue around loosened hip prostheses," *Acta Orthopaedica Scandinavica*, vol. 70, no. 5, pp. 446–451, 1999.

[73] B. Nawrocki, M. Polette, H. Burlet, P. Birembaut, and J.-J. Adnet, "Expression of gelatinase A and its activator MT1-MMP in the inflammatory periprosthetic response to polyethylene," *Journal of Bone and Mineral Research*, vol. 14, no. 2, pp. 288–294, 1999.

[74] R. M. Hembry, M. R. Bagga, J. J. Reynolds, and D. L. Hamblen, "Stromelysin, gelatinase A and TIMP-1 in prosthetic interface tissue: a role for macrophages in tissue remodelling," *Histopathology*, vol. 27, no. 2, pp. 149–159, 1995.

[75] M. Takagi, "Neutral proteinases and their inhibitors in the loosening of total hip prostheses," *Acta orthopaedica Scandinavica. Supplementum*, vol. 271, pp. 3–29, 1996.

[76] Y. Yokohama, T. Matsumoto, M. Hirakawa et al., "Production of matrix metalloproteinases at the bone-implant interface in loose total hip replacements," *Laboratory Investigation*, vol. 73, no. 6, pp. 899–911, 1995.

[77] M. Takagi, Y. T. Konttinen, O. Lindy et al., "Gelatinase/Type IV collagenases in the loosening of total hip replacement endoprostheses," *Clinical Orthopaedics and Related Research*, no. 306, pp. 136–144, 1994.

[78] M. Takagi, Y. T. Konttinen, P. Kemppinen et al., "Tissue inhibitor of metalloproteinase 1, collagenolytic and gelatinolytic activity in loose hip endoprostheses," *Journal of Rheumatology*, vol. 22, no. 12, pp. 2285–2290, 1995.

[79] S. Santavirta, M. Takagi, Y. T. Konttinen, T. Sorsa, and A. Suda, "Inhibitory effect of cephalothin on matrix metalloproteinase activity around loose hip prostheses," *Antimicrobial Agents and Chemotherapy*, vol. 40, no. 1, pp. 244–246, 1996.

[80] N. Ishiguro, T. Ito, K. Kurokouchi et al., "mRNA expression of matrix metalloproteinases and tissue inhibitors of metalloproteinase in interface tissue around implants in loosening total hip arthroplasty," *Journal of Biomedical Materials Research*, vol. 32, pp. 611–617, 1996.

[81] A. Y. Strongin, I. Collier, G. Bannikov, B. L. Marmer, G. A. Grant, and G. I. Goldberg, "Mechanism of cell surface activation of 72-kDa type IV collagenase. Isolation of the activated form of the membrane metalloprotease," *The Journal of Biological Chemistry*, vol. 270, no. 10, pp. 5331–5338, 1995.

[82] S. A. Syggelos, S. C. Eleftheriou, E. Giannopoulou, E. Panagiotopoulos, and A. J. Aletras, "Gelatinolytic and collagenolytic activity in periprosthetic tissues from loose hip endoprostheses," *Journal of Rheumatology*, vol. 28, no. 6, pp. 1319–1329, 2001.

[83] G. I. Goldberg, S. M. Wilhelm, A. Kronberger, E. A. Bauer, G. A. Grant, and A. Z. Eisen, "Human fibroblast collagenase," *The Journal of Biological Chemistry*, vol. 261, pp. 6600–6605, 1986.

[84] K. A. Hasty, J. J. Jeffrey, M. S. Hibbs, and H. G. Welgus, "The collagen substrate specificity of human neutrophil collagenase," *The Journal of Biological Chemistry*, vol. 262, no. 21, pp. 10048–10052, 1987.

[85] T. J. Chambers, J. A. Darby, and K. Fuller, "Mammalian collagenase predisposes bone surfaces to osteoclastic resorption," *Cell and Tissue Research*, vol. 241, no. 3, pp. 671–675, 1985.

[86] L. Nordsletten, L. Buø, M. Takagi et al., "The plasminogen activation system is upregulated in loosening of total hip prostheses," *Acta Orthopaedica Scandinavica*, vol. 67, no. 2, pp. 143–148, 1996.

[87] I. Kazes, F. Delarue, J. Hagège et al., "Soluble latent membrane-type 1 matrix metalloprotease secreted by human mesangial cells is activated by urokinase," *Kidney International*, vol. 54, no. 6, pp. 1976–1984, 1998.

[88] P. Carmeliet, L. Moons, R. Lijnen et al., "Urokinase-generated plasmin activates matrix metalloproteinases during aneurysm formation," *Nature Genetics*, vol. 17, no. 4, pp. 439–444, 1997.

[89] P. Diehl, B. Hantke, M. Hennig et al., "Protein expression of MMP-13, uPA, and PAI-1 in pseudocapsular and interface tissue around implants of loose artificial hip joints and in osteoarthritis," *International Journal of Molecular Medicine*, vol. 13, no. 5, pp. 711–715, 2004.

[90] Y. Nakashima, D.-H. Sun, W. J. Maloney, S. B. Goodman, D. J. Schurman, and R. L. Smith, "Induction of matrix metalloproteinase expression in human macrophages by orthopaedic particulate debris *in vitro*," *Journal of Bone and Joint Surgery B*, vol. 80, no. 4, pp. 694–700, 1998.

[91] P. Laquerriere, A. Grandjean-Laquerriere, S. Addadi-Rebbah et al., "MMP-2, MMP-9 and their inhibitors TIMP-2 and TIMP-1 production by human monocytes *in vitro* in the presence of different forms of hydroxyapatite particles," *Biomaterials*, vol. 25, no. 13, pp. 2515–2524, 2004.

[92] T. Ikeda, M. Kasai, M. Utsuyama, and K. Hirokawa, "Determination of three isoforms of the receptor activator of nuclear factor-κB ligand and their differential expression in bone and thymus," *Endocrinology*, vol. 142, no. 4, pp. 1419–1426, 2001.

[93] W. J. Boyle, W. S. Simonet, and D. L. Lacey, "Osteoclast differentiation and activation," *Nature*, vol. 423, no. 6937, pp. 337–342, 2003.

[94] E. M. Schwarz, A. P. Lu, J. J. Goater et al., "Tumor necrosis factor-α/nuclear transcription factor-κB signaling in periprosthetic osteolysis," *Journal of Orthopaedic Research*, vol. 18, no. 3, pp. 472–480, 2000.

[95] J. C. Clohisy, T. Hirayama, E. Frazier, S.-K. Han, and Y. Abu-Amer, "NF-kB signaling blockade abolishes implant particle-induced osteoclastogenesis," *Journal of Orthopaedic Research*, vol. 22, no. 1, pp. 13–20, 2004.

[96] W. Ren, X. H. Li, B. D. Chen, and P. H. Wooley, "Erythromycin inhibits wear debris-induced osteoclastogenesis by modulation of murine macrophage NF-κB activity," *Journal of Orthopaedic Research*, vol. 22, no. 1, pp. 21–29, 2004.

[97] F. Lamoureux, G. Moriceau, G. Picarda, J. Rousseau, V. Trichet, and F. Rédini, "Regulation of osteoprotegerin pro- or anti-tumoral activity by bone tumor microenvironment," *Biochimica et Biophysica Acta*, vol. 1805, no. 1, pp. 17–24, 2010.

[98] M. Ulrich-Vinther, E. E. Carmody, J. J. Goater, K. Søballe, R. J. O'Keefe, and E. M. Schwarz, "Recombinant adeno-associated virus-mediated osteoprotegerin gene therapy inhibits wear debris-induced osteolysis," *Journal of Bone and Joint Surgery A*, vol. 84, no. 8, pp. 1405–1412, 2002.

[99] J. Jeffrey Goater, R. J. O'Keefe, R. N. Rosier, J. Edward Puzas, and E. M. Schwarz, "Efficacy of ex vivo OPG gene therapy in preventing wear debris induced osteolysis," *Journal of Orthopaedic Research*, vol. 20, no. 2, pp. 169–173, 2002.

[100] L. M. Childs, E. P. Paschalis, L. Xing et al., "*In vivo* RANK signaling blockade using the receptor activator of NF-κB:Fc effectively prevents and ameliorates wear debris-induced osteolysis via osteoclast depletion without inhibiting osteogenesis," *Journal of Bone and Mineral Research*, vol. 17, no. 2, pp. 192–199, 2002.

[101] T. Masui, S. Sakano, Y. Hasegawa, H. Warashina, and N. Ishiguro, "Expression of inflammatory cytokines, RANKL and OPG induced by titanium, cobalt-chromium and polyethylene particles," *Biomaterials*, vol. 26, no. 14, pp. 1695–1702, 2005.

[102] J. Mandelin, T.-F. Li, M. Liljeström et al., "Imbalance of RANKL/RANK/OPG system in interface tissue in loosening of total hip replacement," *Journal of Bone and Joint Surgery B*, vol. 85, no. 8, pp. 1196–1201, 2003.

[103] T. Gehrke, C. Sers, L. Morawietz et al., "Receptor activator of nuclear factor κB ligand is expressed in resident and inflammatory cells in aseptic and septic prosthesis loosening," *Scandinavian Journal of Rheumatology*, vol. 32, no. 5, pp. 287–294, 2003.

[104] D. R. Haynes, T. N. Crotti, A. E. Potter et al., "The osteoclastogenic molecules RANKL and RANK are associated with periprosthetic osteolysis," *Journal of Bone and Joint Surgery B*, vol. 83, no. 6, pp. 902–911, 2001.

[105] M. Horiki, T. Nakase, A. Myoui et al., "Localization of RANKL in osteolytic tissue around a loosened joint prosthesis," *Journal of Bone and Mineral Metabolism*, vol. 22, no. 4, pp. 346–351, 2004.

[106] J. Mandelin, T.-F. Li, M. Hukkanen et al., "Interface tissue fibroblasts from loose total hip replacement prosthesis produce receptor activator of nuclear factor-κB ligand, osteoprotegerin, and cathepsin K," *Journal of Rheumatology*, vol. 32, no. 4, pp. 713–720, 2005.

[107] V. T. Labropoulou, A. D. Theocharis, A. Symeonidis, S. S. Skandalis, N. K. Karamanos, and H. P. Kalofonos, "Pathophysiology and pharmacological targeting of tumor-induced bone disease: current status and emerging therapeutic interventions," *Current Medicinal Chemistry*, vol. 18, no. 11, pp. 1584–1598, 2011.

[108] O. Ishibashi, S. Niwa, K. Kadoyama, and T. Inui, "MMP-9 antisense oligodeoxynucleotide exerts an inhibitory effect on osteoclastic bone resorption by suppressing cell migration," *Life Sciences*, vol. 79, no. 17, pp. 1657–1660, 2006.

[109] W. C. M. Duivenvoorden, H. W. Hirte, and G. Singh, "Use of tetracycline as an inhibitor of matrix metalloproteinase activity secreted by human bone-metastasizing cancer cells," *Invasion and Metastasis*, vol. 17, no. 6, pp. 312–322, 1997.

[110] G. C. N. Franco, M. Kajiya, T. Nakanishi et al., "Inhibition of matrix metalloproteinase-9 activity by doxycycline ameliorates RANK ligand-induced osteoclast differentiation *in vitro* and *in vivo*," *Experimental Cell Research*, vol. 317, no. 10, pp. 1454–1464, 2011.

[111] L. Blavier and J. M. Delaissé, "Matrix metalloproteinases are obligatory for the migration of preosteoclasts to the developing marrow cavity of primitive long bones," *Journal of Cell Science*, vol. 108, no. 12, pp. 3649–3659, 1995.

[112] M. T. Engsig, Q.-J. Chen, T. H. Vu et al., "Matrix metalloproteinase 9 and vascular endothelial growth factor are essential for osteoclast recruitment into developing long bones," *Journal of Cell Biology*, vol. 151, no. 4, pp. 879–889, 2000.

[113] C. C. Lynch, A. Hikosaka, H. B. Acuff et al., "MMP-7 promotes prostate cancer-induced osteolysis via the solubilization of RANKL," *Cancer Cell*, vol. 7, no. 5, pp. 485–496, 2005.

[114] T. J. Wilson, K. C. Nannuru, M. Futakuchi, A. Sadanandam, and R. K. Singh, "Cathepsin G enhances mammary tumor-induced osteolysis by generating soluble receptor activator of nuclear factor-κB ligand," *Cancer Research*, vol. 68, no. 14, pp. 5803–5811, 2008.

[115] S. Thiolloy, J. Halpern, G. E. Holt et al., "Osteoclast-derived matrix metalloproteinase-7, but not matrix metalloproteinase-9, contributes to tumor-induced osteolysis," *Cancer Research*, vol. 69, no. 16, pp. 6747–6755, 2009.

[116] T. Standal, C. Seidel, Ø. Hjertner et al., "Osteoprotegerin is bound, internalized, and degraded by multiple myeloma cells," *Blood*, vol. 100, no. 8, pp. 3002–3007, 2002.

[117] W. S. Simonet, D. L. Lacey, C. R. Dunstan et al., "Osteoprotegerin: a novel secreted protein involved in the regulation of bone density," *Cell*, vol. 89, no. 2, pp. 309–319, 1997.

[118] T. Manon-Jensen, Y. Itoh, and J. R. Couchman, "Proteoglycans in health and disease: the multiple roles of syndecan shedding," *FEBS Journal*, vol. 277, no. 19, pp. 3876–3889, 2010.

[119] Q. Li, P. W. Park, C. L. Wilson, and W. C. Parks, "Matrilysin shedding of syndecan-1 regulates chemokine mobilization and transepithelial efflux of neutrophils in acute lung injury," *Cell*, vol. 111, no. 5, pp. 635–646, 2002.

[120] S. Brule, N. Charnaux, A. Sutton et al., "The shedding of syndecan-4 and syndecan-1 from HeLa cells and human primary macrophages is accelerated by SDF-1/CXCL12 and mediated by the matrix metalloproteinase-9," *Glycobiology*, vol. 16, no. 6, pp. 488–501, 2006.

[121] C. Y. Fears, C. L. Gladson, and A. Woods, "Syndecan-2 is expressed in the microvasculature of gliomas and regulates angiogenic processes in microvascular endothelial cells," *The Journal of Biological Chemistry*, vol. 281, no. 21, pp. 14533–14536, 2006.

[122] K. Endo, T. Takino, H. Miyamori et al., "Cleavage of syndecan-1 by membrane type matrix metalloproteinase-1 stimulates cell migration," *The Journal of Biological Chemistry*, vol. 278, no. 42, pp. 40764–40770, 2003.

[123] C. Vermes, K. A. Roebuck, R. Chandrasekaran, J. G. Dobai, J. J. Jacobs, and T. T. Glant, "Particulate wear debris activates protein tyrosine kinases and nuclear factor κB, which down-regulates type I collagen synthesis in human osteoblasts," *Journal of Bone and Mineral Research*, vol. 15, no. 9, pp. 1756–1765, 2000.

[124] L. Luo, A. Petit, J. Antoniou et al., "Effect of cobalt and chromium ions on MMP-1, TIMP-1, and TNF-α gene expression in human U937 macrophages: a role for tyrosine kinases," *Biomaterials*, vol. 26, no. 28, pp. 5587–5593, 2005.

[125] G.-R. Yan, C.-L. Xiao, G.-W. He et al., "Global phosphoproteomic effects of natural tyrosine kinase inhibitor, genistein, on signaling pathways," *Proteomics*, vol. 10, no. 5, pp. 976–986, 2010.

[126] X. N. Stahtea, O. C. Kousidou, A. E. Roussidis, G. N. Tzanakakis, and N. K. Karamanos, "Small tyrosine kinase inhibitors as key molecules in the expression of metalloproteinases by solid tumors," *Connective Tissue Research*, vol. 49, no. 3-4, pp. 211–214, 2008.

[127] M. H. Ravindranath, S. Muthugounder, N. Presser, and S. Viswanathan, "Anticancer therapeutic potential of soy isoflavone, genistein," *Advances in Experimental Medicine and Biology*, vol. 546, pp. 121–165, 2004.

[128] D. S. Goodsell, "The molecular perspective: VEGF and angiogenesis," *Oncologist*, vol. 7, no. 6, pp. 569–570, 2002.

[129] L. E. Harry and E. M. Paleolog, "From the cradle to the clinic: VEGF in developmental, physiological, and pathological angiogenesis," *Birth Defects Research C*, vol. 69, no. 4, pp. 363–374, 2003.

Extracellular Matrix Degradation and Tissue Remodeling in Periprosthetic Loosening and Osteolysis: Focus on Matrix Metalloproteinases, Their Endogenous Tissue Inhibitors, and the Proteasome

127

[130] J. Waltenberger, L. Claesson-Welsh, A. Siegbahn, M. Shibuya, and C.-H. Heldin, "Different signal transduction properties of KDR and Flt1, two receptors for vascular endothelial growth factor," *The Journal of Biological Chemistry*, vol. 269, no. 43, pp. 26988–26995, 1994.

[131] R. Kunstfeld, S. Hirakawa, Y.-K. Hong et al., "Induction of cutaneous delayed-type hypersensitivity reactions in VEGF-A transgenic mice results in chronic skin inflammation associated with persistent lymphatic hyperplasia," *Blood*, vol. 104, no. 4, pp. 1048–1057, 2004.

[132] M. Nakagawa, T. Kaneda, T. Arakawa et al., "Vascular endothelial growth factor (VEGF) directly enhances osteoclastic bone resorption and survival of mature osteoclasts," *FEBS Letters*, vol. 473, no. 2, pp. 161–164, 2000.

[133] K. Henriksen, M. Karsdal, J.-M. Delaissé, and M. T. Engsig, "RANKL and vascular endothelial growth factor (VEGF) induce osteoclast chemotaxis through an ERK1/2-dependent mechanism," *The Journal of Biological Chemistry*, vol. 278, no. 49, pp. 48745–48753, 2003.

[134] N. Al-Saffar, J. T. L. Mah, Y. Kadoya, and P. A. Revell, "Neovascularisation and the induction of cell adhesion molecules in response to degradation products from orthopaedic implants," *Annals of the Rheumatic Diseases*, vol. 54, no. 3, pp. 201–208, 1995.

[135] J. Y. Wang, B. H. Wicklund, R. B. Gustilo, and D. T. Tsukayama, "Titanium, chromium and cobalt ions modulate the release of bone-associated cytokines by human monocytes/macrophages *in vitro*," *Biomaterials*, vol. 17, no. 23, pp. 2233–2240, 1996.

[136] Y. Niki, H. Matsumoto, Y. Suda et al., "Metal ions induce bone-resorbing cytokine production through the redox pathway in synoviocytes and bone marrow macrophages," *Biomaterials*, vol. 24, no. 8, pp. 1447–1457, 2003.

[137] O. Langholz, D. Röckel, C. Mauch et al., "Collagen and collagenase gene expression in three-dimensional collagen lattices are differentially regulated by α1β1 and α2β1 integrins," *Journal of Cell Biology*, vol. 131, no. 6, pp. 1903–1915, 1995.

[138] L. Ravanti, J. Heino, C. López-Otín, and V. M. Kaharin, "Induction of collagenase-3 (MMP-13) expression in human skin fibroblasts by three-dimensional collagen is mediated by p38 mitogen-activated protein kinase," *The Journal of Biological Chemistry*, vol. 274, pp. 2446–2455, 1999.

[139] Y. Tamaki, Y. Takakubo, K. Goto et al., "Increased expression of toll-like receptors in aseptic loose periprosthetic tissues and septic synovial membranes around total hip implants," *Journal of Rheumatology*, vol. 36, no. 3, pp. 598–608, 2009.

[140] J. A. Gebbia, J. L. Coleman, and J. L. Benach, "Selective induction of matrix metalloproteinases by Borrelia burgdorferi via Toll-like receptor 2 in monocytes," *Journal of Infectious Diseases*, vol. 189, no. 1, pp. 113–119, 2004.

[141] R. A. Lisboa, M. V. Andrade, and J. R. Cunha-Melo, "Toll-like receptor activation and mechanical force stimulation promote the secretion of matrix metalloproteinases 1, 3 and 10 of human periodontal fibroblasts via p38, JNK and NF-kB," *Archives of Oral Biology*, vol. 58, no. 6, pp. 731–739, 2013.

[142] M. H. A. Malik, F. Jury, A. Bayat, W. E. R. Ollier, and P. R. Kay, "Genetic susceptibility to total hip arthroplasty failure: a preliminary study on the influence of matrix metalloproteinase 1, interleukin 6 polymorphisms and vitamin D receptor," *Annals of the Rheumatic Diseases*, vol. 66, no. 8, pp. 1116–1120, 2007.

[143] S. Ye, "Polymorphism in matrix metalloproteinase gene promoters: implication in regulation of gene expression and susceptibility of various diseases," *Matrix Biology*, vol. 19, no. 7, pp. 623–629, 2000.

[144] U. Benbow, J. L. Rutter, C. H. Lowrey, and C. E. Brinckerhoff, "Transcriptional repression of the human collagenase-1 (MMP-1) gene in MDA231 breast cancer cells by all-trans-retinoic acid requires distal regions of the promoter," *British Journal of Cancer*, vol. 79, no. 2, pp. 221–228, 1999.

[145] S. R. Bramhall, A. Rosemurgy, P. D. Brown, C. Bowry, and J. A. C. Buckles, "Marimastat as first-line therapy for patients with unresectable pancreatic cancer: a randomized trial," *Journal of Clinical Oncology*, vol. 19, no. 15, pp. 3447–3455, 2001.

[146] P. C. Ortiz-Lazareno, G. Hernandez-Flores, J. R. Dominguez-Rodriguez et al., "MG132 proteasome inhibitor modulates proinflammatory cytokines production and expression of their receptors in U937 cells: involvement of nuclear factor-κB and activator protein-1," *Immunology*, vol. 124, no. 4, pp. 534–541, 2008.

[147] X. Mao, X. Pan, T. Cheng, and X. Zhang, "Inhibition of titanium particle-induced inflammation by the proteasome inhibitor bortezomib in murine macrophage-like RAW 264.7 cells," *Inflammation*, vol. 35, no. 4, pp. 1411–1418, 2012.

[148] T. Krakauer, "Molecular therapeutic targets in inflammation: cyclooxygenase and NF-κB," *Current Drug Targets*, vol. 3, no. 3, pp. 317–324, 2004.

[149] L. Goffin, Q. Seguin-Estévez, M. Alvarez, W. Reith, and C. Chizzolini, "Transcriptional regulation of matrix metalloproteinase-1 and collagen 1A2 explains the anti-fibrotic effect exerted by proteasome inhibition in human dermal fibroblasts," *Arthritis Research and Therapy*, vol. 12, no. 2, article 73, 2010.

[150] T. Smallie, G. Ricchetti, N. J. Horwood, M. Feldmann, A. R. Clark, and L. M. Williams, "IL-10 inhibits transcription elongation of the human TNF gene in primary macrophages," *The Journal of Experimental Medicine*, vol. 207, no. 10, pp. 2081–2088, 2010.

[151] A. B. Lentsch, T. P. Shanley, V. Sarma, and P. A. Ward, "*In vivo* suppression of NF-κB and preservation of IκBα by interleukin-10 and interleukin-13," *The Journal of Clinical Investigation*, vol. 100, no. 10, pp. 2443–2448, 1997.

[152] N. Giuliani, F. Morandi, S. Tagliaferri et al., "The proteasome inhibitor bortezomib affects osteoblast differentiation *in vitro* and *in vivo* in multiple myeloma patients," *Blood*, vol. 110, no. 1, pp. 334–338, 2007.

[153] E. Terpos, D. J. Heath, A. Rahemtulla et al., "Bortezomib reduces serum dickkopf-1 and receptor activator of nuclear factor-κB ligand concentrations and normalises indices of bone remodelling in patients with relapsed multiple myeloma," *British Journal of Haematology*, vol. 135, no. 5, pp. 688–692, 2006.

[154] I. von Metzler, H. Krebbel, M. Hecht et al., "Bortezomib inhibits human osteoclastogenesis," *Leukemia*, vol. 21, no. 9, pp. 2025–2034, 2007.

[155] V. J. Palombella, O. J. Rando, A. L. Goldberg, and T. Maniatis, "The ubiquitin-proteasome pathway is required for processing the NF-κB1 precursor protein and the activation of NF-κB," *Cell*, vol. 78, no. 5, pp. 773–785, 1994.

[156] E. Reinstein and A. Ciechanover, "Narrative review: potein degradation and human diseases: the ubiquitin connection," *Annals of Internal Medicine*, vol. 145, no. 9, pp. 676–684, 2006.

[157] S. S. Skandalis, A. J. Aletras, C. Gialeli et al., "Targeting the tumor proteasome as a mechanism to control the synthesis

and bioactivity of matrix macromolecules," *Current Molecular Medicine*, vol. 12, pp. 1068–1082, 2012.

[158] M. L. Wang, P. V. Hauschka, R. S. Tuan, and M. J. Steinbeck, "Exposure to particles stimulates superoxide production by human THP-1 macrophages and A vian HD-11EM osteoclasts activated by tumor necrosis factor-α and PMA," *Journal of Arthroplasty*, vol. 17, no. 3, pp. 335–346, 2002.

[159] H.-M. Wu, K.-H. Chi, and W.-W. Lin, "Proteasome inhibitors stimulate activator protein-1 pathway via reactive oxygen species production," *FEBS Letters*, vol. 526, no. 1–3, pp. 101–105, 2002.

[160] B. Catalgol, I. Ziaja, N. Breusing et al., "The proteasome is an integral part of solar ultraviolet a radiation-induced gene expression," *The Journal of Biological Chemistry*, vol. 284, no. 44, pp. 30076–30086, 2009.

[161] S. Fineschi, W. Reith, P. A. Guerne, J.-M. Dayer, and C. Chizzolini, "Proteasome blockade exerts an antifibrotic activity by coordinately down-regulating type I collagen and tissue inhibitor of metalloproteinase-1 and up-regulating metalloproteinase-1 production in human dermal fibroblasts," *FASEB Journal*, vol. 20, no. 3, pp. 562–564, 2006.

[162] H.-J. Kim, B. Barajas, M. Wang, and A. E. Nel, "Nrf2 activation by sulforaphane restores the age-related decrease of TH1 immunity: role of dendritic cells," *Journal of Allergy and Clinical Immunology*, vol. 121, no. 5, pp. 1255–1261, 2008.

[163] L. Sepp-Lorenzino, Z. Ma, D. E. Lebwohl, A. Vinitsky, and N. Rosen, "Herbimycin A induces the 20 S proteasome- and ubiquitin-dependent degradation of receptor tyrosine kinases," *The Journal of Biological Chemistry*, vol. 270, no. 28, pp. 16580–16587, 1995.

[164] D. C. Markel, R. Zhang, T. Shi, M. Hawkins, and W. Ren, "Inhibitory effects of erythromycin on wear debris-induced VEGF/Flt-1 gene production and osteolysis," *Inflammation Research*, vol. 58, pp. 413–421, 2009.

[165] P. J. Millett, M. J. Allen, and M. P. G. Bostrom, "Effects of alendronate on particle-induced osteolysis in a rat model," *Journal of Bone and Joint Surgery A*, vol. 84, no. 2, pp. 236–249, 2002.

[166] M. von Knoch, C. Wedemeyer, A. Pingsmann et al., "The decrease of particle-induced osteolysis after a single dose of bisphosphonate," *Biomaterials*, vol. 26, no. 14, pp. 1803–1808, 2005.

[167] P. G. Dedes, C. Gialeli, A. Tsonis et al., "Expression of matrix macromolecules and functional properties of breast cancer cells are modulated by the bisphosphonate zoledronic acid," *Biochimica et Biophysica Acta*, vol. 1820, pp. 1926–1939, 2012.

[168] A. Ito, T. Nose, S. Takahashi, and Y. Mori, "Cyclooxygenase inhibitors augment the production of pro-matrix metalloproteinase 9 (progelatinase B) in rabbit articular chondrocytes," *FEBS Letters*, vol. 360, no. 1, pp. 75–79, 1995.

[169] J. A. DiBattista, J. Martel-Pelletier, N. Fujimoto, K. Obata, M. Zafarullah, and J.-P. Pelletier, "Prostaglandins E2 and E1 inhibit cytokine-induced metalloprotease expression in human synovial fibroblasts: mediation by cyclic-AMP signalling pathway," *Laboratory Investigation*, vol. 71, no. 2, pp. 270–278, 1994.

[170] S. Takahashi, T. Inoue, M. Higaki, and Y. Mizushima, "Cyclooxygenase inhibitors enhance the production of tissue inhibitor-1 of metalloproteinases (TIMP-1) and pro-matrix metalloproteinase 1 (proMMP-1) in human rheumatoid synovial fibroblasts," *Inflammation Research*, vol. 46, no. 8, pp. 320–323, 1997.

[171] S. K. Lin, C. C. Wang, S. Huang et al., "Induction of dental pulp fibroblast matrix metalloproteinase-1 and tissue inhibitor of metalloproteinase-1 gene expression by interleukin-1alpha and tumor necrosis factor-alpha through a prostaglandin-dependent pathway," *Journal of Endodontics*, vol. 27, no. 3, pp. 185–189, 2001.

[172] R. Yamazaki, S. Kawai, Y. Mizushima et al., "A major metabolite of aceclofenac, 4'-hydroxy aceclofenac, suppresses the production of interstitial pro-collagenase/proMMP-1 and pro-stromelysin- 1/proMMP-3 by human rheumatoid synovial cells," *Inflammation Research*, vol. 49, no. 3, pp. 133–138, 2000.

[173] H. Akimoto, R. Yamazaki, S. Hashimoto, T. Sato, and A. Ito, "4'-Hydroxy aceclofenac suppresses the interleukin-1-induced production of promatrix metalloproteinases and release of sulfated-glycosaminoglycans from rabbit articular chondrocytes," *European Journal of Pharmacology*, vol. 401, no. 3, pp. 429–436, 2000.

[174] S. A. Syggelos, E. Giannopoulou, P. A. Gouvousis, A. P. Andonopoulos, A. J. Aletras, and E. Panagiotopoulos, "In vitro effects of non-steroidal anti-inflammatory drugs on cytokine, prostanoid and matrix metalloproteinase production by interface membranes from loose hip or knee endoprostheses," *Osteoarthritis and Cartilage*, vol. 15, no. 5, pp. 531–542, 2007.

[175] Y. Ishimi, C. Miyaura, C. H. Jin et al., "IL-6 is produced by osteoblasts and induces bone resorption," *Journal of Immunology*, vol. 145, no. 10, pp. 3297–3303, 1990.

[176] N. Kurihara, D. Bertolini, T. Suda, Y. Akiyama, and G. D. Roodman, "IL-6 stimulates osteoclast-like multinucleated cell formation in long term human marrow cultures by inducing IL-1 release," *Journal of Immunology*, vol. 144, no. 11, pp. 4226–4230, 1990.

[177] U. H. Lerner and A. Ohlin, "Tumor necrosis factors α and β can stimulate bone resorption in cultured mouse calvariae by a prostaglandin-independent mechanism," *Journal of Bone and Mineral Research*, vol. 8, no. 2, pp. 147–155, 1993.

[178] J. H. Herman, W. G. Sowder, and E. V. Hess, "Nonsteroidal anti-inflammatory drug modulation of prosthesis pseudomembrane induced bone resorption," *Journal of Rheumatology*, vol. 21, no. 2, pp. 338–343, 1994.

13

HTR8/SVneo Cells Display Trophoblast Progenitor Cell-Like Characteristics Indicative of Self-Renewal, Repopulation Activity, and Expression of "Stemness-" Associated Transcription Factors

Maja Weber,[1] Ilka Knoefler,[1] Ekkehard Schleussner,[1] Udo R. Markert,[1] and Justine S. Fitzgerald[1,2]

[1] *Placenta-Lab, Department of Obstetrics, University Hospital Jena, Germany*
[2] *Abteilung für Geburtshilfe, Placenta-Labor, Universitätsklinikum Jena, Bachstr. 18, 07740 Jena, Germany*

Correspondence should be addressed to Justine S. Fitzgerald; fitzgerald@med.uni-jena.de

Academic Editor: Irma Virant-Klun

Introduction. JEG3 is a choriocarcinoma—and HTR8/SVneo a transformed extravillous trophoblast—cell line often used to model the physiologically invasive extravillous trophoblast. Past studies suggest that these cell lines possess some stem or progenitor cell characteristics. Aim was to study whether these cells fulfill minimum criteria used to identify stem-like (progenitor) cells. In summary, we found that the expression profile of HTR8/SVneo (CDX2+, NOTCH1+, SOX2+, NANOG+, and OCT-) is distinct from JEG3 (CDX2+ and NOTCH1+) as seen only in human-serum blocked immunocytochemistry. This correlates with HTR8/SVneo's self-renewal capacities, as made visible via spheroid formation and multi-passagability in hanging drops protocols paralleling those used to maintain embryoid bodies. JEG3 displayed only low propensity to form and reform spheroids. HTR8/SVneo spheroids migrated to cover and seemingly repopulate human chorionic villi during confrontation cultures with placental explants in hanging drops. We conclude that HTR8/SVneo spheroid cells possess progenitor cell traits that are probably attained through corruption of "stemness-" associated transcription factor networks. Furthermore, trophoblastic cells are highly prone to unspecific binding, which is resistant to conventional blocking methods, but which can be alleviated through blockage with human serum.

1. Introduction

The master regulatory networks of human embryonic stem cell (hESC) transcription factors, OCT4, SOX2, and NANOG, as well as other cell fate determining transcription factors that are implicated in stem cell self-renewal capacities, such as NOTCH1 and STAT3, are expressed not only by embryonic stem cells, but also by a number of cancers [1]. Some of these factors are also expressed in choriocarcinoma (gestational trophoblastic disease) [2]. This has led to the thought that choriocarcinoma may also represent a group of tumors, in which hESC transcription factor deregulation has led to their transformation into cancer stem cells.

In mammalian development, the first cell differentiation step segregates trophoblast and embryonic cell lineages, thus resulting in the formation of the blastocyst's outer lining, the trophectoderm (TE), and its inner cell mass (ICM). The trophectoderm consists of trophoblast stem cells that express CDX2, a homeobox transcription factor, which is required for the emergence of these cells [3]. Physiological invasion is seen during blastocyst implantation, which is mediated through the trophectoderm. Interestingly, both CDX2 and SOX2 deficiency lead to implantation failure of the blastocyst secondary to trophoectoderm differentiation problems [4–6].

The trophectoderm also differentiates into several trophoblast subsets in order to create the placenta of the first trimester pregnancy. Of these subsets, the cytotrophoblast is considered a putative "progenitor cell," which replenishes the outer layer of the villous (syncytiotrophoblast), but which is also able to invade the decidua in a cancer-like manner when necessary and desirable (extravillous trophoblast) [7]. This behaviour is often believed to be driven by hypoxia, and it is a well-orchestrated and closely controlled process, mostly through a network of interaction between the invading trophoblast, the decidua, the maternal endothelium, and the maternal immune system; the detailed description of which would tax the scope of this introduction [8]. The first trimester placenta is especially ample with invasive (cyto)trophoblast, while the term placenta trophoblast loses this capability [8].

The uniqueness of this situation, in which physiologic, spatially (limited to the decidua, first third of the myometrium, and the invasion into maternal spiral arteries), and temporally (limited to the first trimester of pregnancy) regulated invasion (by the trophoblast) and pathologic, deregulated, and malignant invasion (by choriocarcinoma) are set so close together, has drawn the attention of cancer researchers worldwide [8]. However, since isolation of primary trophoblast and choriocarcinoma cells is often cumbersome, in recent years, several trophoblastic cell lines have been utilized as imperfect models for the invasive trophoblast(ic) cell. Some of the most popular cell lines used constitute the immortalized first trimester trophoblast cell line, HTR8/SVneo, and the choriocarcinoma cell line JEG3. HTR8/SVneo cells are often considered a closer model of trophoblast cells, because the HTR8/SVneo cell lines were established by immortalizing a physiologic extravillous trophoblast cell via transfection with a plasmid containing the simian virus 40 large T antigen (SV40) [9], while the JEG3 cell line was cloned from a primary choriocarcinoma strain [10].

Our own recently published data, however, demonstrate that the miRNA profiles of these two cell lines are quite differing, surprisingly with JEG3 encompassing an miRNA profile that is closer to primary first trimester trophoblast cells than that of the HTR8/SVneo cell lines [11]. Villous cytotrophoblast and HTR8/SVneo cells have interestingly also been implicated in producing a "side population" that either demonstrates long-term repopulating properties or expresses classical hESC markers [12, 13].

Following the idea that both JEG3 and HTR8/SVneo are transformed cells and have been proposed to produce cancer stem cell or progenitor (side population) cell populations, we aimed to characterize the putative cancer and trophoblast stem/progenitor cell traits of HTR8/SVneo and JEG3 cells on the basis of general minimum recommendations for identifying cancer stem cells or progenitor cells [14, 15]. This is accomplished first by assessing the capacity of these cells to form spheroid bodies, second by determining the expression of various transcription factors related to progenitor or to cancer stem cell development, and finally by investigating the cells' ability to repopulate trophoblast tissue in a near in vivo model.

For these studies, we phrase SOX2, OCT4, and NANOG as core "stemness-" associated transcription factors [16] and CDX2 [3] as a trophoblast stem/progenitor cell transcription factor. NOTCH1 is included as an often abused, prominent cell-fate transcription factor associated with both cancer stem

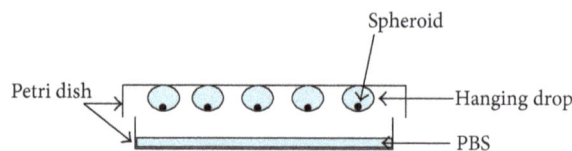

FIGURE 1: Schematic diagram of spheroid formation.

cells and hESC [17–20] and is, henceforth, termed a cell-fate determining transcription factor.

2. Methods

2.1. Spheroid Formation with Hanging Drops. We chose the hanging drops protocol as reviewed by Kurosawa [21].

Briefly, 20 000 cells per 30 μL drop supplemented RPMI (as described later in Section 2.2) were plated onto the lid of two Petri dishes in regular arrays (20 drops/Petri lid). The lid was inverted over the bottom of the PBS-filled Petri dish (see schematic representation Figure 1). The Petri dish with the hanging drops were cultured under standard conditions (37°C, 5% CO_2, humidified atmosphere) for 48 hours. A schematic image of the hanging drop principle is seen in Figure 1. The experiment was carried out in the same manner for HTR8/SVneo cells and for JEG3 cells (40 drops per cell line).

2.2. Cell Culture. HTR-8/SVneo cells (a kind gift from Professor Charles Graham of the Department of Anatomy and Cell Biology at Queen's University, Kingston, ON, Canada) were cultured in RPMI (PAA) and JEG3 cells in F12 Medium. Both media were supplemented with 10% fetal bovine serum (FBS; SIGMA, St. Louis, USA) and 1x penicillin/streptomycin (PAA Laboratories; Pasching, Austria). Cell cultures were maintained under standard culturing conditions (37°C, 5% CO_2, humidified atmosphere).

2.3. Immunocytochemistry (ICH). Cells were trypsinized, centrifuged, and resuspended in 1 mL respective medium. Slides (SuperFrost/Plus slides; Menzel, Germany) were washed and sterilized with ethanol, coated with cells (200 μL), and incubated over night at 37°C. The cells were fixed on the next day with ethanol/methanol and consequently used to perform immunocytochemistry. To inhibit endogenous peroxidise activity, the cells were incubated in methanol/H_2O_2 for 5–10 min and washed for 5 min in phosphate-buffered saline (PBS, pH 7.4), followed by incubation firstly with and without 5% human AB serum (PAA), which corresponds to an approximate Fc-concentration of 0.6 mg/mL, in order to further eliminate the possibility of Fc-receptor cross-reactions (as described in [22]), and secondly with goat serum at room temperature for 20 min (Vector Laboratories) to eliminate regular nonspecific background staining.

Samples were then incubated with the primary antibodies (please refer to Table 1) for 60 min at room temperature. Antibodies were diluted in DAKO Antibody Diluent with Background Reducing Components (DAKO, Denmark).

HTR8/SVneo Cells Display Trophoblast Progenitor Cell-Like Characteristics Indicative of Self-Renewal, Repopulation Activity, and Expression of "Stemness-" Associated Transcription Factors

131

TABLE 1: List of antibodies.

		Immunohistochemistry (IHC)		
Antibody	Clone	Isotype	Concentration IHC	Source
Cdx2	—	Polyclonal Rabbit	1 : 200	Cell Signaling
Sox2	D6D9	Polyclonal Rabbit	1 : 100	Cell Signaling
Notch1	D6F11	Polyclonal Rabbit	1 : 200	Cell Signaling
Nanog	—	Polyclonal Rabbit	1 : 400	Cell Signaling
Oct4A	C52G3	Polyclonal Rabbit	1 : 300	Cell Signaling
Isotyp control	DA1E	Polyclonal Rabbit	1 : 100	Cell Signaling
ABC Elite kit (rabbit IgG)				Vector Laboratories (Lörrach, Germany)

In the next step, our samples were incubated with the biotinylated secondary antibody (Vector Laboratories) for 30 min at room temperature. For a listing of antibodies, please refer also to Table 1. Following incubation with the secondary antibody, an incubation period with ABC-complex (avidin-biotinylated peroxidise; Vector Laboratories) again for 30 min at room temperature was completed. Between each step, all samples were washed profusely with PBS. The peroxidase reaction was achieved with DAB (diaminobenzidine/H_2O_2; 1 mg/mL; DAB; Dako) and after 5 min discontinued with water. Hematoxylin staining was used for cell nuclei staining (2 min). Finally, slides were dehydrated by an ethanol-to-xylene treatment, covered with Histofluid (Paul Marienfeld, Lauda-Königshofen, Germany), and analysis was completed with the Axioplan 2 microscope (Carl Zeiss, Jena, Germany).

A negative control was prepared by replacing the primary antibody with DAKO Antibody Diluent only. Isotype controls were prepared in the same manner as the primary antibody.

Analysis of staining intensity and gross estimation of stained cell numbers was accomplished by eye and by two blinded investigators (criteria similar to standard immunoreactive scoring).

2.4. Immunofluorescence Staining. The cells were cultured on SuperFrost/Plus slides (Menzel, Germany) over night with serum-free media. Cells were fixed on the subsequent day with ethanol/methanol. To reduce nonspecific background staining, all samples were incubated either with goat serum or with 5% human AB serum (PAA, as recommended by [22]) at room temperature for 20 min (Vector Laboratories). Samples were incubated with the primary antibodies (please refer to Table 1 for company names and concentrations used) overnight at 4°C. Antibodies were diluted in DAKO Antibody Diluent with Background Reducing Components (DAKO, Denmark). On the next day, they were incubated with the secondary antibody labeled with Cy3 for 1 h at room temperature. Between each step, all samples were washed profusely with PBS. Sections were counterstained Vectashield Mounting Media with DAPI (VECTOR Laboratories) and then cover slipped.

A negative control was prepared by replacing the primary antibody with DAKO Antibody Diluent only. Isotype controls were prepared in the same manner as the primary antibody. All samples were analyzed with an AxioPlan2 microscope (Carl Zeiss, Jena, Germany).

Assessment was accomplished by eye and by two investigators only in terms of positive or negative expression and pattern of expression.

2.5. Placental Explant Cultures and Confrontation Cultures. Two biopsy-sized "explants" (2 mm diameter) each from villous tissue of five healthy human term placentae (after elective caesarian section) were collected. An approval by the local ethical committee exists. Prior to confrontation cultures, all spheroids were stained by application of 10 nM Mito Tracker dye (fluorescent green; Invitrogen) for 30 min at 37°C and then intensively washed in PBS (Biochrom, Germany). Commencing from the time that the placental "biopsies" or "explants" are placed in culture, these are termed villous explant cultures.

Each explant was then confronted in culture with one spheroid within the respective hanging drop for 48 h. Subsequently, the confronted tissues were incubated with a solution of 10% nonfat milk in PBS containing 1% Triton (AppliChem) to permeabilize cell membranes and to block nonspecific binding sites for 1 h. Following intensive washing step, tissues were incubated with a rat anti-human CD31 (anti-PECAM1; Millipore, Germany) for 2 h, followed by incubation with a goat anti-rat IgG-Cy5 conjugate (Milipore) for 90 min, all within the previously described solution (nonfat milk/PBS/Triton). After staining, descriptive analyses with the tissues and spheroids were accomplished on a confocal laser scanning microscope (Carl-Zeiss, Jena, Germany).

3. Results

3.1. HTR8/SVneo Cells Have a High and JEG3 Cells a Low Propensity to Form Spheroid Bodies Within Hanging Drops. The first step to confirm putative cancer stem/progenitor cell status is to confirm their capacity for self-renewal, which is often accomplished by propagation of these cells as spheroids in stem cell culturing conditions [14]. Cells with self-renewing potential can be disaggregated from the spheroids and passaged multiple times with retention of spheroid-forming ability [15].

Of the 40 hanging drops experiments per cell line, 100% of the incubated HTR8/SVneo cells and only 50% of the incubated JEG3 cells were able to form spheroids (data not demonstrated). The developed HTR8/SVneo spheroids regularly measured a diameter of approximately 700–750 μm,

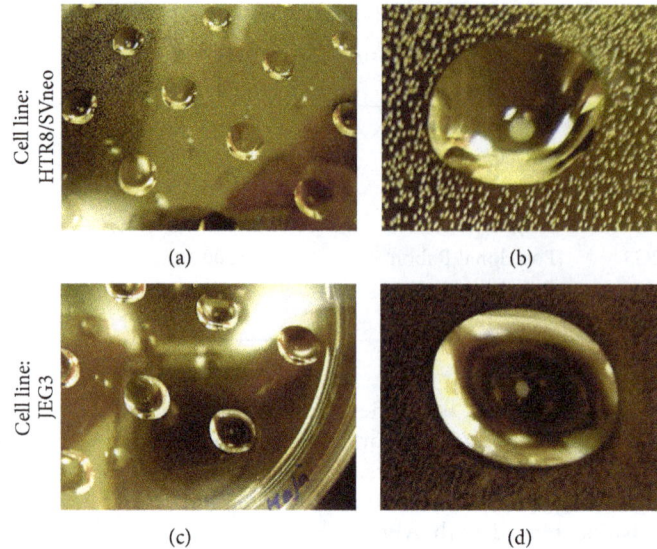

Cell line: HTR8/SVneo

(a)

(b)

Cell line: JEG3

(c)

(d)

FIGURE 2: Spheroid formation of HTR8/SVneo and JEG3 cells via hanging drops. Forty hanging drops per cell line were produced with 20 000 cells per 30 μL drop. The environment of the hanging drop delivers the prerequisite for spheroid formation. After an incubation period of 48 h, 100% of HTR8/SVneo-containing drops formed visible spheroids (a, b), while only 50% of JEG3-containing drops formed spheroids (c). JEG3 spheroids appeared smaller in circumference, less globular (oblong shapes), and rather on the verge of disaggregation (d).

which was also visible by the "naked" eye (Figures 2(a) and 2(b)). On the other hand, JEG3 spheroids were much smaller (as visible even by the "naked" eye) and irregularly shaped (Figures 2(c) and 2(d)). The JEG3 spheroids were also unstable and were disaggregated easily and thus could not easily be pipetted into the shortened tip of a pipette for transportation into, for example, a new hanging drop.

In order to verify that spheroid formation is not secondary to cell aggregation, we proceeded to disaggregate the spheroids, split these cells, and propagate them again in hanging drops to assess their continued ability to reform spheroids (as recommended in [15]). Under this experimental setting, we passaged the HTR8/SVneo spheroids multiple times (5 passages and continuing) without their loss of spheroid-forming abilities. As little as 5000 HTR8/SVneo cells are able to form a spheroid (splitting ongoing). Please note that we have not yet tried to form spheroids with less cells than the mentioned. Only 50% of the initial JEG3 spheroids were able to reform spheroids following disaggregation and splitting; further passages could not be maintained.

3.2. HTR8/SVneo and JEG3 Cells Express the Trophoblast Stem Cell Marker CDX2.

CDX2 is a transcription factor that is necessary for the first differentiation of hESC into the trophoblast stem cell (reviewed in [3]). Loss of CDX2 activity disables the blastocyst to implant correctly during murine pregnancy [23]. Following the hypothesis that HTR8/SVneo or JEG3 cells recapitulate features of a cancer stem cell or a trophoblast(ic) stem or progenitor cell, we sought to investigate whether the classic trophoblast stem cell "marker" is expressed in both cell lines.

Since it has been described that nonspecific binding in trophoblast(ic) cells is extreme and often cannot be alleviated through conventional blocking procedures [22], we initially blocked cell lines with normal goat serum and further blocked them with or without human serum. According to immunofluorescence staining results, both cell lines seem to stain positive for CDX2. However, since there was no alteration in the pattern of staining before and after application of human serum, and since it is rather unlikely that CDX2 is so prominently expressed in the cytoplasm [24, 25], we conclude that unspecific binding is still too high in immunofluorescence stainings to be reliable.

In contrast, ICH results show that HTR8/SVneo cells express CDX2 in the nucleus, although the trophoblast marker expression appeared lower, especially in terms of number of nuclei, after blockage with human serum (Figure 3(b) without human serum versus Figure 3(a) with human serum). In JEG3 cells, a low-intensity CDX2 signal was observable even after blocking with human serum with numbers of positive-stained nuclei rather unchanged (Figures 4(a) and 4(b)). Via immunofluorescence staining procedures, CDX2 expression is made visible, and the signal patterns are again not affected by blocking procedure; thus, this procedure is again deemed unreliable (Figures 3(c), 3(d), 4(c), and 4(d)).

3.3. HTR8/SVneo Cells Express NOTCH1, NANOG, and SOX2, but Not OCT4 according to ICH Analysis.

In theory, cancer stem cells have probably reacquired properties similar to stem cells during transformation into a malignancy [1, 2, 26]. We analyzed the expression of the classic or core "stemness-" associated transcription factors OCT4, NANOG, and SOX2 (reviewed in [16]). We also chose to assess the expression of the cell fate determining transcription factor NOTCH1 due to its recent suggestion as a maintainer of "stemness," as well its ever-emerging role in the invasion potential of tumors (reviewed in [17, 20, 27, 28]).

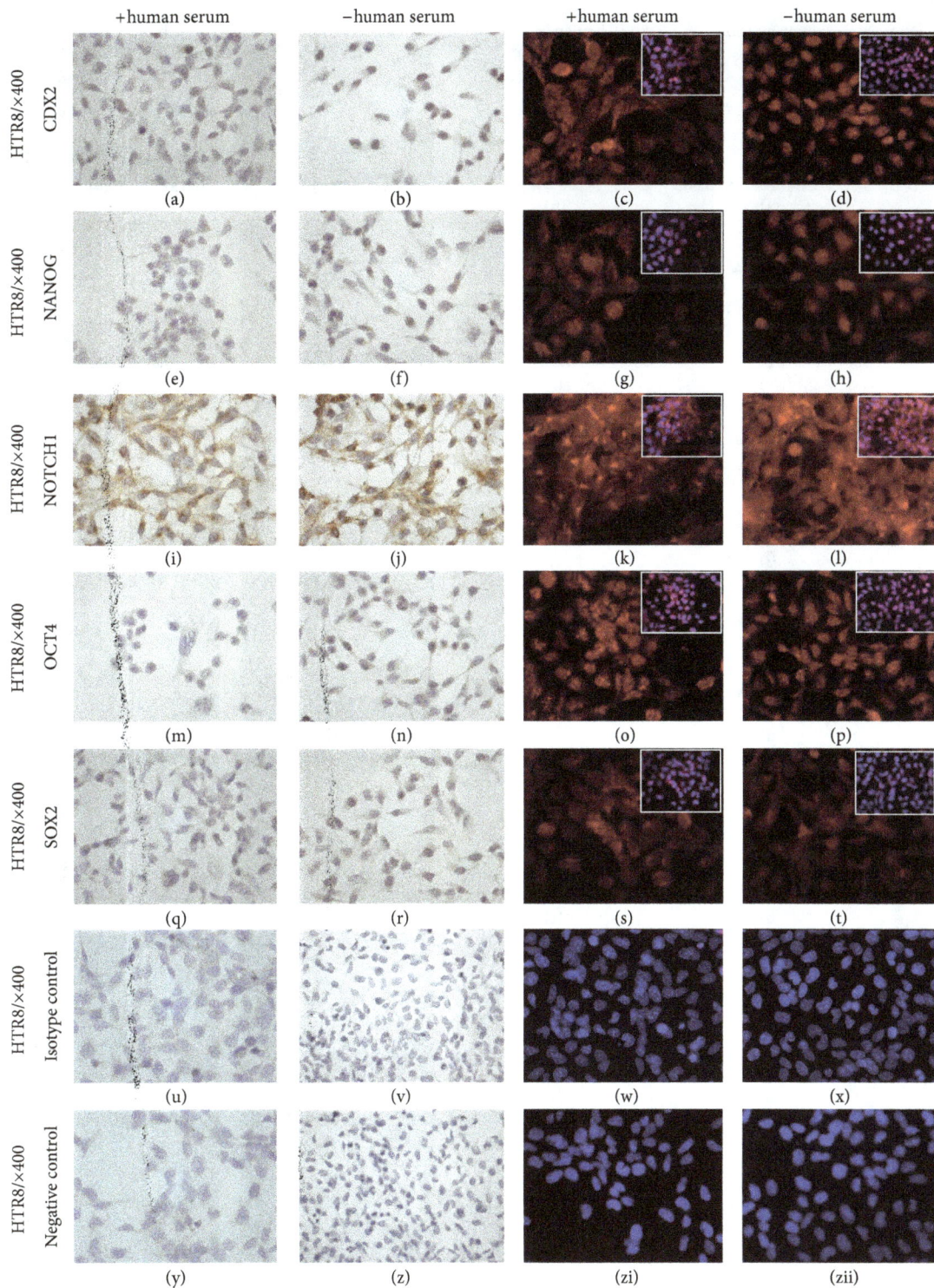

FIGURE 3: Expression of trophoblast (CDX2), cell fate determining (NOTCH1), and core "stemness-" associated (OCT4, NANOG, and SOX2) stem cell transcription factors in HTR8/SVneo cells. HTR8/SVneo cells express CDX2, NOTCH1, NANOG, and SOX2 ((a), (e), (i), and (q)). OCT4 signaling is lost after blocking the samples with human serum ((m) versus (n)). Human serum is needed to further block exceptional, unspecific binding on trophoblastic cells that cannot be eliminated via conventional blocking measures. Immunofluorescent staining seems less specific with or without blockage with human serum ((c), (d), (g), (h), (k), (l), (o), (p), (s), and (t)).

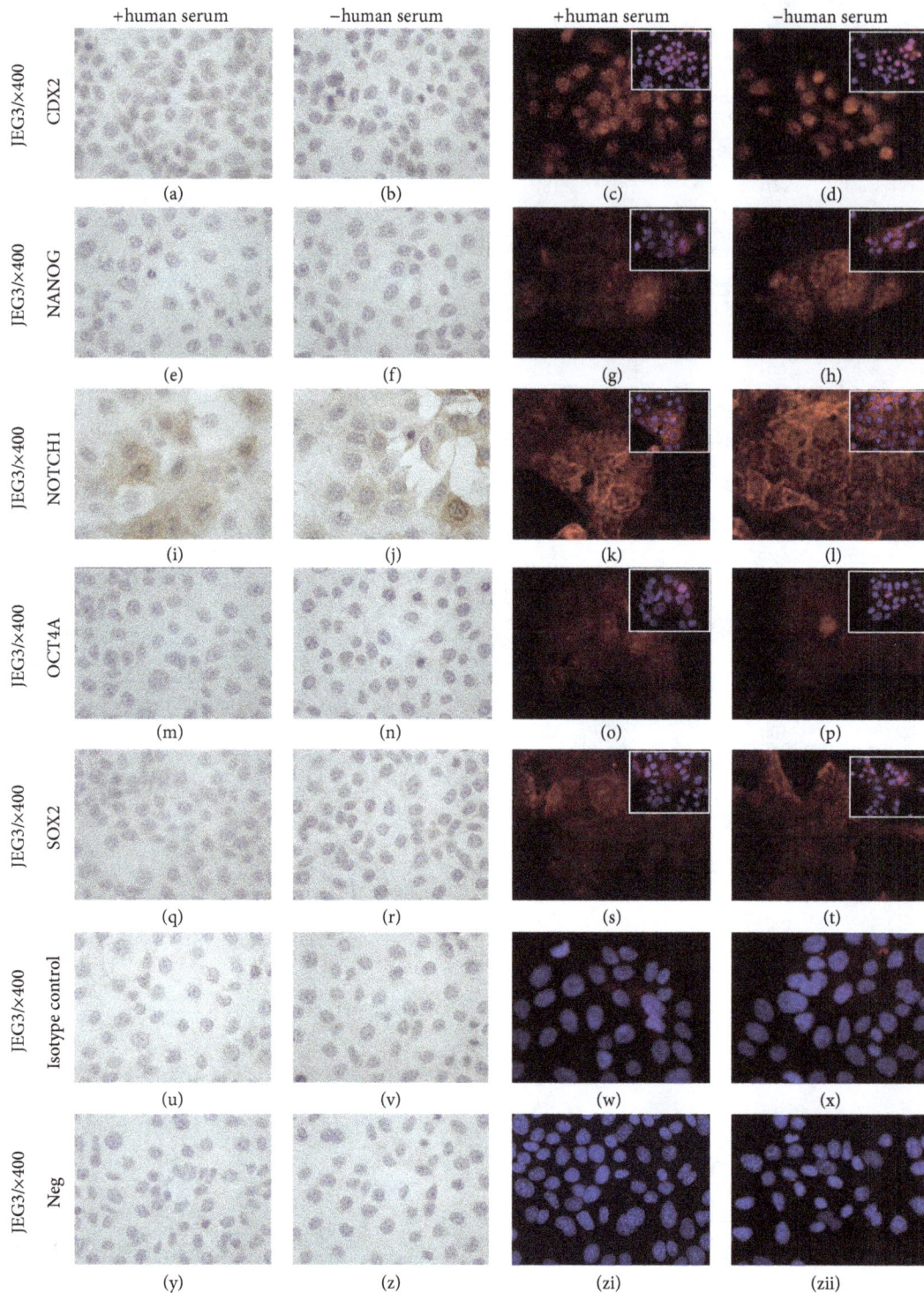

FIGURE 4: Expression of trophoblast (CDX2), cell stypo determining (NOTCH1), and core "stemmness-" associated (OCT4, NANOG, and SOX2) stem cell transcription factors in JEG3 cells. JEG3 cells express CDX2 and NOTCH1 ((a) and (i)). OCT4, NANOG and SOX2 signaling are not visible (with human serum: (e), (m), and (q); without human serum (f), (n), and (r)). Immunofluorescent staining seems less specific with or without blockage with human serum (all positive signals: (c), (d), (g), (h), (k), (l), (o), (p), (s), and (t)).

HTR8/SVneo Cells Display Trophoblast Progenitor Cell-Like Characteristics Indicative of Self-Renewal, Repopulation Activity, and Expression of "Stemness-" Associated Transcription Factors

135

Since it has been described that nonspecific binding in trophoblast(ic) cells is extreme and often cannot be alleviated through conventional blocking procedures [22], we initially blocked cell lines with normal goat serum and further blocked them with or without human serum. Upon blocking of HTR8/SVneo cells with human serum, the positive signals for all investigated transcription factors were reduced, especially in the nuclei, or, as in the case of OCT4a, disappeared altogether (Figure 3: (f), (j), (n), and (r) without human serum versus (e), (i), (m), and (q) with human serum). The expression intensity was highest for NOTCH1, and this expression was only slightly altered after blockage with human serum (Figures 3(i) and 3(j)). In contrast to peroxidase staining, we detected all stem cell markers via fluorescence staining, and there were no visible differences in staining pattern between the blocking procedures (Figure 3, for NANOG: (g) and (h); for NOTCH1: (k) and (l); for OCT4: (o) and (p); for SOX2: (s) and (t)).

Taking the recommendations of Honig et al. [22] into consideration (meaning that we now deem the immunofluorescence procedure unreliable for staining of these transcription factors in these cells), we surmise that HTR8/SVneo cells express all of the investigated transcription factors except for OCT4.

3.4. JEG3 Cells Express NOTCH1, but Not NANOG, SOX2, and OCT4 according to ICH Analysis.

NANOG, OCT4, and SOX2 were not detectable in JEG3 cells regardless of blocking methods (Figure 4: (e), (f), (m), (n), (q), and (r)) after ICH analysis.

JEG3 cells express NOTCH1, and, as in HTR8/SVneo cells, NOTCH1 was detectable with both blocking methods (Figures 4(i) and 4(j)). Interestingly, the NOTCH1 staining signal in JEG3 cells appears less intensive than that in HTR8/SVneo cells (Figure 3(i) versus Figure 4(i)).

Similar to HTR8/SVneo cells, the fluorescent staining of JEG3 cells shows positive signals for NANOG, NOTCH1, OCT4, and SOX2, and it makes no difference in staining pattern if the cells were blocked with or without human serum (Figure 4: (g), (h), (k), (l), (o), (p), (s), and (t)).

3.5. HTR8/SVneo Spheroids Contain Chorionic Villi-Covering Cells Indicative of Repopulation.

To demonstrate repopulation capacity of cancer stem/progenitor cells, the progenitor cell candidates are usually injected in an immunocompromised mouse in the general vicinity of the target organ in which prior elimination of the target population has taken place (as the progenitor cells are supposed to replace them) [29]. If the progenitor cell candidates (e.g., mammary gland stem cell) possess repopulating potential, then they will be found in the stead of the original target cell population (e.g., mammary gland).

It has been observed before that altering in vitro culture standards can cause a side population of HTR8/SVneo cells to differentiate into trophoblast subpopulations, as made visible by ICH analysis of differentiation markers [12]. The HTR8/SVneo capacity to actually repopulate the villous in an in vivo setting has not yet been demonstrated.

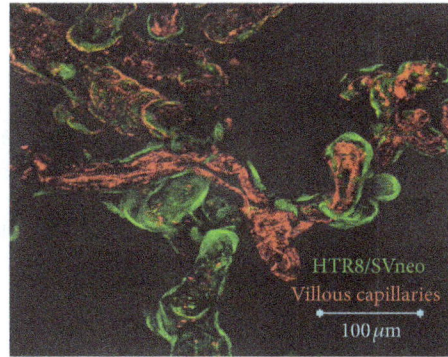

FIGURE 5: HTR8/SVneo spheroid cells repopulate chorionic villi of placental biopsies. HTR8/SVneo spheroids were confronted with placental biopsies derived from healthy pregnancies after elective, ceasarian section. After 48 h, HTR8/SVneo spheroid cells (green) have covered chorionic villi (endothelium of fetal capillaries within the villi are depicted in red).

The syncytiotrophoblast layer within villous explant cultures is known to fully degenerate following 4 h of culture initiation [30].

We sought to determine whether the cells present in the HTR8/SVneo spheroids could repopulate the entire chorionic villous, which according to the previous citation is now replete of the syncytiotrophoblast layer, following confrontation with placental villous explants in an effort to remain in a near in vivo model of the placenta.

We excluded JEG3 from this experiment as the previous investigations suggest that JEG3 cells possess only low prominent cancer stem cell characteristics. Furthermore, as JEG3 spheroids were unstable and easily disaggregated, it was not possible to transfer them to new hanging drops in which coculture experiments were performed. In coculture of HTR8/SVneo with placental explants, spheroids are completely disaggregated after 48 h, and all retrievable HTR8/SVneo cells cover the chorionic villous (Figure 5; HTR8/SVneo spheroid cells in green; endothelium of fetal capillaries within chorionic villi in red). Ten individual confrontation cultures have been performed with qualitatively similar results.

Assuming that the syncytiotrophoblast layer of placental explants truly do degenerate following 4 h cultivation, then the previous situation suggests that the HTR8/SVneo spheroid cells contain the progenitor cell characteristic of repopulating activity in a near in vivo model.

4. Discussion

HTR8/SVneo and JEG3 cells are highly popular transformed cell lines often used as an imperfect model of the trophoblast, with HTR8/SVneo often deemed as closer to the physiologic setting. Both cell types have been proposed to possess "stem-like" or "progenitor-like" characteristics [2, 12].

One aim of this study was to characterize stem/progenitor cell traits of these cell lines according to general minimum standards used to identify putative cancer stem cells

[14, 15]. Another aim of this study was to characterize the HTR8/SVneo and JEG3 cell lines for their expression pattern of various transcription factors associated with "stemness" (OCT4, NANOG, and SOX2), cell fate (NOTCH1), and trophoblast stem cells (CDX2) and to correlate this to their intrinsic progenitor/stem cell capacities, because their transformed character is likely to contribute to altered transcription factor expression, which in turn is likely to be responsible for stem/progenitor-like behavior. This is a first step in defining a cell population as a progenitor or cancer stem cell.

In our hands, HTR8/SVneo cells have a high propensity to form spheroid bodies, while expressing virtually all investigated transcription factors (specifically CDX2, NOTCH1, SOX2, and NANOG, and not OCT4). Furthermore, HTR8/SVneo spheroid cells demonstrate behavior reminiscent of self-renewal and replenishing properties. In contrast, JEG3 cells have only a limited ability to form spheroid bodies, and although they express the trophoblast stem cell marker CDX2 and a cell fate determinant NOTCH1, they did not express the investigated hESC cell markers or so-called "stemness-" associated transcription factors. Taken together, this is indicative of the fact that HTR8/SVneo cells are transformed in a manner that might make them closer in phenotype to a trophoblast progenitor cell than to a differentiated trophoblast cell, such as a syncytio- or extravillous trophoblast. Our experiments reveal traits that speak for JEG3 cells being a form of cancer stem cell as previously proposed.

As mentioned earlier, our own recent results indicate that HTR8/SVneo cells have less in common with the primary extravillous trophoblast cell than JEG3 cells [11]. GeneChip analyses of the expression signatures of primary trophoblast versus choriocarcinoma cell lines (including JEG3) and versus extravillous trophoblast derived cell lines (including HTR8/SVneo) have revealed that all three groups cluster distinctly [31]. Furthermore, the genes that are similarly expressed between EVT and choriocarcinoma cell lines are related to cell motility, signaling, vasculature and tissue development (all functional signs of differentiation), while HTR8/SVneo and EVT similarities are restricted to those genes regulating RNA transport and metabolism (housekeeping characteristics) [31]. Classic hallmark characteristics for primary, differentiated trophoblast cells are its expression of cytokeratin 7 and negative expression of vimentin [32, 33]. HTR8/SVneo cells show an expression profile that is just the opposite [12, 31]. In this aspect, it is interesting that HTR8/SVneo cells express N-cadherin, while their JEG3 counterparts do not [34], meaning that HTR8/SVneo express at least two known epithelial-mesenchymal transition (EMT) markers (N-cadherin and vimentin), which is also a sign of partial dedifferentiation (reviewed in [27]). Finally, HTR8/SVneo cells express HLA-G, a typical extravillous trophoblast differentiation marker [33], only weakly or not at all [12] and secrete β-HCG, a hallmark characteristic of a syncytiotrophoblast cell (as reviewed in [35]), weakly, but more than primary cytotrophoblast, which do not secrete hCG at all [9].

In our study, we also demonstrate the self-renewal properties of HTR8/SVneo cells propagated as spheroids under the same culture conditions as human embryonic stem cell embryoid bodies. HTR8/SVneo cells not only survive multi-passages after spheroid formation, but are continually able to reform spheroids after each passage even with fairly low cell numbers. This is also in line with a recent observation that HTR8/SVneo cells form a side population that displays self-renewal characteristics [12]. We found it somewhat puzzling that all HTR8/SVneo cells incubated in hanging drops formed spheroids in our experiments, while Takao et al. describe only a very exclusive side-population that demonstrates self-renewal properties [12]. However, since we excluded the possibility that HTR8/SVneo spheroids are mere cell aggregates, we propose that the hanging drop environment induces HTR8/SVneo cells to alter its phenotype towards that of a progenitor-like cell.

We have not proven in our own studies that the syncytiotrophoblast layer of placental explants has actually degenerated during the 48 h cultivation period we chose; however, other studies have described that this occurs after 4 h [30]. Following coculture of our placental explants with HTR8/SVneo cells for 48 h, we reveal for the first time that HTR8/SVneo cells have the propensity to "renew" or at least cover villous tissue in a model near to the in vivo situation. We speculate that the syncytiotrophoblast layer is lost, as described, and that HTR8/SVneo cells answer a distress call during the cultivation period, which allows the HTR8/SVneo cells to migrate to this area and cover or indeed replace the villous with syncytiotrophoblast-like cells. This is an indication for a certain degree of plasticity within the HTR8/SVneo cells, which we have not corroborated here with syncytiotrophoblast-specific expression of surface molecules. We believe that this is likely though, since Takao et al. have been able to demonstrate that a side-population of HTR8/SVneo cells is able to differentiate into syncytiotrophoblast and other trophoblast lineages [12]. Future studies are needed in order to unravel which transcription factor or other signal is responsible for regulating this migratory and replenishing characteristic.

Our investigations characterize for the first time the expression of "stemness-" associated cancer and trophoblast stem cell transcription factors in the HTR8/SVneo cell line. The fact that both cell lines produce CDX2, while only HTR8/SVneo shows progenitor cell capacities, could indicate that CDX2 is rather a sign of trophoblast lineage derivation instead of a trophoblast stem cell differentiation alone. However, CDX2 has been shown to be expressed only in first trimester trophoblast, while term placentae lose this capacity. Furthermore, the same study identified that CDX2+ELF5+ cells within the placentae characterize cytotrophoblast populations [36]. This in turn suggests that JEG3 cells have retained a certain cytotrophoblastic identity, which we have not been able to visualize as a progenitor-like function in this study.

The regulatory network between CDX2 and the "stemness-" associated transcription factors is tight and complicated; reiterating this in detail goes beyond the scope of this investigation. Briefly, CDX2 and the "stemness-" associated transcription factors are thought to reciprocate or mutually antagonize each other, also because this is the expression pattern found during blastocyst formation. At least in the mouse,

HTR8/SVneo Cells Display Trophoblast Progenitor Cell-Like Characteristics Indicative of Self-Renewal, Repopulation Activity, and Expression of "Stemness-" Associated Transcription Factors

137

CDX2 is known to downregulate OCT4 and NANOG [37, 38]. However, SOX2 and CDX2 are known to cooperate during trophectoderm formation [4]. In our analysis, while JEG3 cells did not express any of the so-called "stemness-" associated transcription markers, HTR8/SVneo cells expressed most of them (OCT4a^{neg}, NANOG^{weak}, and SOX2^{weak}). Interestingly, a recent characterization of trophoblast progenitor cells derived from first trimester placentae reveals that undifferentiated trophoblast progenitor cells that form embryoid bodies and that are capable of multipassage also display OCT4^{neg}, NANOG^{weak}, and SOX2^{weak} phenotypes [39]. Due to the association of this transcription marker expression profile with the presence of progenitor-like functions in HTR8/SVneo cells and in trophoblast progenitor cells, it is enticing to conclude that NANOG and SOX2 are responsible for the observed functions. Further studies with gain/loss-of-function analyses would, however, be necessary to finalize that conclusion. Furthermore, since CDX2 is coexpressed in HTR8/SVneo cells with certain hESC transcription factors, it is most plausible that the "stemness-" associated transcription factor regulatory network has been corrupted, probably through alterations secondary to transfection of HTR8 parent cells with a plasmid containing the simian virus 40 large T antigen (SV40).

We were surprised to see that NANOG was not expressed in JEG3 cells as described in a recent publication, in which NANOG was especially detected in the nuclear fraction of JEG3 cells [40]. Currently, we cannot explain this discrepancy other than in methodology differences, and further analyses on the mRNA and protein level are under way.

The NOTCH1 signal was remarkably visible in both cell lines. NOTCH1 has been associated with "stemness" properties, as well as with the differentiation of cancer stem cells (tumor stem-like cells) into endothelial progenitor cells [18, 19, 27]. Furthermore, NOTCH1 expression is linked with trophoblast, as well as with malignant types of invasion [17, 20, 28]. In the physiologic placenta, NOTCH1 expression is thought to be vital for placental angiogenesis, while defective NOTCH signaling is thought to contribute in the pathogenesis of preeclampsia (reviewed in [41]). With our current information, it is as yet impossible to conclude whether NOTCH1 expression is a sign of stem-like properties, of EMT or invasion potential. Further functional analyses would be helpful in unraveling this aspect.

In summary, though, all of the previously analyzed characteristics of HTR8/SVneo are in line with the idea that HTR8/SVneo cells share more similarities to a trophoblast progenitor cell, perhaps the cytotrophoblast, than a primary extravillous trophoblast. Our investigation, together with that of Takao et al., is first step in defining HTR8/SVneo progenitor cell characteristics; however, the final step of in vivo testing per xenotransplantation into immuncompromised mice is as yet pending (but seems likely to be successful, since xenografts of human placental explants into immuncompromised mouse is highly successful [42]).

In all finality, we hypothesize that transfection of the HTR8/SVneo cell line has altered the original (extravillous) trophoblast cell in a manner leading to development of the characteristics described in these investigations; thus, investigators should use caution when using the popular HTR8/SVneo cell line as a model for primary extravillous trophoblast cells.

We wish also to direct investigators to the expression alterations (or lack thereof) seen in our immunocytochemistry and immunofluorescence results before and after blockage with human serum. Our results suggest, as with [22], that trophoblastic cells are prone to unspecific binding, and caution should be exercised when interpreting immunostaining results.

Acknowledgments

The authors wish to recognize financial support from the IZKF of Jena (Interdisziplinäres Zentrum für Klinische Forschung; Interdisciplinary Center for Clinical Research in Jena; Project no. J12) and ProChance. They are highly indebted to Professor Maria Wartenberg, Clinic of Internal Medicine I, Cardiology Division of the University Hospital of Jena, for her excellent assistance through teaching and establishing the hanging drops protocol in the Placenta-Labor and kind technical support concerning use of her confocal laser scanning microscope (Carl-Zeiss, Jena, Germany). They furthermore appreciate the gift of HTR8/SVneo cells from Professor Charles Graham of the Department of Anatomy and Cell Biology at Queen's University, Kingston, ON, Canada.

References

[1] N. A. Lobo, Y. Shimono, D. Qian, and M. F. Clarke, "The biology of cancer stem cells," *Annual Review of Cell and Developmental Biology*, vol. 23, pp. 675–699, 2007.

[2] A. N. Y. Cheung, H. J. Zhang, W. C. Xue, and M. K. Y. Siu, "Pathogenesis of choriocarcinoma: clinical, genetic and stem cell perspectives," *Future Oncology*, vol. 5, no. 2, pp. 217–231, 2009.

[3] R. M. Roberts and S. J. Fisher, "Trophoblast stem cells," *Biology of Reproduction*, vol. 84, no. 3, pp. 412–421, 2011.

[4] M. Keramari, J. Razavi, K. A. Ingman et al., "Sox2 is essential for formation of trophectoderm in the preimplantation embryo," *PLoS One*, vol. 5, no. 11, Article ID e13952, 2010.

[5] A. Meissner and R. Jaenisch, "Generation of nuclear transfer-derived pluripotent ES cells from cloned Cdx2-deficient blastocysts," *Nature*, vol. 439, no. 7073, pp. 212–215, 2006.

[6] S. Kyurkchiev, F. Gandolfi, S. Hayrabedyan et al., "Stem cells in the reproductive system," *American Journal of Reproductive Immunology*, vol. 67, pp. 445–462, 2012.

[7] G. E. Lash, T. Ansari, P. Bischof et al., "IFPA meeting 2008 workshops report," *Placenta*, vol. 30, pp. S4–S14, 2009.

[8] J. S. Fitzgerald, T. G. Poehlmann, E. Schleussner, and U. R. Markert, "Trophoblast invasion: The role of intracellular cytokine signalling via signal transducer and activator of transcription 3 (STAT3)," *Human Reproduction Update*, vol. 14, no. 4, pp. 335–344, 2008.

[9] C. H. Graham, T. S. Hawley, R. G. Hawley et al., "Establishment and characterization of first trimester human trophoblast cells with extended lifespan," *Experimental Cell Research*, vol. 206, no. 2, pp. 204–211, 1993.

[10] P. O. Kohler and W. E. Bridson, "Isolation of hormone-producing clonal lines of human choriocarcinoma," *Journal of Clinical Endocrinology and Metabolism*, vol. 32, no. 5, pp. 683–687, 1971.

[11] D. M. Morales-Prieto, W. Chaiwangyen, S. Ospina-Prieto et al., "MicroRNA expression profiles of trophoblastic cells," *Placenta*, vol. 33, pp. 725–734, 2012.

[12] T. Takao, K. Asanoma, K. Kato et al., "Isolation and characterization of human trophoblast side-population (SP) cells in primary villous Cytotrophoblasts and HTR-8/SVneo cell line," *PLoS One*, vol. 6, no. 7, Article ID e21990, 2011.

[13] P. Spitalieri, G. Cortese, A. Pietropolli et al., "Identification of multipotent cytotrophoblast cells from human first trimester chorionic Villi," *Cloning and Stem Cells*, vol. 11, no. 4, pp. 535–556, 2009.

[14] A. Fabian, M. Barok, G. Vereb, and J. Szöllosi, "Die hard: are cancer stem cells the bruce willises of tumor biology?" *Cytometry A*, vol. 75, no. 1, pp. 67–74, 2009.

[15] B. M. Boman and M. S. Wicha, "Cancer stem cells: a step toward the cure," *Journal of Clinical Oncology*, vol. 26, no. 17, pp. 2795–2799, 2008.

[16] D. Pei, "Regulation of pluripotency and reprogramming by transcription factors," *Journal of Biological Chemistry*, vol. 284, no. 6, pp. 3365–3369, 2009.

[17] R. T. K. Pang, C. O. Leung, T. M. Ye et al., "MicroRNA-34a suppresses invasion through downregulation of Notch1 and Jagged1 in cervical carcinoma and choriocarcinoma cells," *Carcinogenesis*, vol. 31, no. 6, pp. 1037–1044, 2010.

[18] V. L. Bautch, "Cancer: tumour stem cells switch sides," *Nature*, vol. 468, no. 7325, pp. 770–771, 2010.

[19] S. M. Prasad, M. Czepiel, C. Cetinkaya et al., "Continuous hypoxic culturing maintains activation of Notch and allows long-term propagation of human embryonic stem cells without spontaneous differentiation," *Cell Proliferation*, vol. 42, no. 1, pp. 63–74, 2009.

[20] B. B. Hafeez, V. M. Adhami, M. Asim et al., "Targeted knockdown of notch1 inhibits invasion of human prostate cancer cells concomitant with inhibition of matrix metalloproteinase-9 and urokinase plasminogen activator," *Clinical Cancer Research*, vol. 15, no. 2, pp. 452–459, 2009.

[21] H. Kurosawa, "Methods for inducing embryoid body formation: in vitro differentiation system of embryonic stem cells," *Journal of Bioscience and Bioengineering*, vol. 103, no. 5, pp. 389–398, 2007.

[22] A. Honig, L. Rieger, M. Kapp, J. Dietl, and U. Kämmerer, "Immunohistochemistry in human placental tissue—pitfalls of antigen detection," *Journal of Histochemistry and Cytochemistry*, vol. 53, no. 11, pp. 1413–1420, 2005.

[23] K. Chawiengsaksophak, R. James, V. E. Hammond, F. Köntgen, and F. Beck, "Homeosis and intestinal tumours in Cdx2 mutant mice," *Nature*, vol. 386, no. 6620, pp. 84–87, 1997.

[24] T. M. Erb, C. Schneider, S. E. Mucko et al., "Paracrine and epigenetic control of trophectoderm differentiation from human embryonic stem cells: the role of bone morphogenic protein 4 and histone deacetylases," *Stem Cells and Development*, vol. 20, pp. 1601–1614, 2011.

[25] E. Wydooghe, L. Vandaele, J. Beek et al., "Differential apoptotic staining of mammalian blastocysts based on double immunofluorescent CDX2 and active caspase-3 staining," *Analytical Biochemistry*, vol. 416, pp. 228–230, 2011.

[26] P. C. Hermann, S. L. Huber, T. Herrler et al., "Distinct populations of cancer stem cells determine tumor growth and metastatic activity in human pancreatic cancer," *Cell Stem Cell*, vol. 1, no. 3, pp. 313–323, 2007.

[27] A. Pannuti, K. Foreman, P. Rizzo et al., "Targeting Notch to target cancer stem cells," *Clinical Cancer Research*, vol. 16, no. 12, pp. 3141–3152, 2010.

[28] Z. Wang, S. Banerjee, Y. Li, K. M. W. Rahman, Y. Zhang, and F. H. Sarkar, "Down-regulation of Notch-1 inhibits invasion by inactivation of nuclear factor-κB, vascular endothelial growth factor, and matrix metalloproteinase-9 in pancreatic cancer cells," *Cancer Research*, vol. 66, no. 5, pp. 2778–2784, 2006.

[29] M. J. Liao, C. C. Zhang, B. Zhou et al., "Enrichment of a population of mammary gland cells that form mammospheres and have in vivo repopulating activity," *Cancer Research*, vol. 67, no. 17, pp. 8131–8138, 2007.

[30] J. L. James, P. R. Stone, and L. W. Chamley, "Cytotrophoblast differentiation in the first trimester of pregnancy: evidence for separate progenitors of extravillous trophoblasts and syncytiotrophoblast," *Reproduction*, vol. 130, no. 1, pp. 95–103, 2005.

[31] M. Bilban, S. Tauber, P. Haslinger et al., "Trophoblast invasion: assessment of cellular models using gene expression signatures," *Placenta*, vol. 31, no. 11, pp. 989–996, 2010.

[32] I. T. Manyonda, G. S. J. Whitley, and J. E. Cartwright, "Trophoblast cell lines: a response to the workshop report by King et al.," *Placenta*, vol. 22, no. 2-3, pp. 262–263, 2001.

[33] D. W. Morrish, G. S. J. Whitley, J. E. Cartwright, C. H. Graham, and I. Caniggia, "In vitro models to study trophoblast function and dysfunction—a workshop report," *Placenta*, vol. 23, no. 1, pp. S114–S118, 2002.

[34] J. N. Bulmer, G. J. Burton, S. Collins et al., "IFPA Meeting 2011 workshop report II: angiogenic signaling and regulation of fetal endothelial function, placental and fetal circulation and growth, spiral artery remodeling," *Placenta*, vol. 33, pp. S9–S14, 2012.

[35] L. A. Cole, "hCG, the wonder of today's science," *Reproductive Biology and Endocrinology*, vol. 10, article 24, 2012.

[36] M. Hemberger, R. Udayashankar, P. Tesar, H. Moore, and G. J. Burton, "ELF5-enforced transcriptional networks define an epigenetically regulated trophoblast stem cell compartment in the human placenta," *Human Molecular Genetics*, vol. 19, no. 12, pp. 2456–2467, 2010.

[37] A. Nishiyama, L. Xin, A. A. Sharov et al., "Uncovering early response of gene regulatory networks in ESCs by systematic induction of transcription factors," *Cell Stem Cell*, vol. 5, no. 4, pp. 420–433, 2009.

[38] L. Chen, A. Yabuuchi, S. Eminli et al., "Cross-regulation of the nanog and Cdx2 promoters," *Cell Research*, vol. 19, no. 9, pp. 1052–1061, 2009.

[39] O. Genbacev, M. Donne, M. Kapidzic et al., "Establishment of human trophoblast progenitor cell lines from the chorion," *Stem Cells*, vol. 29, pp. 1427–1436, 2011.

[40] M. K. Y. Siu, E. S. Y. Wong, H. Y. Chan, H. Y. S. Ngan, K. Y. K. Chan, and A. N. Y. Cheung, "Overexpression of NANOG in gestational trophoblastic diseases: effect on apoptosis, cell invasion, and clinical outcome," *American Journal of Pathology*, vol. 173, no. 4, pp. 1165–1172, 2008.

[41] W. X. Zhao and J. H. Lin, "Notch signaling pathway and human placenta," *International Journal of Medical Sciences*, vol. 9, pp. 447–452, 2012.

[42] T. Poehlmann, S. Bashar, U. R. Markert et al., "Letter to the editors," *Placenta*, vol. 25, no. 4, pp. 357–358, 2004.

14

Bodyweight Assessment of Enamelin Null Mice

Albert H.-L. Chan,[1] **Rangsiyakorn Lertlam,**[2] **James P. Simmer,**[2]
Chia-Ning Wang,[3] **and Jan C. C. Hu**[2]

[1] *Department of Periodontics and Oral Medicine, University of Michigan School of Dentistry, 1011 North University Avenue, Ann Arbor, MI 48109-1078, USA*

[2] *Department of Biologic and Materials Sciences, University of Michigan School of Dentistry, 1210 Eisenhower Place, Ann Arbor, MI 48108, USA*

[3] *Department of Biostatistics, University of Michigan School of Public Health, 109 Observatory Street, 1700 SPH I, Ann Arbor, MI 48109-2029, USA*

Correspondence should be addressed to Jan C. C. Hu; janhu@umich.edu

Academic Editor: L. Brian Foster

The *Enam* null mice appear to be smaller than wild-type mice, which prompted the hypothesis that enamel defects negatively influence nutritional intake and bodyweight gain (BWG). We compared the BWG of $Enam^{-/-}$ and wild-type mice from birth (D0) to Day 42 (D42). Wild-type (WT) and $Enam^{-/-}$ (N) mice were given either hard chow (HC) or soft chow (SC). Four experimental groups were studied: WTHC, WTSC, NHC, and NSC. The mother's bodyweight (DBW) and the average litter bodyweight (ALBW) were obtained from D0 to D21. After D21, the pups were separated from the mother and provided the same type of food. Litter bodyweights were measured until D42. ALBW was compared at 7-day intervals using one-way ANOVA, while the influence of DBW on ALBW was analyzed by mixed-model analyses. The ALBW of $Enam^{-/-}$ mice maintained on hard chow (NHC) was significantly lower than the two WT groups at D21 and the differences persisted into young adulthood. The ALBW of $Enam^{-/-}$ mice maintained on soft chow (NSC) trended lower, but was not significantly different than that of the WT groups. We conclude that genotype, which affects enamel integrity, and food hardness influence bodyweight gain in postnatal and young adult mice.

1. Introduction

Enamelin is the largest (~200 kDa) but the least abundant (3–5%) of the three major secretary stage enamel matrix proteins. Twelve mutations in the enamelin gene (*ENAM*, 4q13.2) have been published associated with autosomal dominant amelogenesis imperfecta (ADAI) [1–16]. Clinical features of affected teeth showed thin enamel, with either severe or localized hypoplasia. The enamel phenotype is dose-dependent and ranges from small, well-circumscribed enamel pits to enamel agenesis (when both *ENAM* alleles are mutated) [7]. None of these *ENAM* mutation studies reported a phenotype outside the dentition. In mice, mutations in the *Enam* gene have been induced with the mutagen N-ethyl-N-nitrosourea (ENU), and four separate point mutations

have been identified: p.Ser55Ile, p.Glu57Gly, a splice donor site in exon 4, and a premature stop codon in exon 8 (p.Gln176*) [17, 18]. Heterozygous mice exhibited rough and pitted enamel while the null mice showed enamel agenesis. Enamelin null mice were generated by replacing the *Enam* coding sequence from the translation initiation site through exon 7 with a *lacZ* reporter gene [19]. The enamel defects were dose-dependent. The enamel layer was completely absent in *Enam* null mice compared to the mild enamel phenotype in the heterozygotes ($Enam^{+/-}$). A thin, highly irregular, easily abraded mineralized crust over the dentin was observed in the *Enam* null mice. The affected teeth showed significant wear and were generally chalky-white. The histologic, morphometric, and protein/mineral analyses demonstrated that enamelin is essential for proper enamel matrix organization and mineralization. Serum calcium, phosphate, alkaline phosphatase, and glucose levels overlapped normal ranges.

There is considerable evidence that enamelin is an enamel-specific protein. The *lacZ* knock-in in the *Enam* $^{-/-}$ only detected enamelin expression in ameloblasts. Although not all ages and organs were surveyed, *Enam* tooth-specific expression is consistent with the human and mouse expressed sequence tag EST profiles. The human (Hs.667018) lists only three *ENAM* transcripts per million for healthy (nondental) tissues. The mouse (Mm.8014) lists four *Enam* transcripts per million in embryonic tissue (which has developing teeth), 1414 per million in molars, and zero in all other tissues. Humans with *ENAM* defects only show an enamel phenotype. Among the reports of 16 kindreds with 12 different amelogenesis imperfecta (AI) causing *ENAM* mutations, none revealed a history of systemic problems associated with the genetic condition. Finally, *Enam* is consistently found to be pseudogenized in vertebrates that have lost the ability to make teeth during evolution [20–23]. These findings support the conclusion that *Enam* defects are unlikely to directly cause a significant reduction in animal bodyweight (BW). Perhaps absence of enamel covering the dentin surface causes pain that discourages eating, especially if the food is hard and requires mastication.

The relationship between poor oral health and lower bodyweight has been observed in human studies. Acs et al. showed that three-year-old children with dental caries and with at least one pulp-involved tooth weighed 1 kg less than the counterparts [24]. A follow-up study reported that there was significant "catch-up" in growth following complete dental rehabilitation of the children who suffered severe dental caries with pulpal involvement. There are several plausible mechanisms for which dental caries may contribute to underweight and poor growth in young children and one of them is pain and discomfort from dental caries reducing nutritional intake because eating is painful [25]. Severe forms of amelogenesis imperfecta are definitely associated with increased dental pain and can lead to behavioral compensations. When both *ENAM* alleles are defective, the chief complaint typically includes dental pain, particularly thermal sensitivity [7, 12]. Anterior open-bites, which are often associated with tongue-thrust and thumb sucking behaviors, are observed with increased frequency in AI patients, regardless of the genetic etiology, including AI caused by *ENAM* [6], *KLK4* [26], *MMP20* [27], *FAM83H* [28], or *AMELX* [29] mutations.

Enamelin null mice have virtually no enamel [19]. Given that genetic (AI) or environmental (caries and pulp involvement) factors can cause pain upon chewing in humans, it is reasonable to deduce that mice with severe enamel defects might experience eating difficulties that reduce BW gain, especially following the transition from milk to chow. If true, soft chow may prove more tolerable than hard chow, given the lack of enamel. The primary aim of this study was to compare the patterns of BW gain in *Enam* null and wild-type mice during the first 6 weeks of postnatal life, when maintained separately on hard or soft chow. As the mothers of the *Enam* null mice also had no enamel covering the crown of their teeth (which could affect their nutrition and the quantity of their milk), a secondary aim was to evaluate whether maternal bodyweight (DBW) correlated with ALBW among the four experimental groups in the preweaning period.

2. Materials and Methods

2.1. Animal Protocol. All procedures involving animals were reviewed and approved by the UCUCA Committee at the University of Michigan.

2.2. Animal Breeding and Bodyweight Measurement. Wild-type (strain C57BL/6) and *Enam* null (strain C57BL/6) females at the age of 4 to 6 months were mated to males with the same *Enam* genotype and genetic background. The *Enam* null and wild-type mice were maintained separately on soft or hard chow throughout the experiment. The chow was LabDiet 5001 (Purina Mills, St. Louis, MO, USA). Four experimental groups were generated: wild-type hard chow (WTHC), wild-type soft chow (WTSC), null hard chow (NHC), and null soft chow (NSC). The soft chow was water-moistened hard chow, and therefore both contained the same nutritional value. Signs of pregnancy were checked once a day following breeding and each pregnant mouse was transferred to a new cage with minimal disturbance. The mother mice were checked twice a day to determine the accurate birth time (D0) for each litter. Each litter's bodyweight was measured daily as a group at the same time of day using a digital analytical balance (Denver Instrument, Denver, CO, USA) from D0 to D42. The measured value for the weight of the litter divided by the number of pups in the litter gave the average litter bodyweight (ALBW). Each litter was housed with the mother from D0 to D21, after which the litter was weaned and separated by gender. From D22 to D42, the mean litter weight of males and females in the same litter was measured separately on a daily basis. When a mouse died, the average bodyweight (BW) of the remaining litter was used. Dam bodyweight (DBW) was measured daily from D0 to D21. The young adult male mice (at the age of 4-5 months) that were not sacrificed for morphologic examination were weighed, and the mean BW was compared among the four groups.

2.3. Morphologic Examination. After the BW analyses, mice at the age of 8 weeks were sacrificed by inhalation anesthesia using isoflurane (Sigma, St. Louis, MO, USA) and perfused with ice-cold 4% paraformaldehyde. Mandibles were dissected from the skull, fixed in 4% paraformaldehyde for 24 h, rinsed with phosphate buffered saline, and stored in 70% ethanol-diethyl pyrocarbonate. Mandibles were separated into halves by incision in the symphysis using a no. 11 scalpel blade. The right hemimandibles were photographed at 3x magnification (SMZ1000, Nikon) for morphologic evaluation under a stereomicroscope.

2.4. Statistic Analysis. Descriptive analysis of BW (in grams) was presented as the mean ± SD. ALBW on D0 and on subsequent days at an interval of 7 days until D42 was compared among the four experimental groups by one-way ANOVA and the Tukey test for pairwise group comparisons. DBW was compared on D0, D7, D14, and D21 as well as the mean BW of young adult males using the same method. Subsequently, the mixed model (SPSS 16.0) was implemented to assess the association of ALBW with DBM. Other independent variables include day, group (WHC, WSC, NHC, and NSC), litter size, and combinations. "Day" was treated as a categorical factor. All statistical analyses were conducted by consulting the Center for Statistical Consultation and Research, University of Michigan Ann Arbor, MI, USA.

3. Results

Four litters in the wild-type hard chow (WTHC) and wild-type soft chow (WTSC) groups, six litters in the null hard chow (NHC) group and five litters in the null soft chow (NSC) group were available for data analyses. One mother in the WTSC group died for an unknown reason on D17. A foster mother of the same genotype nursed her litter until D21, and the data from this litter was included in the analysis. The foster mother was not also nursing her own pups at that time.

Plots of the average pup weights (litter weight/pups per litter) ±2 standard errors for each group at specific time points are shown in Figures 1 and 2 and the ANOVA analysis in Table 1. The average bodyweight (BW) of pups from all four groups was not significantly different until D21, when the NHC group was significantly lighter than both of the wild-type groups (Figure 1 and Table 1). In the postweaning period (after D21) when male and female pups were weighed separately, the mean male average litter bodyweight (ALBW) in the NHC group was significantly lower than the other three groups at D35 and D42 (Figure 2(a) and Table 1). The same pattern was observed for female mean ALBW at D28 and D35 (Figure 2(b) and Table 1). These data are consistent with the interpretation that null pups, especially null pups fed hard chow, were gaining weight more slowly because they were eating less due to the lack of enamel on their teeth.

Table 2 shows the mothers' average bodyweights (DBW) of the four groups starting at the birth of their pups (day 0) and again at days 7, 14, and 21. The mothers of the two wild-type groups were significantly heavier than the NSC group on days 0 and 7. On day 14, the mothers of the two null mice groups weighed significantly less than the mothers of the wild-type groups. However, on day 21, no statistically significant differences in maternal bodyweight (DBW) among the four groups were observed, although the null mothers' bodyweights trended lower. Perhaps a larger sample size would have demonstrated significance at day 21 or other variables besides genetic background affected the analysis. The mixed model analyses (Table 3) demonstrated that ALBW was significantly related to the group ($P = 0.01$), litter size ($P < 0.00$), day ($P < 0.00$) but not DBW ($P = 0.72$). We suspect that nursing mothers may exhibit adaptation behaviors in managing dietary intake.

One-way ANOVA analysis of adult male mouse bodyweight and *post hoc* comparison of four experimental groups revealed that the mean bodyweight in the NHC group was still significantly lower than the other three groups (Table 4).

Although the average bodyweight of NSC group was 7.8% lower than the average weight of the WTSC, there was no statistical difference between the two groups, which suggests that maintaining *Enam* null mice with tooth defects on soft chow may allow them to obtain adequate nutrition for proper weight gain.

Morphologic evaluation of the hemimandibles (Figure 3) showed that the appearance of molars was identical to what was described in previous work [19]. The color of molars was chalky-white and the surface was rough in the null mice. There was no apparent enamel layer, occlusal wear was apparent, and periodontal defects associated food and debris impaction was observed between molars with open interproximal contacts. All findings suggested that the enamel layer of null mice was defective and chewing function may have been compromised.

4. Discussion

The ANOVA analyses showed the BW of NHC was significantly different from the two wild-type groups after day 21. In contrast, the BW of NSC was not significantly different from the wild-type groups throughout the 6-week experimental period, even though it was consistently lower than that of wild-type. This result indicated that the food hardness is important for null mice as a factor in determining BW. As soon as the eyes of young mice open, they start to consume solid food [30]. The timing of opening their eyes, around day 14, approximates the timing of BW deviation that we observed. It is plausible that when the young null mice tried to eat solid food as a supplement, they had difficulties because of their defective enamel. Furthermore, they had more trouble gaining weight when they were given only hard chow. The impeded ability to eat hard chow effectively among the null mice resulted in lower bodyweight gain.

The mean BW of pups decreased by 0.3 gm with each additional pup in the litter on day 21 (the results of the mixed model analyses). The negative relationship between litter size and average litter BW is in accordance with previous studies. The study evaluated the effect of litter size on average pup weight in rats using regression analysis showed that the relation was negative and increased in magnitude from birth to weaning (3 wk) [31]. The effect persisted into the postweaning period; although a compensatory growth spurt seemed to occur during wk 3 to 5. It was noted that when litter size was not adjusted in the mixed model, there was only marginal difference in bodyweight ($P = 0.08$) among groups. The average litter size (average of D0 to D21) in the null groups was about 1 to 2 less than wild-type groups. There were an average of 6.2 pups per litter in null groups, regardless of food type; 7 in WTSC and 8 in WTHC mice. Although no statistic analysis was performed to compare litter size among groups, the fact that the litter size was smaller in null mice suggested that the attrition of pups might have occurred. It may be due to the mother selectively nurse healthier pups and/or smaller/weaker pups in the same litter were outcompeted. Both indicated that the nutrition from

TABLE 1: One-way ANOVA analyses of ALBW at days 0, 7, 14, 21, 28, 35, and 42. The average litter bodyweights (g) with standard deviations in parentheses are presented. Each litter weight was divided by the number of pups in the litter. Statistically significant ($P < 0.05$) differences between groups within each time point are noted.

Day	Gender	WTHC	WTSC	NHC	NSC
0	M + F	1.4 (0.1)	1.4 (0.1)	1.3 (0.1)	1.3 (0.1)
7	M + F	4.1 (0.5)	3.9 (0.4)	3.4 (0.3)	3.8 (0.2)
14	M + F	6.8 (0.6)	6.8 (1.0)	5.7 (0.8)	6.4 (0.6)
21	M + F	9.0[*] (0.8)	8.6[*] (1.2)	6.4[**] (1.2)	7.9 (0.4)
28	M	14.1[*] (1.1)	13.4[*] (1.3)	9.1[**] (1.8)	11.6 (1.6)
	F	12.5[*] (0.7)	12.2[*] (1.3)	7.8[**‡] (2.1)	11.2[‡] (1.1)
35	M	18.6[*] (1.4)	18.2[*] (1.1)	11.5[**‡] (3.8)	16.5[‡] (1.7)
	F	15.8[*] (0.7)	15.6[*] (1.6)	10.4[**‡] (2.7)	14.3[‡] (1.0)
42	M	20.3[*] (1.1)	20.1[*] (0.5)	14.6[**‡] (3.7)	18.9[‡] (0.9)
	F	17.0 (0.5)	17.2 (1.2)	13.8 (2.3)	15.4 (1.2)

Statistically significant differences (SSDs) between NHC and WTHC ([*]); NHC and WTSC ([*]); NHC and NSC ([‡]). Until day 21, the males and females of each litter were weighed together. There were no statistically significant differences among the groups until day 21, suggesting that bodyweight differences might be associated with the transition to eating chow.

TABLE 2: One-way ANOVA analyses of dam bodyweights on litter days 0, 7, 14, and 21. The average bodyweights (g) of the mothers with standard deviations in parentheses are presented. Statistically significant ($P < 0.05$) differences between groups within each time point are noted.

Group	WTHC	WTSC	NHC	NSC
N	4	4	6	5
D0	31.94[†*] (0.84)	30.27[‖] (1.74)	27.92[*] (1.37)	26.48[†‖] (1.59)
D7	35.33[†*] (1.70)	33.77[‖] (1.53)	30.74[*] (1.80)	29.61[†‖] (2.33)
D14	36.56[†*] (3.54)	36.68[*‖] (2.59)	31.73[**] (2.18)	30.89[†‖] (1.99)
D21	34.00 (4.77)	34.23 (4.71)	31.38 (2.52)	31.12 (2.15)

Statistically significant differences (SSD) between NHC and WTHC ([*]); NHC and WTSC ([*]); NSC and WTHC ([†]); NSC and WTSC ([‖]); N: number of litters. This data shows that the null mothers fed hard chow did not show significant differences in bodyweight with null mothers fed soft chow. The null mothers' bodyweights were smaller than those of the wild-type mothers.

TABLE 3: Mixed-model analyses of independent variables and ALBW. The potential for independent variables to influence the average litter bodyweight was assessed. The intercept is the ALBW at D21. A P value <0.05 was accepted as significant.

Source	Numerator df	Denominator df	F	P
Intercept	1	87.32	405.12	<0.00
Day	21	108.72	14.67	<0.00
Group	3	16.04	5.45	0.01
Litter size	1	74.81	22.00	<0.00
DBW	1	225.85	0.13	0.72
Day-group	63	103.78	1.70	0.01
Day-litter size	21	109.79	3.51	<0.00
Day-DBW	21	110.17	0.85	0.66

df: degrees of freedom. The table shows that factors such as litter size could influence the average litter bodyweight, while the mother's bodyweight (DBW) did not.

TABLE 4: One-way ANOVA analyses comparing adult male mouse bodyweights. Statistically significant ($P < 0.05$) differences between groups at time point of 4-5 months are noted.

Group	WTHC	WTSC	NHC	NSC
N	9	13	10	11
Mean BW (g)	25.64[*] (1.55)	28.03[*] (1.08)	21.94[**‡] (4.54)	25.85[‡] (2.09)

Statistically significant differences (SSD) between NHC and WTHC ([*]); NHC and WTSC ([*]); NHC and NSC ([‡]). The bodyweight of NHC adult male mice was significantly lower than that of the other three groups.

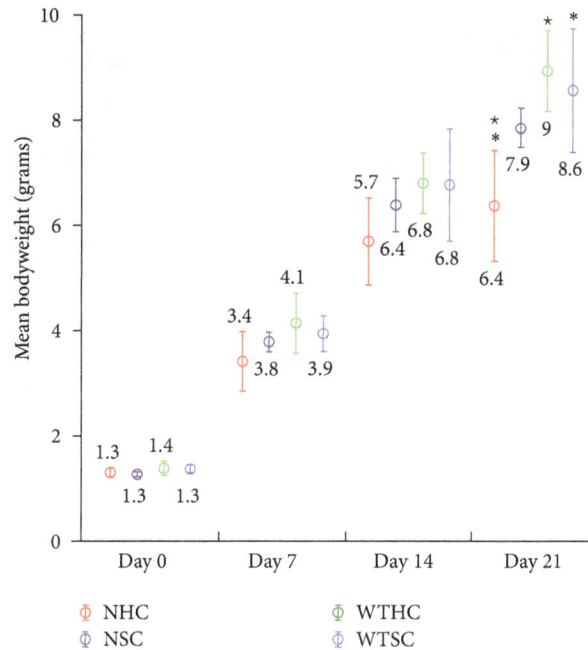

FIGURE 1: Mean bodyweights of pups at days 0, 7, 14, and 21 (prior to weaning). Four groups of mice were measured and compared wild-type hard chow (WTHC), null hard chow (NHC), wild-type soft chow (WTSC), and null soft chow (NSC). At birth, the mean bodyweight of mice in all groups was similar and showed little variance. Over time, the mean bodyweights showed increasing variance within and between groups. The only statistically significant differences were found at 21 days where the mean bodyweight of null mice provided with hard chow was lower than wild-type mice. Statistically significant differences (SSD) between NHC and WTSC (∗) or WTHC (⋆).

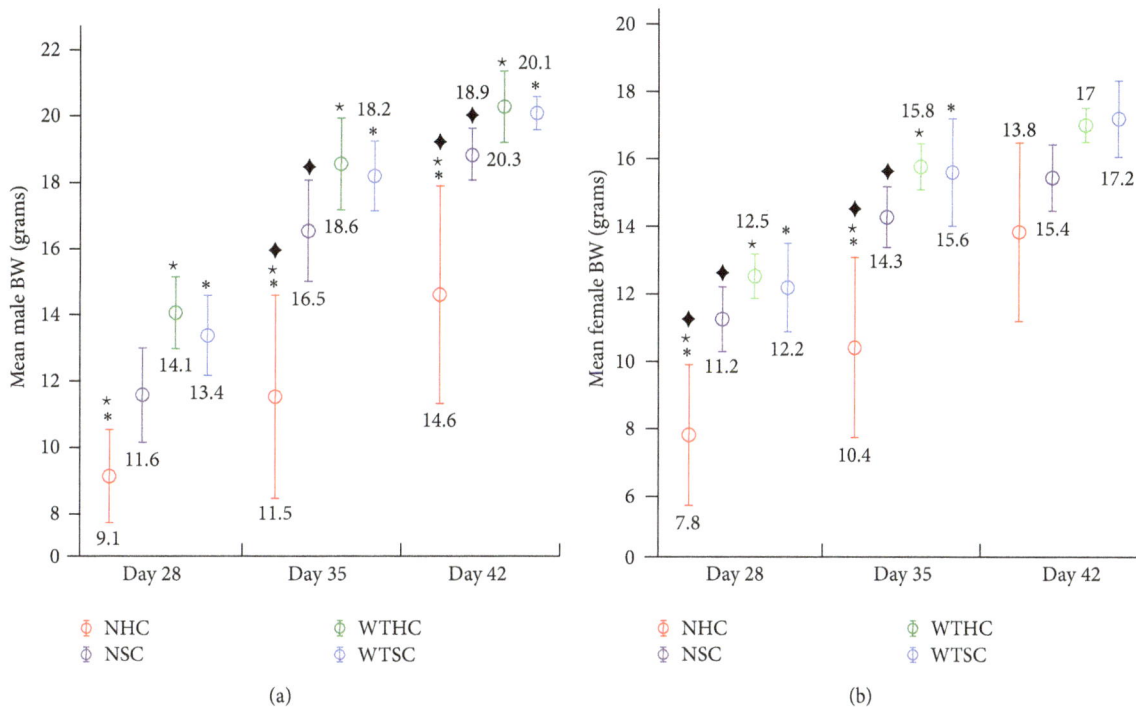

(a)

(b)

FIGURE 2: Mean bodyweights of male (a) and female (b) pups after weaning (days 28, 35, and 42). Statistically significant differences (SSD) between NHC and WTSC (∗), WTHC (⋆), or NSC (◆). The null mice on hard chow had a lower mean bodyweight than null mice on soft chow and wild-type mice on hard or soft chow. Although the mean bodyweight of null mice on soft chow tended to have a lower average bodyweight than wild-type mice, the differences were small and not statistically significant. The mean bodyweight of null mice on hard chow varied more than mice in other groups, suggesting differences in how individual mice adapted to eating with defective dentitions.

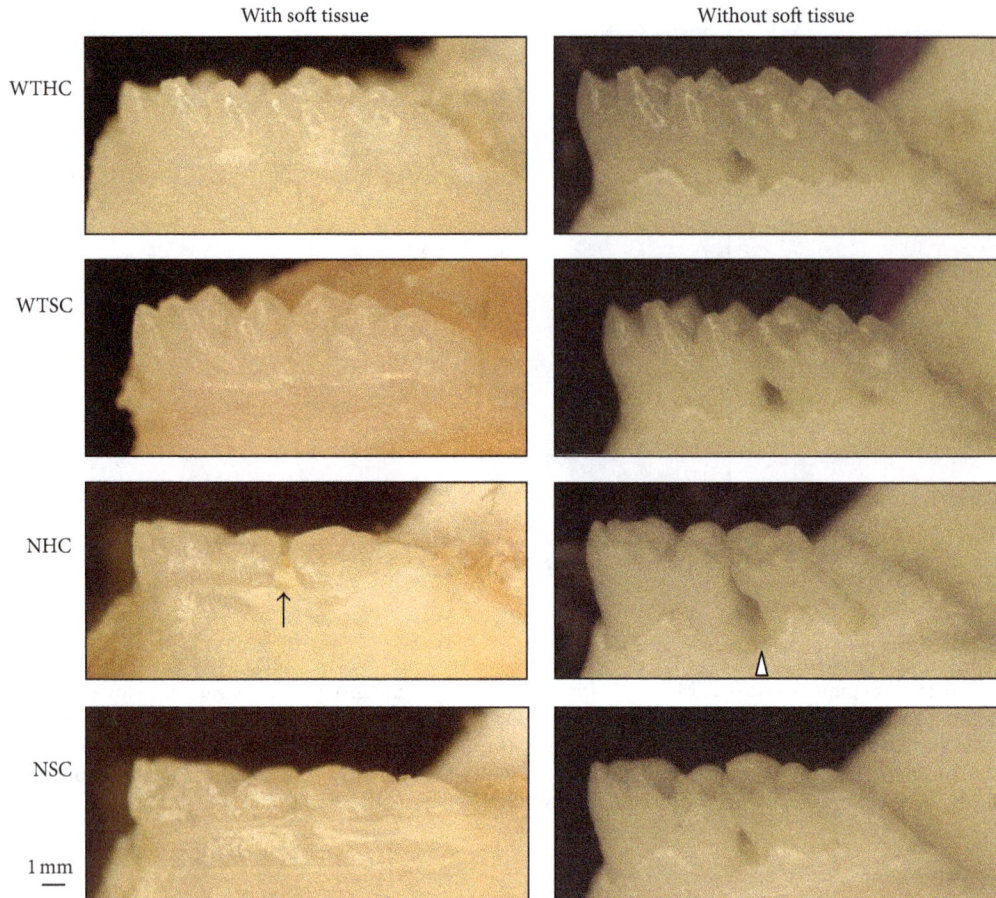

FIGURE 3: Morphology of mandibular molars at 8 weeks. Normal cuspal morphology, coronal crown contour, and supporting bone structure can be observed among wild-type mice maintained on hard chow (WTHC) or soft chow (WTSC). Flatten cusps and smaller occlusal table are consistent features of null mouse molars (NHC and NSC). Food impaction (arrow) between molars was often observed in the null mice and was associated with bony defects (arrowhead).

null mother mice might not be sufficient compared to the wild-type groups with the same litter size.

It was surprising that there was no correlation between DBW and ALBW. The BW of the null mother mice was consistently lower than that of the wild-type mothers; however, the pattern of litter weight gain did not correspond to that of maternal BW. The ALBW was generally similar among groups until the end of preweaning period. After that time point wild-type pups showed higher BW gain than null pups with soft chow, which in time was heavier than null pups with hard chow. Although not statistically significant, the parallel correlation of DBW and ALBW did not exclude the possibility that the nutritional status of null dams was a factor influencing the observed lower BW of their litters. During the experiment, we found that the values of DBW fluctuated frequently and the magnitude can be as high as 5 grams between two consecutive days. The high BW variation of the individual dam may have contributed toward the finding of no significant difference on ALBW among the experimental groups. One study evaluated the genetic and maternal effects of mice using the cross-fostering technique [32]. Two females littering within a 12-hour period were paired and a random half of pups of each sex were then reciprocally switched

within the pair. The BW was taken every 3 days from birth to 6 weeks and weekly weight thereafter until 84 days. The variance due to nurse dams peaked at 12 days, accounting for about 70% of the phenotype variance in D12 weights.

The growth and development of mice in various developmental stages and how they were bred has been studied extensively [30]. In standard husbandry of mice, pups are kept with their mother for the first three weeks after birth, during which their nutrition primarily comes from mother's milk. Young mice start to consume solid food as soon as their eyes open roughly around day 14. They are weaned on day 21, signaling the end of their dependence on their mother. At the age of 6 weeks, they are considered young adult and can be active in reproduction. In a study assessing the impact of inherited enamel defects on bodyweight gain of amelogenin (*Amelx*) null mice [33], it was reported that the average BW of the null mice was less than that of the wild-type mice on each day of the 3-week preweaning period. It is possible that the result is due partially to decreased nutritional intake starting from the affected mother mice. It would be interesting to determine if the BW difference would be even more significant after the young mice start to eat with their defective teeth independently from their mother.

Recently, a phenotyping screening of two enamelin-mutant mouse lines originating from an ENU mutagenesis project for dominant mutations on a C3HeB/FeJ genetic background observed dominant effects of heterozygous enamelin mutations on bone and energy metabolism, as well as on clinical chemistry and hematological parameters, suggesting that enamelin plays a critical role in other organs besides developing teeth [34]. The first mouse, with the p.Gln176* mutation, had previously been characterized without noting systemic effects [18]. The second mouse Enam mutation (c.A382T; p.Lys128*) was dominant so that mice heterozygous and homozygous for the mutation both showed the same level of defective "whitish" enamel. They concluded that the heterozygous enamelin truncations likely resulted in defective energy metabolism possibly from subtle changes in liver and/or pancreas function. This conclusion is hard to reconcile with the fact that not a single enamelin transcript is listed among the 111,391 (mouse) or 205,232 (human) ESTs for liver and 106,259 (mouse) or 213,410 (human) ESTs for pancreas. Enamelin is a member of the secretory calcium-binding phosphoproteins (SCPPs) encoded by a family of genes clustered on chromosome 4q13 in humans and chromosome 5 in mice [35, 36] that are critical for many processes, such as bone and tooth biomineralization, saliva, and lactation. Perhaps, there are undetected mutations affecting these linked genes that account for their findings. The results of this study cast doubt upon the claim that their mice with severe enamel defects ate as much hard chow as wild-type mice. If there are no differences in food intake by mice with and without enamel, what is the selective pressure that maintains the enamel layer during evolution?

In this study, we first tested the hypothesis that Enam null mice have lower bodyweights than wild-type mice. Our results showed that Enam null mice have significantly lower bodyweights than wild-type mice starting at day 21, when the mice are weaning. We also showed that weaned null mice fed on hard chow tend to have lower bodyweights than null mice fed on soft chow, revealing that at least part of the lower bodyweight can be explained by eating difficulties secondary to the lack of enamel on the dentition. Our results, in conjunction with evidence that enamelin is expressed specifically by ameloblasts and there is an absence of selection pressure to maintain Enam in vertebrates that have lost the ability to make enamel during evolution, support the conclusion that enamelin does not perform necessary functions outside of dental enamel formation.

Acknowledgments

The authors thank Dr. Yuanyuan Hu for her technical consultations on mouse breeding and Mr. Joe Kazemi at the Center for Statistical Consultation and Research, University of Michigan, for providing statistical consultations. The study was supported by NIDCR Grant DE011301.

References

[1] M. H. Rajpar, K. Harley, C. Laing, R. M. Davies, and M. J. Dixon, "Mutation of the gene encoding the enamel-specific protein, enamelin, causes autosomal-dominant amelogenesis imperfecta," Human Molecular Genetics, vol. 10, no. 16, pp. 1673–1677, 2001.

[2] C. K. Mårdh, B. Bäckman, G. Holmgren, J. C. C. Hu, J. P. Simmer, and K. Forsman-Semb, "A nonsense mutation in the enamelin gene causes local hypoplastic autosomal dominant amelogenesis imperfecta (AIH2)," Human Molecular Genetics, vol. 11, no. 9, pp. 1069–1074, 2002.

[3] M. Kida, T. Ariga, T. Shirakawa, H. Oguchi, and Y. Sakiyama, "Autosomal-dominant hypoplastic form of amelogenesis imperfecta caused by an enamelin gene mutation at the exon-intron boundary," Journal of Dental Research, vol. 81, no. 11, pp. 738–742, 2002.

[4] P. S. Hart, M. D. Michalec, W. K. Seow, T. C. Hart, and J. T. Wright, "Identification of the enamelin (g.8344delG) mutation in a new kindred and presentation of a standardized ENAM nomenclature," Archives of Oral Biology, vol. 48, no. 8, pp. 589–596, 2003.

[5] T. C. Hart, P. S. Hart, M. C. Gorry et al., "Novel ENAM mutation responsible for autosomal recessive amelogenesis imperfecta and localised enamel defects," Journal of Medical Genetics, vol. 40, no. 12, pp. 900–906, 2003.

[6] J. W. Kim, F. Seymen, B. P. J. Lin et al., "ENAM mutations in autosomal-dominant amelogenesis imperfecta," Journal of Dental Research, vol. 84, no. 3, pp. 278–282, 2005.

[7] D. Ozdemir, P. S. Hart, E. Firatli, G. Aren, O. H. Ryu, and T. C. Hart, "Phenotype of ENAM mutations is dosage-dependent," Journal of Dental Research, vol. 84, no. 11, pp. 1036–1041, 2005.

[8] J. W. Kim, J. P. Simmer, B. P. L. Lin, F. Seymen, J. D. Bartlett, and J. C. C. Hu, "Mutational analysis of candidate genes in 24 amelogenesis imperfecta families," European Journal of Oral Sciences, vol. 114, supplement 1, pp. 3–12, 2006.

[9] S. J. Gutierrez, M. Chaves, D. M. Torres, and I. Briceño, "Identification of a novel mutation in the enamalin gene in a family with autosomal-dominant amelogenesis imperfecta," Archives of Oral Biology, vol. 52, no. 5, pp. 503–506, 2007.

[10] A. Pavlič, M. Petelin, and T. Battelino, "Phenotype and enamel ultrastructure characteristics in patients with ENAM gene mutations g.13185-13186insAG and 8344delG," Archives of Oral Biology, vol. 52, no. 3, pp. 209–217, 2007.

[11] H. Y. Kang, F. Seymen, S. K. Lee et al., "Candidate gene strategy reveals ENAM mutations," Journal of Dental Research, vol. 88, no. 3, pp. 266–269, 2009.

[12] H. C. Chan, L. Mai, A. Oikonomopoulou et al., "Altered enamelin phosphorylation site causes amelogenesis imperfecta," Journal of Dental Research, vol. 89, no. 7, pp. 695–699, 2010.

[13] R. G. Lindemeyer, C. W. Gibson, and T. J. Wright, "Amelogenesis imperfecta due to a mutation of the enamelin gene: clinical case with genotype-phenotype correlations," Pediatric Dentistry, vol. 32, no. 1, pp. 56–60, 2010.

[14] J. T. Wright, M. Torain, K. Long et al., "Amelogenesis imperfecta: genotype-phenotype studies in 71 families," Cells Tissues Organs, vol. 194, pp. 279–283, 2011.

[15] Y. L. Song, C. N. Wang, C. Z. Zhang, K. Yang, and Z. Bian, "Molecular characterization of amelogenesis imperfecta in Chinese patients," Cells Tissues Organs, vol. 13, article 13, 2012.

[16] S. Simmer, N. Estrella, R. Milkovich, and J. Hu, "Autosomal dominant amelogenesis imperfecta associated with ENAM frameshift mutation p.Asn36Ilefs56," Clinical Genetics. In press.

[17] H. Masuya, K. Shimizu, H. Sezutsu et al., "Enamelin (Enam) is essential for amelogenesis: ENU-induced mouse mutants as

models for different clinical subtypes of human amelogenesis imperfecta (AI)," *Human Molecular Genetics*, vol. 14, no. 5, pp. 575–583, 2005.

[18] H. Seedorf, M. Klaften, F. Eke, H. Fuchs, U. Seedorf, and M. Hrabe De Angelis, "A mutation in the enamelin gene in a mouse model," *Journal of Dental Research*, vol. 86, no. 8, pp. 764–768, 2007.

[19] J. C. C. Hu, Y. Hu, C. E. Smith et al., "Enamel defects and ameloblast-specific expression in Enam knock-out/lacZ knock-in mice," *Journal of Biological Chemistry*, vol. 283, no. 16, pp. 10858–10871, 2008.

[20] T. A. Deméré, M. R. McGowen, A. Berta, and J. Gatesy, "Morphological and molecular evidence for a stepwise evolutionary transition from teeth to baleen in mysticete whales," *Systematic Biology*, vol. 57, no. 1, pp. 15–37, 2008.

[21] R. W. Meredith, J. Gatesy, W. J. Murphy, O. A. Ryder, and M. S. Springer, "Molecular decay of the tooth gene enamelin (ENAM) mirrors the loss of enamel in the fossil record of placental mammals," *PLoS Genetics*, vol. 5, no. 9, Article ID e1000634, 2009.

[22] N. Al-Hashimi, A. G. Lafont, S. Delgado, K. Kawasaki, and J. Y. Sire, "The enamelin genes in lizard, crocodile, and frog and the pseudogene in the chicken provide new insights on enamelin evolution in tetrapods," *Molecular Biology and Evolution*, vol. 27, no. 9, pp. 2078–2094, 2010.

[23] R. W. Meredith, J. Gatesy, J. Cheng, and M. S. Springer, "Pseudogenization of the tooth gene enamelysin (MMP20) in the common ancestor of extant baleen whales," *Proceedings of Biological Sciences*, vol. 278, no. 1708, pp. 993–1002, 2011.

[24] G. Acs, G. Lodolini, S. Kaminsky, and G. J. Cisneros, "Effect of nursing caries on body weight in a pediatric population," *Pediatric Dentistry*, vol. 14, no. 5, pp. 302–305, 1992.

[25] A. Sheiham, "Dental caries affects body weight, growth and quality of life in pre-school children," *British Dental Journal*, vol. 201, no. 10, pp. 625–626, 2006.

[26] P. S. Hart, T. C. Hart, M. D. Michalec et al., "Mutation in kallikrein 4 causes autosomal recessive hypomaturation amelogenesis imperfecta," *Journal of Medical Genetics*, vol. 41, no. 7, pp. 545–549, 2004.

[27] J. W. Kim, J. P. Simmer, T. C. Hart et al., "MMP-20 mutation in autosomal recessive pigmented hypomaturation amelogenesis imperfecta," *Journal of Medical Genetics*, vol. 42, no. 3, pp. 271–275, 2005.

[28] P. S. Hart, S. Becerik, D. Cogulu et al., "Novel FAM83H mutations in Turkish families with autosomal dominant hypocalcified amelogenesis imperfecta," *Clinical Genetics*, vol. 75, no. 4, pp. 401–404, 2009.

[29] J. W. Kim, J. P. Simmer, Y. Y. Hu et al., "Amelogenin p.M1T and p.W4S mutations underlying hypoplastic X-linked amelogenesis imperfecta," *Journal of Dental Research*, vol. 83, no. 5, pp. 378–383, 2004.

[30] H. A. el-Oksh, T. M. Sutherland, and J. S. Williams, "Prenatal and postnatal maternal influence on growth in mice," *Genetics*, vol. 57, no. 1, pp. 79–94, 1967.

[31] S. M. Azzam, M. K. Nielsen, and G. E. Dickerson, "Postnatal litter size effects on growth and reproduction in rats," *Journal of Animal Science*, vol. 58, no. 6, pp. 1337–1342, 1984.

[32] J. J. Rutledge, O. W. Robison, E. J. Eisen, and J. E. Legates, "Dynamics of genetic and maternal effects in mice," *Journal of Animal Science*, vol. 35, no. 5, pp. 911–918, 1972.

[33] Y. Li, Z. A. Yuan, M. A. Aragon, A. B. Kulkarni, and C. W. Gibson, "Comparison of body weight and gene expression in amelogenin null and wild-type mice," *European Journal of Oral Sciences*, vol. 114, supplement 1, pp. 190–193, 2006.

[34] H. Fuchs, S. Sabrautzki, H. Seedorf et al., "Does enamelin have pleiotropic effects on organs other than the teeth? Lessons from a phenotyping screen of two enamelin-mutant mouse lines," *European Journal of Oral Sciences*, vol. 120, pp. 269–277, 2012.

[35] K. Kawasaki and K. M. Weiss, "Mineralized tissue and vertebrate evolution: the secretory calcium-binding phosphoprotein gene cluster," *Proceedings of the National Academy of Sciences of the United States of America*, vol. 100, no. 7, pp. 4060–4065, 2003.

[36] K. Kawasaki, A. V. Buchanan, and K. M. Weiss, "Gene duplication and the evolution of vertebrate skeletal mineralization," *Cells Tissues Organs*, vol. 186, no. 1, pp. 7–24, 2007.

15

An Immunohistochemistry Study of Sox9, Runx2, and Osterix Expression in the Mandibular Cartilages of Newborn Mouse

Hong Zhang,[1,2] Xiaopeng Zhao,[3] Zhiguang Zhang,[4] Weiwei Chen,[2,5] and Xinli Zhang[2]

[1] Department of Orthodontics, Guanghua School of Stomatology, Sun Yat-Sen University, Guangzhou 510055, China
[2] Dental and Craniofacial Research Institute, School of Dentistry, University of California, Los Angeles, CA, 90095, USA
[3] Department of Oral and Maxillofacial Surgery, Sun Yat-sen Memorial Hospital, Sun Yat-Sen University, Guangzhou 510120, China
[4] Department of Oral and Maxillofacial Surgery, Guanghua School of Stomatology, Sun Yat-Sen University, Guangzhou 510055, China
[5] Institute for Medical Biology, College of Life Sciences, South-Central University for Nationalities, Wuhan 430074, China

Correspondence should be addressed to Zhiguang Zhang; drzhangzg@163.com and Xinli Zhang; xzhang@dentistry.ucla.edu

Academic Editor: Chad M. Novince

The purpose of this study is to investigate the spacial expression pattern and functional significance of three key transcription factors related to bone and cartilage formation, namely, Sox9, Runx2, and Osterix in cartilages during the late development of mouse mandible. Immunohistochemical examinations of Sox9, Runx2, and Osterix were conducted in the mandibular cartilages of the 15 neonatal C57BL/6N mice. In secondary cartilages, both Sox9 and Runx2 were weakly expressed in the polymorphic cell zone, strongly expressed in the flattened cell zone and throughout the entire hypertrophic cell zone. Similarly, both transcriptional factors were weakly expressed in the uncalcified Meckel's cartilage while strongly expressed in the rostral cartilage. Meanwhile, Osterix was at an extremely low level in cells of the flattened cell zone and the upper hypertrophic cell zone in secondary cartilages. Surprisingly, Osterix was intensely expressed in hypertrophic chondrocytes in the center of the uncalcified Meckel's cartilage while moderately expressed in part of hypertrophic chondrocytes in the rostral process. Consequently, it is suggested that Sox9 is a main and unique positive regulator in the hypertrophic differentiation process of mandibular secondary cartilages, in addition to Runx2. Furthermore, Osterix is likely responsible for phenotypic conversion of Meckel's chondrocytes during its degeneration.

1. Introduction

The development of cartilages plays a pivotal role in the development and growth of the mandible. Mandibular cartilages are derived from ectomesenchymal cells of the first pharyngeal arch, but their characteristics differ. Meckel's cartilage is a fetal cartilaginous skeleton in the mandible. Although it is classified as primary cartilage similar to limb bud cartilage, it contains four distinct portions, each having a different fate. The anterior, intermediate, and proximal portions convert to intramandibular symphysis, sphenomandibular ligament, and the inner ear ossicles, respectively. The posterior portion of intramandibular Meckel's cartilage facing the developing molar buds undergoes developmental events similar to endochondral ossification, but the degradation mechanisms of this portion are distinct from those in endochondral ossification [1]. Independent of the chondroskeleton, four secondary cartilages including the condylar, coronoid, angular, and symphyseal cartilage strongly influence the further development of the mandible. These secondary cartilages differ from the primary cartilage in embryonic origin, morphological and biochemical organization. They are derived from the periosteum of intramembranous bone after (secondary to) bone formation [2, 3]. Furthermore, they display a unique mode of cell proliferation and differentiation. The condylar cartilage, as a principle secondary cartilage, does not form columns of proliferating chondrocytes and grows multidirectionally to adapt to the mandibular fossa of the temporal bone [4].

Recent studies showed that the three master transcription factors of Sox9, Runx2, and Osterix are involved in the formation of Meckel's cartilage and mandibular condylar cartilage [3, 10]. Sox9, Runx2, and Osterix are key transcription factors, which are necessary in skeletal cell fate decision [11]. Sox9 (SRY-box containing gene 9) is an essential15 and nonredundant factor of chondrogenesis. Analyses in genetically modified mice revealed that Sox9 promotes the early stage, but suppresses the terminal stage of chondrocyte differentiation in limb bud cartilage [12–14]. On the contrary, the multifunctional transcription factor Runx2, which is expressed in prehypertrophic and hypertrophic chondrocytes, is a main positive regulator of hypertrophic differentiation in late chondrogenesis of the limb buds [5, 15, 16]. New in vitro data demonstrated that Sox9 negatively regulates Runx2 by enhancing Bapx1 expression, which leads to the inhibition of terminal chondrocyte differentiation [17]. Osterix, which acts downstream of Runx2 during bone formation, is expressed in chondrocyte progenitors and prehypertrophic chondrocytes in rib, spine, and limb cartilages, suggesting that Osterix may play a critical role during the primary cartilage maturation in combination with Runx2 and Sox9 [6, 7].

However, the transcriptional control of the later development of mandibular cartilages remains poorly understood. At birth, the rostral process of intramandibular Meckel's cartilage is undergoing endochondral ossification, while the posterior portion of intramandibular Meckel's cartilage is degenerating [18–20]. Meanwhile, four secondary cartilages, especially the condylar cartilage, were not well documented in terms of their developmental characteristics, although they function mainly as a growth cartilage similar to limb bud cartilage. At present, transcription factors are attracting increasing clinical attention because of their roles in the etiology and pathogenesis of malformations and growth disorders, degenerative diseases, and in regenerative and repair processes [21, 22]. The findings that Runx2-deficient mice lack mandibular condylar cartilage and had deformed Meckel's cartilage indicate that Runx2 is essential for the formation of the mandibular cartilages [23]. In many cleidocranial dysplasia (CCD) patients who were link to Runx2 deficent, however, there are no abnormal findings in the mandible, in spite of cases of condylar malformation, persistent symphysis, or a narrow coronoid process being also known [24, 25]. These investigations provided a hint that Runx2 may be just one of essential biological factors influencing the development and growth of mandibular cartilages. The present study is to examine tissue distribution of Sox9, Runx2, and Osterix in newborn mice mandibular cartilages using immunohistochemistry technique and investigate whether these transcription factors have similar functions to those in limb bud cartilage which will contribute to current understanding of mechanisms of the development of mandible and the possible pathogenesis of some craniofacial anomalies involving mandible.

2. Materials and Methods

All animals were housed and handled in accordance with guidelines of the Chancellor's Animal Research Committee of the Office for Protection of Research Subjects at the University of California, Los Angeles, CA, USA.

2.1. Tissue Preparation. A total of 15 newborn C57BL/6N mice were collected in 2 hours right after being delivered and used for this study. The mandibles were then removed and immersed in 4% paraformaldehyde (0.1 M phosphate buffer, pH 7.4) for 1 day at 4°C. The specimens were decalcified with 10% EDTA for 5 days at 4°C and then embedded in paraffin using standard procedures. Sections (5 μm) were cut in the plane parallel to the ascending ramus of the mandible, all the mandibular cartilages being in one section. For general morphology, deparaffinized sections were stained with hematoxylin and eosin. The skeletal staining with Alizarin Red-Alcian Blue was performed for preparation of gross specimen of mandible as reported previously [26].

2.2. Immunohistochemistry. Five-micron-thick paraffin sections were dewaxed in xylenes and rehydrated in ethanol baths. Endogenous peroxidases were blocked by incubating sections in 3% hydrogen peroxide for 20 min at room temperature. Sections were incubated with anti-Runx2, anti-Osterix, and anti-Sox9 primary antibodies (Santa Cruz Biotechnology, CA, USA) (dilution 1 : 100) and biotinylated anti-rabbit or anti-goat IgG secondary antibody (Vector Laboratories, Burlingame, CA) for 1 h at room temperature. Positive immunoreactivity was detected using Vectastain ABC kit (Vector Laboratories, Burlingame, CA, USA) and AEC chromogenic substrate (Dako, Carpiteria, CA, USA) with red positive staining. A negative control was performed by replacing primary antibody solutions with PBS. Sections were counterstained with hematoxylin for 30 sec followed by rinsing 5 min in running water. Photomicrographs were acquired using an Olympus BX51. Image-pro Plus 6.0 software was used to calculate stained area and Integrated Optical Density (IOD). The average optical density (mean density) represented the intensity of protein expression and was counted in 4 random fields (×20 objective) of each cartilage area and trabecular bone area per section. The mean density is equal to (IOD SUM)/area. For exact analysis, three sections were prepared at similar plane for each sample. ANOVA was used for multiple groups' comparison, and Student's t-test was used for comparison between any two groups. Statistical significance defined as $P < 0.05$.

3. Results

3.1. Histological Analysis of Cartilages in Newborn Mouse Mandible. Mandibular cartilages in newborn mouse included the portions of Meckel's cartilage, condylar, angular, and symphyseal secondary cartilage, while cartilage was not present in the coronoid process of the newborn mouse (Figure 1(a)). On the basis of the cellular morphological changes, the mandibular condylar and angular cartilages were histologically composed of four different cell zones: a thin fibrous cell zone, a polymorphic cell zone, a wider flattened cell zone, and a broad hypertrophic cell zone occupied the lower half of the organ (Figures 1(b) and 1(c)). Almost all of the intramandibular bar of Meckel's cartilage had ossified completely, but a small amount of Meckel's cartilage remained in a limited portion of the rostral region and at the mylohyoid groove between the condylar and angular processes. At the posterior portion of intramandibular Meckel's cartilage, HE staining pattern of the matrix changed from the intense

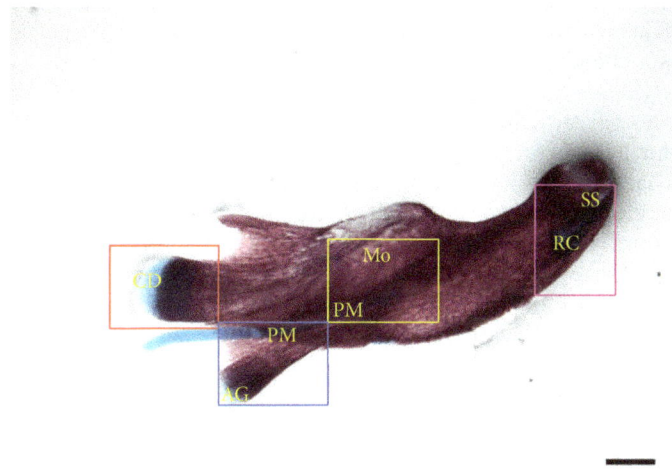

(a)

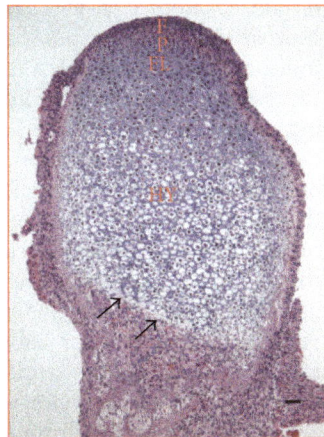

(b)

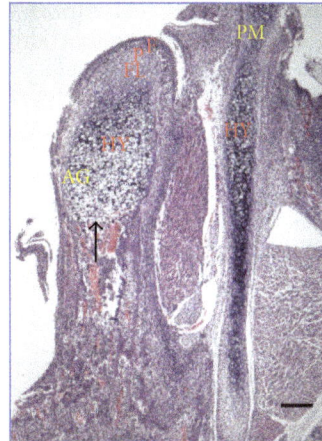

(c)

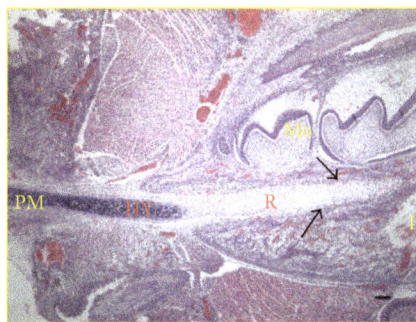

(d)

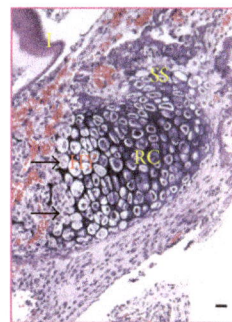

(e)

FIGURE 1: Histological analysis of mandibular cartilages of mice at newborn stage. (a) Lingual view of mandible by Alizarin Red and Alcian Blue staining; (b) hematoxylin-eosin sections of condylar cartilage, and (c) hematoxylin-eosin sections of angular cartilage (AG) similarly displaying four different cell zones: a thin fibrous cell zone (F), a polymorphic cell zone (P), a wider flattened cell zone (FL), and a broad hypertrophic cell zone (HY); (d) hematoxylin-eosin staining pattern of the matrix of the posterior portion of intramandibular Meckel's cartilage (PM) changed from the intense hematoxylin to the light eosin in the resorption (R) facing the molar buds (Mo) and incisor (I); (e) hematoxylin-eosin sections of the endochondral ossification rostral process of Meckel's cartilage (RC) and symphyseal secondary cartilage (SS) facing the incisor (I). Scale bar: 100 μm for (a) and (b), 250 μm for (c) and (d), and 50 μm for (e).

hematoxylin to the light eosin in resorption area, which indicated the degradation of Meckel's cartilage matrix during development (Figure 1(d)). Furthermore, the endochondral ossification rostral process of Meckel's cartilage and symphyseal secondary cartilage serve as a chondrogenic mandibular symphysis of newborn mice (Figure 1(e)). The opened chondrocytic lacunae and disconnected cartilaginous matrix (arrows in Figures 1(b), 1(c), and 1(e)) were clearly found in the resorption area of condylar, angular, and symphyseal secondary cartilage, in addition to rostral cartilage, while the appearing perichondrium and the eosinophilic cartilage erode on the lateral sides (arrows in Figure 1(d)) were observed in the posterior portion of intramandibular Meckel's cartilage. These results indicated that the degradation of cartilaginous matrix in the posterior portion of intramandibular Meckel's cartilage is distinct from others among mandibular cartilages.

3.2. Immunohistochemical Analysis of Sox9, Rux2, and Osterix in Cartilages of the Newborn Mouse Mandible.

Interestingly, transcription factors Sox9 and Runx2 showed similar expression level and tissue distribution patterns throughout all the mandibular cartilages of newborn mice. In secondary cartilages, both Sox9 and Runx2 were weakly expressed by cells in the polymorphic cell zone, strongly in the flattened cell zone and throughout the entire hypertrophic cell zone. To quantitatively measure changes in expression of transcriptional factors critical for chondrogenic differentiation, Sox9 (Figure 2(f)) and Runx2 (Figure 2(h)) immunohistochemistry in hypertrophic zones of mandibular cartilages at newborn stage were quantitated by average optical density of positive staining. The expression levels of both transcriptional factors in the degrading posterior portion of intramandibular Meckel's cartilage (Figures 2(b) and 2(d)) exhibited a significantly decrease, compared with others among mandibular cartilages. Meanwhile, cells in the rostral cartilage (Figures 2(e) and 2(g)) and cells in extramandibular Meckel's cartilage (Figures 2(f) and 2(h)) similarly expressed both transcriptional factors more than in the degrading posterior portion of intramandibular Meckel's cartilage. Unexpectedly, Sox9, as Runx2, was expressed in all the terminal chondrocytes of the mandibular cartilages in newborn mice (Figures 2(a), 2(b), and 2(e)), contrary to the express pattern of Sox9 in limb bud cartilage [8]. This spatial distribution pattern indicated Sox9's requirement in the terminal stage of mandibular chondrocyte differentiation.

Runx2 and Osterix are involved in the formation of Meckel's cartilage and mandibular condylar cartilage [3, 10]. Thus, we correlated the expression patterns of Runx2 (Figures 2(c), 2(d), and 2(g)) and Osterix (Figure 3) in mandibular cartilages at newborn stages. Results showed that Osterix was at an extremely low level in part of cells of the flattened cell zone and the upper hypertrophic cell zone in condylar cartilage and angular cartilage, independent on Runx2. Unlike Sox9, the spatial pattern of Osterix in condylar cartilage and angular cartilage was consistent with that in limb bud cartilage [7, 9]. Notably, Osterix was intensely expressed only in hypertrophic chondrocytes of the center of the uncalcified Meckel's cartilage containing the strong basophilic matrix,

while it was entirely absent in hypertrophic chondrocytes in the resorption area containing the light eosinophilic matrix (Figures 3(b), 1(c), and 1(d)). Additionally, the expression level of Osterix in the hypertrophic chondrocytes of Meckel's cartilage (Figure 3(b)) was significantly higher compared with that in condylar cartilage and angular cartilage which have only few positive cells in the flattened cell zone and the upper hypertrophic cell zone (Figures 3(a), 3(b), and 3(d)). Moreover, Osterix was moderately expressed in part of hypertrophic chondrocytes in the rostral process (Figure 3(c)), while it was absent in extramandibular Meckel's cartilage. At present, the mechanisms of Osterix regulation of chondrocyte differentiation and function are still under investigation, whereas the significantly intense immunohistochemistry of Osterix in hypertrophic chondrocytes of the center of the uncalcified Meckel's cartilage provided evidence of Osterix's role in the degradation of the posterior portion of uncalcified intramandibular Meckel's cartilage.

3.3. Comparison of Expression Intensity of Runx2 and Osterix in the Chondrocytes with That in the Osteoblasts of the Newborn Mouse Mandible.

Since Runx2 and Osterix are indispensable for osteoblast differentiation and known to be expressed in osteoblasts, we first confirmed the positive staining of osteoblasts using the same sections of cartilages with Runx2 and Osterix positive staining, which also validated our IHC approach to be highly reliable. Then, we compared the expression intensity of the two key transcriptional factors related to bone formation in chondrocytes with that in mandibular osteoblasts to further confirm the significance of both during the development of mandibular cartilages. Similar to the condylar subchondral bone in 56-day-old rats [27], Runx2 protein which was expressed in secondary hypertrophic chondrocytes was not localized in the cells gathering in the erosive front of all the mandibular secondary cartilages (Figure 4(a)), but in some osteoblasts surrounding the trabecular bone and some osteocytes buried in the trabecular bone in the mandible (Figure 4(b)). Thus, we quantitatively analyzed the expression of Runx2 protein in osteoblasts and osteocytes (Figure 4(e)), comparing with that in condylar cartilage and the posterior portion of intramandibular Meckel's cartilage. In the present study, the expression of Runx2 protein in condylar hypertrophic chondrocytes was the most intense (Figures 2(h) and 4(e)), being statistically significant difference from that in osteoblasts, which indicated an important role of Runx2 in secondary chondrocyte maturation, in addition to that in chondrocyte maturation of growth cartilage and osteoblast differentiation. Expectedly, Osterix was localized in some osteoblasts and bone marrow cells in sub-chondral bone area (Figure 4(c)). Furthermore, more positive osteoblasts and osteocytes were visualized in the trabecular bone area (Figure 4(d)). Interestingly, the immunohistochemistry of Osterix in hypertrophic chondrocytes of the center of the uncalcified Meckel's cartilage is still significantly more intense than that in the osteoblasts (Figure 4(f)). This pointed out that Osterix highly likely performed a regulatory effect on the degradation of the posterior portion of uncalcified intramandibular Meckel's cartilage.

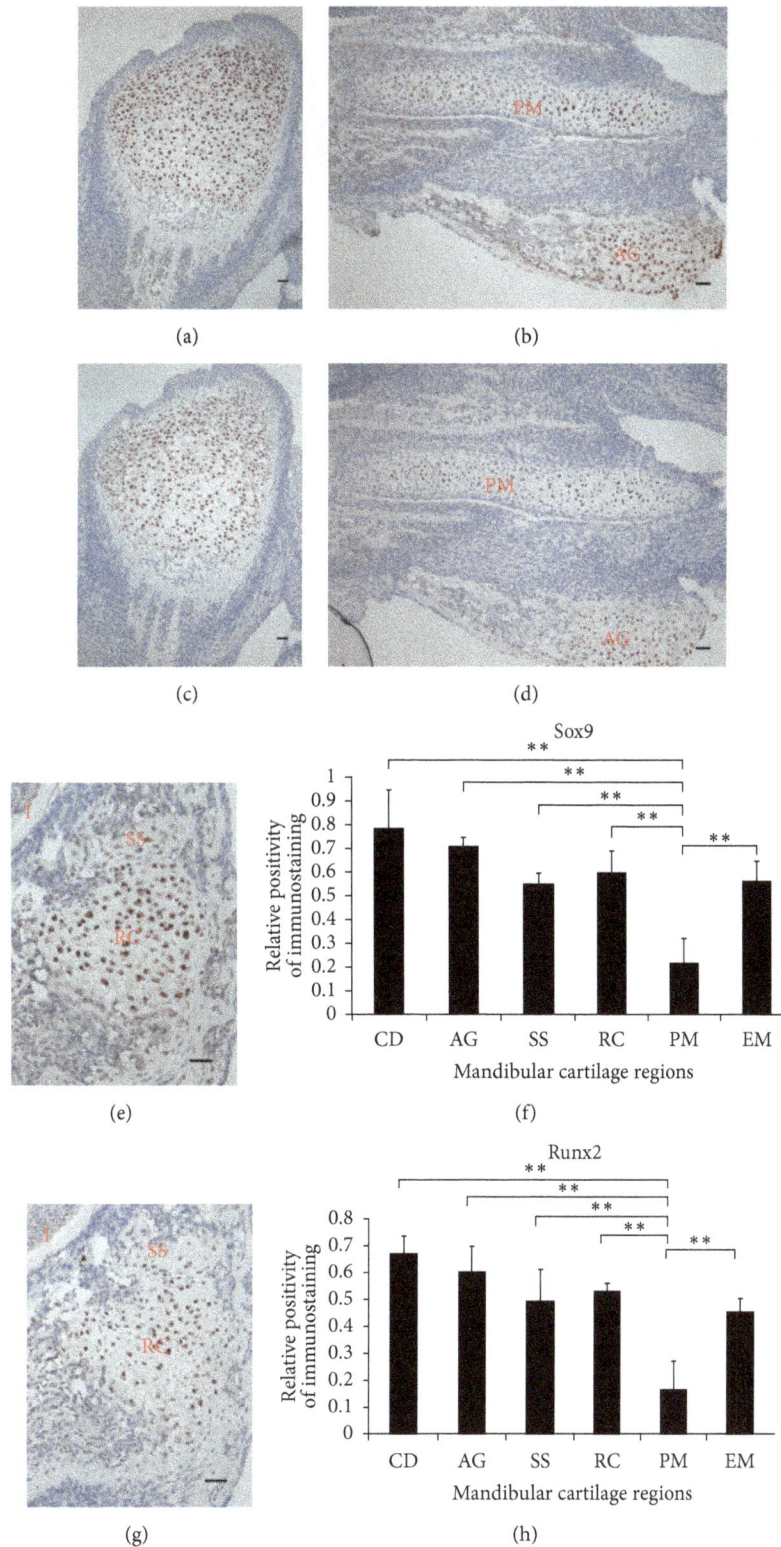

(a)

(b)

(c)

(d)

(e)

(f)

(g)

(h)

FIGURE 2: Immunohistochemistry of Sox9 and Runx2 in mandibular cartilages of mice at newborn stage. Sox9 (a, b, and e) and Runx2 (c, d, and g) showed similar expression patterns throughout all the mandibular cartilages. In condylar cartilage (a and c), angular cartilage (b and d), and symphyseal secondary cartilage and rostral cartilage (e and g), both Sox9 and Runx2 were strongly expressed by cells entire hypertrophic cell zone. Scale bar: 100 μm for (a, b, c, and d) and 50 μm for (e and g). Results of Sox9 (f) and Runx2 (h) immunohistochemistry in hypertrophic zones of mandibular cartilages including condylar (CD), angular (AG), and symphyseal secondary cartilage (SS), and rostral process (RC), posterior Meckel's (PM), and extramandibular Meckel's (EM) cartilage were quantitated by average optical density of positive staining per 200 field ($^{**}P < 0.001$). The expression levels of both transcriptional factors in the posterior portion of uncalcified intramandibular Meckel's cartilage (b and d) were significantly reduced than in other cartilages.

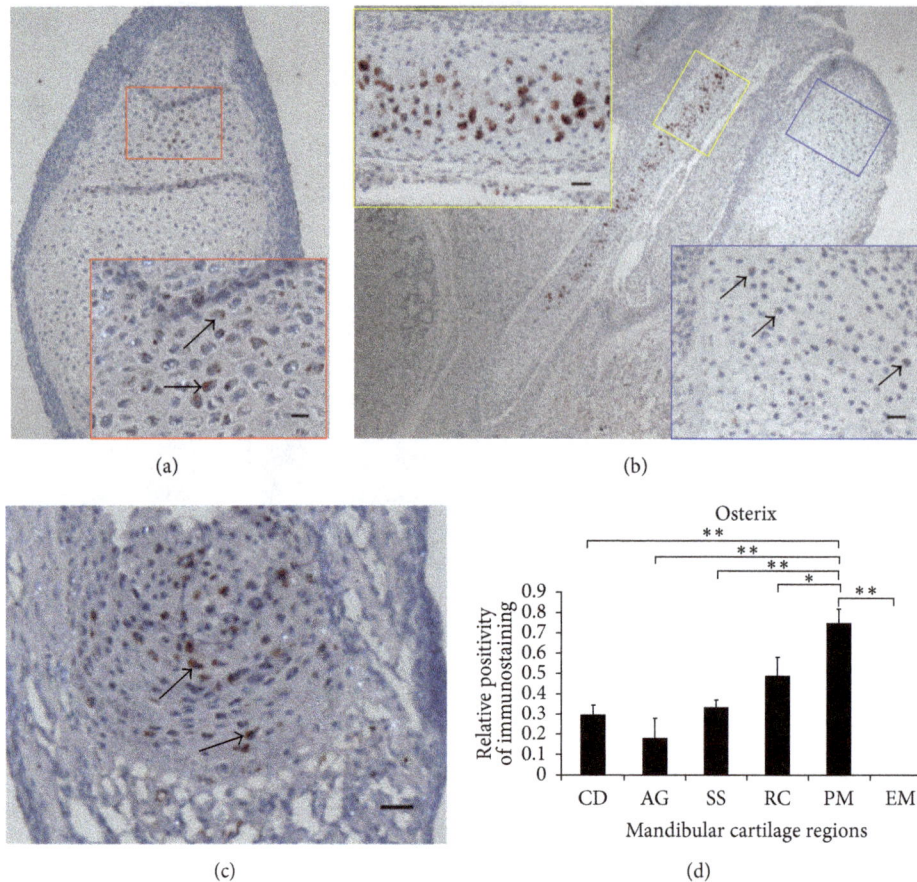

(a)

(b)

(c)

(d)

FIGURE 3: Immunohistochemistry of Osterix in mandibular cartilages of mice at newborn stage. Osterix was at a extremely low level in condylar cartilage (in the red box of (a)) and angular cartilage (in the blue box of (b)) while intensely expressed in the center of the Meckel's cartilage containing the strong basophilic matrix (in the yellow box of (b) and Figure 1(c)). Further, Osterix was moderately expressed in the rostral process (c). Scale bar: 25 μm for (a', b', and b'') and 50 μm for (c). Results of Osterix (d) immunohistochemistry in mandibular cartilages including condylar (CD), angular (AG), and symphyseal secondary cartilage (SS) and rostral process (RC), posterior Meckel's (PM), and extramandibular Meckel's (EM) cartilage were quantitated by average optical density of positive staining per 200 field ($^*P < 0.05$, $^{**}P < 0.001$). Immunohistochemistry of Osterix in the posterior portion of intramandibular Meckel's cartilage was significantly stronger and had more positive cells than in other cartilages.

4. Discussion

The majority of in vivo studies on cartilage differentiation are carried out using the appendicular skeleton as a model system, with the implicit assumption that chondrogenesis is equivalent throughout the body. However, Eames directly tested that the programs of chick head chondrogenesis are unique by comparing the neural crest-derived pharyngeal arch skeleton to that of the mesoderm-derived limb, due to the fact that each skeleton forms from unique embryonic populations [28]. Meckel's cartilage and mandibular secondary cartilages are markedly distinguished from limb bud cartilage in their embryonic origin. The mechanisms that regulate the diverse developmental programs in Meckel's cartilage and mandibular secondary cartilages remain to be discovered. The present study investigated the expression of the essential transcription factors related to chondrogenesis in these cartilages during the later development of mandibular cartilages.

The accumulated studies confirmed that Sox9 accelerates chondrocyte differentiation in proliferating chondrocytes but inhibits the terminal stages of chondrocyte differentiation in limb bud cartilage [29, 30]. However, few investigations focused on the mechanism of Sox9 in secondary chondrocyte differentiation [31]. Our findings clearly demonstrated that the key transcription factor Sox9 was strongly expressed at the whole hypotrophic cell zone of condylar cartilage and angular cartilage in newborn mice, which was different from the expression pattern of Sox9 in limb bud cartilage (Figure 5(a)) [8]. Moreover, Rabie et al. have demonstrated that Sox9 were expressed at the hypertrophic cell zone of the condylar cartilage in 36-day-old rats and continued to be expressed throughout the examined period until day 52 [32]. Conversely, in limb bud cartilage, both of the Sox9 transcripts and protein were absent or at very diminished levels in hypertrophic chondrocytes [8, 9]. The recent data demonstrated that Sox9 is a major negative regulator of cartilage vascularization, bone marrow formation, and endochondral

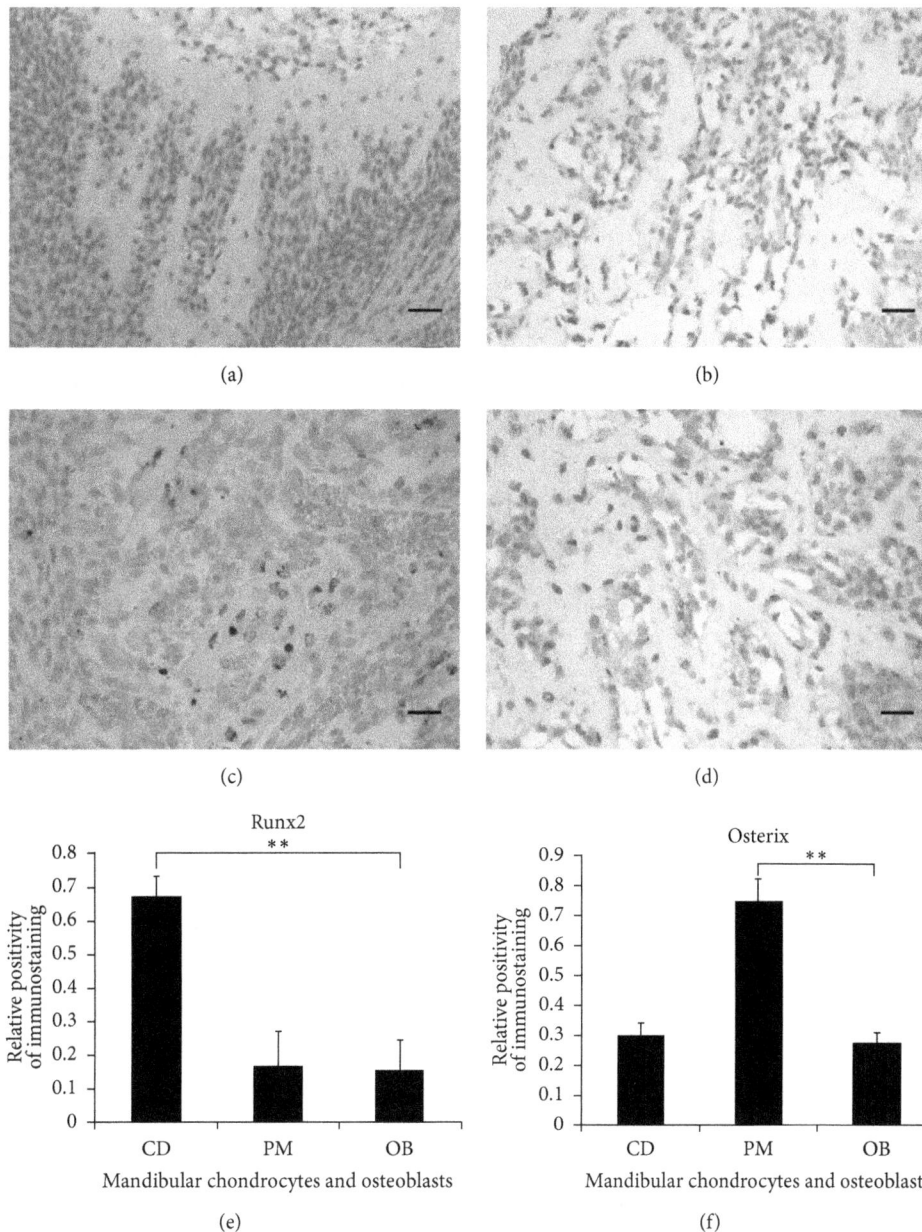

(a)

(b)

(c)

(d)

(e)

(f)

FIGURE 4: Immunohistochemistry of Runx2 and Osterix in mandibular osteoblasts of mice at newborn stage. Runx2 protein which was expressed in secondary hypertrophic chondrocytes was not localized in the cells gathering in the erosive front of mandibular secondary cartilages (a), but in some osteoblasts surrounding the trabecular bone and some osteocytes buried in the trabecular bone in the mandible (b). The expression of Runx2 protein in condylar hypertrophic chondrocytes was significantly stronger than that in osteoblasts (e). Meanwhile, Osterix was localized in some osteoblasts and bone marrow cells in subchondral bone area (c), while more positive osteoblasts and osteocytes were visualized in the trabecular bone area (d). The immunohistochemistry of Osterix in hypertrophic chondrocytes of the center of the uncalcified Meckel's cartilage is still significantly more intense than that in the osteoblasts (f). ($^{**}P < 0.001$) CD: condylar cartilage; PM: the posterior portion of intramandibular Meckel's cartilage; and OB: osteoblasts. Scale bar: 25 μm for (a, b, c, and d).

ossification [33]. Despite this observation, our investigations indicate that Sox9 downregulation is not necessary in the terminal stage of secondary cartilage development. We speculated that the transcription factor Sox9 may be a main positive regulator in the secondary cartilage terminal maturation, contrary to its function in later differentiation of limb bud cartilage, based on strong expression of Sox9 in the mandibular secondary hypertrophic cell zone.

In the present study, surprisingly, transcription factors Sox9 and Runx2 were similarly expressed at mandibular secondary cartilages in newborn mice, suggesting that Sox9 and Runx2 may coregulate secondary chondrocyte differentiation. In avian secondary cartilage formation, Buxton reported that Runx2-expressing preosteoblasts exit from the cell cycle and rapidly differentiate into hypertrophic chondrocytes, which is correlated with the up-regulation of Sox9

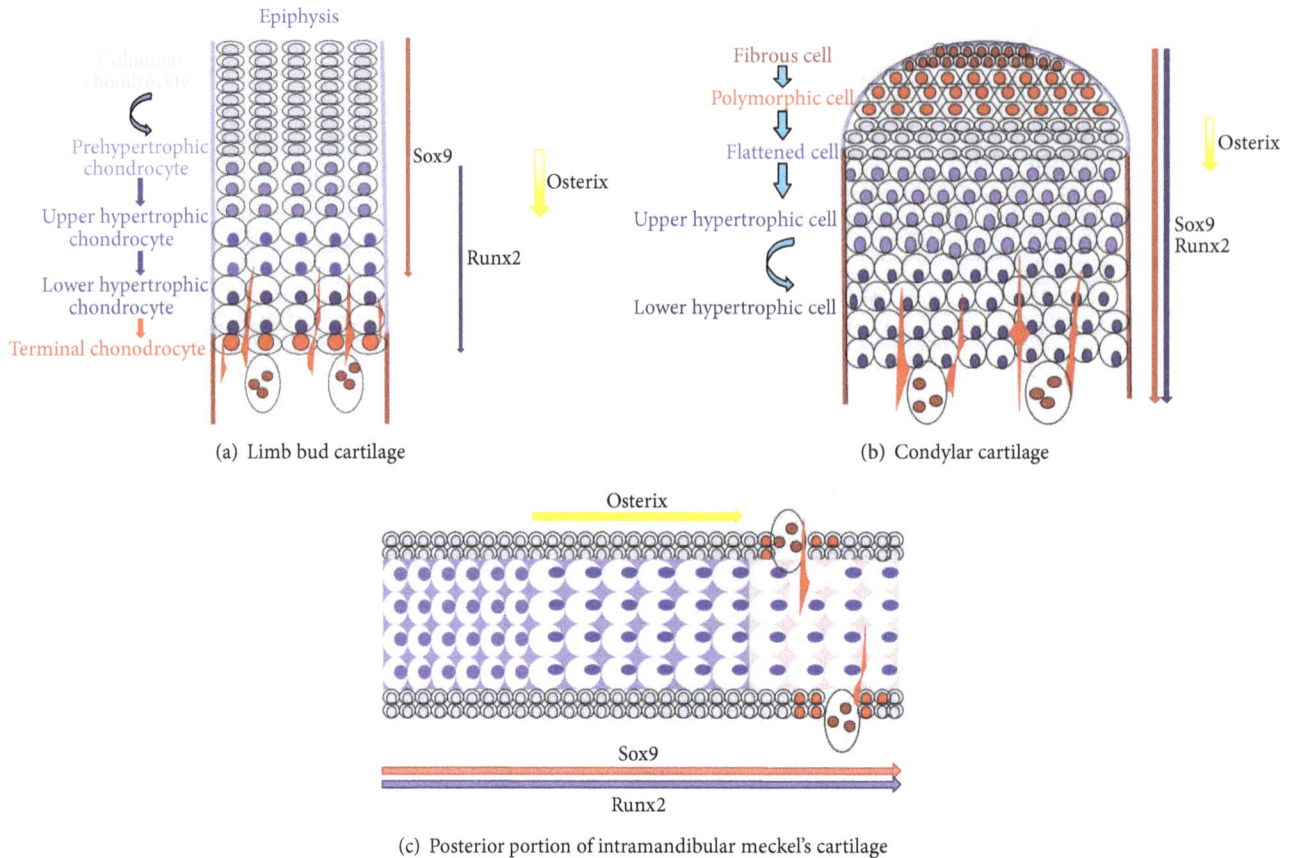

(a) Limb bud cartilage

(b) Condylar cartilage

(c) Posterior portion of intramandibular meckel's cartilage

FIGURE 5: Schematic representations of the expression pattern of three key transcription factors in the different types of cartilage during the newborn stage. The expression pattern of Sox9, Runx2, and Osterix in limb bud cartilage is based on previous reports of Kim et al. [5], Kaback et al. [6], Nishimura et al. [7], Ng et al. [8], and Dy et al. [9]. Furthermore, the expression patterns of Sox9 (red), Runx2 (blue), and Osterix (yellow) in condylar cartilage and Meckel's cartilage are based on the present findings. Long arrows indicate the expressing cell zones of transcription factors in cartilage. (a) limb bud cartilage, (b) condylar cartilage, and (c) the posterior portion of intramandibular Meckel's cartilage.

[31]. In addition, Buxton described two routes to chondrocyte hypertrophy and had postulated that precursors expressing Sox9 differentiate into prehypertrophic/hypertrophic chondrocytes mediated by the up-regulation of Runx2 in primary cartilage formation. Whereas, preosteoblasts expressing Runx2 differentiate into prehypertrophic/hypertrophic chondrocytes mediated by the upregulation of Sox9 in secondary cartilage formation [31]. Mammalian mandibular secondary cartilages are a heterogeneous tissue containing cells at various stages of chondrocyte maturation [34]. Moreover, these secondary cartilages manifest a unique zone-like packing of maturing chondrocytes [35]. Shibata and Yokohama-Tamaki recently demonstrated that the mandibular secondary cartilage anlages are derived from Runx2 mRNA expressing mandibular anlage [10]. Thus, our observations on the overlapping expression of Sox9 and Runx2 at mandibular secondary cartilages in newborn mice support Buxton's proposed concept in principle. The up-regulation of Sox9 from the polymorphic cell zone to the hypertrophic cell zone might act as a trigger for subsequent mammalian secondary chondrocyte differentiation. This can be interpreted as evidence of a unique differentiation pathway: the formation of secondary

hypertrophic chondrocytes from osteoblast precursors, with the help of the positive regulator Sox9.

More unexpectedly, our finding that Sox9 and Runx2 were coexpressed in the hypertrophic cell zone of the rostral region is not in line with the analyses of endochondral ossification in limb bud cartilage. The previous studies revealed that Sox9 inhibits the hypertrophic chondrocyte differentiation through suppression of Runx2 in endochondral ossification of limb bud cartilage [17]. Furthermore, Sox9 protein needs to be degraded to allow chondrocyte terminal maturation in limb bud cartilage [13]. However, Eames et al. had proposed that a unique combination of Sox9 and Runx2 may drive the expression of the major marker of hypertrophic chondrocytes, Col10, based on the analysis of Sox9 and Runx2 functions in primary cartilage differentiation of the avian cranial skeleton [36]. In the present study, the overlapping expression pattern of Sox9 and Runx2 in the hypertrophic cell zone of the rostral region of Meckel's cartilage provides clear evidence that Runx2 can drive the chondrocyte terminal differentiation in the presence of Sox9 protein. Additionally, our data that Sox9 and Runx2 were similarly expressed less in the hypertrophic cell zone of the posterior portion of intramandibular Meckel's cartilage has reinforced the notion that degeneration of Meckel's cartilage represents a different process from endochondral ossification.

Normally, Osterix is present at an extremely low level in prehypertrophic chondrocytes of limb bud cartilage, compared to osteoblasts [9]. To our knowledge, the present study is the first to demonstrate the expression of Osterix protein in mandibular cartilages. Osterix protein is faintly expressed in prehypertrophic chondrocytes of secondary cartilages, similar to limb bud cartilage, which suggests that Osterix plays similar roles during the two types of cartilage development. By contrast, Osterix protein is intensely expressed in hypertrophic chondrocytes in the central zone of the bars of intramandibular Meckel's cartilage, while Osterix protein is not present in cells around light eosinophilic matrix dynamically changed from the strong basophilic matrix in the front of the degrading Meckel's cartilage. The light eosinophilic matrix in front of the degrading Meckel's cartilage might display the calcified cartilage matrix [37]. A great amount of in vitro data demonstrated that the chondrocytes of Meckel's cartilage can transdifferentiate to osteogenic cells as characterized by production of type I collagen [38–40]. Furthermore, the previous in vivo investigations revealed that the extracellular matrix of intramandibular portion of the Meckel's cartilage is replaced gradually by type I collagen secreted by chondrocytes during the development of Meckel's cartilage [41]. We speculated that Osterix may be relevant to phenotypic conversion of Meckel's chondrocytes. The enhanced expression of Osterix in mature chondrocytes might be an explanation of type I collagen synthesis by chondrocytes in Meckel's cartilage. Further studies are needed to elucidate the exact role of Osterix during the late development of Meckel's cartilage. On the other hand, the disparity in the expression pattern between Osterix and Runx2 in chondrocytes in the present study, suggested that Osterix might perform its regulation and function in mandibular cartilage development, independent of Runx2. Moreover, with respect to the more remarkable expression of Runx2 in the condylar cartilage and Osterix in intramandibular degrading Meckel's cartilage relative to those in osteoblasts in the present study we speculated that Runx2 or Osterix could need much more intense expression in the chondrocytes than in the osteoblasts, in order to play a functional role during the development of mandibular cartilages.

Cartilage is a complex and developmentally important tissue type. Transcriptional factors are crucial to the development of cartilages. The differential expression of key transcriptional factors in several types of cartilages will dictate the distinct cellular events during the development of the cartilages. The present data provide insights into the similar roles that master transcriptional factors Sox9 and Runx2 play during the later development of mandibular cartilages, which is different from that in limb bud cartilage. It is necessary to investigate in further detail whether the differences in cellular events between ectomesenchymal chondrocytes and mesodermal chondrocytes involve the derivation of the cells. Furthermore, Osterix is likely responsible for phenotypic conversion of Meckel's chondrocytes during its degeneration, based on its intensive expression in hypertrophic chondrocytes of the degrading Meckel's cartilage of newborn mice. Human mandibular anomaly appears to be a common malformation and appears in multiple congenital birth defect syndromes, ranging from agnathia (agenesis of the jaw) to micrognathia to patterning malformations. These malformations are particularly devastating, as our faces are our identity [42]. The regeneration of complex facial structures requires precision and specificity. A much more thorough understanding of the mechanism of master transcriptional factors in mandibular chondrogenesis lay the important foundation for the application of targeted interventions at the molecular level, endogenous tissue engineering, and cell-based therapies in mandibular anomalies.

5. Conclusions

Our study demonstrated similar tissue distribution of Sox9 and Runx2 in newborn mice mandibular cartilages, which is distinguished from that in limb bud cartilage. It is speculated that Sox9 is a main and unique positive regulator in the hypertrophic differentiation process of mandibular secondary cartilages, in addition to Runx2. Moreover, the distinct expression pattern of osterix in degenerating posterior portion of Meckel's cartilage suggests that Osterix may be relevant to phenotypic conversion of Meckel's chondrocytes.

Conflict of Interests

The authors declare that they have no conflict of interests.

Acknowledgment

The authors thank the support of the National Natural Science Foundation of China (no. 10972242).

References

[1] Y. Sakakura, Y. Hosokawa, E. Tsuruga, K. Irie, M. Nakamura, and T. Yajima, "Contributions of matrix metalloproteinases toward Meckel's cartilage resorption in mice: immunohistochemical studies, including comparisons with developing endochondral bones," *Cell and Tissue Research*, vol. 328, no. 1, pp. 137–151, 2007.

[2] S. Shibata, K. Fukada, S. Suzuki, and Y. Yamashita, "Immunohistochemistry of collagen types II and X, and enzyme-histochemistry of alkaline phosphatase in the developing condylar cartilage of the fetal mouse mandible," *Journal of Anatomy*, vol. 191, no. 4, pp. 561–570, 1997.

[3] S. Shibata, N. Suda, S. Suzuki, H. Fukuoka, and Y. Yamashita, "An in situ hybridization study of Runx2, Osterix, and Sox9 at the onset of condylar cartilage formation in fetal mouse mandible," *Journal of Anatomy*, vol. 208, no. 2, pp. 169–177, 2006.

[4] M. S. Kim, S. Y. Jung, J. H. Kang et al., "Effects of bisphosphonate on the endochondral bone formation of the mandibular condyle," *Anatomia, Histologia, Embryologia*, vol. 38, no. 5, pp. 321–326, 2009.

[5] I. S. Kim, F. Otto, B. Zabel, and S. Mundlos, "Regulation of chondrocyte differentiation by Cbfa1," *Mechanisms of Development*, vol. 80, no. 2, pp. 159–170, 1999.

[6] L. A. Kaback, D. Y. Soung, A. Naik et al., "Osterix/Sp7 regulates mesenchymal stem cell mediated endochondral ossification," *Journal of Cellular Physiology*, vol. 214, no. 1, pp. 173–182, 2008.

[7] R. Nishimura, M. Wakabayashi, K. Hata et al., "Osterix regulates calcification and degradation of chondrogenic matrices through matrix metalloproteinase 13 (MMP13) expression in association with transcription factor Runx2 during endochondral ossification," *The Journal of Biological Chemistry*, vol. 287, no. 40, pp. 33179–33190, 2012.

[8] L. J. Ng, S. Wheatley, G. E. O. Muscat et al., "SOX9 binds DNA, activates transcription, and coexpresses with type II collagen during chondrogenesis in the mouse," *Developmental Biology*, vol. 183, no. 1, pp. 108–121, 1997.

[9] P. Dy, W. Wang, P. Bhattaram et al., "Sox9 directs hypertrophic maturation and blocks osteoblast differentiation of growth plate chondrocytes," *Developmental Cell*, vol. 22, no. 3, pp. 597–609, 2012.

[10] S. Shibata and T. Yokohama-Tamaki, "An in situ hybridization study of Runx2, Osterix, and Sox9 in the anlagen of mouse mandibular condylar cartilage in the early stages of embryogenesis," *Journal of Anatomy*, vol. 213, no. 3, pp. 274–283, 2008.

[11] L. Zou, X. Zou, H. Li et al., "Molecular mechanism of osteo-chondroprogenitor fate determination during bone formation," *Advances in Experimental Medicine and Biology*, vol. 585, no. 7, pp. 431–441, 2006.

[12] W. Bi, J. M. Deng, Z. Zhang, R. R. Behringer, and B. de Crombrugghe, "Sox9 is required for cartilage formation," *Nature Genetics*, vol. 22, no. 1, pp. 85–89, 1999.

[13] H. Akiyama, M. C. Chaboissier, J. F. Martin, A. Schedl, and B. de Crombrugghe, "The transcription factor Sox9 has essential roles in successive steps of the chondrocyte differentiation pathway and is required for expression of Sox5 and Sox6," *Genes & Development*, vol. 16, no. 21, pp. 2813–2828, 2002.

[14] H. Akiyama, J. P. Lyons, Y. Mori-Akiyama et al., "Interactions between Sox9 and β-catenin control chondrocyte differentiation," *Genes & Development*, vol. 18, no. 9, pp. 1072–1087, 2004.

[15] M. Mikasa, S. Rokutanda, H. Komori et al., "Regulation of Tcf7 by Runx2 in chondrocyte maturation and proliferation," *Journal of Bone and Mineral Metabolism*, vol. 29, no. 3, pp. 291–299, 2011.

[16] M. Ding, Y. Lu, S. Abbassi et al., "Targeting Runx2 expression in hypertrophic chondrocytes impairs endochondral ossification during early skeletal development," *Journal of Cellular Physiology*, vol. 227, no. 10, pp. 3446–3456, 2012.

[17] S. Yamashita, M. Andoh, H. Ueno-Kudoh, T. Sato, S. Miyaki, and H. Asahara, "Sox9 directly promotes Bapx1 gene expression to repress Runx2 in chondrocytes," *Experimental Cell Research*, vol. 315, no. 13, pp. 2231–2240, 2009.

[18] H. Sugito, Y. Shibukawa, T. Kinumatsu et al., "Ihh signaling regulates mandibular symphysis development and growth," *Journal of Dental Research*, vol. 90, no. 5, pp. 625–631, 2011.

[19] F. Tsuzurahara, S. Soeta, T. Kawawa, K. Baba, and M. Nakamura, "The role of macrophages in the disappearance of Meckel's cartilage during mandibular development in mice," *Acta Histochemica*, vol. 113, no. 2, pp. 194–200, 2011.

[20] Y. Sakakura, Y. Hosokawa, E. Tsuruga, K. Irie, and T. Yajima, "In situ localization of gelatinolytic activity during development and resorption of Meckel's cartilage in mice," *European Journal of Oral Sciences*, vol. 115, no. 3, pp. 212–223, 2007.

[21] B. Rath-Deschner, N. Daratsianos, S. Dührr et al., "The significance of RUNX2 in postnatal development of the mandibular condyle," *Journal of Orofacial Orthopedics*, vol. 71, no. 1, pp. 17–31, 2010.

[22] S. Matsushima, N. Isogai, R. Jacquet et al., "The nature and role of periosteum in bone and cartilage regeneration," *Cells Tissues Organs*, vol. 194, no. 2–4, pp. 320–325, 2011.

[23] S. Shibata, N. Suda, S. Yoda et al., "Runx2-deficient mice lack mandibular condylar cartilage and have deformed Meckel's cartilage," *Anatomy and Embryology*, vol. 208, no. 4, pp. 273–280, 2004.

[24] K. Ishii, I. L. Nielsen, and K. Vargervik, "Characteristics of jaw growth in cleidocranial dysplasia," *The Cleft Palate-Craniofacial Journal*, vol. 35, no. 2, pp. 161–1666, 1998.

[25] B. L. Jensen, "Cleidocranial dysplasia: craniofacial morphology in adult patients," *Journal of Craniofacial Genetics and Developmental Biology*, vol. 14, no. 3, pp. 163–176, 1994.

[26] X. L. Zhang, K. Ting, C. M. Bessette et al., "Nell-1, a key functional mediator of Runx2, partially rescues calvarial defects in Runx2+/- mice," *Journal of Bone and Mineral Research*, vol. 26, no. 4, pp. 777–791, 2011.

[27] A. B. M. Rabie, G. H. Tang, and U. Hägg, "Cbfa1 couples chondrocytes maturation and endochondral ossification in rat mandibular condylar cartilage," *Archives of Oral Biology*, vol. 49, no. 2, pp. 109–118, 2004.

[28] B. F. Eames and J. A. Helms, "Conserved molecular program regulating cranial and appendicular skeletogenesis," *Developmental Dynamics*, vol. 231, no. 1, pp. 4–13, 2004.

[29] M. Wuelling and A. Vortkamp, "Transcriptional networks controlling chondrocyte proliferation and differentiation during endochondral ossification," *Pediatric Nephrology*, vol. 25, no. 4, pp. 625–631, 2010.

[30] V. Lefebvre and P. Smits, "Transcriptional control of chondrocyte fate and differentiation," *Birth Defects Research C*, vol. 75, no. 3, pp. 200–212, 2005.

[31] P. G. Buxton, B. Hall, C. W. Archer, and P. Francis-West, "Secondary chondrocyte-derived lhh stimulates proliferation of periosteal cells during chick development," *Development*, vol. 130, no. 19, pp. 4729–4739, 2003.

[32] A. B. M. Rabie, T. T. She, and U. Hägg, "Functional appliance therapy accelerates and enhances condylar growth," *American Journal of Orthodontics and Dentofacial Orthopedics*, vol. 123, no. 1, pp. 40–48, 2003.

[33] T. Hattori, C. Müller, S. Gebhard et al., "SOX9 is a major negative regulator of cartilage vascularization, bone marrow formation and endochondral ossification," *Development*, vol. 137, no. 6, pp. 901–911, 2010.

[34] J. Chen, A. Utreja, Z. Kalajzic, T. Sobue, D. Rowe, and S. Wadhwa, "Isolation and characterization of murine mandibular condylar cartilage cell populations," *Cells, Tissues, Organs*, vol. 195, no. 3, pp. 232–243, 2012.

[35] G. Shen and M. A. Darendeliler, "The adaptive remodeling of condylar cartilage—a transition from chondrogenesis to osteogenesis," *Journal of Dental Research*, vol. 84, no. 8, pp. 691–699, 2005.

[36] B. F. Eames, P. T. Sharpe, and J. A. Helms, "Hierarchy revealed in the specification of three skeletal fates by Sox9 and Runx2," *Developmental Biology*, vol. 274, no. 1, pp. 188–200, 2004.

[37] K. Ishizeki, H. Saito, T. Shinagawa, N. Fujiwara, and T. Nawa, "Histochemical and immunohistochemical analysis of the mechanism of calcification of Meckel's cartilage during mandible development in rodents," *Journal of Anatomy*, vol. 194, no. 2, pp. 265–277, 1999.

[38] K. Ishizeki, N. Takahashi, and T. Nawa, "Formation of the sphenomandibular ligament by Meckel's cartilage in the mouse:

possible involvement of epidermal growth factor as revealed by studies in vivo and in vitro," *Cell and Tissue Research*, vol. 304, no. 1, pp. 67–80, 2001.

[39] K. Ishizeki, T. Kagiya, N. Fujiwara, K. Otsu, and H. Harada, "Expression of osteogenic proteins during the intrasplenic transplantation of Meckel's chondrocytes: a histochemical and immunohistochemical study," *Archives of Histology and Cytology*, vol. 72, no. 1, pp. 1–12, 2009.

[40] K. Ishizeki, "Imaging analysis of osteogenic transformation of Meckel's chondrocytes from green fluorescent protein-transgenic mice during intrasplenic transplantation," *Acta Histochemica*, vol. 114, no. 6, pp. 608–619, 2012.

[41] Y. Harada and K. Ishizeki, "Evidence for transformation of chondrocytes and site-specific resorption during the degradation of Meckel's cartilage," *Anatomy and Embryology*, vol. 197, no. 6, pp. 439–450, 1998.

[42] Y. Chai and R. E. Maxson Jr., "Recent advances in craniofacial morphogenesis," *Developmental Dynamics*, vol. 235, no. 9, pp. 2353–2375, 2006.

16

The Role of Changes in Extracellular Matrix of Cartilage in the Presence of Inflammation on the Pathology of Osteoarthritis

Maricela Maldonado[1] and Jin Nam[1,2]

[1] Department of Bioengineering, University of California, 900 University Avenue, Riverside, CA 92521, USA
[2] Center for Bioengineering Research, University of California, Riverside, CA 92521, USA

Correspondence should be addressed to Jin Nam; jnam@engr.ucr.edu

Academic Editor: Martin Götte

Osteoarthritis (OA) is a degenerative disease that affects various tissues surrounding joints such as articular cartilage, subchondral bone, synovial membrane, and ligaments. No therapy is currently available to completely prevent the initiation or progression of the disease partly due to poor understanding of the mechanisms of the disease pathology. Cartilage is the main tissue afflicted by OA, and chondrocytes, the sole cellular component in the tissue, actively participate in the degeneration process. Multiple factors affect the development and progression of OA including inflammation that is sustained during the progression of the disease and alteration in biomechanical conditions due to wear and tear or trauma in cartilage. During the progression of OA, extracellular matrix (ECM) of cartilage is actively remodeled by chondrocytes under inflammatory conditions. This alteration of ECM, in turn, changes the biomechanical environment of chondrocytes, which further drives the progression of the disease in the presence of inflammation. The changes in ECM composition and structure also prevent participation of mesenchymal stem cells in the repair process by inhibiting their chondrogenic differentiation. This review focuses on how inflammation-induced ECM remodeling disturbs cellular activities to prevent self-regeneration of cartilage in the pathology of OA.

1. Introduction

Osteoarthritis (OA) is a debilitating disease, which primarily affects joints, especially load-bearing areas such as hips and knees. It is characterized by pain and degenerative changes in the tissues surrounding those areas. There are no current therapies which can completely prevent the progression of the disease. Some of the main factors that drive the progression of OA are chronic inflammation and gradual structural changes within the joint tissues [1]. Unlike the general concept of OA being a degenerative disease, the remodeling processes are highly active throughout each stage of the disease [2]. During the active remodeling, however, the quality of extracellular matrix (ECM) is compromised due to the quick turnover rate and atypical composition of the newly synthesized ECMs [3]. Among many factors, inflammatory cytokines and proteases are main contributors which mediate the changes in the quality of ECM [2]. As a consequence of the microenvironmental changes, the altered ECM synthesis in the presence of

inflammation, in turn, further disturbs the functions of the cells. Therefore, there is a constant cycle of evolution between the cells and their newly synthesized ECM, forming a positive feedback loop, which drives the progression of OA. In this review, we will focus on the interplay between ECM and cellular functions under inflammation, and how these factors are responsible for the progression of OA. An understanding of the complexity of the interplay between the cells and their microenvironment may provide a sound basis for developing suitable therapies to treat osteoarthritis.

2. Changes in Extracellular Matrix Synthesis during Osteoarthritis

Progression of OA can be characterized by changes in ECM composition and structure. Natural, healthy cartilage matrix is mainly composed of collagen type II which provides tensile support for the tissue. Aggrecan, a negatively charged proteoglycan that attracts water molecules, provides the compressive resistant and shock absorbing capability of cartilage under loading [2]. It has been shown that during OA, there are sequential events that affect the integrity of homeostatic ECM; aggrecan content is decreased, while collagen content is increased [2, 3, 5]. This change in ECM composition predisposes the tissue for mechanical fault resulting in significantly altered mechanical environments of the cells within the cartilage matrix.

In the initial stages of OA, proliferative chondrocytes form clusters in order to adjust to the changing microenvironments [2]. This alteration of cellular configuration also changes the quantity and composition of the ECM secreted by the cells. It has been shown that there is a significant downregulation of aggrecan gene expression at the onset of OA in a rat model [1], and this finding agrees with markedly low proteoglycan synthesis, observed in human OA samples with normal appearance [6]. The changes of aggrecan, which exists in a nonaggregated form in OA, alter the permeability and thus mechanical compliance of the matrix [2, 7]. The reduced proteoglycan content decreases compressive modulus of cartilage and consequently exposes the tissue to greater strains when exposed to mechanical stress.

Unlike the decreased production of proteoglycan, collagen synthesis rate increases in the early stages of OA and remains elevated [8]. In addition to the increased ratio of collagen/aggrecan synthesis, the composition of collagen type has been also shown to change from collagen type II to type I [9]. Healthy cartilage matrix mainly contains collagen type II, while collagen type I is mainly found in subchondral bone tissue [2, 3, 10]. The compositional change affects the mechanical stability of the ECM network [10]. Compared to collagen type I, type II chains contain a higher content of hydroxylysine as well as glucosyl and galactosyl residues which mediate the interaction with proteoglycans [11]. Therefore, the decreased collagen type II content during OA inevitably undermines the integrity of ECM networks formed by collagen and proteoglycan. Furthermore, Silver et al. showed that the elastic modulus, due to shortened collagen fibril lengths, decreases with an increased extent of OA [12]. As a result of these changes, the osteoarthritic cartilaginous tissues exhibit a reduced ability to store elastic energy, and this, in turn, leads to fibrillation and fissure formation [12]. Figure 1 shows the structural and compositional changes in cartilage in a monoiodoacetate- (MIA-) induced arthritis model in rats. Although the animal model induces significantly accelerated cartilage degeneration as compared to typical human osteoarthritis, it depicts similar structural and compositional changes in cartilage exhibited in the pathogenesis of OA [13]. On day 11 post-MIA injection, the overall cartilage damage was assessed at Grade 2-3 according to Osteoarthritis Research Society International's (OARSI's) histopathology grading system showing cartilage lesion formation, articular surface fissurization, subchondral bone advancement, and bone marrow edema/cyst [14]. An area exhibiting chondrocyte disorientation without vertical fissure development was chosen to observe changes in cartilage matrix. In this area, nonchondrocytic collagen type I

is present in the cartilage matrix of the OA tissue, whereas it is negligible in the control (Figure 1(B)). These changes in the structure and composition of ECM progressively alter the biological and mechanical microenvironments that significantly modulate cellular activities as described later in this review.

3. Inflammation-Induced Extracellular Matrix Changes in Osteoarthritis

ECM changes in cartilage can be attributed to multiple factors during the progression of OA. Among them, inflammation plays an active role affecting both quantity and quality of ECMs. Mechanical damage and/or age-related wear/tear are thought to trigger systematic inflammatory responses in all tissues surrounding the joint including articular cartilage, synovial membrane, subchondral bone, and ligaments [2, 15]. Chondrocytes, the only cell type residing in cartilage, respond to such inflammatory conditions and participate in the catabolic activities that ultimately lead to the degradation of cartilaginous ECM [16]. An animal model of MIA-induced arthritis showed that the sequential upregulation of inflammatory genes is associated with all levels of cartilage damage throughout the progression of OA [1]. These upregulated inflammatory genes form a positive feedback loop, mainly through the NF-κB signaling pathway, as the severity of the cartilage damage progresses [17]. In fact, it was observed that chondrocytes in human arthritic cartilages also constitutively exhibit elevated activities of NF-κB [18]. Factors that contribute to the catabolic processes in OA include interleukin 1β (IL-1β), tissue necrosis factor-α (TNF-α), IL-12, IL-15, and various associated chemokines [19–23]. These inflammatory factors were shown to significantly increase the expression of matrix degrading proteins including matrix metalloproteinases (MMPs) (i.e., MMP-1 and MMP-13) and various types of a disintegrin and metalloproteinase with a thrombospondin type 1 motif (ADAMTS) (i.e., ADAMTS 1,4,5) in chondrocytes [1, 24–30]. For example, an increase in cell clustering, a typical morphological feature of chondrocytes in the early stage of OA, was observed with an increase in MMP-13 expression [31]. The receptor for advanced glycation end products (RAGE), which is increased in OA articular chondrocytes, was also shown to stimulate MAP kinase and NF-κB activities that, in turn, increased the production of MMP-13 and propagated the catabolism of the cartilage matrix [32, 33].

The degenerative activities of matrix degrading proteins are intensified by the elevated level of nitric oxide (NO), a molecule which is also upregulated by inflammatory proteins in chondrocytes. NO, upregulated by the transcriptional activity of NF-κB, perpetuates the chronic inflammation that enhances matrix degradation and mediates apoptosis of chondrocytes by creating oxidative environments [34–36]. In a canine model of OA, the use of a NO inhibitor reduced the degenerative changes in cartilage, possibly demonstrating the critical role of NO in the progression of OA [34].

Concurrently with matrix degradation, the inflammation-mediated downregulation of chondrogenic growth/

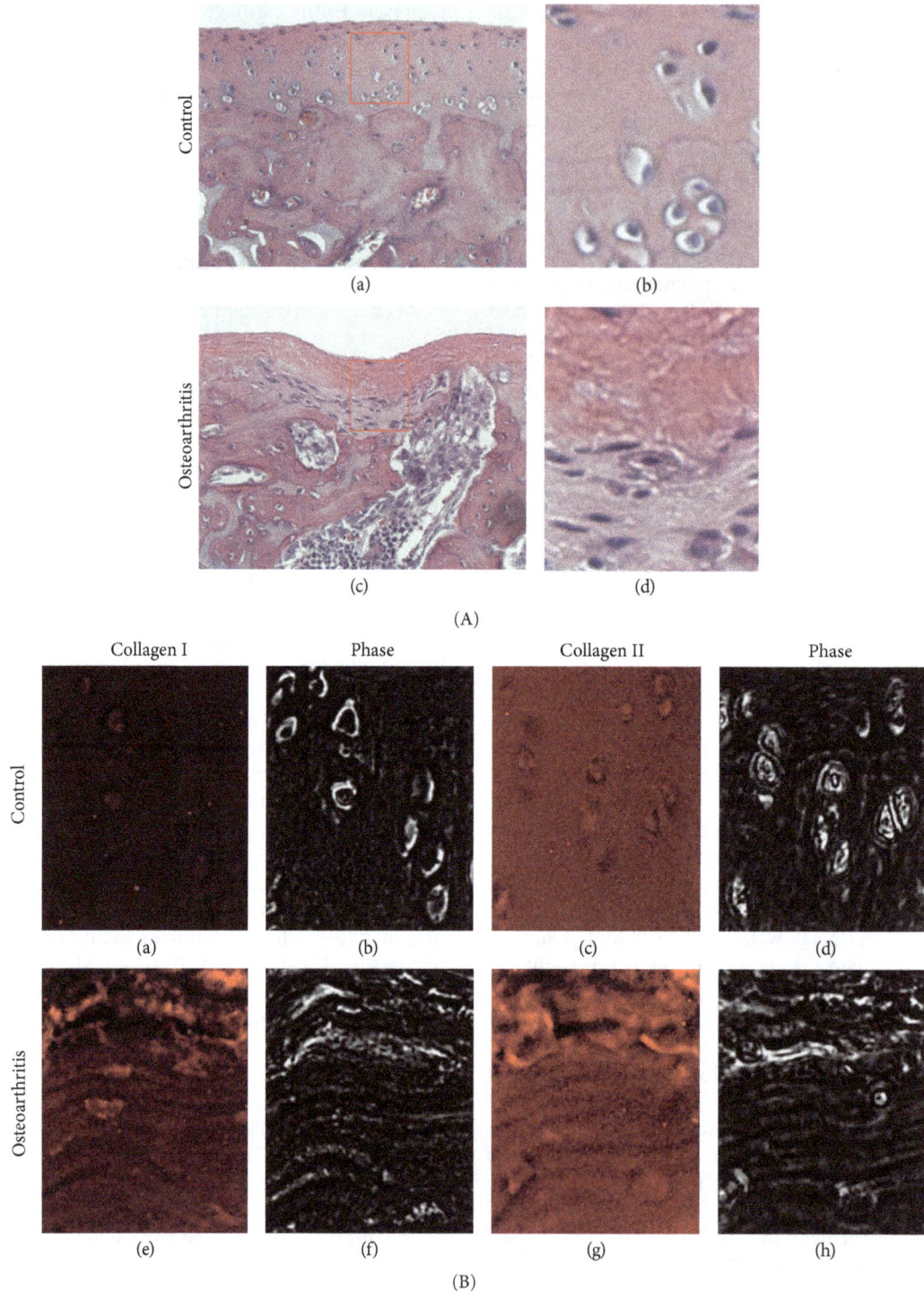

FIGURE 1: Changes in the extracellular matrix structure and composition of cartilage afflicted by osteoarthritis (OA). Experimental OA was induced by intra-articular injection of monoiodoacetate (MIA) similar to the previously described protocol using a rat model [4]. OA induced rats were sacrificed at day 11, and the medial condyles of the arthritic knees (A (c-d); B (e–h)) were histologically (H&E staining (A)) and immunohistologically (collagen type I (B (a) and (e)) and type II (Figure B (c) and (g)) compared to that of the saline-injected sham control ((A (a-b); B (a–d)). (A) Microscopic features of OA cartilage (grade 2-3) show cartilage lesion formation, articular surface fissurization, subchondral bone advancement, and bone marrow edema/cyst. In addition, cell clustering and fibrocartilage formation is apparent in OA samples. (A) (b and d) are magnified images of the area indicated in (A) (a and c), respectively, to reveal the changes in cellular morphology. (B) Consecutive sections of the healthy and OA cartilages were stained using monoclonal antibodies for collagen type I or type II. An increase in intensity for collagen type I is observed in the OA cartilage, while it is not present in the control cartilage. Collagen type II is readily observed for both the healthy and OA cartilage. (B) (b, d, f, and h) are phase-contrast images of (B) (a, c, e, and g), respectively, to reveal tissue morphologies.

transcription factors that mediate chondrocytic ECM synthesis, such as transforming growth factor β (TGF-β), sex determining region Y-box 9 (SOX9), insulin-like growth factor (IGF), and connective tissue growth factor (CTGF), is also responsible for suppressing the anabolic activities of chondrocytes [1, 37, 38]. Taken together, these results demonstrate the significant influence of inflammatory mediators in the progression of OA by altering the homeostasis of cartilage ECM.

Another matrix component which is found in increased concentrations in synovial fluid during OA is Tenascin-C (TN-C), an ECM glycoprotein. Elevated levels of TN-C have been suggested to induce inflammatory mediators and promote ECM degradation in OA patients [39]. Although TN-C is highly expressed during embryogenesis, its presence is minimal in healthy adult tissues. Its expression during OA is, however, highly upregulated [40, 41]. The elevated concentration of TN-C causes a significant effect in the catabolism of the cartilage, resulting in degradation of ECM [39, 40]. Additionally, biglycan fragments in articular cartilage and meniscus and fibronectin fragments in hip and knee synovia have also been found in elevated levels as OA progresses [42–44]. Both fragmented biglycan and fibronectin exhibit proinflammatory effects through the activation of toll-like receptors [45, 46]. Overall, the combination of inflammation-induced upregulation of matrix-degrading proteins, downregulation of chondrocytic ECM synthesis, and accelerated matrix degradation due to fragmented inflammatory ECMs, promotes the progression of disease.

4. Alteration in Biomechanical Environments during Osteoarthritis

The changes in altered ECM synthesis and elevated activities of matrix degrading proteins drastically change the mechanical properties of cartilage, which further intensifies the destructive processes associated with OA [47]. Initially, an increase in cartilage thickness is observed by hyperproliferative chondrocytes before noticeable surface fibrillation occurs [48]. The highly proliferating chondrocytes produce greater amount of aggrecan that leads to cartilage thickening in dimensions as well as softening of extracellular matrix [2]. At this stage, a lower shear modulus was observed in the cartilage from an OA model when compared to normal articular cartilage [49, 50]. In a mouse model, a reduction in tensile stiffness in articular cartilage is also accompanied by the tentative cartilage thickening [51]. These biomechanical changes expose chondrocytes to an environment more susceptible to greater strains, as compared to physiological levels, thus altering their cellular functions.

As the disease progresses, however, the tissue gradually loses aggrecan content, which has provided compliance of local mechanical environments due to its ability to interact with water molecules. In addition to aggrecan loss, it has been recently shown that collagen fibril stiffens in osteoarthritic cartilage [52]. Furthermore, another possible mechanism through which the mechanical microenvironment changes is the accumulation of advanced glycation end products (AGEs)

which can crosslink to the collagen network [53]. In vitro, the increased AGE crosslinking to the collagen network was shown to increase the stiffness of human adult articular cartilage [53]. The combination of aggrecan loss and collagen network stiffening results in increased overall stiffness of the tissue. Consequently, as OA advances, the cartilage layer becomes thinner and stiffer transmitting greater load to the underlying subchondral bones. The change in mechanical conditions induces the advancement of subchondral bones towards the articular surface leading to the development of bone marrow edema/subchondral bone cysts and the propagation of periarticular osteophytes [2, 54, 55]. Recent studies suggest that these changes in subchondral bone structure may precede the articular cartilage thinning [56].

Nevertheless, due to changes in the mechanical properties of the cartilage via altered homeostasis of ECM, its residing cells, chondrocytes, are exposed to vastly different biomechanical microenvironments that further intensify the progression of OA by altering cellular behaviors. Ultimately, this leads to the formation of fibrocartilaginous tissues that exhibit more bone-like properties replacing the completely degenerated cartilage in addition to osteophyte formation at the periphery of the articular surface [2, 54].

5. The Effects of Inflammation on Cartilage Extracellular Matrix Homeostasis by Articular Chondrocytes

Global inflammation in synovium during OA affects chondrocytes that are responsible for ECM turnover and thus cartilage homeostasis [57]. Inflammation which is persistent in OA has shown to directly induce the catabolic activities of chondrocytes. IL-1β, a highly upregulated cytokine during OA, has shown to induce upregulation of matrix degrading enzymes such as MMP-1, 3, and 13 in chondrocytes [58]. Dozin et al. also showed that when exposed to inflammatory cytokines, chondrocytes, regardless of patient age or OA status of human donors, enhance their production of proinflammatory cytokines such as IL-6 and IL-8 [59]. TNF-α, another critical cytokine that is highly upregulated in OA, has been shown to induce MMP-13 expression, mediated by ERK, p38, JNK MAP kinases, and AP-1 and NF-κB transcriptions factors [24, 60, 61]. At the same time, the presence of inflammatory cytokine IL-1β has been shown to play a role in suppressed ECM synthesis through downregulation of SOX9 [62]. This, in turn, decreases the expression of collagen type II and aggrecan in articular chondrocytes. The activation/suppression of such signaling cascades autoregulates chondrocytes to further upregulate the synthesis of matrix degrading enzymes and downregulate the production of chondrocytic ECMs [63]. Nitric oxide (NO) and cyclooxygenase-2 (COX-2), two components which have active roles in perpetuating inflammation, were also endogenously expressed at high levels in chondrocytes from OA tissues even when cultured in vitro in the absence of inflammatory cytokines [64, 65]. These changes in metabolism may demonstrate a possible permanent phenotypical change in the OA chondrocytes.

In this regard, one notable alteration of chondrocytes in arthritic joints is their production of nonchondrocytic ECM. In addition to the increase in the production of collagen type I replacing type II as previously described, chondrocytes isolated from OA diseased tissues have shown to produce collagen type X, a marker for hypertrophic chondrocytes, as compared to undetectable expression of the protein in healthy cartilage [66]. Collagen type X is typically synthesized by hypertrophic chondrocytes that also produce collagen type I. The emergence of these nonarticular chondrocytic proteins may indicate the change of phenotype in chondrocytes as the disease progresses. The morphological change of chondrocytes with abnormal nonround morphology in arthritic cartilages could be related to a phenotypical change such as an increase in IL-1β production and a decrease in pericellular collagen type VI synthesis [67]. When the cells from arthritic knees are subject to a chondrogenic in vitro culture condition, they are not able to fully recover normal tissue phenotype as evident by low cellularity and decreased chondrocytic ECM production as compared to chondrocytes from healthy joints [66, 67]. This demonstrates that damages in OA cartilage may not be able to be fully recovered by autologous chondrocytes.

One possible cause of the phenotype change of OA chondrocytes is inflammation as inflammatory synovial fluid has shown to activate chondrocytes and dramatically affect the normal processes of the cells. When healthy chondrocytes are subjected to inflammation, simulated by inflammatory cytokines such as IL-1β, TNF-α, CXCL1, or 8, all of which are upregulated during OA, the cells exhibit hypertrophic differentiation [68]. This differentiation is shown to be mediated by RAGE signaling through the p38 MAPK pathway [69]. Interestingly, the activation of the p38 MAPK signaling pathway has also shown to promote the synthesis of MMP-13 possibly linking the change in phenotype to the facilitated rate of matrix turnover [32]. In addition to the synthesis of nonchondrocytic ECM and enhancement in matrix degradation, chronic inflammation also induces cell death. When healthy chondrocytes were subject to synovial fluids from osteoarthritic patients, the cells not only upregulated the expression of cytokines, such as IL-6, IL-8, monocyte chemotactic protein-1 (MCP-1), and vascular endothelial growth factor (VEGF), but also underwent apoptosis [16].

6. The Effects of Changes in Extracellular Matrix on Articular Chondrocytes

The altered microenvironments by ECM changes, in the presence of inflammation, further drive catabolic/nonreparative activities of chondrocytes, ultimately leading to cartilage destruction/achondrocytic ECM formation. As previously described, the mechanical properties of cartilage are dynamically altered during the progression of OA due to imbalanced matrix turnover (greater matrix degradation versus synthesis) and noncartilaginous ECM formation. The increase in local matrix stiffness due to changes in ECM appears to suppress chondrocytic activities of the cells. Recent studies show that chondrocytes sense the stiffness of the matrix and differentially respond to it by altering their phenotype,

resulting in production of different types of ECM (i.e., ratio of collagen type II to type I) [70–72]. An optimal stiffness has been shown to promote greater SOX9, COL2A1, and aggrecan gene expression in chondrocytes and either above or below this stiffness induced dedifferentiation of the cells towards fibrochondrocytic phenotype [70]. This effect of matrix stiffness on modulating chondrogenic phenotype has been shown to occur through the regulation of the TGF-β signaling pathway [70]. In addition, the mechanosensitive behavior of chondrocytes may explain the fact that typical in vitro 2D culture of chondrocytes on stiff tissue culture plastics results in the dedifferentiation of the cells [73–75].

The changes in matrix composition during OA not only affect the mechanical environments of chondrocytes but also alter interactions of matrix proteins with the cells. Matrilin-3 (MATN3) is a matrix protein that is highly upregulated during OA [76, 77]. Although the protein is a part of healthy cartilage matrix, the soluble form of MATN3 is upregulated and released to synovial fluid in OA [78]. When human chondrocytes were cultured in the presence of soluble MATN3, there was a decrease in ECM anabolism and increased catabolism only at concentrations higher than those found in OA patients. On the other hand, when soluble MATN3 was immobilized, ECM synthesis and accumulation was enhanced [78]. These results show how MATN3, which is found in synovial fluid of OA patients, can change the behavior of chondrocytes, demonstrating the direct involvement of ECM in the progression of OA by interacting with the cells as well as indirectly by changing the cells' mechanical environments.

The presence of calcium crystals in cartilage has been shown to increase with severity of OA, and these changes have a strong correlation with hypertrophic chondrocyte differentiation [79]. Interestingly, bovine articular chondrocytes within cartilage explants, when exposed to basic calcium phosphate crystals, had significant increases in intracellular calcium content, which is correlated with cartilage matrix degradation [80]. Another ECM component that affects chondrocyte metabolism is fibronectin, which showed a significant positive correlation between chondrocyte apoptosis and fibronectin content [81]. Overall, these multifaceted effects by changes in ECM, including dysregulation of matrix synthesis (reduction in collagen type II and aggrecan, increase in collagen type I and X), upregulation of matrix degradation, and induction of cell apoptosis, promote the progression of OA by altering the cellular behaviors of chondrocytes.

7. The Effects of Inflammation on Chondrogenic Differentiation of Mesenchymal Stem Cells during Osteoarthritis

The mechanisms involving the initiation of OA are still elusive as some argue it is mechanical damage-induced and others inflammation-induced. Nevertheless, once the disease is initiated, the degeneration of cartilage matrix progresses due to the combination of chronic inflammation and altered mechanical loading as discussed earlier. A part of the progressive degenerative processes is due to the limited regenerative capability of chondrocytes. These cells are typically quiescent in healthy cartilage [2]. When they are exposed to proliferating conditions to repair the cartilage damage, they often dedifferentiate to a phenotype that produces nonchondrocytic ECM [2]. This atypical ECM synthesis further drives chondrocyte dedifferentiation and nonhomeostatic ECM synthesis by altered mechanical environments. In addition to chondrocytes, the repair of the damaged tissue is attempted by another cell type, mesenchymal stem cell (MSC), that can differentiate to all mesenchymal lineage cells including chondrocyte, osteoblast, and adipocyte [82]. MSCs often participate in the repair of bone damage since they constitute bone marrow. Due to its close proximity to the cartilage layer in the subchondral marrow and their ability to differentiate into chondrocytes, MSCs have been considered as a possible cell source involved in cartilage repair.

For this reason, microfracture (or microperforation) surgery is often used to treat a localized cartilage lesion. Small fractures are created in the subchondral bone, and this causes new cartilage formation mainly due to the regenerative activities of MSCs from the bone marrow [83]. Although this technique has shown some benefits repairing damaged cartilage, the neotissue contains fibrocartilage that exhibits different mechanical properties, leading to question its long-term stability [84, 85]. These studies may provide clues for why endogenous MSCs cannot fully rescue damaged cartilage during the progression of osteoarthritis, unlike the positive healing response after bone fractures. Typically, subchondral bone advances towards the cartilage surface as the articular surface degrades [86]. In this condition, MSCs are subjected to a milieu of inflammation, altered ECM composition, and vastly different mechanical loading profiles in the injured cartilage, all of which affect the differentiation of MSCs to chondrocytes.

As described earlier, the native cartilage is exposed to chronic inflammation conditions by increased levels of inflammatory mediators including IL-1β, TNF-α, and prostaglandin E$_2$ (PGE$_2$) [87, 88]. These inflammatory cytokines not only affect the homeostatic functions of residential chondrocytes but also impact the chondrogenic differentiation of MSCs [87, 89–91]. Treatment of IL-1β during chondrogenic differentiation of bone marrow-derived MSCs suppresses Sox9 expression, a critical transcription factor that controls chondrogenesis [90]. The suppression of Sox9 subsequently leads to a decrease in collagen type II and aggrecan expression. In addition, TNF-α, in combination with IL-1β, has been shown to transform embryonic chondroprogenitor cells into fibroblast-like cells, further suggesting the inhibitory effects of inflammatory cytokines on chondrogenesis [87]. Similarly, when human MSCs are exposed to conditioned medium derived from osteoarthritic synovium, chondrogenesis is inhibited [92]. These antichondrogenic effects of inflammatory cytokines were shown to be caused by the activation of the NF-κB signaling pathway [93]. Overall, inflammatory conditions present in OA cartilage prevent chondrocytic differentiation of MSCs, thus inhibiting

regeneration of damaged cartilage with appropriate chondrocytic ECMs.

8. The Effects of Changes in Extracellular Matrix on Chondrocytic Differentiation of Mesenchymal Stem Cell

The changes in the composition of ECM also affect chondrogenic differentiation of MSCs. In a study by Bosnakovski et al., MSCs cultured in collagen type II hydrogels exhibited greater gene expression levels of chondrocytic markers as compared to those cultured on typical tissue culture plates [94]. As OA progresses, residential chondrocytes start to produce collagen type I instead of type II. This change can affect the subsequent chondrogenesis of MSCs as it has been shown that collagen type II favors chondrogenic induction by modulating cell shape, as compared to collagen type I [95]. It was demonstrated that collagen type II promotes a more rounded cell shape, similar to that of the native chondrocyte in healthy cartilage, through the β1 integrin-mediated Rho A/Rock signaling pathway.

In addition to the compositional effect, mechanical changes of ECMs (become stiffer due to the loss of hydrating aggrecan in OA) affect chondrogenesis of MSCs by regulating cell morphology [96]. A softer mechanical environment enhances chondrogenesis of MSCs, evident by greater gene and protein expression of chondrogenic markers including SOX9, collagen type II, and aggrecan by inhibiting stress fiber formation, as compared to the stiffer environment. Similarly, using polyacrylamide hydrogels with varying stiffnesses, Xue et al. showed that human mesenchymal stem cells are differentiated towards a chondrocytic phenotype on softer gels, regardless of initial cell seeding density [97]. The study highlights the importance of cell-matrix interactions during chondrogenic differentiation of MSCs.

Along with the direct influence of local stiffness change on MSC differentiation, the altered mechanical profiles under loading also affect the differentiation process. Bone marrow derived MSCs seeded onto fibrin hydrogels developed a spread out morphology and differentiated towards a myogenic lineage [98]. In the presence of long-term, dynamic compression, myogenic differentiation was inhibited, while markers for chondrogenic phenotype were upregulated. However, the magnitude of loading is an important factor determining chondrocytic differentiation of MSCs and thus synthesis of proper ECMs. Under the same loading regimen, a stiffer ECM induces less strain on the cells. In this regard, Michalopoulos et al. have recently shown that physiological compressive loading (15% strain) on MSC-laden scaffolds induces greater chondrogenesis as compared to a smaller strain of 10% that led to greater osteogenesis [99]. Similarly, stiffer agarose gels inhibited cartilage matrix production and gene expression of MSCs under hydrostatic pressure as compared to those in softer microenvironments [100]. These studies demonstrate that changes in the mechanical properties of cartilage during OA may favor the differentiation of MSCs towards nonchondrocytic lineages further intensifying the degeneration of cartilage. Overall, altered environments

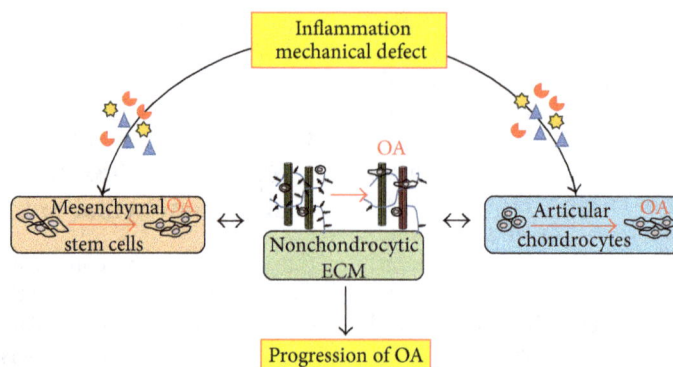

FIGURE 2: Schematic of the interplay between the extracellular matrix and cellular activities under inflammation during the progression of osteoarthritis (OA). Wear and tear or trauma induces inflammation and mechanical defects in cartilage, which initiate OA. These altered microenvironments affect the residential chondrocytes to produce nonchondrocytic extracellular matrix (ECM) that, in turn, further drives the dedifferentiation of the chondrocytes. The changes in microenvironments also negatively affect the chondrogenic differentiation of mesenchymal stem cells that originate from subchondral bone marrow, preventing the self-regeneration of cartilage. The positive feedback loop between mal-formed ECM and cellular activities drives the progression of OA.

in ECM composition and mechanical properties during the progression of OA significantly limit the chondrogenesis of MSCs inhibiting the regeneration process of cartilage damage.

9. Summary

Both inflammatory factors and compositional/structural changes of ECM drive the progression of OA by affecting residential articular chondrocytes as well as MSCs that migrate from bone marrow in the underlying subchondral bone to repair the cartilage defect (Figure 2). Due to chronic inflammation and altered microenvironments, chondrocytes change their phenotype towards more hypertrophic cells resulting in achondrocytic ECM synthesis. These changes in ECM, in combination with cartilage matrix degradation under inflammation, further fuel the degeneration process resulting in the alteration of biomechanical conditions, which disturb the surrounding tissues in the joint. The ECM changes in the presence of inflammation also negatively affect chondrogenic differentiation of MSCs, limiting self-regeneration of cartilage. Overall, the interplay between changes in ECM and changes in cellular function under inflammation forms a positive feedback loop that drives the pathology of OA.

References

[1] J. Nam, P. Perera, J. Liu et al., "Sequential alterations in catabolic and anabolic gene expression parallel pathological changes during progression of monoiodoacetate-induced arthritis," *PLoS ONE*, vol. 6, no. 9, Article ID e24320, 2011.

[2] A. D. Pearle, R. F. Warren, and S. A. Rodeo, "Basic science of articular cartilage and osteoarthritis," *Clinics in Sports Medicine*, vol. 24, no. 1, pp. 1–12, 2005.

[3] J. Martel-Pelletier, C. Boileau, J. Pelletier, and P. J. Roughley, "Cartilage in normal and osteoarthritis conditions," *Best Practice and Research*, vol. 22, no. 2, pp. 351–384, 2008.

[4] J. Nam, P. Perera, J. Liu et al., "Transcriptome-wide gene regulation by gentle treadmill walking during the progression of

monoiodoacetate-induced arthritis," *Arthritis & Rheumatism*, vol. 63, no. 6, pp. 1613–1625, 2011.

[5] M. W. Lark, E. K. Bayne, J. Flanagan et al., "Aggrecan degradation in human cartilage: evidence for both matrix metalloproteinase and aggrecanase activity in normal, osteoarthritic, and rheumatoid joints," *The Journal of Clinical Investigation*, vol. 100, no. 1, pp. 93–106, 1997.

[6] P. Lorenzo, M. T. Bayliss, and D. Heinegård, "Altered patterns and synthesis of extracellular matrix macromolecules in early osteoarthritis," *Matrix Biology*, vol. 23, no. 6, pp. 381–391, 2004.

[7] Y. Sun, D. R. Mauerhan, J. S. Kneisl et al., "Histological examination of collagen and proteoglycan changes in osteoarthritic menisci," *The Open Rheumatology Journal*, vol. 6, pp. 24–32, 2012.

[8] T. Videman, I. Eronen, and T. Candolin, "[3H]proline incorporation and hydroxyproline concentration in articular cartilage during the development of osteoarthritis caused by immobilization. A study in vivo with rabbits," *Biochemical Journal*, vol. 200, no. 2, pp. 435–440, 1981.

[9] A. Lahm, E. Mrosek, H. Spank et al., "Changes in content and synthesis of collagen types and proteoglycans in osteoarthritis of the knee joint and comparison of quantitative analysis with photoshop-based image analysis," *Archives of Orthopaedic and Trauma Surgery*, vol. 130, no. 4, pp. 557–564, 2010.

[10] A. R. Poole, M. Kobayashi, T. Yasuda et al., "Type II collagen degradation and its regulation in articular cartilage in osteoarthritis," *Annals of the Rheumatic Diseases*, vol. 61, supplement 2, pp. ii78–ii81, 2002.

[11] K. Gelse, E. Pöschl, and T. Aigner, "Collagens—structure, function, and biosynthesis," *Advanced Drug Delivery Reviews*, vol. 55, no. 12, pp. 1531–1546, 2003.

[12] F. H. Silver, G. Bradica, and A. Tria, "Elastic energy storage in human articular cartilage: estimation of the elastic modulus for type II collagen and changes associated with osteoarthritis," *Matrix Biology*, vol. 21, no. 2, pp. 129–137, 2002.

[13] R. E. Guzman, M. G. Evans, S. Bove, B. Morenko, and K. Kilgore, "Mono-iodoacetate-induced histologic changes in subchondral bone and articular cartilage of rat femorotibial joints: an animal model of osteoarthritis," *Toxicologic Pathology*, vol. 31, no. 6, pp. 619–624, 2003.

[14] K. P. H. Pritzker, S. Gay, S. A. Jimenez et al., "Osteoarthritis cartilage histopathology: grading and staging," *Osteoarthritis and Cartilage*, vol. 14, no. 1, pp. 13–29, 2006.

[15] M. B. Goldring, "Articular cartilage degradation in osteoarthritis," *HSS Journal*, vol. 8, no. 1, pp. 7–9, 2012.

[16] P. Hoff, F. Buttgereit, G. R. Burmester et al., "Osteoarthritis synovial fluid activates pro-inflammatory cytokines in primary human chondrocytes," *International Orthopaedics*, vol. 37, no. 1, pp. 145–151, 2013.

[17] J. A. Roman-Blas and S. A. Jimenez, "NF-κB as a potential therapeutic target in osteoarthritis and rheumatoid arthritis," *Osteoarthritis and Cartilage*, vol. 14, no. 9, pp. 839–848, 2006.

[18] S. C. Rosa, F. Judas, M. C. Lopes, and A. F. Mendes, "Nitric oxide synthase isoforms and NF-κB activity in normal and osteoarthritic human chondrocytes: regulation by inducible nitric oxide," *Nitric Oxide*, vol. 19, no. 3, pp. 276–283, 2008.

[19] D. Pfander, N. Heinz, P. Rothe, H.-D. Carl, and B. Swoboda, "Tenascin and aggrecan expression by articular chondrocytes is influenced by interleukin 1β: a possible explanation for the changes in matrix synthesis during osteoarthritis," *Annals of the Rheumatic Diseases*, vol. 63, no. 3, pp. 240–244, 2004.

[20] I. Meulenbelt, A. B. Seymour, M. Nieuwland, T. W. J. Huizinga, C. M. van Duijn, and P. E. Slagboom, "Association of the interleukin-1 gene cluster with radiographic signs of osteoarthritis of the hip," *Arthritis & Rheumatism*, vol. 50, no. 4, pp. 1179–1186, 2004.

[21] L. I. Sakkas, N. A. Johanson, C. R. Scanzello, and C. D. Platsoucas, "Interleukin-12 is expressed by infiltrating macrophages and synovial lining cells in rheumatoid arthritis and osteoarthritis," *Cellular Immunology*, vol. 188, no. 2, pp. 105–110, 1998.

[22] C. R. Scanzello, E. Umoh, F. Pessler et al., "Local cytokine profiles in knee osteoarthritis: elevated synovial fluid interleukin-15 differentiates early from end-stage disease," *Osteoarthritis and Cartilage*, vol. 17, no. 8, pp. 1040–1048, 2009.

[23] G. M. Campo, A. Avenoso, A. D'Ascola et al., "Hyaluronan in part mediates IL-1beta-induced inflammation in mouse chondrocytes by up-regulating CD44 receptors," *Gene*, vol. 494, no. 1, pp. 24–35, 2012.

[24] A. Liacini, J. Sylvester, W. Q. Li et al., "Induction of matrix metalloproteinase-13 gene expression by TNF-α is mediated by MAP kinases, AP-1, and NF-κB transcription factors in articular chondrocytes," *Experimental Cell Research*, vol. 288, no. 1, pp. 208–217, 2003.

[25] L. Troeberg and H. Nagase, "Proteases involved in cartilage matrix degradation in osteoarthritis," *Biochimica et Biophysica Acta*, vol. 1824, no. 1, pp. 133–145, 2012.

[26] R. Song, M. D. Tortorella, A. Malfait et al., "Aggrecan degradation in human articular cartilage explants is mediated by both ADAMTS-4 and ADAMTS-5," *Arthritis & Rheumatism*, vol. 56, no. 2, pp. 575–585, 2007.

[27] K. Imai, S. Ohta, T. Matsumoto et al., "Expression of membrane-type 1 matrix metalloproteinase and activation of progelatinase A in human osteoarthritic cartilage," *American Journal of Pathology*, vol. 151, no. 1, pp. 245–256, 1997.

[28] R. C. Billinghurst, L. Dahlberg, M. Ionescu et al., "Enhanced cleavage of type II collagen by collagenases in osteoarthritic articular cartilage," *The Journal of Clinical Investigation*, vol. 99, no. 7, pp. 1534–1545, 1997.

[29] A. J. Freemont, V. Hampson, R. Tilman, P. Goupille, Y. Taiwo, and J. A. Hoyland, "Gene expression of matrix metalloproteinases 1, 3, and 9 by chondrocytes in osteoarthritic human knee articular cartilage is zone and grade specific," *Annals of the Rheumatic Diseases*, vol. 56, no. 9, pp. 542–549, 1997.

[30] A. Kaspiris, L. Khaldi, T. B. Grivas et al., "Subchondral cyst development and MMP-1 expression during progression of osteoarthritis: an immunohistochemical study," *Orthopaedics & Traumatology*, 2013.

[31] A. Hasegawa, H. Nakahara, M. Kinoshita, H. Asahara, J. Koziol, and M. K. Lotz, "Cellular and extracellular matrix changes in anterior cruciate ligaments during human knee aging and osteoarthritis," *Arthritis Research & Therapy*, vol. 15, no. 1, article R29, 2013.

[32] R. F. Loeser, R. R. Yammani, C. S. Carlson et al., "Articular chondrocytes express the receptor for advanced glycation end products: potential role in osteoarthritis," *Arthritis & Rheumatism*, vol. 52, no. 8, pp. 2376–2385, 2005.

[33] S. Frank, M. A. Peters, C. Wehmeyer et al., "Regulation of matrixmetalloproteinase-3 and matrixmetalloproteinase-13 by SUMO-2/3 through the transcription factor NF-kappaB," *Annals of the Rheumatic Diseases*. In press.

[34] A. R. Amin, P. E. Di Cesare, P. Vyas et al., "The expression and regulation of nitric oxide synthase in human osteoarthritis-affected chondrocytes: evidence for up-regulated neuronal nitric oxide synthase," *Journal of Experimental Medicine*, vol. 182, no. 6, pp. 2097–2102, 1995.

[35] K. Kühn, A. R. Shikhman, and M. Lotz, "Role of nitric oxide, reactive oxygen species, and p38 MAP kinase in the regulation of human chondrocyte apoptosis," *Journal of Cellular Physiology*, vol. 197, no. 3, pp. 379–387, 2003.

[36] A. Karan, M. A. Karan, P. Vural et al., "Synovial fluid nitric oxide levels in patients with knee osteoarthritis," *Clinical Rheumatology*, vol. 22, no. 6, pp. 397–399, 2003.

[37] E. N. B. Davidson, E. L. Vitters, P. M. van der Kraan, and W. B. van den Berg, "Expression of transforming growth factor-β (TGFβ) and the TGFβ signalling molecule SMAD-2P in spontaneous and instability-induced osteoarthritis: role in cartilage degradation, chondrogenesis and osteophyte formation," *Annals of the Rheumatic Diseases*, vol. 65, no. 11, pp. 1414–1421, 2006.

[38] E. N. B. Davidson, A. Scharstuhl, E. L. Vitters, P. M. van der Kraan, and W. B. van den Berg, "Reduced transforming growth factor-beta signaling in cartilage of old mice: role in impaired repair capacity," *Arthritis Research & Therapy*, vol. 7, no. 6, pp. R1338–R1347, 2005.

[39] L. Patel, W. Sun, S. S. Glasson, E. A. Morris, C. R. Flannery, and P. S. Chockalingam, "Tenascin-C induces inflammatory mediators and matrix degradation in osteoarthritic cartilage," *BMC Musculoskeletal Disorders*, vol. 12, article 164, 2011.

[40] D. M. Salter, "Tenascin is increased in cartilage and synovium from arthritic knees," *British Journal of Rheumatology*, vol. 32, no. 9, pp. 780–786, 1993.

[41] E. J. Mackie and L. I. Murphy, "The role of tenascin-C and related glycoproteins in early chondrogenesis," *Microscopy Research and Technique*, vol. 43, no. 2, pp. 102–110, 1998.

[42] J. Melrose, E. S. Fuller, P. J. Roughley et al., "Fragmentation of decorin, biglycan, lumican and keratocan is elevated in degenerate human meniscus, knee and hip articular cartilages compared with age-matched macroscopically normal and control tissues," *Arthritis Research and Therapy*, vol. 10, no. 4, article R79, 2008.

[43] X. Chevalier, N. Groult, and W. Hornebeck, "Increased expression of the Ed-B-containing fibronectin (an embryonic isoform

of fibronectin) in human osteoarthritic cartilage," *British Journal of Rheumatology*, vol. 35, no. 5, pp. 407–415, 1996.

[44] X. Chevalier, P. Claudepierre, N. Groult, L. Zardi, and W. Hornebeck, "Presence of ED-A containing fibronectin in human articular cartilage from patients with osteoarthritis and rheumatoid arthritis," *Journal of Rheumatology*, vol. 23, no. 6, pp. 1022–1030, 1996.

[45] L. Schaefer, A. Babelova, E. Kiss et al., "The matrix component biglycan is proinflammatory and signals through Toll-like receptors 4 and 2 in macrophages," *The Journal of Clinical Investigation*, vol. 115, no. 8, pp. 2223–2233, 2005.

[46] L. Fan, Q. Wang, R. Liu et al., "Citrullinated fibronectin inhibits apoptosis and promotes the secretion of pro-inflammatory cytokines in fibroblast-like synoviocytes in rheumatoid arthritis," *Arthritis Research & Therapy*, vol. 14, no. 6, article R266, 2012.

[47] F. H. Silver, G. Bradica, and A. Tria, "Do changes in the mechanical properties of articular cartilage promote catabolic destruction of cartilage and osteoarthritis?" *Matrix Biology*, vol. 23, no. 7, pp. 467–476, 2004.

[48] H. E. Panula, M. M. Hyttinen, J. P. A. Arokoski et al., "Articular cartilage superficial zone collagen birefringence reduced and cartilage thickness increased before surface fibrillation in experimental osteoarthritis," *Annals of the Rheumatic Diseases*, vol. 57, no. 4, pp. 237–245, 1998.

[49] R. C. Appleyard, D. Burkhardt, P. Ghosh et al., "Topographical analysis of the structural, biochemical and dynamic biomechanical properties of cartilage in an ovine model of osteoarthritis," *Osteoarthritis and Cartilage*, vol. 11, no. 1, pp. 65–77, 2003.

[50] F. Guilak, A. Ratcliffe, N. Lane, M. P. Rosenwasser, and V. C. Mow, "Mechanical and biochemical changes in the superficial zone of articular cartilage in canine experimental osteoarthritis," *Journal of Orthopaedic Research*, vol. 12, no. 4, pp. 474–484, 1994.

[51] L. Xu, C. M. Flahiff, B. A. Waldman et al., "Osteoarthritis-like changes and decreased mechanical function of articular cartilage in the joints of mice with the chondrodysplasia gene (cho)," *Arthritis & Rheumatism*, vol. 48, no. 9, pp. 2509–2518, 2003.

[52] C. Y. Wen, C. B. Wu, B. Tang et al., "Collagen fibril stiffening in osteoarthritic cartilage of human beings revealed by atomic force microscopy," *Osteoarthritis and Cartilage*, vol. 20, no. 8, pp. 916–922, 2012.

[53] N. Verzijl, J. DeGroot, Z. C. Ben et al., "Crosslinking by advanced glycation end products increases the stiffness of the collagen network in human articular cartilage: a possible mechanism through which age is a risk factor for osteoarthritis," *Arthritis & Rheumatism*, vol. 46, no. 1, pp. 114–123, 2002.

[54] M. B. Goldring and S. R. Goldring, "Articular cartilage and subchondral bone in the pathogenesis of osteoarthritis," *Annals of the New York Academy of Sciences*, vol. 1192, pp. 230–237, 2010.

[55] P. M. van der Kraan and W. B. van den Berg, "Osteophytes: relevance and biology," *Osteoarthritis and Cartilage*, vol. 15, no. 3, pp. 237–244, 2007.

[56] O. Stannus, G. Jones, F. Cicuttini et al., "Circulating levels of IL-6 and TNF-α are associated with knee radiographic osteoarthritis and knee cartilage loss in older adults," *Osteoarthritis and Cartilage*, vol. 18, no. 11, pp. 1441–1447, 2010.

[57] M. J. Benito, D. J. Veale, O. FitzGerald, W. B. van den Berg, and B. Bresnihan, "Synovial tissue inflammation in early and late osteoarthritis," *Annals of the Rheumatic Diseases*, vol. 64, no. 9, pp. 1263–1267, 2005.

[58] J. T. Tung, C. E. Arnold, L. H. Alexander et al., "Evaluation of the influence of prostaglandin E2 on recombinant equine interleukin-1 β-stimulated matrix metalloproteinases 1, 3, and 13 and tissue inhibitor of matrix metalloproteinase 1 expression in equine chondrocyte cultures," *American Journal of Veterinary Research*, vol. 63, no. 7, pp. 987–993, 2002.

[59] B. Dozin, L. Camardella, R. Cancedda, and A. Pietrangelo, "Response of young, aged and osteoarthritic human articular chondrocytes to inflammatory cytokines: molecular and cellular aspects," *Matrix Biology*, vol. 21, no. 5, pp. 449–459, 2002.

[60] A. Liacini, J. Sylvester, W. Q. Li, and M. Zafarullah, "Inhibition of interleukin-1-stimulated MAP kinases, activating protein-1 (AP-1) and nuclear factor kappa B (NF-κB) transcription factors down-regulates matrix metalloproteinase gene expression in articular chondrocytes," *Matrix Biology*, vol. 21, no. 3, pp. 251–262, 2002.

[61] C. T. G. Appleton, S. E. Usmani, J. S. Mort, and F. Beier, "Rho/ROCK and MEK/ERK activation by transforming growth factor-α induces articular cartilage degradation," *Laboratory Investigation*, vol. 90, no. 1, pp. 20–30, 2010.

[62] L. Dai, X. Zhang, X. Hu et al., "Silencing of microRNA-101 prevents IL-1beta-induced extracellular matrix degradation in chondrocytes," *Arthritis Research & Therapy*, vol. 14, no. 6, article R268, 2012.

[63] J. R. Kammermann, S. A. Kincaid, P. F. Rumph, D. K. Baird, and D. M. Visco, "Tumor necrosis factor-α (TNF-α) in canine osteoarthritis: immunolocalization of TNF-α, stromelysin and TNF receptors in canine osteoarthritis cartilage," *Osteoarthritis and Cartilage*, vol. 4, no. 1, pp. 23–34, 1996.

[64] W. Lee, S. Yu, S. Cheong, J. Sonn, and S. Kim, "Ectopic expression of cyclooxygenase-2-induced dedifferentiation in articular chondrocytes," *Experimental and Molecular Medicine*, vol. 40, no. 6, pp. 721–727, 2008.

[65] A. R. Amin, M. Attur, R. N. Patel et al., "Superinduction of cyclooxygenase-2 activity in human osteoarthritis-affected cartilage. Influence of nitric oxide," *The Journal of Clinical Investigation*, vol. 99, no. 6, pp. 1231–1237, 1997.

[66] K. G. A. Yang, D. B. F. Saris, R. E. Geuze et al., "Altered in vitro chondrogenic properties of chondrocytes harvested from unaffected cartilage in osteoarthritic joints," *Osteoarthritis and Cartilage*, vol. 14, no. 6, pp. 561–570, 2006.

[67] M. M. Temple, W. C. Bae, M. Q. Chen et al., "Age- and site-associated biomechanical weakening of human articular cartilage of the femoral condyle," *Osteoarthritis and Cartilage*, vol. 15, no. 9, pp. 1042–1052, 2007.

[68] D. Merz, R. Liu, K. Johnson, and R. Terkeltaub, "IL-8/CXCl8 and growth-related oncogene α/CXCL1 induce chondrocyte hypertrophic differentiation," *Journal of Immunology*, vol. 171, no. 8, pp. 4406–4415, 2003.

[69] D. L. Cecil, K. Johnson, J. Rediske, M. Lotz, A. M. Schmidt, and R. Terkeltaub, "Inflammation-induced chondrocyte hypertrophy is driven by receptor for advanced glycation end products," *Journal of Immunology*, vol. 175, no. 12, pp. 8296–8302, 2005.

[70] J. L. Allen, M. E. Cooke, and T. Alliston, "ECM stiffness primes the TGFbeta pathway to promote chondrocyte differentiation," *Molecular Biology of the Cell*, vol. 23, no. 18, pp. 3731–3742, 2012.

[71] E. Schuh, J. Kramer, J. Rohwedel et al., "Effect of matrix elasticity on the maintenance of the chondrogenic phenotype," *Tissue Engineering A*, vol. 16, no. 4, pp. 1281–1290, 2010.

[72] E. Schuh, S. Hofmann, K. S. Stok, H. Notbohm, R. Müller, and N. Rotter, "The influence of matrix elasticity on chondrocyte

behavior in 3D," *Journal of Tissue Engineering and Regenerative Medicine*, vol. 6, no. 10, pp. e31–e42, 2012.

[73] S. Marlovits, M. Hombauer, M. Truppe, V. Vècsei, and W. Schlegel, "Changes in the ratio of type-I and type-II collagen expression during monolayer culture of human chondrocytes," *Journal of Bone and Joint Surgery B*, vol. 86, no. 2, pp. 286–295, 2004.

[74] R. H. J. Das, H. Jahr, J. A. N. Verhaar, J. C. van der Linden, G. J. V. M. van Osch, and H. Weinans, "In vitro expansion affects the response of chondrocytes to mechanical stimulation," *Osteoarthritis and Cartilage*, vol. 16, no. 3, pp. 385–391, 2008.

[75] Z. Lin, J. B. Fitzgerald, J. Xu et al., "Gene expression profiles of human chondrocytes during passaged monolayer cultivation," *Journal of Orthopaedic Research*, vol. 26, no. 9, pp. 1230–1237, 2008.

[76] O. Pullig, G. Weseloh, A. R. Klatt, R. Wagener, and B. Swoboda, "Matrilin-3 in human articular cartilage: increased expression in osteoarthritis," *Osteoarthritis and Cartilage*, vol. 10, no. 4, pp. 253–263, 2002.

[77] J.-B. Vincourt, P. Gillet, A.-C. Rat et al., "Measurement of matrilin-3 levels in human serum and synovial fluid using a competitive enzyme-linked immunosorbent assay," *Osteoarthritis and Cartilage*, vol. 20, no. 7, pp. 783–786, 2012.

[78] J. Vincourt, S. Etienne, L. Grossin et al., "Matrilin-3 switches from anti- to pro-anabolic upon integration to the extracellular matrix," *Matrix Biology*, vol. 31, no. 5, pp. 290–298, 2012.

[79] M. Fuerst, J. Bertrand, L. Lammers et al., "Calcification of articular cartilage in human osteoarthritis," *Arthritis & Rheumatism*, vol. 60, no. 9, pp. 2694–2703, 2009.

[80] C. Nguyen, M. Lieberherr, C. Bordat et al., "Intracellular calcium oscillations in articular chondrocytes induced by basic calcium phosphate crystals lead to cartilage degradation," *Osteoarthritis and Cartilage*, vol. 20, no. 11, pp. 1399–1408, 2012.

[81] C. M. Thomas, R. Murray, and M. Sharif, "Chondrocyte apoptosis determined by caspase-3 expression varies with fibronectin distribution in equine articular cartilage," *International Journal of Rheumatic Diseases*, vol. 14, no. 3, pp. 290–297, 2011.

[82] A. I. Caplan, "Mesenchymal stem cells," *Journal of Orthopaedic Research*, vol. 9, no. 5, pp. 641–650, 1991.

[83] R. J. Williams III and H. W. Harnly, "Microfracture: indications, technique, and results," *Instructional Course Lectures*, vol. 56, pp. 419–428, 2007.

[84] D. K. Bae, K. H. Yoon, and S. J. Song, "Cartilage healing after microfracture in osteoarthritic knees," *Arthroscopy*, vol. 22, no. 4, pp. 367–374, 2006.

[85] P. Orth, M. Cucchiarini, G. Kaul et al., "Temporal and spatial migration pattern of the subchondral bone plate in a rabbit osteochondral defect model," *Osteoarthritis and Cartilage*, vol. 20, no. 10, pp. 1161–1169, 2012.

[86] Y.-S. Qui, B. F. Shahgaldi, W. J. Revell, and F. W. Heatley, "Observations of subchondral plate advancement during osteochondral repair: a histomorphometric and mechanical study in the rabbit femoral condyle," *Osteoarthritis and Cartilage*, vol. 11, no. 11, pp. 810–820, 2003.

[87] H. Mohamed-Ali, "Influence of interleukin-1β, tumour necrosis factor alpha and prostaglandin E2 on chondrogenesis and cartilage matrix breakdown in vitro," *Rheumatology International*, vol. 14, no. 5, pp. 191–199, 1995.

[88] F. Kojima, H. Naraba, S. Miyamoto, M. Beppu, H. Aoki, and S. Kawai, "Membrane-associated prostaglandin E synthase-1 is upregulated by proinflammatory cytokines in chondrocytes from patients with osteoarthritis," *Arthritis Research & Therapy*, vol. 6, no. 4, pp. R355–R365, 2004.

[89] T. Felka, R. Schäfer, B. Schewe, K. Benz, and W. K. Aicher, "Hypoxia reduces the inhibitory effect of IL-1β on chondrogenic differentiation of FCS-free expanded MSC," *Osteoarthritis and Cartilage*, vol. 17, no. 10, pp. 1368–1376, 2009.

[90] M. K. Majumdar, E. Wang, and E. A. Morris, "BMP-2 and BMP-9 promote chondrogenic differentiation of human multipotential mesenchymal cells and overcome the inhibitory effect of IL-1," *Journal of Cellular Physiology*, vol. 189, no. 3, pp. 275–284, 2001.

[91] S. Boeuf, F. Graf, J. Fischer, B. Moradi, C. B. Little, and W. Richter, "Regulation of aggrecanases from the ADAMTS family and aggrecan neoepitope formation during in vitro chondrogenesis of human mesenchymal stem cells," *European Cells and Materials*, vol. 23, pp. 320–332, 2012.

[92] G. T. H. Heldens, E. N. Blaney Davidson, E. L. Vitters et al., "Catabolic factors and osteoarthritis-conditioned medium inhibit chondrogenesis of human mesenchymal stem cells," *Tissue Engineering A*, vol. 18, no. 1-2, pp. 45–54, 2012.

[93] N. Wehling, G. D. Palmer, C. Pilapil et al., "Interleukin-1β and tumor necrosis factor α inhibit chondrogenesis by human mesenchymal stem cells through NF-κB-dependent pathways," *Arthritis & Rheumatism*, vol. 60, no. 3, pp. 801–812, 2009.

[94] D. Bosnakovski, M. Mizuno, G. Kim, S. Takagi, M. Okumura, and T. Fujinaga, "Chondrogenic differentiation of bovine bone marrow mesenchymal stem cells (MSCs) in different hydrogels: influence of collagen type II extracellular matrix on MSC chondrogenesis," *Biotechnology and Bioengineering*, vol. 93, no. 6, pp. 1152–1163, 2006.

[95] Z. Lu, B. Z. Doulabi, C. Huang, R. A. Bank, and M. N. Helder, "Collagen type II enhances chondrogenesis in adipose tissue-derived stem cells by affecting cell shape," *Tissue Engineering A*, vol. 16, no. 1, pp. 81–90, 2010.

[96] J. Nam, J. Johnson, J. J. Lannutti, and S. Agarwal, "Modulation of embryonic mesenchymal progenitor cell differentiation via control over pure mechanical modulus in electrospun nanofibers," *Acta Biomaterialia*, vol. 7, no. 4, pp. 1516–1524, 2011.

[97] R. Xue, J. Y. Li, Y. Yeh et al., "Effects of matrix elasticity and cell density on human mesenchymal stem cells differentiation," *Journal of Orthopaedic Research*, vol. 31, no. 9, pp. 1360–1365, 2013.

[98] S. D. Thorpe, C. T. Buckley, A. J. Steward, and D. J. Kelly, "European Society of Biomechanics S.M. Perren Award 2012: the external mechanical environment can override the influence of local substrate in determining stem cell fate," *Journal of Biomechanics*, vol. 45, no. 15, pp. 2483–2492, 2012.

[99] E. Michalopoulos, R. L. Knight, S. Korossis, J. N. Kearney, J. Fisher, and E. Ingham, "Development of methods for studying the differentiation of human mesenchymal stem cells under cyclic compressive strain," *Tissue Engineering C*, vol. 18, no. 4, pp. 252–262, 2012.

[100] A. J. Steward, D. R. Wagner, and D. J. Kelly, "The pericellular environment regulates cytoskeletal development and the differentiation of mesenchymal stem cells and determines their response to hydrostatic pressure," *European Cells and Materials*, vol. 25, pp. 167–178, 2013.

17

Craniosynostosis-Associated Fgfr2^{C342Y} Mutant Bone Marrow Stromal Cells Exhibit Cell Autonomous Abnormalities in Osteoblast Differentiation and Bone Formation

J. Liu,[1] **T.-G. Kwon,**[2] **H. K. Nam,**[1] **and N. E. Hatch**[1]

[1] *Department of Orthodontics and Pediatric Dentistry, School of Dentistry, University of Michigan, Ann Arbor, MI 48109-1078, USA*

[2] *Department of Oral and Maxillofacial Surgery, School of Dentistry, Kyungpook National University, Jung Gu, Daegu, Republic of Korea*

Correspondence should be addressed to N. E. Hatch; nhatch@umich.edu

Academic Editor: Zhao Lin

We recently reported that cranial bones of Fgfr2$^{C342Y/+}$ craniosynostotic mice are diminished in density when compared to those of wild type mice, and that cranial bone cells isolated from the mutant mice exhibit inhibited late stage osteoblast differentiation. To provide further support for the idea that craniosynostosis-associated Fgfr mutations lead to cell autonomous defects in osteoblast differentiation and mineralized tissue formation, here we tested bone marrow stromal cells isolated from Fgfr2$^{C342Y/+}$ mice for their ability to differentiate into osteoblasts. Additionally, to determine if the low bone mass phenotype of Crouzon syndrome includes the appendicular skeleton, long bones were assessed by micro CT. Fgfr2$^{C342Y/+}$ cells showed increased osteoblastic gene expression during early osteoblastic differentiation but decreased expression of alkaline phosphatase mRNA and enzyme activity, and decreased mineralization during later stages of differentiation, when cultured under 2D *in vitro* conditions. Cells isolated from Fgfr2$^{C342Y/+}$ mice also formed less bone when allowed to differentiate in a 3D matrix *in vivo*. Cortical bone parameters were diminished in long bones of Fgfr2$^{C342Y/+}$ mice. These results demonstrate that marrow stromal cells of Fgfr2$^{C342Y/+}$ mice have an autonomous defect in osteoblast differentiation and bone mineralization, and that the Fgfr2^{C342Y} mutation influences both the axial and appendicular skeletons.

1. Introduction

Craniosynostosis is a debilitating pediatric condition characterized by the premature fusion of cranial bones. This fusion leads to high intracranial pressure and abnormal skull and facial shapes presumably resulting from limited growth at fused craniofacial sutures with compensating overgrowth at nonfused cranial sutures [1–5]. Untreated craniosynostosis can lead to blindness, seizures, and death [6–10]. Current treatment options for craniosynostosis and its associated craniofacial abnormalities are limited to surgery with genetic counseling, orthodontic, medical, and social support [11]. Notably, even with an appropriately early and accurate diagnosis, craniosynostosis can carry high morbidity, with some patients requiring multiple surgeries throughout infancy and childhood for treatment of recurring craniosynostosis and normalization of skull and facial shapes [12].

It has been known for over a decade that craniosynostosis occurs in association with mutations in the genes for fibroblast growth factor receptors (Fgfr's). Mutations in Fgfr2 cause Apert, Crouzon, Jackson-Weiss, and Pfeiffer craniosynostosis syndromes, while mutations in Fgfr1 cause Pfeiffer syndrome and mutations in Fgfr3 cause Muenke craniosynostosis syndrome and Crouzon syndrome with acanthosis nigricans [13–17]. Numerous prior studies have shown that these craniosynostosis-associated Fgfr mutations lead to a gain of function in terms of Fgfr signaling [18–22]. Yet, despite our extensive knowledge in the genetics

Craniosynostosis-Associated Fgfr2^{C342Y} Mutant Bone Marrow Stromal Cells Exhibit Cell Autonomous Abnormalities in Osteoblast Differentiation and Bone Formation

169

underlying syndromic craniosynostosis, the pathogenesis by which mutations in Fgfr's lead to craniosynostosis has yet to be fully elucidated. Because Fgf/Fgfr signaling is an important form of intercellular communication, it has been proposed that craniosynostosis-associated mutations in Fgf receptors lead to craniosynostosis by causing inappropriate signaling to cranial cells from neighboring tissues such as the dura mater [23–25]. In contrast, it has also been proposed that craniosynostosis-associated mutations in Fgf receptors lead to intrinsic defects in the behavior of cranial bone cells and tissues [26–28]. Importantly, while the biologic pathogenesis of craniosynostosis remains unknown, the only treatment for craniosynostosis will remain that of surgical intervention.

To advance our understanding of mechanisms that lead to craniosynostosis, we are investigating the Fgfr2$^{C342Y/+}$ mouse model of Crouzon syndrome. Common features of Crouzon syndrome in humans include coronal suture synostosis with rare pansynostosis, hypertelorism, severe ocular proptosis, strabismus, hypoplastic maxilla, and relative mandibular prognathism [11, 29]. Notably, high rates of stylohyoid ligament calcification and vertebral fusions, as well as the occasional fusion of limb joints, have also been reported [30–32]. Mice carrying the classic Crouzon syndrome associated Fgfr2^{C342Y} mutation were initially reported to have characteristics similar to those of Crouzon syndrome patients including a dome-shaped skull, wide set and proptotic eyes, premature fusion of cranial sutures, and a shortened maxilla [32]. Notably, fusions were also evident between the femur and tibia and between cervical vertebral arches in homozygous mutant mice. The homozygous mutant mice also lacked vertebral body ossification. These findings indicate that the Crouzon syndrome phenotype involves both the axial and appendicular skeletons. The findings also suggest that the bony fusions of Crouzon syndrome occur in the context of diminished eutopic bone formation.

We recently reported that the frontal bones of Crouzon Fgfr2$^{C342Y/+}$ mice are diminished in bone volume and density when compared to those of wild type mice, and that while frontal bone cells isolated from the mutant mice exhibit increased osteoblastic gene during early stages of osteoblast differentiation, the cells are also diminished in their ability to differentiate into mature osteoblasts in vitro [5]. To provide further support for the idea that the Crouzon syndrome associated Fgfr2^{C342Y} mutation leads to intrinsic changes in osteoblast differentiation and bone formation, we isolated cells from Crouzon mice and assayed their ability to differentiate into osteoblasts and form mineralized tissue, as compared to cells isolated from wild type mice. Because cell culture in a three-dimensional matrix supports a more physiologically relevant environment than more traditional two-dimensional cell culture methods [33], we assayed cells in traditional in vitro monolayer culture and in a three-dimensional collagenous matrix in vivo. Additionally, to determine if the appendicular skeleton is altered in Fgfr2$^{C342Y/+}$ mice, we used microcomputed tomography to quantify parameters of tibial bone quality and quantity. Here we report that the Fgfr2^{C342Y} mutation enhances expression of osteoblastic genes during early stages of differentiation

while inhibiting expression of alkaline phosphatase enzyme (AP/Tnap/Alpl/Akp2) and mineralization during later differentiation, in two-dimensional in vitro cell culture. Here we also show that the Fgfr2^{C342Y} mutation inhibits the ability of these cells to express Tnap enzyme and form mineralized tissue when allowed to differentiate in a three-dimensional matrix in vivo. Additionally, here we report for the first time that the long bones of Fgfr2$^{C342Y/+}$ mice have significantly diminished cortical bone quality and quantity, when compared to those of wild type mice. Together, these results demonstrate that the Crouzon syndrome associated Fgfr2^{C342Y} mutation causes cell autonomous abnormalities in osteoblast differentiation that include enhanced early differentiation but inhibited later expression of Tnap enzyme and mineralization by bone marrow stromal cells. Results also show that Crouzon syndrome is associated with significantly diminished appendicular bone volume and density.

2. Experimental Procedures

2.1. Animals. Fgfr2$^{C342Y/+}$ and wild type mice were genotyped as previously described [29]. Briefly, DNA from tail digests was amplified by polymerase chain reaction using 5'-gagtaccatgctgactgcatgc-3' and 5'-ggagaggcatctctgtttcaagacc-3' primers. The reaction product was resolved by gel electrophoresis, yielding a 200 base pair band for wild type Fgfr2 and a 300 base pair band for Fgfr2^{C342Y}. Fgfr2$^{C342Y/+}$ and wild type mice on the Balb/C genetic background were utilized for cell isolations and for micro-CT analyses. NIH III nude mice were obtained from Charles River Laboratories International, Inc. (Wilmington, MA) and utilized as donor mice for subcutaneous implant experiments. All animal procedures were performed according to University of Michigan's University Committee on Use and Care of Animals.

2.2. Microcomputed Tomography. Tibias of Fgfr2$^{C342Y/+}$ ($n = 14$) and wild type ($n = 14$) four-week-old mice were embedded in 1% agarose and scanned at the proximal metaphysis and the mid-diaphysis using a microcomputed tomography imaging system (Scanco μCT100, Scanco Medical, Bassersdorf, Switzerland). Scan settings were voxel size 12 μm, 70 kVp, 114 μA, 0.5 mm AL filter, and integration time of 500 ms. Density measurements were calibrated to the manufacturer's hydroxyapatite phantom. Analysis was performed using the manufacturer's evaluation software (Scanco μCT100, Scanco Medical, Bassersdorf, Switzerland) using a fixed global threshold of 28% (280 on a grayscale of 0–1000) to segment bone from nonbone. Micro-CT bone data was analyzed and is reported in accordance with the recommendations of Bouxsein et al., 2010 [34]. Statistical significance between groups was established by use of the Student's t-test.

2.3. Primary Cell Isolation. Bone marrow stromal cells were isolated from the long bones of four-week-old Crouzon mice and wild type littermates, as previously described [35]. Briefly, marrow cells were aspirated using a 25-gauge needle and a 5 mL syringe containing media. Marrow was flushed and cells were then dispersed by aspirating several times through a 22-gauge needle. Cells were cultured in αMEM supplemented with 20% fetal bovine serum (FBS) and 10,000 μg/mL penicillin/streptomycin for several days. Media was changed every three days until all suspension cells were removed and adherent cells were confluent.

2.4. In Vitro Osteoblast Differentiation. Cells were induced to differentiate *in vitro* by culture in media containing 50 μg/mL ascorbate for the indicated number of days. RNA was isolated using Trizol reagent (Invitrogen) following manufacturer protocols. mRNA levels were assayed by reverse transcription and real-time PCR. Real-time PCR was performed utilizing the murine tissue nonspecific alkaline phosphatase (TNAP) primer/probe set Mm00475834_m1, the murine bone sialoprotein (BSP) primer/probe set Mm00492555_m1, the murine osteocalcin (OCN) primer/probe set Mm03413826_mH, the murine runt related transcription factor 2 (Runx2) primer/probe set Mm00501578_m1, the murine collagen type I, alpha 1 (col1a1) primer/probe set Mm00801666_g1, the murine hypoxanthine phosphoribosyltransferase-1 (Hprt1) primer/probe set Mm01545399_m1, and Taqman Universal PCR Master Mix (Applied Biosystems). Real-time PCR was performed on a GeneAmp 7700 thermocycler (Applied Biosystems) and quantified by comparison to a standard curve. mRNA levels are reported after normalization to Hprt1 mRNA levels. Cells were induced to form mineral by addition of 10 mM β-glycerophosphate. Mineralized nodules were stained by Von Kossa. Briefly, cells were rinsed with phosphate-buffered saline, fixed with 100% ethanol and rehydrated in a graded ethanol series. Cells were then incubated in 5% AgNO$_3$, rinsed with dH$_2$O and exposed to light for 1 hour. For quantification, wells were scanned and densitometry was measured using *NIH Image* software. Tissue non-specific alkaline phosphatase (AP) enzyme activity was assayed using the colorimetric substrate, NBT/BCIP (Sigma). Cells were fixed in 70% ethanol for 10 minutes at room temperature, air dried, and incubated with substrate for 1 hour at 37C. Cells were then rinsed with dH$_2$O, air dried, and visualized macroscopically for evidence of staining. For quantification, wells were scanned and densitometry was measured using *NIH Image* software. Statistical significance between genotypes for mRNA levels, quantification of mineralization, and quantification of AP enzyme activity was established by use of the Student's *t*-test.

2.5. Subcutaneous Implant Preparation and Analysis of Ossicles. Subcutaneous implants were prepared as previously described [36]. 4×10^7 bone marrow stromal cells were mixed with 0.01% NaOH in phosphate buffered saline and 1 ml of rat tail collagen solution (BD Biosciences, San Jose, CA) on ice. The solution was aliquoted into glass tissue culture wells (*Lab-Tek* 16 well chamber slide system; Nalge Nunc International, Rochester, NY) and then incubated at 37°C for 1 hour to allow for gel hardening. Midline longitudinal incisions were made along the dorsal surface of each six-week-old NIH III mouse, and subcutaneous pockets were formed laterally, by gentle blunt dissection. A single implant was placed into

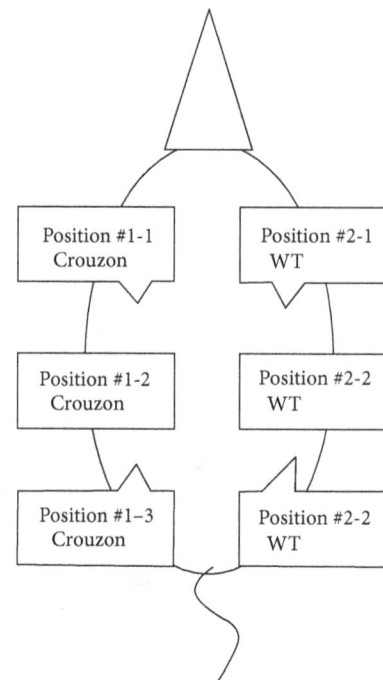

FIGURE 1: Subcutaneous implant placement. This schematic shows the position of six implants that were placed subcutaneously, on the dorsal surface of immunodeficient mice.

each subcutaneous pocket, for a total of six implants per animal (Figure 1). Implants were removed eight weeks after implantation and analyzed for mineralized tissue formation by radiography (Faxitron MX-20, Faxitron Bioptics LLC, Tucson, AZ). All implants were imaged on the same film. Mineralized tissue of twelve implants from each genotype was quantified by densitometry (ImageJ, NIH). Statistical significance between genotypes was established by use of the Student's *t*-test. Implants were then homogenized for alkaline phosphatase measurements or decalcified and embedded in paraffin for histologic analysis by trichrome or hematoxylin and eosin staining. Alkaline phosphatase enzyme activity of implants was measured by homogenizing the implants in a solution containing 1.6 M MgCl$_2$, 0.2 M Tris-Cl pH 8.1, and 1% triton X-100 followed by incubation of lysate with 7.5 mM of 4-nitrophenyl-phosphate at room temperature for 1 hour. Absorbance at 405 nm was measured and results were normalized to DNA content. Statistical significance between genotypes was established by use of the Student's *t*-test.

3. Results

3.1. Animals. The Crouzon Fgfr2$^{C342Y/+}$ mutant mice show a phenotype similar to that which was previously reported [5, 32, 37]. The mutant mice are slightly smaller in body size than their wild type littermates and exhibit craniofacial abnormalities associated with craniosynostosis including a dome-shaped skull, wide set and proptotic eyes, and severe midface hypoplasia.

Craniosynostosis-Associated Fgfr2^{C342Y} Mutant Bone Marrow Stromal Cells Exhibit Cell Autonomous Abnormalities in Osteoblast Differentiation and Bone Formation

171

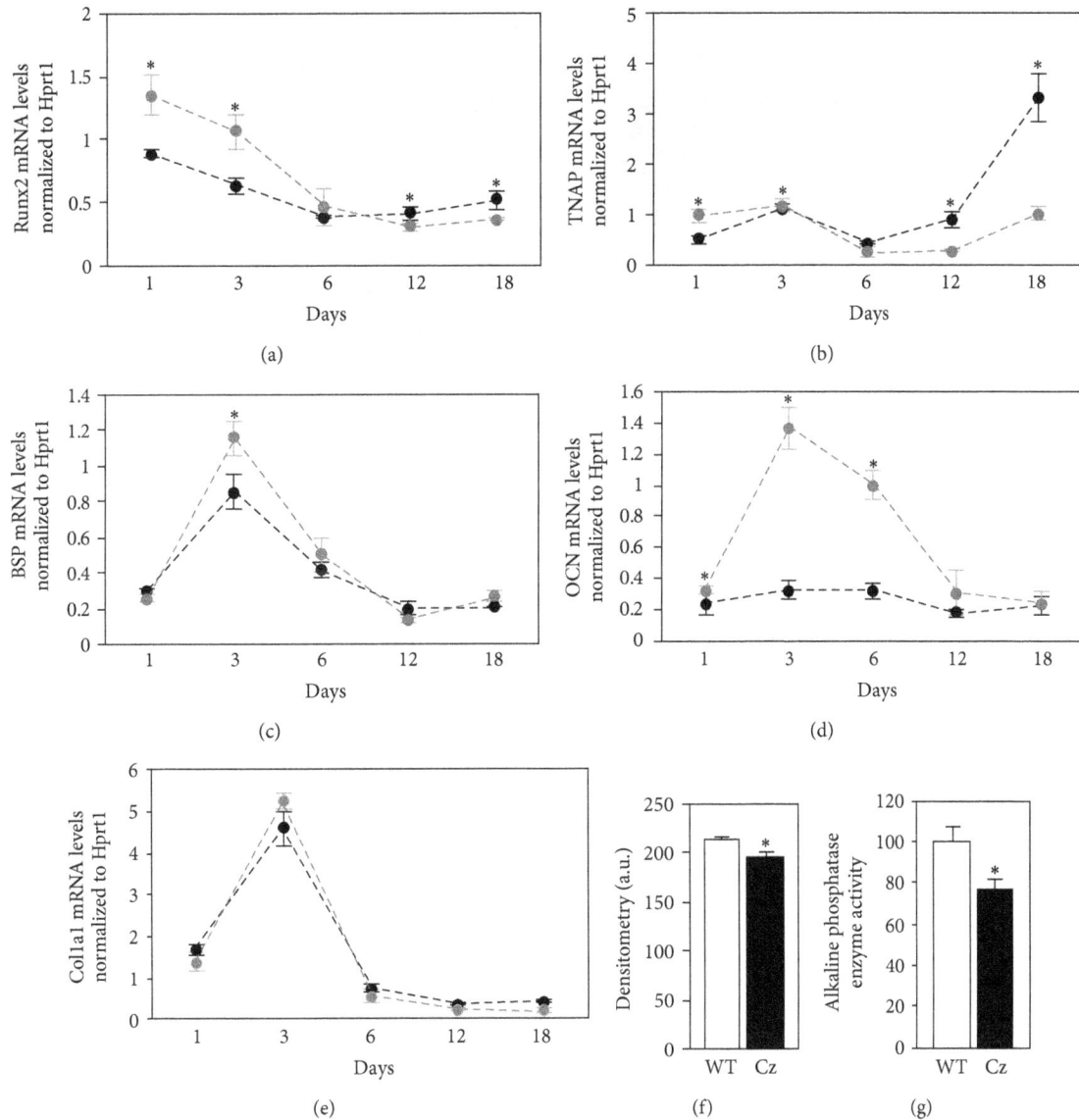

FIGURE 2: The Fgfr2^{C342Y} mutant bone marrow stromal cells exhibit abnormal osteoblastic gene expression and diminished mineralization *in vitro*. Bone marrow stromal cells were isolated from Crouzon Fgfr2$^{C342Y/+}$ (Cz) and wild type (WT) littermates and then cultured with ascorbate for the indicated number of days to induce osteoblast differentiation. Runx2, bone sialoprotein (BSP), osteocalcin (OCN), and tissue non-specific alkaline phosphatase (TNAP) and collagen type 1 alpha1 (col1a1) mRNA levels were measured by real-time PCR. Black lines represent wild type; grey lines represent Crouzon (a, b, c, d, and e). Results are presented as normalized to Hprt1. Cells were cultured with ascorbate and β-glycerophosphate to induce mineralization for 18 days (f). Mineralized nodules were stained by Von Kossa and quantified by densitometry. Cells were cultured with ascorbate for 18 days, and alkaline phosphatase (Tnap/Alpl/Akp2) enzyme activity was quantified by incubation of cells with a colorimetric substrate. Enzyme activity was quantified by densitometry (g). Results shown are means ± standard deviations from triplicate experiments for all data shown. $^*P < .05$ between genotypes.

3.2. *In Vitro Osteoblast Differentiation.* Analysis of mRNA levels demonstrates that bone marrow stromal cells isolated from Crouzon mice express significantly higher levels of Runx2 and tissue non-specific alkaline phosphatase enzyme (AP) at days 1 and 3 of differentiation, significantly higher levels of bone sialoprotein at day 3 of differentiation, and significantly higher levels of osteocalcin at days 1, 3, and 6 of differentiation (Figures 2(a), 2(b), 2(c), and 2(d)). This data indicates that the Fgfr2^{C342Y} mutation enhances the expression of osteoblastic genes in bone marrow stromal cells during early stages of osteoblast differentiation. In contrast, Runx2 and tissue non-specific alkaline phosphatase mRNA levels were significantly lower in cells isolated from Crouzon mice at days 12 and 18 of differentiation (Figures 2(a) and 2(b)). Cells isolated from Crouzon mice also exhibited less alkaline phosphatase enzyme activity at day 18 of differentiation and mineralized to a diminished extent than cells isolated from wild type littermates (Figures 2(f) and 2(g)). In combination,

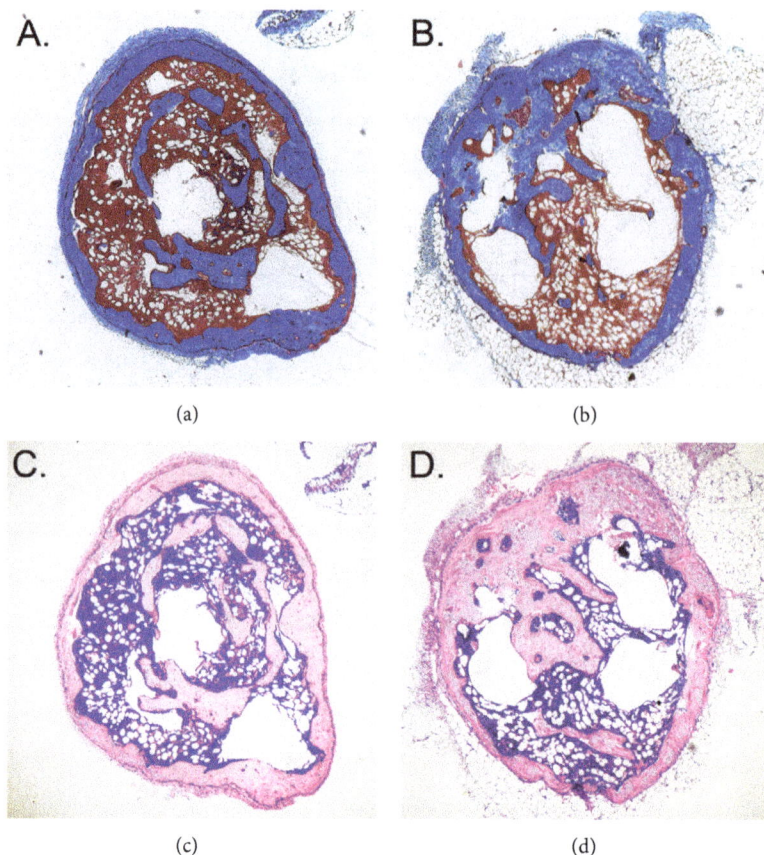

FIGURE 3: Histologic staining of implants. Ossicles formed eight weeks after subcutaneous implantation of cells mixed with a collagenous matrix were stained by trichrome (a and b) or by hematoxylin and eosin (c and d). Note the greater amount of deep blue (a) and light pink (c) staining in ossicles formed by wild type cells, indicative of greater bone formation by these cells. In comparison, note the greater amount of light blue (b) and greyish-pink (d) staining in ossicles formed by Crouzon cells, indicative of greater amounts of original implanted collagen. Also note that ossicles formed by either wild type or mutant cells contain bone marrow.

this data suggests that the Fgfr2^{C342Y} mutation may inhibit later stages of osteoblast differentiation and inhibit mineralization. Notably, collagen type 1, alpha 1 mRNA expression levels were not different between Crouzon and wild type cells at any stages of differentiation. This data suggests that the diminished mineralization by Crouzon cells is not likely due to diminished collagen expression. Together, these results indicate that while the Fgfr2^{C342Y} mutation may enhance early osteoblast differentiation, it also appears to inhibit expression of tissue non-specific alkaline phosphatase and mineralization by more differentiated cells. The findings also indicate that the Fgfr2^{C342Y} mutation induces autonomous abnormalities in osteoblast differentiation when bone marrow stromal cells are cultured in a conventional two-dimensional monolayer system *in vitro*.

3.3. In Vivo Mineralized Tissue Formation.

Because a three-dimensional matrix promotes a more physiologically relevant cellular state than conventional *in vitro* monolayer cell culture [33, 38–40], we next assayed the cells when allowed to differentiate in a collagenous matrix *in vivo*. In accordance with the *in vitro* data, bone marrow stromal cells isolated

from Crouzon mice formed significantly less mineralized tissue than cells isolated from wild type mice, when allowed to differentiate in a three-dimensional collagenous matrix *in vivo*. Both trichrome (bone tissue stains deep blue and implanted collagen stains lighter blue) and H&E histologic stains (bone tissue stains pink and implanted collagen stains greyish-pink) indicate that the ossicles formed by Crouzon cells may contain less mineralized tissue than ossicles formed by wild type cells (Figure 3). Notably, the marrow of implants prepared using either Crouzon or wild type cells contains both hematopoietic and adipocytic cells (small empty ovals are adipocyte ghosts). This may indicate that the Fgfr2^{C342Y} mutation does not influence hematopoietic or adipocytic cell differentiation, although more comprehensive analyses of these cells types in both heterozygous and homozygous mutant mice are required to definitely determine if this is the case. Radiographs of representative implants dissected eight weeks after implantation reveal apparently diminished radiodensity of ossicles prepared using cells isolated from the mutant, as compared to the wild type mice (Figure 4(a)). Quantification of mineralized tissue confirms that the mutant cell implants have significantly less mineralized tissue than the wild type cell implants (Figure 4(b)). Finally, alkaline

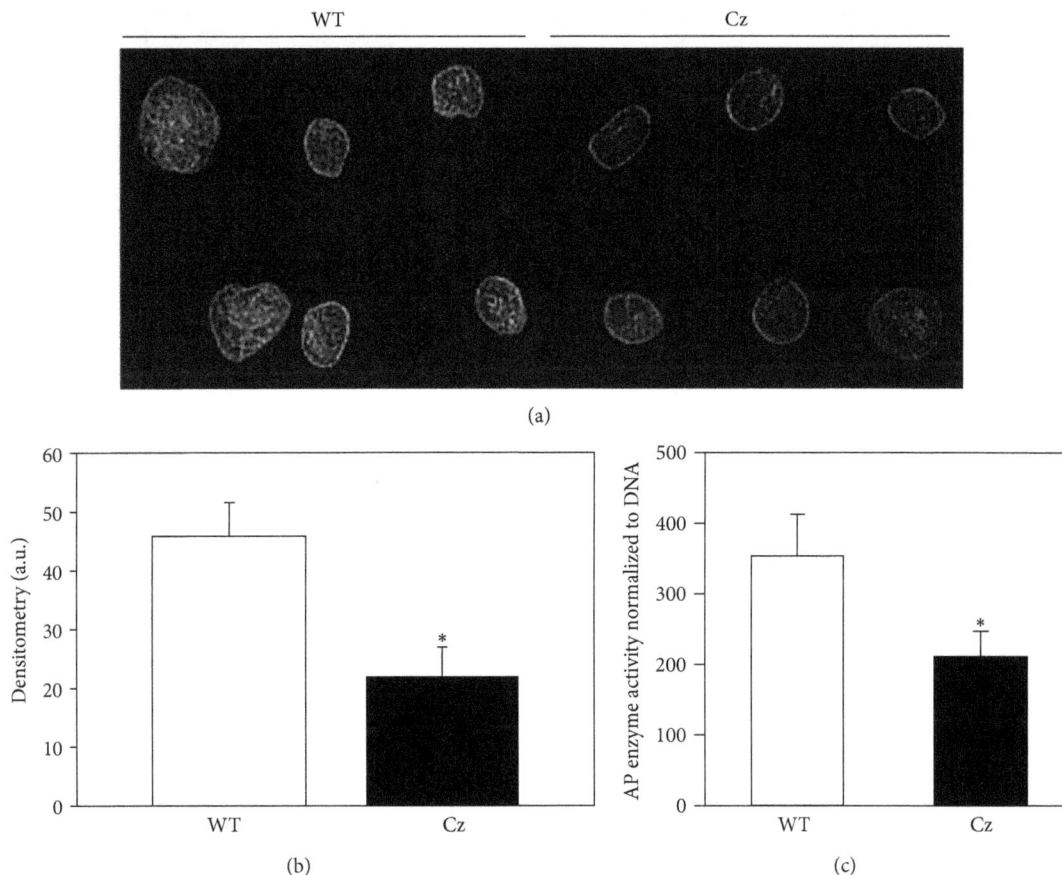

FIGURE 4: The Fgfr2^{C342Y} mutation inhibits mineralized tissue formation *in vivo*. (a) Radiographic image of representative implants shows increased radiodensity of ossicles formed by wild type (WT), as compared to those formed by Crouzon (Cz) cells. (b) Quantification of radiodense tissue by densitometry confirms that implanted wild type cells formed significantly greater amounts of mineralized tissue compared to mutant cells ($n = 12$ implants per genotype). $^*P < .05$ versus WT. (c) Homogenized implants formed by wild type cells also have significantly greater levels of alkaline phosphatase enzyme expression than homogenized implants formed by Crouzon cells ($n = 3$ implants per genotype). $^*P < .05$ versus WT.

phosphatase enzyme expression, which is essential for mineral deposition in collagenous tissues [41] was found to be significantly lower in implants prepared using mutant than wild type cells (Figure 4(c)). Together, these results demonstrate that the Fgfr2^{C342Y} mutation inhibits alkaline phosphatase enzyme expression and the formation of mineralized tissue by osteoprogenitor cells when allowed to differentiate in a three-dimensional matrix *in vivo*.

3.4. Microcomputed Tomographic Analysis of Long Bones.

We recently showed that the frontal cranial bones of Fgfr2$^{C342Y/+}$ mice are diminished in bone volume and density when compared to those of wild type mice [5]. To determine if bones of the appendicular skeleton are similarly affected, we utilized microcomputed tomography to analyze parameters of bone quality and quantity in tibias of four-week-old Fgfr2$^{C342Y/+}$ and Fgfr2$^{+/+}$ mice. Results demonstrate that Fgfr2$^{C342Y/+}$ mice have significantly diminished tibial cortical bone volume/total volume, bone mineral density, and tissue mineral density (Figure 5), when compared to FGFR2$^{+/+}$ mice.

Trabecular measures of tibial bone volume/total volume, bone mineral density, and tissue mineral density were not different between Fgfr2$^{C342Y/+}$ and Fgfr2$^{+/+}$ mice (Figure 6). These results demonstrate that the Fgfr2$^{C342Y/+}$ mutation is associated with decreased bone volume and density that is not limited to the craniofacial skeleton.

4. Discussion

The pathogenesis of craniosynostosis remains unknown and until this knowledge has been realized, the only treatment for craniosynostosis will remain that of surgical intervention. While it is not known if craniosynostosis results primarily from abnormalities in cranial bone cells, cranial suture cells and/or abnormalities in other cell types, mounting evidence does indicate that craniosynostosis occurs in the context of diminished cranial bone quantity and quality. Fgfr2^{S250W} Apert mice and Fgfr3^{P244R} Muenke mice were both previously shown to exhibit craniosynostosis in combination with lower levels of bone formation and/or mineralization when

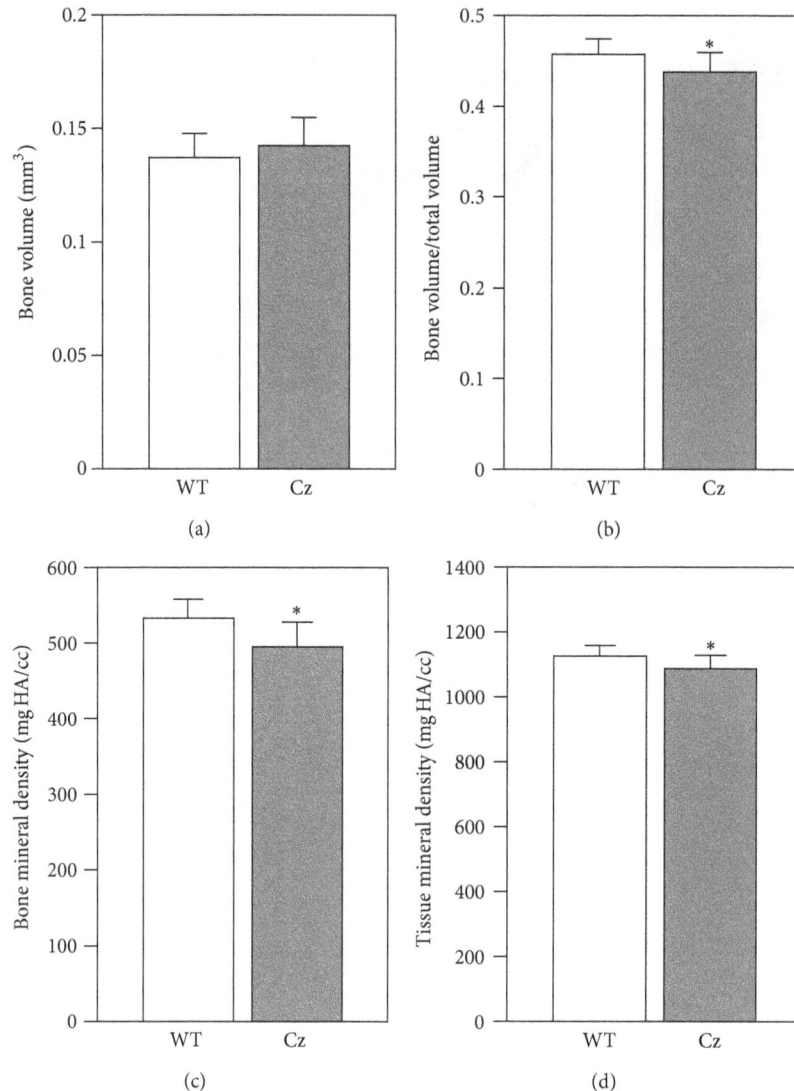

FIGURE 5: Diminished cortical bone volume and density in the long bones of Fgfr2$^{C342Y/+}$ mice. Micro-CT analyses demonstrate significantly diminished cortical bone volume/total volume, bone mineral density, and tissue mineral density in tibias of Fgfr2$^{C342Y/+}$ (Cz) mice as compared to wild type (WT) mice. $^{*}P < .05$ between genotypes.

compared to wild type littermates [42, 43]. As noted previously, Fgfr2$^{C342Y/C342Y}$ Crouzon mice were originally characterized as having decreased ossification of vertebral bodies [32]. In addition, we recently reported that Fgfr2$^{C342Y/+}$ mice on a BALB/c genetic background have diminished cranial bone volume and density when compared to wild type mice [5]. Abnormal BMP signaling in neural crest cells was also recently shown to cause diminished cranial bone volume and density in combination with craniosynostosis [44]. Taken together, these results appear to indicate that craniosynostosis is an abnormality involving excessive ectopic mineralization (bone formation at a temporally and spatially inappropriate location, such as the juvenile cranial suture) and not a disorder of excessive eutopic bone formation. In fact, craniosynostosis associated with several distinct genetic mutations (cited earlier) appears to occur in combination with diminished eutopic bone mass. This distinction is critical for the future development of biologically based therapeutics for the prevention and/or treatment of craniosynostosis.

Here we report that bone marrow stromal cells isolated from Crouzon Fgfr2$^{C342Y/+}$ mice express significantly higher levels of some osteoblastic genes during early stages of differentiation but significantly lower levels of tissue non-specific alkaline phosphatase mRNA and enzyme activity, as well as diminished mineralization when cells are further differentiated in a 2D *in vitro* cell culture system. These results are in accordance with our previous report which showed that frontal bone cells isolated from Fgfr2$^{C342Y/+}$ mice exhibited enhanced early osteoblastic differentiation but diminished later stage differentiation and a decreased tendency to form mineralized tissue, when compared to cells isolated from wild type mice *in vitro* [5]. That the Fgfr2^{C342Y} mutation stimulates early osteoblast differentiation while inhibiting later maturation into fully functional osteoblasts could explain

Craniosynostosis-Associated Fgfr2^{C342Y} Mutant Bone Marrow Stromal Cells Exhibit Cell Autonomous Abnormalities in Osteoblast Differentiation and Bone Formation

175

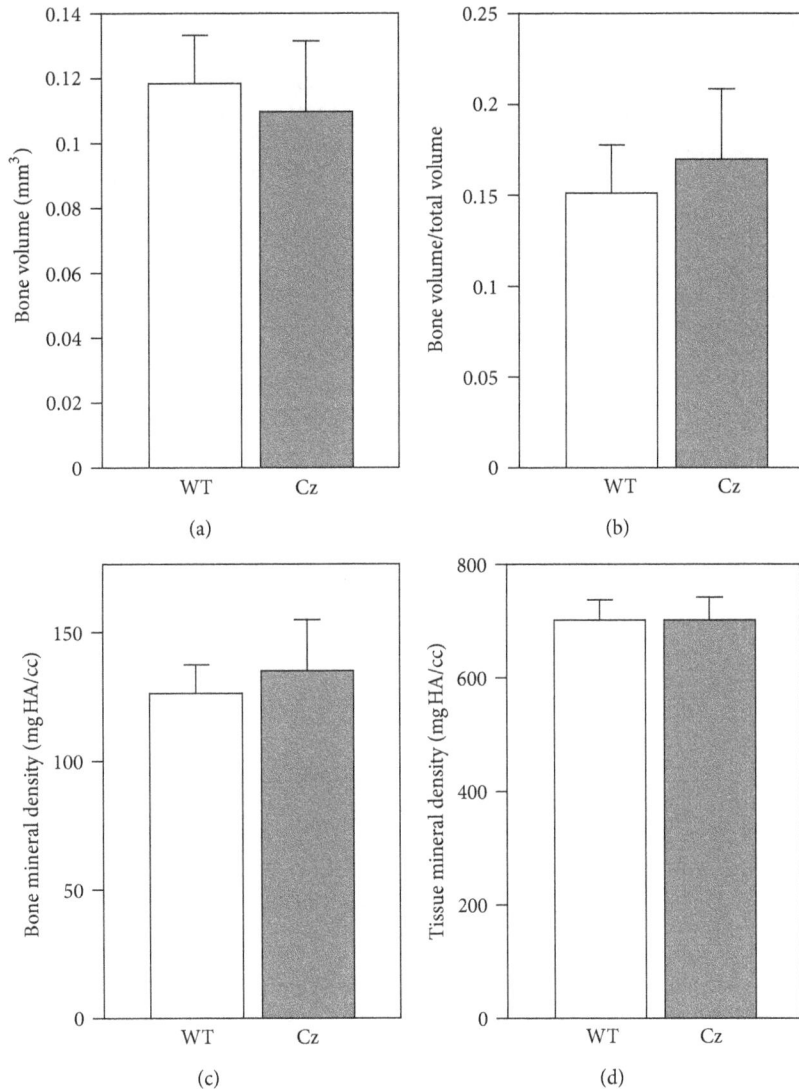

FIGURE 6: Similar trabecular bone volume and density in the long bones of Fgfr2$^{C342Y/+}$ mice. Micro-CT analyses demonstrate no significant differences in bone volume, bone volume/total volume, bone mineral density, and tissue mineral density in tibias of Fgfr2$^{C342Y/+}$ (Cz) mice as compared to wild type (WT) mice.

the apparently discrepant results found in the literature regarding effects of craniosynostosis-associated FGFR mutations on osteoblast differentiation. Our *in vitro* data, showing that Crouzon Fgfr2^{C342Y} marrow stromal cells express higher levels of multiple osteoblastic genes than wild type cells during early stages of differentiation, is in accordance with previous reports which showed increased osteoblastic gene expression in cranial suture tissues. Previous *in vivo* analyses of the Fgfr2$^{P253R/+}$ and Fgfr2$^{S252W/+}$ mouse models of Apert syndrome revealed increased osteoblastic gene expression around the coronal suture [45, 46]. Mice carrying the P250R mutation in Fgfr1 associated with Pfeiffer syndrome were also previously shown to exhibit enhanced osteoblastic differentiation of cells within the sagittal suture [47]. Finally, newborn Fgfr2$^{C342Y/+}$ mice were also previously shown to have increased Runx2 mRNA levels around the coronal

suture, as compared to wild type littermates [32]. Increased expression of osteoblastic genes in suture tissues is well reconciled with our results showing increased osteoblastic gene expression in Fgfr2$^{C342Y/+}$ bone marrow stromal cells during early stages of differentiation. In contrast, our data showing that Fgfr2^{C342Y} marrow stromal cells express significantly lower levels of Runx2 and tissue non-specific alkaline phosphatase mRNA, as well as significantly diminished alkaline phosphatase enzyme activity and mineralization, is in accordance with previous studies showing that S252W, C342Y, and P253R craniosynostosis-associated mutations in FGFR2 inhibit osteoblast differentiation [48–50]. The data can also potentially explain the diminished eutopic bone mass that is seen in multiple mouse models of craniosynostosis, including the Fgfr2^{C342Y} mouse model of Crouzon syndrome. Taken together, it appears that craniosynostosis syndrome-associated mutations in Fgfr's enhance or inhibit osteoblast differentiation in a cell type, environment, and differentiation stage dependent manner.

Importantly, two-dimensional cell culture on plastic does not well represent the environmental conditions that cells find in physiologic tissues, and it has been suggested that three-dimensional cell culture helps to "bridge the gap" between cultured cells and the *in vivo* environment [33]. Therefore, to increase confidence in our *in vitro* findings, we also mixed bone marrow stromal cells in a three dimensional collagenous matrix and allowed them to differentiate when implanted into donor mice. Results of these experiments showed that the Crouzon Fgfr2^{C342Y} mutation inhibited bone formation and alkaline phosphatase enzyme expression, again supporting the idea that the Fgfr2^{C342Y} mutation inhibits later stage osteoblast differentiation and bone formation. While our results also show significantly diminished bone volume and density in the long bones of Fgfr2^{C342Y} mice, future studies are required to definitively establish that Crouzon bone marrow stromal cells are deficient in their ability to differentiate into fully functional osteoblasts and form bone *in vivo*. If correct, our results suggest that patients carrying craniosynostosis syndrome associated Fgfr mutations may be at higher risk for osteoporosis and/or slow repair of long bone fractures.

While the diminished cranial bone formation in these Fgfr-associated mouse models of craniosynostosis has not been previously considered as contributing to the development of craniosynostosis, it is interesting to note that craniosynostosis is also known to occur in other disorders of low bone mineralization. Mutations in the phosphate regulating protein Phex cause X-linked hypophosphatemic rickets involving low serum phosphate, defective bone mineralization and also craniosynostosis [51–53]. It is unknown how Phex mutations lead to craniosynostosis, but, similar to studies of human patients with Fgfr2-associated craniosynostosis [30], these patients also commonly have paradoxical heterotopic calcification of normally nonmineralizing tissues, such as tendons and ligaments [53]. Craniosynostosis is also seen in infantile hypophosphatasia due to inactivating mutations in the enzyme, tissue non-specific alkaline phosphatase (TNAP) [54–56]. These patients have severely deficient bone mineralization [57]. TNAP is an enzymatic generator of inorganic phosphate and an established essential promoter of tissue mineralization, but it is again unknown how diminished TNAP activity leads to craniosynostosis [41, 58]. Finally, craniosynostosis was also previously reported to occur secondary to antacid-induced infantile hypophosphatemia [59]. This data indicates that craniosynostosis occurs in multiple disorders involving dysregulated phosphate homeostasis and bone mineralization. Notably, signaling through Fgfr's was also previously shown to regulate expression of enzymes that control the local production of inorganic phosphate [35], and here we show that cells isolated from Crouzon mice express significantly lower levels of tissue non-specific alkaline phosphatase mRNA and significantly diminished alkaline phosphatase enzyme expression. Together, these results make it tempting to hypothesize that craniosynostosis

may be promoted by abnormal tissue levels of phosphate leading to aberrant tissue mineralization.

In conclusion, this study demonstrates that the Crouzon syndrome associated C342Y mutation in Fgfr2 enhances early osteoblast differentiation but inhibits later differentiation of bone marrow stromal cells into fully functional osteoblasts when cultured in a conventional *in vitro* monolayer system and when allowed to differentiate in a three-dimensional matrix *in vivo*. This study also demonstrates that the long bones of Fgfr2^{C342Y} mice have significantly diminished bone volume and density when compared to wild type littermates. Taken together, our results indicate that Crouzon cells have an intrinsic or cell autonomous defect in osteoblast differentiation and bone formation that includes cells of the appendicular skeleton. Future studies are required to determine if Crouzon syndrome patients are at increased risk for osteoporosis and/or poor repair of bony fractures due to this abnormal cell behavior.

Conflict of Interests

The authors have no relationship with the aforementioned companies and so have no conflict of interests to report.

Acknowledgment

This work was supported by NIDCR Grant R03DE021082.

References

[1] D. Renier, E. Lajeunie, E. Arnaud, and D. Marchac, "Management of craniosynostoses," *Child's Nervous System*, vol. 16, no. 10-11, pp. 645–658, 2000.

[2] M. Seruya, A. Oh, M. J. Boyajian, J. C. Posnick, and R. F. Keating, "Treatment for delayed presentation of sagittal synostosis: challenges pertaining to occult intracranial hypertension—clinical article," *Journal of Neurosurgery*, vol. 8, no. 1, pp. 40–48, 2011.

[3] G. M. Morriss-Kay and A. O. M. Wilkie, "Growth of the normal skull vault and its alteration in craniosynostosis: insights from human genetics and experimental studies," *Journal of Anatomy*, vol. 207, no. 5, pp. 637–653, 2005.

[4] S. Kreiborg, "Craniofacial growth in plagiocephaly and Crouzon syndrome," *Scandinavian Journal of Plastic and Reconstructive Surgery*, vol. 15, no. 3, pp. 187–197, 1981.

[5] J. Liu, H. K. Na, E. Wang, and N. E. Hatch, "Further analysis of the crouzon mouse: effects of the FGFR2^{C342Y} mutation are cranial bone dependent," *Calcified Tissue International*, vol. 92, no. 5, pp. 451–466, 2013.

[6] K. Okajima, L. K. Robinson, M. A. Hart et al., "Ocular anterior chamber dysgenesis in craniosynostosis syndromes with a fibroblast growth factor receptor 2 mutation," *American Journal of Medical Genetics*, vol. 85, no. 2, pp. 160–170, 1999.

[7] P. Stavrou, S. Sgouros, H. E. Willshaw, J. H. Goldin, A. D. Hockley, and M. J. C. Wake, "Visual failure caused by raised intracranial pressure in craniosynostosis," *Child's Nervous System*, vol. 13, no. 2, pp. 64–67, 1997.

[8] H. Abe, T. Ikota, M. Akino, K. Kitami, and M. Tsuru, "Functional prognosis of surgical treatment of craniosynostosis," *Child's Nervous System*, vol. 1, no. 1, pp. 53–61, 1985.

Craniosynostosis-Associated Fgfr2^{C342Y} Mutant Bone Marrow Stromal Cells Exhibit Cell Autonomous Abnormalities in Osteoblast Differentiation and Bone Formation

177

[9] P. S. Shah, K. Siriwardena, G. Taylor et al., "Sudden infant death in a patient with FGFR3 P250R mutation," *American Journal of Medical Genetics A*, vol. 140, no. 24, pp. 2794–2796, 2006.

[10] M. M. Cohen and S. Kreiborg, "Upper and lower airway compromise in the Apert syndrome," *American Journal of Medical Genetics*, vol. 44, no. 1, pp. 90–93, 1992.

[11] S. A. Rasmussen, M. M. Yazdy, J. L. Frías, and M. A. Honein, "Priorities for public health research on craniosynostosis: summary and recommendations from a Centers for Disease Control and Prevention-sponsored meeting," *American Journal of Medical Genetics A*, vol. 146, no. 2, pp. 149–158, 2008.

[12] J. K. Williams, S. R. Cohen, F. D. Burstein, R. Hudgins, W. Boydston, and C. Simms, "A longitudinal, statistical study of reoperation rates in craniosynostosis," *Plastic and Reconstructive Surgery*, vol. 100, no. 2, pp. 305–310, 1997.

[13] W. Reardon, R. M. Winter, P. Rutland, L. J. Pulleyn, B. M. Jones, and S. Malcolm, "Mutations in the fibroblast growth factor receptor 2 gene cause Crouzon syndrome," *Nature Genetics*, vol. 8, no. 1, pp. 98–103, 1994.

[14] U. Schell, A. Hehr, G. J. Feldman et al., "Mutations in FGFR1 and FGFR2 cause familial and sporadic Pfeiffer syndrome," *Human Molecular Genetics*, vol. 4, no. 3, pp. 323–328, 1995.

[15] A. O. M. Wilkie, S. F. Slaney, M. Oldridge et al., "Apert syndrome results from localized mutations of FGFR2 and is allelic with Crouzon syndrome," *Nature Genetics*, vol. 9, no. 2, pp. 165–172, 1995.

[16] O. A. Ibrahimi, F. Zhang, A. V. Eliseenkova, R. J. Linhardt, and M. Mohammadi, "Proline to arginine mutations in FGF receptors 1 and 3 result in Pfeiffer and Muenke craniosynostosis syndromes through enhancement of FGF binding affinity," *Human Molecular Genetics*, vol. 13, no. 1, pp. 69–78, 2004.

[17] N. E. Hatch, "FGF signaling in craniofacial biological control and pathological craniofacial development," *Critical Reviews in Eukaryotic Gene Expression*, vol. 20, no. 4, pp. 295–311, 2010.

[18] J. Andersen, H. D. Burns, P. Enriquez-Harris, A. O. M. Wilkie, and J. K. Heath, "Apert syndrome mutations in fibroblast growth factor receptor 2 exhibit increased affinity for FGF ligand," *Human Molecular Genetics*, vol. 7, no. 9, pp. 1475–1483, 1998.

[19] K. Yu, A. B. Herr, G. Waksman, and D. M. Ornitz, "Loss of fibroblast growth factor receptor 2 ligand-binding specificity in Apert syndrome," *Proceedings of the National Academy of Sciences of the United States of America*, vol. 97, no. 26, pp. 14536–14541, 2000.

[20] K. M. Neilson and R. E. Friesel, "Constitutive activation of fibroblast growth factor receptor-2 by a point mutation associated with Crouzon syndrome," *Journal of Biological Chemistry*, vol. 270, no. 44, pp. 26037–26040, 1995.

[21] V. Shukla, X. Coumoul, R. H. Wang, H. S. Kim, and C. X. Deng, "RNA interference and inhibition of MEK-ERK signaling prevent abnormal skeletal phenotypes in a mouse model of craniosynostosis," *Nature Genetics*, vol. 39, no. 9, pp. 1145–1150, 2007.

[22] V. P. Eswarakumar, F. Özcan, E. D. Lew et al., "Attenuation of signaling pathways stimulated by pathologically activated FGF-receptor 2 mutants prevents craniosynostosis," *Proceedings of the National Academy of Sciences of the United States of America*, vol. 103, no. 49, pp. 18603–18608, 2006.

[23] G. M. Cooper, E. L. Durham, J. J. Cray, M. I. Siegel, J. E. Losee, and M. P. Mooney, "Tissue interactions between craniosynostotic dura mater and bone," *Journal of Craniofacial Surgery*, vol. 23, no. 3, pp. 919–924, 2012.

[24] B. U. Ang, R. M. Spivak, H. D. Nah, and R. E. Kirschner, "Dura in the pathogenesis of syndromic craniosynostosis: fibroblast growth factor receptor 2 mutations in dural cells promote osteogenic proliferation and differentiation of osteoblasts," *Journal of Craniofacial Surgery*, vol. 21, no. 2, pp. 462–467, 2010.

[25] B. J. Slater, M. D. Kwan, D. M. Gupta, R. R. Amasha, D. C. Wan, and M. T. Longaker, "Dissecting the influence of regional dura mater on cranial suture biology," *Plastic and Reconstructive Surgery*, vol. 122, no. 1, pp. 77–84, 2008.

[26] K. Mangasarian, Y. Li, A. Mansukhani, and C. Basilico, "Mutation associated with Crouzon syndrome causes ligand-independent dimerization and activation of FGF receptor-2," *Journal of Cellular Physiology*, vol. 172, no. 1, pp. 117–125, 1997.

[27] A. Lomri, J. Lemonnier, M. Hott et al., "Increased calvaria cell differentiation and bone matrix formation induced by fibroblast growth factor receptor 2 mutations in Apert syndrome," *Journal of Clinical Investigation*, vol. 101, no. 6, pp. 1310–1317, 1998.

[28] G. Holmes, G. Rothschild, U. B. Roy, C. X. Deng, A. Mansukhani, and C. Basilico, "Early onset of craniosynostosis in an Apert mouse model reveals critical features of this pathology," *Developmental Biology*, vol. 328, no. 2, pp. 273–284, 2009.

[29] Online Mendelian Inheritance in Man and OMIM, MIM Number 123500, Johns Hopkins University, Baltimore, Md, USA, 2011, http://omim.org/entry/123500.

[30] S. Kreiborg, "Crouzon Syndrome. A clinical and roentgencephalometric study," *Scandinavian Journal of Plastic and Reconstructive Surgery and Hand Surgery*, vol. 18, pp. 1–198, 1981.

[31] T. W. Proudman, M. H. Moore, A. H. Abbott, and D. J. David, "Noncraniofacial manifestations of Crouzon's disease," *Journal of Craniofacial Surgery*, vol. 5, no. 4, pp. 218–222, 1994.

[32] V. P. Eswarakumar, M. C. Horowitz, R. Locklin, G. M. Morriss-Kay, and P. Lonai, "A gain-of-function mutation of Fgfr2c demonstrates the roles of this receptor variant in osteogenesis," *Proceedings of the National Academy of Sciences of the United States of America*, vol. 101, no. 34, pp. 12555–12560, 2004.

[33] F. Pampaloni, E. G. Reynaud, and E. H. K. Stelzer, "The third dimension bridges the gap between cell culture and live tissue," *Nature Reviews Molecular Cell Biology*, vol. 8, no. 10, pp. 839–845, 2007.

[34] M. L. Bouxsein, S. K. Boyd, B. A. Christiansen, R. E. Guldberg, K. J. Jepsen, and R. Müller, "Guidelines for assessment of bone microstructure in rodents using micro-computed tomography," *Journal of Bone and Mineral Research*, vol. 25, no. 7, pp. 1468–1486, 2010.

[35] N. E. Hatch, Y. Li, and R. T. Franceschi, "FGF2 stimulation of the pyrophosphate-generating enzyme, PC-1, in pre-osteoblast cells is mediated by RUNX2," *Journal of Bone and Mineral Research*, vol. 24, no. 4, pp. 652–662, 2009.

[36] P. H. Krebsbach, K. Gu, R. T. Franceschi, and R. B. Rutherford, "Gene therapy-directed osteogenesis: BMP-7-transduced human fibroblasts form bone *in vivo*," *Human Gene Therapy*, vol. 11, no. 8, pp. 1201–1210, 2000.

[37] C. A. Perlyn, V. B. DeLeon, C. Babbs et al., "The craniofacial phenotype of the Crouzon mouse: analysis of a model for syndromic craniosynostosis using three-dimensional microCT," *Cleft Palate-Craniofacial Journal*, vol. 43, no. 6, pp. 740–747, 2006.

[38] A. Birgersdotter, R. Sandberg, and I. Ernberg, "Gene expression perturbation *in vitro*—a growing case for three-dimensional (3D) culture systems," *Seminars in Cancer Biology*, vol. 15, no. 5, pp. 405–412, 2005.

[39] U. A. Gurkan, V. Kishore, K. W. Condon, T. M. Bellido, and O. Akkus, "A scaffold-free multicellular three-dimensional *in vitro* model of osteogenesis," *Calcified Tissue International*, vol. 88, no. 5, pp. 388–401, 2011.

[40] P. R. Baraniak and T. C. McDevitt, "Scaffold-free culture of mesenchymal stem cell spheroids in suspension preserves multilineage potential," *Cell and Tissue Research*, vol. 347, no. 3, pp. 701–711, 2012.

[41] M. Murshed, D. Harmey, J. L. Millán, M. D. McKee, and G. Karsenty, "Unique coexpression in osteoblasts of broadly expressed genes accounts for the spatial restriction of ECM mineralization to bone," *Genes and Development*, vol. 19, no. 9, pp. 1093–1104, 2005.

[42] L. Chen, D. Li, C. Li, A. Engel, and C. X. Deng, "A Ser250Trp substitution in mouse fibroblast growth factor receptor 2 (Fgfr2) results in craniosynostosis," *Bone*, vol. 33, no. 2, pp. 169–178, 2003.

[43] S. R. F. Twigg, C. Healy, C. Babbs et al., "Skeletal analysis of the $Fgfr3^{P244R}$ mouse, a genetic model for the muenke craniosynostosis syndrome," *Developmental Dynamics*, vol. 238, no. 2, pp. 331–342, 2009.

[44] Y. Komatsu, P. B. Yu, N. Kamiya et al., "Augmentation of Smad-dependent BMP signaling in neural crest cells causes craniosynostosis in mice," *Journal of Bone and Mineral Research*, 2012.

[45] G. Holmes, G. Rothschild, U. B. Roy, C. X. Deng, A. Mansukhani, and C. Basilico, "Early onset of craniosynostosis in an Apert mouse model reveals critical features of this pathology," *Developmental Biology*, vol. 328, no. 2, pp. 273–284, 2009.

[46] L. Yin, X. Du, C. Li et al., "A Pro253Arg mutation in fibroblast growth factor receptor 2 (Fgfr2) causes skeleton malformation mimicking human Apert syndrome by affecting both chondrogenesis and osteogenesis," *Bone*, vol. 42, no. 4, pp. 631–643, 2008.

[47] Y. X. Zhou, X. Xu, L. Chen, C. Li, S. G. Brodie, and C. X. Deng, "A Pro250Arg substitution in mouse Fgfr1 causes increased expression of Cbfa1 and premature fusion of calvarial sutures," *Human Molecular Genetics*, vol. 9, no. 13, pp. 2001–2008, 2000.

[48] A. Fragale, M. Tartaglia, S. Bernardini et al., "Decreased proliferation and altered differentiation in osteoblasts from genetically and clinically distinct craniosynostotic disorders," *American Journal of Pathology*, vol. 154, no. 5, pp. 1465–1477, 1999.

[49] A. Mansukhani, P. Bellosta, M. Sahni, and C. Basilico, "Signaling by fibroblast growth factors (FGF) and fibroblast growth factor receptor 2 (FGFR2)-activating mutations blocks mineralization and induces apoptosis in osteoblasts," *Journal of Bone and Mineral Research*, vol. 149, pp. 1297–1308, 2000.

[50] Online Mendelian Inheritance in Man and OMIM, MIM Number 307800, Johns Hopkins University, Baltimore, Md, USA, 2011, http://omim.org/entry/307800.

[51] W. A. Roy, R. J. Iorio, and G. A. Meyer, "Craniosynostosis in vitamin D-resistant rickets. A mouse model," *Journal of Neurosurgery*, vol. 55, no. 2, pp. 265–271, 1981.

[52] Y. Sabbagh, A. O. Jones, and H. S. Tenenhouse, "PHEXdb, a locus-specific database for mutations causing X-linked hypophosphatemia," *Human Mutation*, vol. 16, pp. 1–6, 2000.

[53] G. Liang, L. D. Katz, K. L. Insogna, T. O. Carpenter, and C. M. MacIca, "Survey of the enthesopathy of X-linked hypophosphatemia and its characterization in Hyp mice," *Calcified Tissue International*, vol. 85, no. 3, pp. 235–246, 2009.

[54] Online Mendelian Inheritance in Man and OMIM, MIM Number 171760, Johns Hopkins University, Baltimore, Md, USA, 2009, http://omim.org/entry/171760.

[55] D. Wenkert, M. Benigno, K. Mack, W. McAlister, S. Mumm, and M. Whyte, "Hypophosphatasia: prevalence of clinical problems in 175 pediatric patients," in *Proceedings of the American Society for Bone and Mineral Research 31st Annual Meeting*, Denver, Colo, USA, 2009, Abstract A09001674.

[56] M. P. Whyte, "Physiological role of alkaline phosphatase explored in hypophosphatasia," *Annals of the New York Academy of Sciences*, vol. 1192, pp. 190–200, 2010.

[57] E. Mornet, "Hypophosphatasia," *Orphanet Journal of Rare Diseases*, vol. 2, p. 40, 2007.

[58] S. Narisawa, N. Fröhlander, and J.L. Millán, "Inactivation of two mouse alkaline phosphatase genes and establishment of a model of infantile hypophosphatasi," *Developmental Dynamics*, vol. 208, pp. 432–446, 1997.

[59] A. K. Shetty, T. Thomas, J. Rao, and A. Vargas, "Rickets and secondary craniosynostosis associated with long-term antacid use in an infant," *Archives of Pediatrics and Adolescent Medicine*, vol. 152, no. 12, pp. 1243–1245, 1998.

18

Mineral and Matrix Changes in Brtl/+ Teeth Provide Insights into Mineralization Mechanisms

Adele L. Boskey,[1] **Kostas Verdelis,**[2] **Lyudmila Spevak,**[1] **Lyudmila Lukashova,**[1] **Elia Beniash,**[3] **Xu Yang,**[3] **Wayne A. Cabral,**[4] **and Joan C. Marini**[4]

[1] *Musculoskeletal Integrity Program, Hospital for Special Surgery, 535 E 70th Street, New York, NY 10021, USA*
[2] *Department of Endodontics, School of Dental Medicine, University of Pittsburgh, 3501 Terrace Street, Pittsburgh, PA 15261, USA*
[3] *Department of Oral Biology, School of Dental Medicine, University of Pittsburgh, 3501 Terrace Street, Pittsburgh, PA 15261, USA*
[4] *Bone & Extracellular Matrix Branch, NIH/ NICHD, Bethesda, MD 20892, USA*

Correspondence should be addressed to Adele L. Boskey; boskeya@hss.edu

Academic Editor: Yong-Hee P. Chun

The Brtl/+ mouse is a knock-in model for osteogenesis imperfecta type IV in which a Gly349Cys substitution was introduced into one COL1A1 allele. To gain insight into the changes in dentin structure and mineral composition in these transgenic mice, the objective of this study was to use microcomputed tomography (micro-CT), scanning electron microscopy (SEM), and Fourier transform infrared imaging (FTIRI) to analyze these structures at 2 and 6 months of age. Results, consistent with the dental phenotype in humans with type IV OI, showed decreased molar volume and reduced mineralized tissue volume in the teeth without changes in enamel properties. Increased acid phosphate content was noted at 2 and 6 months by FTIRI, and a trend towards altered collagen structure was noted at 2 but not 6 months in the Brtl/+ teeth. The increase in acid phosphate content suggests a delay in the mineralization process, most likely associated with the defect in the collagen structure. It appears that in the Brtl/+ teeth slow maturation of the mineralized structures allows correction of altered mineral content and acid phosphate distribution.

1. Introduction

Osteogenesis imperfecta (OI) and dentinogenesis imperfecta (DGI) are rare genetic diseases associated, for the most part, with abnormalities in collagen structure, production, or processing, which result in fragile (brittle) mineralized tissues [1–4]. In addition to the four types of OI originally described by Sillence et al. [5], additional types have been identified in humans [6], and their underlying molecular origins for the most part have been determined. The reasons for the variation in mineralization and mechanical properties in OI and DGI are less certain. The Brtl/+ mouse is a knock-in model for moderately severe OI in which a Gly349Cys substitution identified in a child with type IV OI was introduced into one COL1A1 allele, resulting in a phenotype representative of type IV OI [7]. Changes in the whole bone mechanical strength and collagen D-spacing in these Brtl/+ mice have previously been reported [8, 9] along with variations in their osteoblast and osteoclast activities [10]. The animals were also noted, based on light microscopy, to have a DGI phenotype at 2 months due to changes in the pulp cavities of their teeth [7]. The purpose of the present study was to gain further insight into the mineralization and collagen defects in Brtl/+ mouse teeth using a combination of Fourier transform infrared microscopic imaging (FTIRI), scanning electron microscopy (SEM), and microcomputed tomography (micro-CT) and to use this information to elucidate the mechanism underlying these defects.

2. Materials and Methods

2.1. Animals. The Brtl/+ mice were derived as described [7] and sacrificed for other purposes under an NIH-approved IACUC approval at two and six months. The mice were fed a standard (NIH 31) chow and were given sterilized fluoridated tap water, ad libitum. Brtl/+ mice have a mixed background of Sv129/CD-1/C57BL/6S and are bred by crossing heterozygous

Brtl/+ with WT [7]. Based on preliminary analyses using GraphPad StatMate 2.0 (GraphPad Corp, San Diego, CA, USA) it was determined that to detect a significant difference with a power of 80% in mineral/matrix content by FTIRI, six teeth would be needed at 2 months of age. Jaws (n = 6 per genotype), provided frozen, were fixed in ethanol for micro-CT analyses. Jaws were then embedded in polymethyl methacrylate (PMMA) for backscattered SEM and Fourier transform infrared imaging (FTIRI).

2.2. Micro-CT. Three-dimensional architecture and geometry of both the crowns and roots of first and second molars were determined by microcomputed tomography using a Scanco μCT35 (Scanco Medical, Basselrsdorf, Switzerland) system with a 12 μm voxel size at 55 kVp as detailed elsewhere [11, 12]. Crown and root regions of interest were generated by defining the respective contours on sequential reconstructed volume slices [12, 13]. Because of clear and consistent respective interfaces, the segmentation of dentin from the background and enamel from dentin was based on 0.6 and 1.7 g/cc, respectively, global thresholds. The parameters evaluated were tissue volume (TV-representing the total crown or root volume including the pulp or root canal space), enamel + dentin volume (BV), the volumes of Enamel and Dentin separately (EV and DV), the mineral density for each fraction (TMD (E) and TMD (D)), and the tissue volume fraction (BV/TV). All parameters were evaluated on the molar 3D volumes using the crown and root regions of interest through the Scanco evaluation software operating in an open VMS environment. Mineral density for dentin was calculated on basis of the dentin/background segmentation value as a minimum threshold and the dentin/enamel one as a maximum threshold, while the enamel mineral density was calculated using only the dentin/enamel segmentation value as a minimum threshold with maximum threshold kept as the highest available in the scale. The Scanco system uses built-in beam energy-specific algorithms for the conversion of attenuation coefficient into mg/cm^3 values.

2.3. Backscattered Scanning Electron Microscopy (BS SEM). First and second mandibular molars were embedded in PMMA resin and polished on a MiniMet polisher (Buehler, Lake Bluff, IL, USA) down to 0.25 μm using series of MetaDi diamond suspensions, as previously described [14]. Three WT and 3 Brtl/+ samples were studied in JEOL 6335F Field Emission SEM equipped with a backscattered electron detector. The microscopy was conducted at 10 KV and at a working distance of 14 mm.

2.4. Fourier Transform Infrared Imaging. Multiple sections (1-2 microns) of nondecalcified molars, embedded in PMMA, were examined by Fourier transform infrared spectroscopic imaging (Perkin Elmer model Spotlight 100 imaging system) as detailed elsewhere [15, 16]. In brief, dentin spectral images from the whole crown (sectioned along the sagittal plane) or root areas (the mesial or distal root, imaged separately) from first and second molars were examined at 4 μm spatial resolution. ISYS software (Spectral Dimensions, Olney, MD,

USA) was used to process the data, including a subtraction of water vapor and PMMA. Parameters calculated and exhibited as images were (i) mineral-to-matrix ratios (Min/Mat, a comparison of the relative ratios of the integrated intensities of the v1, v3 phosphate band, ~900–1200 cm^{-1} to that of the protein amide I band (centered at 1660 cm^{-1})), (ii) carbonate (855–890 cm^{-1}) to phosphate band area ratio (CO_3/PO_4, indicating carbonate substitution for phosphate in the mineral) (iii) crystallinity (XST, an estimate of crystallite size and perfection, based on the proportion of stoichiometric and non-stoichiometric apatite), (iv) collagen maturity (XLR, a peak height ratio of subbands in the collagen amide I peak), and (v) acid phosphate content HPO_4, (1128 cm^{-1}/1096 cm^{-1}) an estimate of the amount of acid phosphate substitution in the mineral lattice. The validation of these parameters is described in detail elsewhere [15–17].

2.5. Statistics. The mean values and standard deviations for each tissue type and each parameter in each animal were calculated and the values were compared by ANOVA between Brtl/+ and WT crown and root areas, followed by a Bonferrni multiple comparison test. Due to the limited number of animals available for analysis no correction for animal sex was made. Where multiple comparisons of data were involved, a $P < (0.05/n)$, where n is the number of comparisons, was considered significant.

3. Results

The molars of the Brtl/+ animals differed from those of the WT when examined by micro-CT (Figures 1(a) and 1(b)) with small but significant ($P < 0.01$) decrease in the first and second molar Brtl/+ crowns and root total volume and dentin volumes relative to WT at 2 months and 6 months (Table 1). This decrease of dentin thickness was most pronounced in the root. The mineralized tissue volume fraction (BV/TV) was decreased in the root of the Brtl/+ molars at both 2 and 6 months. There was a small, not consistent increase in the first and second molars, crowns and root mineral density in the Brtl/+ molars at 2 months. As previously noted [7], both first and second molars had widened pulp spaces and a disorganized bone structure around the molar roots.

No major morphological difference between Brtl/+ (Figure 2) and wild type (Figure 3) mandibular molars was observed by backscattered electron imaging, based on the analysis of 3 samples per group. The density and organization of dentinal tubules were similar in WT and Brtl/+ molars. The reduced dentin thickness and increased pulp chamber size in Brtl/+ molars observed in the micro-CT 3-dimensional volumes were not noticeable in the SEM images. Only one tooth had an abnormal mesial root with thin root dentin and a wide root canal. Interestingly, the presence of globular dentin was observed in the cervical portions of the roots of all Brtl/+ molars (Figure 2), while no globular dentin was detected in wild type molars (Figure 3). At higher magnifications, however, significant abnormalities in dentin mineralization in Brtl/+ were observed. Specifically, noticeable variations in

TABLE 1: Micro-CT parameters measured in Brtl/+ and wild type molars at 2 and 6 months.

| | Two months | | | | Six months | | | |
| | First molars | | Second molars | | First molars | | Second molars | |
	Brtl/+	WT	Brtl/+	WT	Brtl/+	WT	Brtl/+	WT
Crown								
BV (D + E) mm^3	0.552 ± 0.0002^a	0.578 ± 0.0003	0.286 ± 0.0003^a	$0.342 \pm 6.71E - 05$	0.516 ± 0.0015	0.552 ± 0.0004	$0.293 \pm 1.44E - 05^a$	0.332 ± 0.0010
BV (D) mm^3	0.404 ± 0.0001^a	0.435 ± 0.0005	0.200 ± 0.0001^a	0.247 ± 0.0001	0.368 ± 0.0007^a	0.411 ± 0.0002	$0.195 \pm 1.98E - 05^a$	0.235 ± 0.0007
BV (E) mm^3	$0.148 \pm 7.69E - 05$	$0.143 \pm 7.58E - 05$	$0.086 \pm 5.35E - 05^a$	$0.095 \pm 2.99E - 05$	0.148 ± 0.0002	$0.141 \pm 2.64E - 05$	$0.098 \pm 3.76E - 05$	$0.098 \pm 8.37E - 05$
TMD (D) mg/cm^3	1327 ± 336^a	1266 ± 304	1275 ± 1121	1264.9 ± 271	1337 ± 1080	1306 ± 267	1328 ± 1907	1299 ± 295
TMD (E) mg/cm^3	2024 ± 627	2005 ± 329	1996 ± 382	1995.6 ± 347	2008 ± 1436	2007 ± 244	2002 ± 1346	2006 ± 96
Root								
TV mm^3	0.482 ± 0.0005^a	0.544 ± 0.0004	0.243 ± 0.0001^a	0.309 ± 0.0002	0.557 ± 0.0014^a	0.610 ± 0.0005	0.294 ± 0.0008^a	0.355 ± 0.0004
BV mm^3	0.362 ± 0.0007^a	0.460 ± 0.0004	0.111 ± 0.0014^a	0.253 ± 0.0002	0.430 ± 0.0034^a	0.530 ± 0.0004	0.211 ± 0.025^a	0.303 ± 0.0004
BV/TV%	0.752 ± 0.003^a	0.845 ± 0.0002	0.454 ± 0.021^a	0.820 ± 0.0001	0.773 ± 0.086^a	$0.870 \pm 8.03E - 05$	0.712 ± 0.018^a	0.853 ± 0.0003
TMD mg/cm^3	1146 ± 416	1162 ± 406	1056 ± 2187^a	1131 ± 704	1177 ± 2709	1227 ± 684	1141 ± 2945	1199.4 ± 644

[a] $P < 0.0125$ versus WT of same age and same tooth type, $n = 6$/genotype.

2-month old

(a)

6-month old

(b)

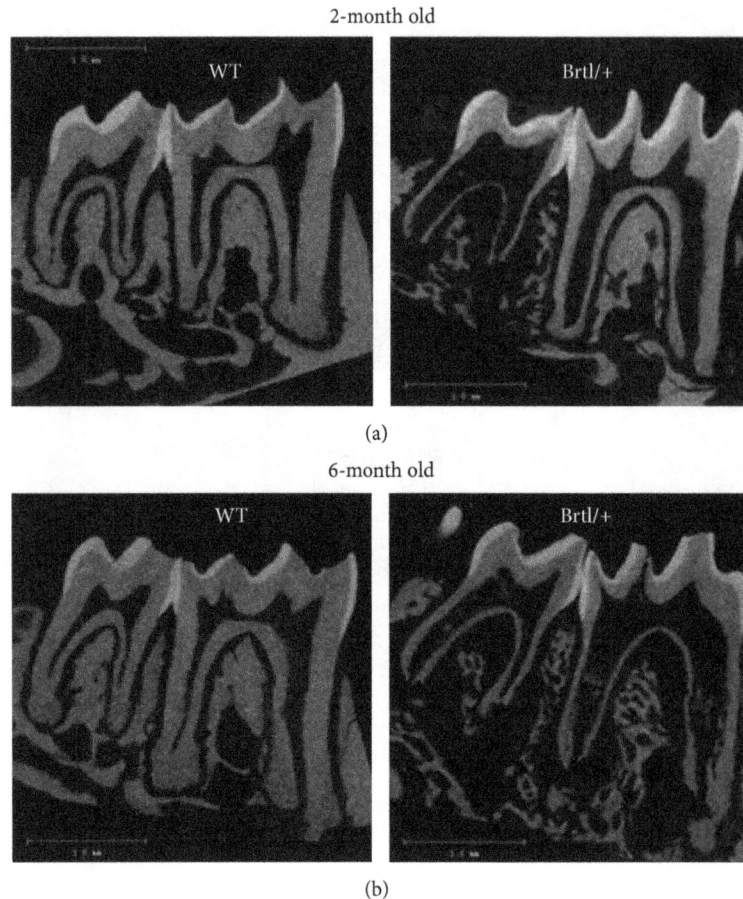

FIGURE 1: Sagittal views from micro-CT images of Brtl/+ and WT first and second molars at (a) 2 and (b) 6 months. Brtl/+ images show wide pulp spaces and decreased dentin thickness, more pronounced in the root, compared to age matched WT. Note the disorganized trabecular jaw bone around the Brtl/+ roots.

the dentin density were observed in the ~50 μm thick proximal dentin layer adjacent to the pulp cavity (Figure 2(c)). Additional studies are needed to further investigate this phenomenon.

Fourier transform infrared imaging (FTIRI) of the molars showed the average mineral/matrix ratio not being significantly altered in the Brtl/+ molars at both 2 and 6 months (Table 2). In fact, the only significant differences in the mean values of the FTIRI images of the teeth were in the acid phosphate content, which was increased in the Brtl/+ roots and crowns at both ages. The distribution of mineral/matrix ratio, crystallinity, and acid phosphate content looked different in typical images of the molars and roots (Figure 4). These "chemical photographs" [18] show, as is typical for mouse molar sections, a broad distribution of mineral properties with visibly increased mineral/matrix ratio in the Brtl/+ molars, as shown in terms of standard deviation in Table 2. The sharp lines and small holes in the section (indicated by arrows) are an artifact due to folds and actual holes made during sectioning and were not included when the average values were calculated. The increased acid phosphate accumulation was apparent in the images (far right column, Figure 4). In the 2-month-old animals the mean crystallinity tended to be reduced in the Brtl/+ crowns and roots, and at

6 months there were some differences in CO_3/PO_4 ratio, but the most significant differences were in the acid phosphate substitution.

4. Discussion

Dentin and enamel mineralizations occur on different types of matrices. The enamel matrix mineralization is influenced by globular proteins such as amelogenin and enamelin [19, 20] which constitute a small fraction of the final mineralized matrix. Dentin mineralization, by contrast, is controlled by the fibrous protein collagen and its associated noncollagenous proteins [21] which persist throughout the lifetime of the tissue. Many of the noncollagenous proteins are intrinsically disordered and assume conformations to match their binding partners [22], such as collagen and hydroxyapatite mineral. In the Brtl/+ molars, the defect in the collagen structure [9] may affect the binding and the concentration of these noncollagenous proteins, their synthesis by collagen producing cells, and their ability to regulate mineralization, as occurs in both Brtl/+ stem cells and bone [23]. This is seen in the data at both 2 and 6 months, where the teeth are smaller and the mineral properties of the dentin, but not the enamel, are distinct from those of the wild type.

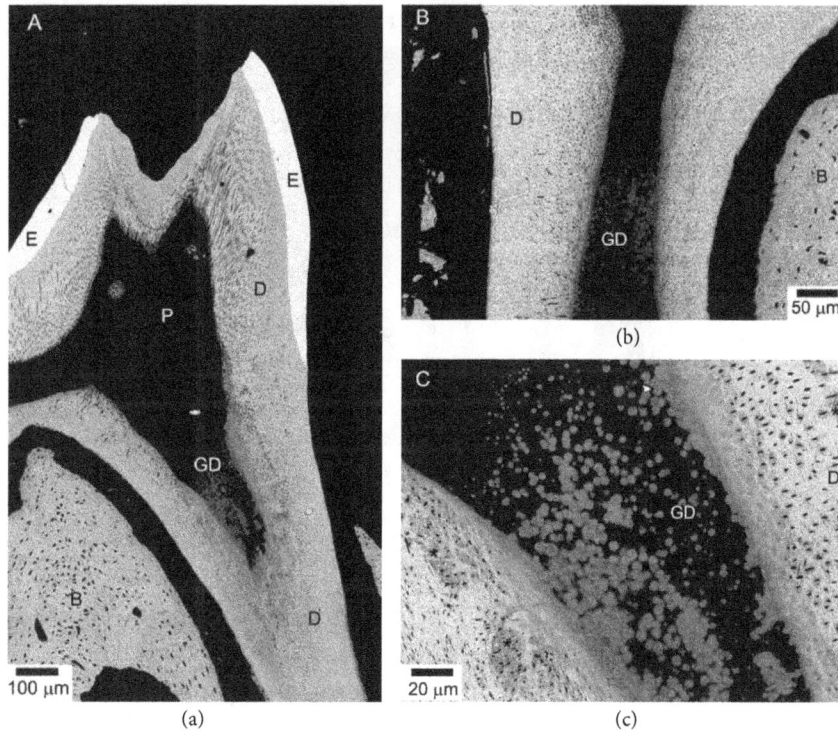

FIGURE 2: BS SEM micrographs of Brtl/+ mandibular molars. (a) Low resolution micrograph showing overall morphology of the tooth. Note an area of globular dentin (GD) in the cervical root. (b) A micrograph of a molar from a different animal also featuring GD. (c) Closeup of the area containing GD from (a). Note the variations in the degree of mineralization. B: bone, D: dentin, E: enamel, and P: pulp. Asterisks mark undermineralized areas.

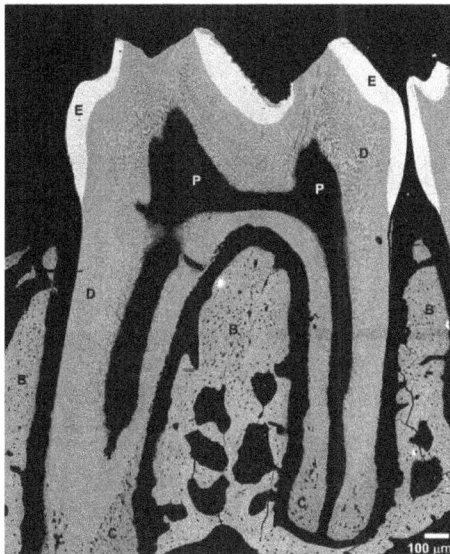

FIGURE 3: BS SEM micrograph of WT mandibular molars. Abbreviations are as in Figure 2.

There is only one report of changes in mouse molar micro-CT parameters with age, in which fibromodulin-deficient mice at 3 weeks and 10 weeks were compared to wild type mice, and a reversal of phenotypic differences was found [24]. It is not clear from only two time points in each of the fibromodulin studies whether older animals begin to show degradation of their teeth or whether the KO animal's molars simply begin to approach the WT values as they age. This is one of the limitations of our Brtl/+ study as we only studied animals at 2 and 6 months. It might have been interesting if we had included teeth from Brtl/+ mice at a younger age. The increases in mineralization of teeth with age, along with the altered collagen structure, imply that the hypermineralization in this variant of classical OI is associated with an abnormal collagen matrix and provide one possible mechanism for the way these abnormal collagen matrices could affect initial mineralization. In the mice studied here, the 2-month-old Brtl/+ mice have completed their tooth development and have been weaned for about a month. Thus they are chewing their food, and the phenotypic difference between their less mineralized teeth and those of the wild type animals may be more apparent than those which are noted at 6 months when age-related physiologic effects (grinding of the teeth, change in curvature due to alveolar osteoclasts, etc. [25]) may mask the Brtl/+ phenotype.

Our data represent the first quantitative evaluations of the mineral properties of the teeth in Brtl/+ mice. Earlier studies had reported the presence of widened and infected pulp cavities characteristic of dentinogenesis imperfecta in patients with osteogenesis imperfecta [7]. Although age-dependent changes were noted in both WT and Brtl/+ molars, the absence of studies monitoring age-dependent change in healthy mice teeth during development made it

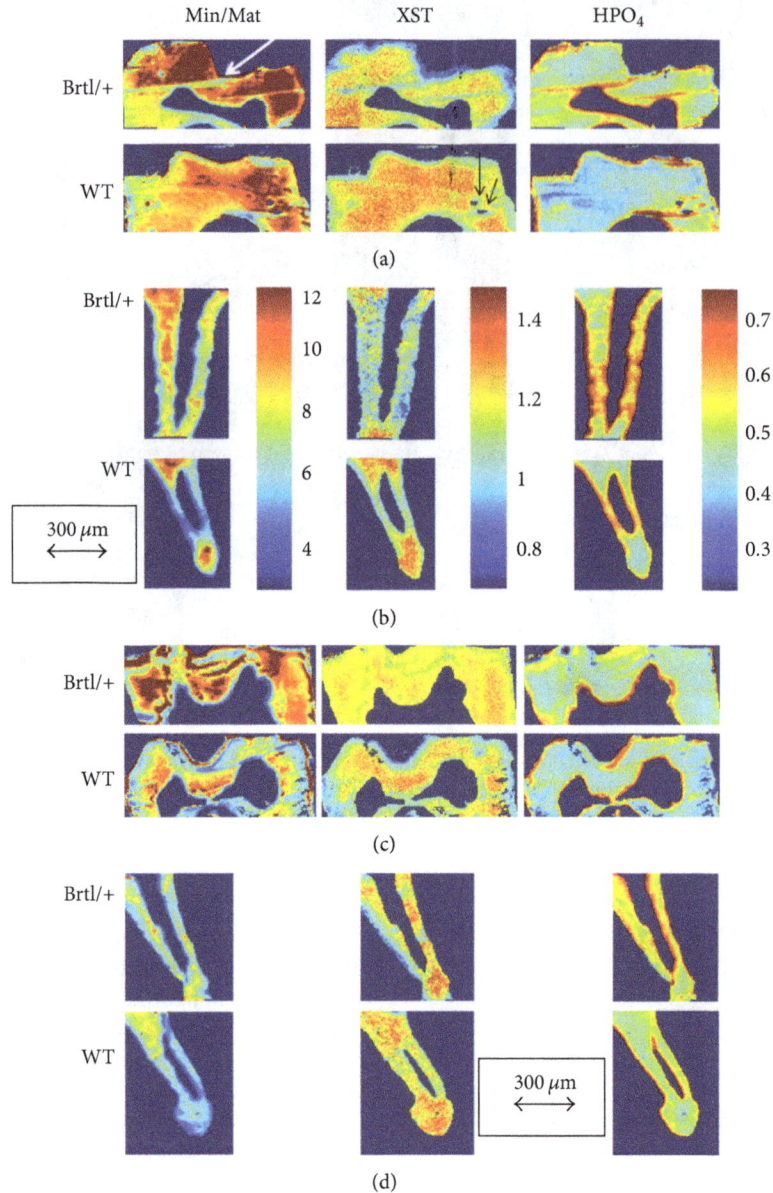

FIGURE 4: Typical FTIRI images for Brtl/+ and WT molars showing mineral/matrix ratio (Min/Mat), crystal size and perfection (XST), and acid phosphate content (HPO$_4$) in (a) first molar crowns at 2 months; (b) second molar root at 2 months; (c) first molar crowns at 6 months; and (d) second molar root at 6 months. The arrows indicate the distance corresponding to 300 μm. The color bars shown for (b) are the same for all similar parameters in the figure. White arrow indicates folds in specimen while black arrows show holes in the replicate images. These areas were masked when calculating numeric data.

difficult to comment on the effect of the transgene in this study. However there is definitely a suggestion of impaired matrix formation which in turn leads to the persistence of a more acid phosphate containing mineral phase in the Brtl/+ molars at both ages. Acid phosphate is present to a greater extent in newly deposited mineral [16]. The observation that no significant change in mineral/matrix ratio was detected in the FTIRI analysis of the molars agreed with the finding of no change in dentin or enamel mineral density but may indicate that these tissues were already fully mineralized by 2 months as suggested by a comparison of the mineral/matrix ratios of comparable sites in wild type teeth at 2 and 6 months.

The presence of globular dentin in the cervical areas of the roots and density variations in the proximal dentin of Brtl/+ detected by SEM is suggestive of mild dysplasia associated with dentinogenesis imperfecta type I (DGI-I), a dental manifestation of osteogenesis imperfecta [26–28]. Although a globular mineralization pattern can be found in normal teeth, where interglobular areas eventually mineralize, extensive areas of globular dentin are often associated with biomineralization abnormalities such as vitamin D deficient X-linked hypophosphatemia, vitamin D deficiency [29–31], and DGI-I.

Brtl/+ mice demonstrate bone fragility, a moderately deformed skeleton, and a low ductility phenotype, accurately

TABLE 2: FTIRI analysis of Brtl/+ and WT molars (all values are dimensionless ratios).

	Min/Mat	CO_3/PO_4	XLR	XST	HPO_4
			2 months		
			First molar crown		
WT	10.6 ± 1.0	0.0052 ± 0.0003	4.06 ± 0.20	1.184 ± 0.05	0.44 ± 0.04
BRTL/+	11.0 ± 2.0	0.0051 ± 0.0007	4.07 ± 0.37	1.13 ± 0.03	0.48 ± 0.03
P	NS	NS	NS	0.06	0.07
			Second molar crown		
WT	10.3 ± 1.0	0.0050 ± 0.0004	4.1 ± 0.29	1.171 ± 0.06	0.450 ± 0.03
BRTL/+	10.6 ± 3.0	0.0050 ± 0.0004	4.2 ± 0.37	1.133 ± 0.04	0.469 ± 0.04
P	NS	NS	NS	NS	NS
			First molar mesial root		
WT	7.99 ± 0.94	0.0053 ± 0.0004	3.97 ± 0.19	1.175 ± 0.041	0.46 ± 0.03
BRTL/+	7.28 ± 1.54	0.0047 ± 0.0007	4.21 ± 0.39	1.14 ± 0.025	0.53 ± 0.05
P	NS	NS	0.08	NS	0.009
			Second molar distal root		
WT	6.78 ± 1.0	0.0048 ± 0.0004	4.38 ± 0.89	1.15 ± 0.076	$0.51 \pm .04$
BRTL/+	6.54 ± 0.75	0.0048 ± 0.0006	4.97 ± 0.85	1.09 ± 0.087	0.61 ± 0.05
P	NS	NS	NS	0.10	0.001
			6 months		
			First molar crown		
WT	10.9 ± 2.2	0.0066 ± 0.0005	4.1 ± 0.3	1.1 ± 0.02	0.45 ± 0.02
Brtl/+	13.0 ± 2.6	0.0064 ± 0.0008	4.4 ± 0.7	1.1 ± 0.01	0.49 ± 0.01
P	NS	NS	NS	NS	0.015
			Second molar crown		
WT	9.0 ± 0.2	0.0063 ± 0.0006	4.1 ± 0.2	1.1 ± 0.01	0.48 ± 0.04
Brtl/+	9.8 ± 1.5	0.0064 ± 0.0007	4.2 ± 1.1	1.06 ± 0.51	0.51 ± 0.05
P	NS	NS	NS	NS	NS
			Second molar mesial root		
WT	7.1 ± 0.8	0.0062 ± 0.0003	4.2 ± 0.3	1.08 ± 0.03	0.53 ± 0.02
Brtl/+	6.7 ± 1.5	0.0053 ± 0.0004	4.4 ± 0.4	1.09 ± 0.02	0.61 ± 0.15
P	NS	0.0009	NS	NS	NS
			Second molar distal root		
WT	7.9 ± 0.5	0.0065 ± 0.0004	4.5 ± 0.3	1.08 ± 0.03	0.49 ± 0.03
Brtl/+	7.1 ± 1.5	0.0065 ± 0.0006	4.6 ± 0.4	1.07 ± 0.03	0.59 ± 0.09
P	NS	NS	NS	NS	0.07

representing the biomechanical phenotype of OI as a disease [8]. Their bones also sustain more microdamage than the WT [32]. The respective failure of the odontoblasts to make the physiologic type I collagen trimer results in dental abnormalities and tooth discolorations in patients with all types of the classic OI types [33]. Since we did not observe any enamel wear, we think that the mouse diet had no effect on the 2-month or 6-month molar phenotype.

The overall mechanism of biologic hydroxyapatite formation is similar in dentin and bone [21, 34, 35]; however, the dental and bone phenotypes can be expected to differ, to some extent because bone is remodeled, and dentin, in general, is not. Future TEM or AFM studies using a fluid cell to provide a source of ions should examine age dependent changes in Brtl/+ teeth and bones to separate effects of mineral nucleation and mineral propagation [36]. Thus the increased

remodeling caused by increased RANKL expression in the Brtl/+ mice bones [10] may accentuate the bone phenotype compared to the dentin, but it may also affect alveolar bone remodeling and similarly affect the dentin. It is important to note that the RANKL/OPG ratio in the Brtl/+ mice is normal. Study of the dentin phenotype indicates an impaired matrix mineralization and a slower correction of the phenotype with age, implying that both the collagen matrix and the noncollagenous proteins that regulate the function of that matrix may be altered in the Brtl/+ teeth.

This study has several limitations. Firstly, the Brtl/+ mice were heterozygous rather than homozygous for the mutant allele, resulting in greater heterogeneity of matrix collagen including forms with no, one, or two mutant $\alpha 1(I)$ alleles. This heterogeneity may contribute to greater variation in the data for mutant as opposed to wild type mice. Nonetheless, significant differences were seen in some critical parameters that can help explain the phenotype. Secondly, animals were studied at 2 and 6 months; thus, initial developmental time points were missed. However, all the earlier studies on the Brtl/+ mice bones used 2- [8–10, 18] and sometimes both 2- and 6-month-old animals [8, 10], and the FTIRI and micro-CT data for the teeth were limited to jaws from these earlier studies. The use of frozen jaws precluded histologic evaluation beyond that already published [7]. The use of jaws from previous studies also prevented observation of the molars as they erupted during their early development.

5. Conclusions

In conclusion, this imaging study of Brtl/+ teeth demonstrated decreased molar volume and reduced mineralized tissue volume in the teeth with the mutant collagen. The Brtl/+ molar dentin was also thinner. As expected there were no changes in enamel properties demonstrating the different mechanisms involved in collagen-based dentin mineralization and collagen-free enamel mineralization. Increased acid phosphate content was noted at 2 months by FTIRI, and altered collagen structure was noted at 2 but not 6 months in the Brtl/+ teeth.

Conflict of Interests

None of the authors have conflicts of interests.

Acknowledgments

This work is supported by NIH Grants DE04141 and AR046121 (to Adele L. Boskey) and by the NICHD Intramural Funding to Joan C. Marini.

References

[1] J. C. Marini, A. Forlino, W. A. Cabral et al., "Consortium for osteogenesis imperfecta mutations in the helical domain of type I collagen: regions rich in lethal mutations align with collagen binding sites for integrins and proteoglycans," *Human Mutation*, vol. 28, no. 3, pp. 209–221, 2007.

[2] F. Rauch, L. Lalic, P. Roughley, and F. H. Glorieux, "Genotype-phenotype correlations in nonlethal osteogenesis imperfecta caused by mutations in the helical domain of collagen type I," *European Journal of Human Genetics*, vol. 18, no. 6, pp. 642–647, 2010.

[3] R. D. Blank and A. L. Boskey, "Genetic collagen diseases: influence of collagen mutations on structure and mechanical behavior," in *Collagen: Structure and Mechanics*, P. Fratzl, Ed., Chapter 16, pp. 447–474, Springer Science, Business Media, 2008.

[4] M. J. Barron, S. T. McDonnell, I. MacKie, and M. J. Dixon, "Hereditary dentine disorders: dentinogenesis imperfecta and dentine dysplasia," *Orphanet Journal of Rare Diseases*, vol. 3, no. 1, article 31, 2008.

[5] D. O. Sillence, A. Senn, and D. M. Danks, "Genetic heterogeneity in osteogenesis imperfecta," *Journal of Medical Genetics*, vol. 16, no. 2, pp. 101–116, 1979.

[6] A. Forlino, W. A. Cabral, A. M. Barnes, and J. C. Marini, "New perspectives on osteogenesis imperfecta: invited review," *Nature Reviews Endocrinology*, vol. 7, pp. 540–557, 2011.

[7] A. Forlino, F. D. Porter, L. Eric J, H. Westphal, and J. C. Marini, "Use of the Cre/lox recombination system to develop a non-lethal knock-in murine model for osteogenesis imperfecta with an α1(I) G349C substitution. Variability in phenotype in BrtlIV mice," *Journal of Biological Chemistry*, vol. 274, no. 53, pp. 37923–37931, 1999.

[8] K. M. Kozloff, A. Carden, C. Bergwitz et al., "Brittle IV mouse model for osteogenesis imperfecta IV demonstrates postpubertal adaptations to improve whole bone strength," *Journal of Bone and Mineral Research*, vol. 19, no. 4, pp. 614–622, 2004.

[9] J. M. Wallace, B. G. Orr, J. C. Marini, and M. M. B. Holl, "Nanoscale morphology of Type I collagen is altered in the Brtl mouse model of Osteogenesis Imperfecta," *Journal of Structural Biology*, vol. 173, no. 1, pp. 146–152, 2011.

[10] T. E. Uveges, P. Collin-Osdoby, W. A. Cabral et al., "Cellular mechanism of decreased bone in Brtl mouse model of OI: imbalance of decreased osteoblast function and increased osteoclasts and their precursors," *Journal of Bone and Mineral Research*, vol. 23, no. 12, pp. 1983–1994, 2008.

[11] A. L. Boskey, M. F. Young, T. Kilts, and K. Verdelis, "Variation in mineral properties in normal and mutant bones and teeth," *Cells Tissues Organs*, vol. 181, no. 3-4, pp. 144–153, 2006.

[12] L. G. Sloofman, K. Verdelis, L. Spevak et al., "Effect of HIP/ribosomal protein L29 deficiency on mineral properties of murine bones and teeth," *Bone*, vol. 47, no. 1, pp. 93–101, 2010.

[13] K. Verdelis, L. Lukashova, E. Atti et al., "MicroCT morphometry analysis of mouse cancellous bone: intra- and inter-system reproducibility," *Bone*, vol. 49, no. 3, pp. 580–587, 2011.

[14] M. Baldassarri, H. C. Margolis, and E. Beniash, "Compositional determinants of mechanical properties of enamel," *Journal of Dental Research*, vol. 87, no. 7, pp. 645–649, 2008.

[15] S. Gourion-Arsiquaud, P. A. West, and A. L. Boskey, "Fourier transform-infrared microspectroscopy and microscopic imaging," *Methods in Molecular Biology*, vol. 455, pp. 293–303, 2008.

[16] A. L. Boskey and R. Mendelsohn, "Infrared spectroscopic characterization of mineralized tissues," *Vibrational Spectroscopy*, vol. 38, no. 1-2, pp. 107–114, 2005.

[17] L. Spevak, C. Flach, T. Hunter, R. Mendelsohn, and A. L. Boskey, "FTIRI parameters describing acid phosphate substitution in biologic hydroxyapatite," *Calcified Tissue International*, vol. 92, pp. 418–428, 2013.

[18] S. G. Kazarian and K. L. A. Chan, "'Chemical photography' of drug release," *Macromolecules*, vol. 36, pp. 9866–9872, 2003.

[19] J. P. Simmer, A. S. Richardson, Y. Y. Hu, C. E. Smith, and J. Ching-Chun Hu, "A post-classical theory of enamel biomineralization...and why we need one," *International Journal of Oral Science*, vol. 4, pp. 129–134, 2012.

[20] J. Moradian-Oldak, "Protein-mediated enamel mineralization," *Frontiers in Bioscience*, vol. 17, pp. 1996–2023, 2012.

[21] M. Goldberg, A. B. Kulkarni, M. Young, and A. Boskey, "Dentin: structure, composition and mineralization," *Frontiers in Bioscience*, vol. 3, pp. 711–735, 2011.

[22] A. L. Boskey, B. Christensen, H. Taleb, and E. S. Sørensen, "Post-translational modification of osteopontin: effects on in vitro hydroxyapatite formation and growth," *Biochemical and Biophysical Research Communications*, vol. 419, pp. 333–338, 2012.

[23] R. Gioia, C. Panaroni, R. Besio et al., "Impaired osteoblastogenesis in a murine model of dominant osteogenesis imperfecta: a new target for osteogenesis imperfecta pharmacological therapy," *Stem Cells*, vol. 30, pp. 1465–1476, 2012.

[24] M. Goldberg, A. Marchadier, C. Vidal et al., "Differential effects of fibromodulin deficiency on mouse mandibular bones and teeth: a micro-CT time course study," *Cells Tissues Organs*, vol. 194, no. 2-4, pp. 205–210, 2011.

[25] N. L. Leong, J. M. Hurng, S. I. Djomehri, S. A. Gansky, M. I. Ryder, and S. P. Ho, "Age-related adaptation of bone-PDL-tooth complex: Rattus-Norvegicus as a model system," *PLoS One*, vol. 7, Article ID e35980, 2012.

[26] J. W. Kim and J. P. Simmer, "Hereditary dentin defects," *Journal of Dental Research*, vol. 86, no. 5, pp. 392–399, 2007.

[27] T. Murayama, R. Iwatsubo, S. Akiyama, A. Amano, and I. Morisaki, "Familial hypophosphatemic vitamin d-resistant rickets: dental findings and histologic study of teeth," *Oral Surgery, Oral Medicine, Oral Pathology, Oral Radiology, and Endodontics*, vol. 90, no. 3, pp. 310–316, 2000.

[28] D. Pallos, P. S. Hart, J. R. Cortelli et al., "Novel COL1A1 mutation (G599C) associated with mild osteogenesis imperfecta and dentinogenesis imperfecta," *Archives of Oral Biology*, vol. 46, no. 5, pp. 459–470, 2001.

[29] T. Boukpessi, D. Septier, S. Bagga, M. Garabedian, M. Goldberg, and C. Chaussain-Miller, "Dentin alteration of deciduous teeth in human hypophosphatemic rickets," *Calcified Tissue International*, vol. 79, no. 5, pp. 294–300, 2006.

[30] G. Hillmann and W. Geurtsen, "Pathohistology of undecalcified primary teeth in vitamin D-resistant rickets Review and report of two cases," *Oral Surgery, Oral Medicine, Oral Pathology, Oral Radiology, and Endodontics*, vol. 82, no. 2, pp. 218–224, 1996.

[31] H. Fong, E. Y. Chu, K. A. Tompkins et al., "Aberrant cementum phenotype associated with the hypophosphatemic Hyp mouse," *Journal of Periodontology*, vol. 80, no. 8, pp. 1348–1354, 2009.

[32] M. S. Davis, B. L. Kovacic, J. C. Marini, A. J. Shih, and K. M. Kozloff, "Increased susceptibility to microdamage in Brtl/+ mouse model for osteogenesis imperfecta," *Bone*, vol. 50, pp. 784–791, 2012.

[33] A. Majorana, E. Bardellini, P. C. Brunelli, M. Lacaita, A. P. Cazzolla, and G. Favia, "Dentinogenesis imperfecta in children with osteogenesis imperfecta: a clinical and ultrastructural study," *International Journal of Paediatric Dentistry*, vol. 20, no. 2, pp. 112–118, 2010.

[34] A. Veis, C. Sfeir, and C. B. Wu, "Phosphorylation of the proteins of the extracellular matrix of mineralized tissues by casein kinase-like activity," *Critical Reviews in Oral Biology and Medicine*, vol. 8, no. 4, pp. 360–379, 1997.

[35] A. Veis and J. R. Dorvee, "Biomineralization mechanisms: a new paradigm for crystal nucleation in organic matrices," *Calcified Tissue International*, 2012.

[36] D. Li, M. H. Nielsen, J. R. Lee, C. Frandsen, J. F. Banfield, and J. J. De Yoreo, "Direction-specific interactions control crystal growth by oriented attachment," *Science*, vol. 336, pp. 1014–1018, 2012.

19

Estradiol Synthesis and Release in Cultured Female Rat Bone Marrow Stem Cells

Dalei Zhang,[1] **Bei Yang,**[1] **Weiying Zou,**[1] **Xiaying Lu,**[1,2] **Mingdi Xiong,**[1]
Lei Wu,[1] **Jinglei Wang,**[1] **Junhong Gao,**[3] **Sifan Xu,**[4] **and Ting Zou**[1]

[1] *Department of Physiology, Medical College of Nanchang University, Nanchang, Jiangxi 330006, China*
[2] *Department of Physiology, Gannan Medical University, Ganzhou, Jiangxi 341000, China*
[3] *Institute of Acupuncture and Moxibustion, China Academy of Chinese Medical Sciences, Beijing 100700, China*
[4] *Institute of Chinese Minority Traditional Medicine, Minzu University of China, Bejing 100081, China*

Correspondence should be addressed to Ting Zou; zouttzou@yahoo.com

Academic Editor: Thomas Skutella

Bone marrow stem cells (BMSCs) have the capacity to differentiate into mature cell types of multiple tissues. Thus, they represent an alternative source for organ-specific cell replacement therapy in degenerative diseases. In this study, we demonstrated that female rat BMSCs could differentiate into steroidogenic cells with the capacity for *de novo* synthesis of Estradiol-17β (E2) under high glucose culture conditions with or without retinoic acid (RA). The cultured BMSCs could express the mRNA and protein for P450arom, the enzyme responsible for estrogen biosynthesis. Moreover, radioimmunoassay revealed that BMSCs cultured in the present culture system produced and secreted significant amounts of testosterone, androstenedione, and E2. In addition, RA promoted E2 secretion but did not affect the levels of androgen. These results indicate that BMSCs can synthesize and release E2 and may contribute to autologous transplantation therapy for estrogen deficiency.

1. Introduction

Steroid hormones play important regulatory roles in female reproduction, in which estrogen is essential for folliculogenesis beyond the antral stage and is necessary to maintain the female phenotype of ovarian somatic cells [1–4]. Estradiol-17β (E2), a product of androgen aromatization, is the principal estrogen and is secreted in large amounts by the large preovulatory follicles in the ovary [5]. Although the ovaries are the principal source of systemic oestrogen in the premenopausal nonpregnant woman, a number of extragonadal sites of oestrogen biosynthesis, including mesenchymal cells of the adipose tissue and skin, osteoblasts, vascular endothelial, aortic smooth-muscle cells, and brain, become the major sources of oestrogen beyond menopause. However, the total amount of oestrogen synthesized by these extragonadal sites may be small. Within these sites, E2 is probably biologically active only at local tissue level in a paracrine or intracrine fashion without significantly affecting circulating levels [6–8].

The reduction of estrogen production in the ovary may cause menopausal symptoms. In addition, premature ovarian failure may be caused by any process which reduces the number of oocytes within the ovary [9]. For example, chemotherapy can reduce ovarian reserve and affect ovarian stromal function to produce less estrogen [10]. Although estrogen replacement therapy has been established and is recommended for postmenopausal women or patients with hypogonadism, due to its beneficial effects, follicular estrogen production is regulated by a complex set of signals that synergize to produce optimal steroidogenesis [11]. Still, it is difficult to provide an optimal therapeutic dose for long-term estrogen replacement therapy. Furthermore, it is associated with a substantial risk for cardiovascular disease and breast cancer [12]. For this reason, alternative therapies such as steroidogenic cell transplantation may have advantages over

HRT for hypogonadism. It should allow control of hormone levels in nature by hypothalamus-anterior pituitary axis.

Several earlier studies have suggested that stem cells can serve as an alternative source for various steroid hormones [13–17]. Bone marrow stem cells (BMSCs) are thought to be multipotent cells, which can replicate as undifferentiated cells and have the potential to differentiate into mature cell types of multiple tissues [18, 19]. In the present study, we investigated whether female rat BMSCs could produce steroidogenic cells with the capacity for the synthesis of E2.

2. Materials and Methods

2.1. Isolation and Culture of BMSCs. Female SD rats, weighing 80 to 100 g, were obtained from Center of Laboratory Animals of Nanchang University and used in accordance with a protocol approved by the Nanchang University Animal Care and Use Committee. The bone marrow cells were isolated from femurs and tibias of female rats by flushing the shaft with phosphate-buffered saline (PBS) using needles, and the cells were further dispersed several times by gentle, repeated pipetting with a sterile pipet. The dissociated cells were seeded in 75 cm^2 culture flasks for the primary culture in the high glucose (4.5 g/L) Dulbecco's modified Eagle's medium (DMEM, Hyclone, Utah) supplemented with 10% fetal bovine serum (FBS, Stem Cell Tech Inc., Canada) and incubated at 37°C in a water-saturated atmosphere of 95% air and 5% CO_2. The nonadherent cells were removed by washing with PBS and replacing the fresh complete medium every 3 or 4 days. The adherent cells were passaged every 7 days by harvesting the cells with 0.25% trypsin/0.02% EDTA, and replating at a 1 : 4 dilution.

2.2. Induction of BMSC Differentiation In Vitro. Cultured cells at passage 3 were recovered and used in these experiments. BMSCs were reincubated in 12-well culture plates (Nunc, Denmark) at a density of 2×10^5/well in the high glucose (4.5 g/L) DMEM containing 10% FBS supplemented with or without 10^{-5} mol/L all-trans retinoic acid (RA, Sigma) for 4 days. Differentiated cells from the BMSCs were analyzed by immunocytochemical staining or RT-PCR analysis for expression of aromatase cytochrome P450 (P450arom). The levels of testosterone (T), androstenedione (ASD), and E2 in culture media were measured by radioimmunoassay (RIA).

2.3. Immunocytochemistry of P450arom. BMSCs were fixed in 4% acetone at 4°C for 15 minutes and washed 3 times with PBS. Endogenous peroxidase was quenched by incubating the fixed cells with 3% H_2O_2 in methanol for 20 minutes. After being washed with PBS for 15 min (5 min ×3 times), cells were incubated for 20 minutes with 10% normal goat serum; then with a rabbit polyclonal antibody to aromatase (Boster Co., Wuhan, China), they were diluted 1 : 100 in PBS overnight at 4°C. The negative control was prepared in an identical manner except that the primary antibody was replaced with normal serum. After washing with PBS, cells were incubated with horseradish peroxidase-conjugated

goat anti-rabbit IgG for 1 hour at room temperature. After washing, the immunoreaction was detected by using DAB system.

2.4. Real-Time PCR Analysis. Total RNA was extracted from cultured BMSCs using Trizol reagent (Sigma, St. Louis, MO) and was reverse transcribed into cDNA using the First-Strand cDNA synthesis kit. Real-time PCR was performed to quantify the samples' cDNA copies using SYBR premix ExTaqTM fluorescent quantitation PCR kit (TaKaRa, Japan). The *CYP19* primers forward: 5$'$-GCTTCTCATCGCAGSGTAT-3$'$, reverse: 5$'$-CAAGGGTAAATTCATTGGG-3$'$. The β-actin primers forward: 5$'$-GGAAATCGTGCGTGACATTAAA-3$'$, reverse: 5$'$-TGCGGCAGTGGCCATC-3$'$. Conditions for PCR were 40 cycles of 95°C for 5 seconds and 60°C for 34 seconds. The cycle threshold (Ct) was set up at the level that reflected the best kinetic PCR parameters, and melting curves were acquired and analyzed. The $2^{-\Delta\Delta Ct}$ method of relative quantification was used to estimate the copy numbers in *CYP19* gene.

2.5. T, ASD and E2 Measurement. Before culture (defined as time 0) and 1, 2, 3, or 4 days after culture, cell culture medium was centrifuged and collected, and the levels of T, ASD and E2 were measured by Beijing Sino-UK Institute of Biological Technology.

2.6. Statistical Analysis. The experiment was repeated a minimum of three times. All data were expressed as the mean ± SD and analyzed by ANOVA and Ducan's multiple range test using the SAS 8.0 software. $P < 0.05$ was considered significantly different.

3. Results

3.1. RT-PCR Analysis for CYP19 mRNA Expression. RT-PCR analysis showed that there was expression of aromatase gene *CYP19* in BMSCs cultured for 4 days in a high glucose DMEM, and the expression was significantly higher than that in 0 day cells. Furthermore, we investigated the effects of RA on BMSC differentiation in vitro. The result showed that the expression of *CYP19* mRNA was not further elevated by RA treatment at a concentration of 10^{-5} mol/L (Figure 1).

3.2. Immunocytochemical Analysis of P450arom. To elucidate the capacity of BMSCs to generate E2, we examined the expression of P450arom protein, the enzyme responsible for estrogen biosynthesis, by immunocytochemical staining. The results showed that P450arom was expressed in BMSCs cultured in a high glucose culture condition alone or in combination with RA treatment, with a positive labeling in part of the cells, and it was primarily immunolocalized in the cytoplasm (Figure 2).

3.3. RIA for E2. To evaluate E2 biosynthesis and release in cultured BMSCs, the levels of E2 in culture medium were

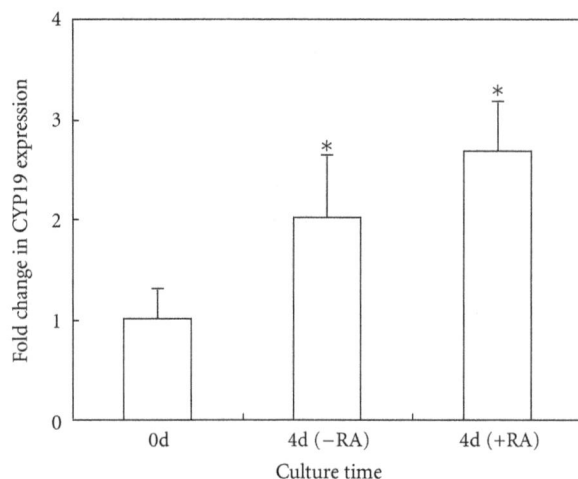

FIGURE 1: Real-time PCR analysis of CYP19 expression in rat BMSCs. The expression of CYP19 was determined relative to the β-actin expression. Expression data ($n = 5$) were reported as fold change ($2^{-\Delta\Delta Ct}$). BMSCs cultured in high glucose DMEM alone (control) or in combination with RA treatment (10^{-5} mol/L) for 4 days as compared with 0 day cells. The experiment was repeated three times. Statistical significance was determined by a t-test. $^*P < 0.05$, versus 0 day.

measured by RIA. Before incubation, the culture medium of BMSCs at passage 3 contained a low concentration of E2. After exposure to a high glucose condition for 1 day, E2 levels were increased significantly. However, E2 content was not obviously altered after prolonged culture time (2–4 days). Similar results were observed in cultured BMSCs that were given a combined treatment with high glucose and RA. The maximal effect of RA was observed in BMSCs cultured for 48 hours, and the release of E2 significantly increased compared with the high glucose medium alone (Figure 3).

3.4. RIA for T and ASD. To investigate *de novo* synthesis of E2, we measured the levels of T and ASD in medium by RIA. After BMSC culture for 4 days and 1 day in high glucose medium alone or together with RA, the release of T and ASD significantly increased, respectively. However, there were no obvious differences in the levels of T and ASD between the high glucose group and combination group (Figures 4 and 5). In addition, prolonged culture time (2–4 d) had no effect on the release of ASD (Figure 5).

4. Discussion

The high degree of stem cell plasticity provides a promising strategy for cell replacement therapy. During the past several years, a great deal of attention has been focused on the plasticity of BMSCs. Since BMSCs have tremendous differentiative potential, they can differentiate in vitro and in vivo into mature cells of the heart, liver, kidney, lungs, GI tract, skin, bone, muscle, cartilage, fat, endothelium, and brain. These BMSC-derived cells have been shown to contribute to clinical treatment of genetic disease or tissue repair [20–29]. In the

present study, we investigated the ability of the BMSCs to generate steroidogenic cells and release E2 in vitro. Our study revealed that BMSCs cultured in high glucose DMEM with or without RA were capable of differentiating into cells that produced and secreted significant amounts of E2.

Under physiological conditions, E2 is produced *de novo* from cholesterol and synthesized by the ovary in a sequential manner. Steroidogenic granulosa and theca cells cooperate under gonadotropin control to produce estrogens by stimulating synthesis of steroidogenic enzyme messenger RNAs [11]. In the theca, under the influence of LH, cholesterol is converted to pregnenolone and metabolized through a series of substrates ending in androgen production. Theca cell-derived androgens transported to the granulosa cells of developing follicles, where they are aromatized to oestrogens by P450arom, the product of the *CYP19* gene, which is responsible for conversion of C19 steroids to estrogen [2, 6, 30]. In our culture systems, P450arom mRNA and protein were expressed in BMSCs, which also produced and released T and ASD. These results suggested that BMSCs could produce steroidogenic cells with the capacity for the synthesis of E2.

In recent years, some studies found that transfection of BMSCs from human and murine with steroidogenic factor 1 (SF-1, an essential factor for differentiation of the pituitary-gonadal axis) can transform BMSCs into steroidogenic cells, which produce various steroid hormones, including E2, and expressed mRNA for P450arom [16, 17]. When transplanted into immature rat testes, adherent marrow-derived cells were found to be engrafted and differentiated into steroidogenic cells that were indistinguishable from Leydig cells [17]. These results provided evidence that BMSCs were capable of differentiating into steroidogenic cells and represented a useful source of stem cells for cell transplantation therapy. In this study, without forced expression of SF-1, we demonstrated the ability of the BMSCs to spontaneously form steroidogenic cells and secrete E2 under a high glucose condition.

RA is well known as the biologically active form of vitamin A and has been shown to play an important role in normal embryonic development and maintenance of differentiation in the adult organism [31]. Previous studies showed that RA could induce BMSCs to differentiate into male germ cells [32] and stimulate E2 and T synthesis in rat hippocampal slice cultures [33]. Therefore, we examined the effects of RA on E2 and T biosynthesis in cultured female rat BMSCs. In the present study, BMSCs were incubated in the absence or presence of RA to investigate RA-induced differentiation of BMSCs to steroidogenic cells in vitro. Compared with high glucose medium alone, E2 secretion was stimulated by RA treatment without any increase in the levels of androgen, suggesting that RA, at least at a concentration of 10^{-5} mol/L, may promote the differentiation of BMSCs to estrogen-producing cells.

A great deal of efforts had been directed at understanding what role stem cells may play in the physiology and pathology of the mammalian female gonads [34]. Over the past few years, some studies found that bone marrow transplantation (BMT) generated immature oocytes and rescued long-term

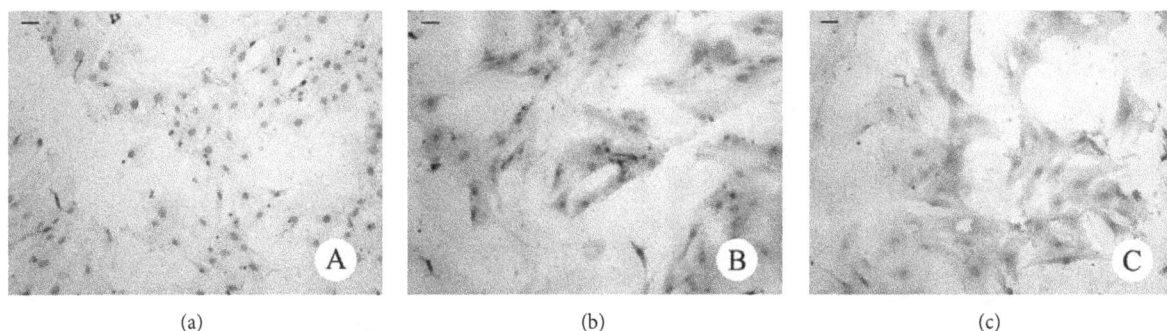

(a) (b) (c)

FIGURE 2: Immunocytochemical staining of P450arom in rat BMSCs cultured for 4 days. (a), negative staining; (b), control group; (c), RA treatment group. P450arom was expressed in BMSCs cultured in a high glucose culture condition alone (control group) or in combination with RA treatment. Scale bar: 20 μm.

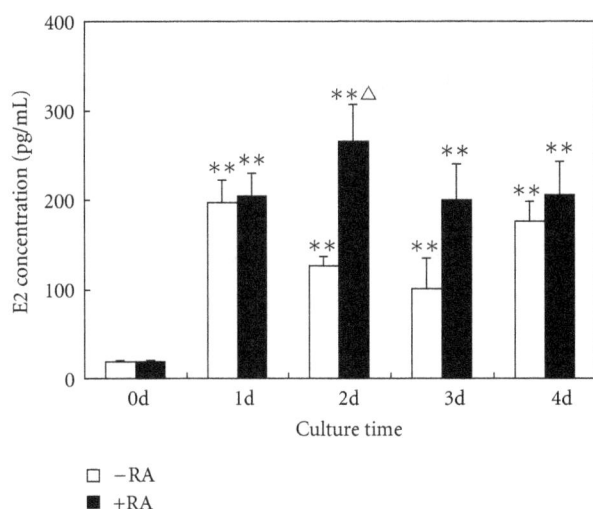

□ −RA
■ +RA

FIGURE 3: Measurement of E2 concentrations in culture media by RIA. Compared with 0 day cells, the release of E2 significantly increased in rat BMSCs cultured in a high glucose DMEM or in combination with RA for 1–4 days. The experiment was repeated three times. Values represent means ± SD ($n = 5$). $^{**}P < 0.01$, versus 0 day; $^{\triangle}P < 0.01$, versus high glucose DMEM.

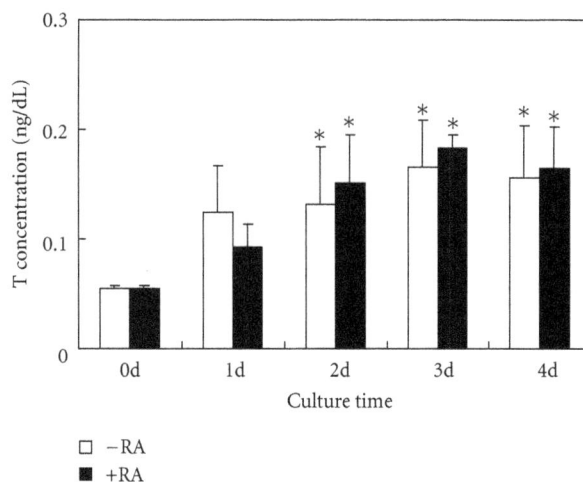

□ −RA
■ +RA

FIGURE 4: Measurement of T concentrations in culture media by RIA. Compared with 0 day cells, the release of T significantly increased in rat BMSCs cultured in a high glucose DMEM for 2–4 days. RA treatment did not obviously increase the levels of ASD. The experiment was repeated three times. Values represent means ± SD ($n = 5$). $^{*}P < 0.05$, versus 0 day; $^{**}P < 0.01$, versus 0 day.

fertility in a preclinical mouse model of chemotherapy-induced premature ovarian failure. Although all offspring were derived from the recipient germline, donor-derived oocytes were generated in ovaries of recipients after BMT [35]. Furthermore, MSC transplantation can improve ovarian function and structure damaged by chemotherapy, and the paracrine mediators secreted by MSC might be involved in the repair of damaged ovaries [36]. These results suggested that the potential of BMSCs for ameliorating female reproductive function was involved in reversal of both ovarian germline and somatic cell insufficiency.

In a previous report, mouse embryonic stem cells in culture developed into oogonia that could enter meiosis, recruit adjacent cells to form follicle-like structures, which expressed aromatase and secreted E2 [15]. In addition, mouse-induced pluripotent stem cells cocultured with ovarian granulosa cells in vitro could form granulosa cell-like cells and secret E2

[37]. BMSCs in our culture systems did not form follicle-like structures, nor did exhibit the morphology of mature ovarian cells. However, they expressed P450arom, suggesting that BMSCs have the ability to synthesize and to release E2, which may contribute to autologous transplantation therapy of BMSCs for hypogonadism.

5. Conclusion

In this study, we showed that female rat BMSCs cultured in high glucose DMEM with or without RA could express CYP19 and P450arom, and excrete T, ASD, and E2. These results indicated that the cultured BMSCs could produce steroidogenic cells with the capacity for E2 synthesis. This study would help to provide basis for clinical application of BMSCs in autologous cell transplantation therapy for patients with estrogen deficiency.

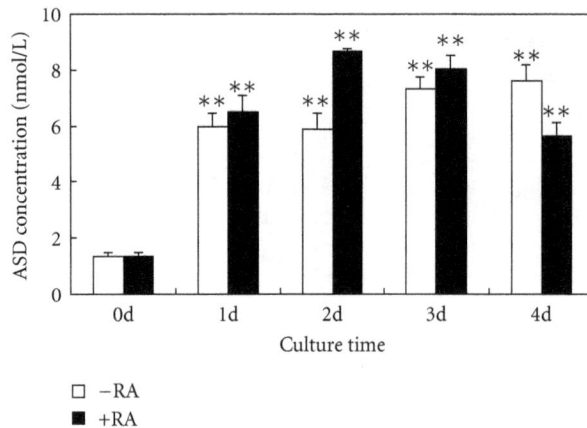

FIGURE 5: Measurement of ASD concentrations in culture media by RIA. Compared with 0 day cells, the release of ASD significantly increased in rat BMSCs cultured in a high glucose DMEM for 1–4 days. RA treatment did not obviously increase the levels of ASD. The experiment was repeated three times. Values represent means ± SD ($n = 5$). $^*P < 0.05$, versus 0 day; $^{**}P < 0.01$, versus 0 day.

Conflict of Interests

The authors declare that they have no conflict of interests.

Authors' Contribution

D. Zhang, B. Yang, and W. Zou equally contributed to this work.

Acknowledgment

This work was supported by the National Natural Science Foundation of China (no. 30960409 and 81060056).

References

[1] E. Y. Adashi and A. J. W. Hsueh, "Estrogens augment the stimulation of ovarian aromatase activity by follicle-stimulating hormone in cultured rat granulosa cells," *Journal of Biological Chemistry*, vol. 257, no. 11, pp. 6077–6083, 1982.

[2] A. E. Drummond, "The role of steroids in follicular growth," *Reproductive Biology and Endocrinology*, vol. 4, article 16, 2006.

[3] F. Otsuka, R. K. Moore, X. Wang, S. Sharma, T. Miyoshi, and S. Shimasaki, "Essential role of the oocyte in estrogen amplification of follicle-stimulating hormone signaling in granulosa cells," *Endocrinology*, vol. 146, no. 8, pp. 3362–3367, 2005.

[4] L. Z. Zhuang, E. Y. Adashi, and A. J. Hsuch, "Direct enhancement of gonadotropin-stimulated ovarian estrogen biosynthesis by estrogen and clomiphene citrate.," *Endocrinology*, vol. 110, no. 6, pp. 2219–2221, 1982.

[5] C. P. Channing and S. P. Coudert, "Contribution of granulosa cells and follicular fluid to ovarian estrogen secretion in the rhesus monkey in vivo," *Endocrinology*, vol. 98, no. 3, pp. 590–597, 1976.

[6] E. R. Simpson, Y. Zhao, V. R. Agarwal et al., "Aromatase expression in health and disease," *Recent Progress in Hormone Research*, vol. 52, pp. 185–214, 1997.

[7] E. R. Simpson and S. R. Davis, "Minireview: aromatase and the regulation of estrogen biosynthesis—some new perspectives," *Endocrinology*, vol. 142, no. 11, pp. 4589–4594, 2001.

[8] E. R. Simpson, "Role of aromatase in sex steroid action," *Journal of Molecular Endocrinology*, vol. 25, no. 2, pp. 149–156, 2000.

[9] G. S. Conway, "Premature ovarian failure," *British Medical Bulletin*, vol. 56, no. 3, pp. 643–649, 2000.

[10] O. Oktem and K. Oktay, "Quantitative assessment of the impact of chemotherapy on ovarian follicle reserve and stromal function," *Cancer*, vol. 110, no. 10, pp. 2222–2229, 2007.

[11] Z. Shoham and M. Schachter, "Estrogen biosynthesis—regulation, action, remote effects, and value of monitoring in ovarian stimulation cycles," *Fertility and Sterility*, vol. 65, no. 4, pp. 687–701, 1996.

[12] H. P. Women, "Risks and benefits of estrogen plus progestin in healthy postmenopausal women: principal results from the women's health initiative randomized controlled trial," *The Journal of the American Medical Association*, vol. 288, pp. 321–333, 2002.

[13] P. A. Crawford, Y. Sadovsky, and J. Milbrandt, "Nuclear receptor steroidogenic factor 1 directs embryonic stem cells toward the steroidogenic lineage," *Molecular and Cellular Biology*, vol. 17, no. 7, pp. 3997–4006, 1997.

[14] S. Gondo, T. Yanase, T. Okabe et al., "SF-1/Ad4BP transforms primary long-term cultured bone marrow cells into ACTH-responsive steroidogenic cells," *Genes to Cells*, vol. 9, no. 12, pp. 1239–1247, 2004.

[15] K. Hübner, G. Fuhrmann, L. K. Christenson et al., "Derivation of oocytes from mouse embryonic stem cells," *Science*, vol. 300, no. 5623, pp. 1251–1256, 2003.

[16] T. Tanaka, S. Gondo, T. Okabe et al., "Steroidogenic factor 1/adrenal 4 binding protein transforms human bone marrow mesenchymal cells into steroidogenic cells," *Journal of Molecular Endocrinology*, vol. 39, no. 5-6, pp. 343–350, 2007.

[17] T. Yazawa, T. Mizutani, K. Yamada et al., "Differentiation of adult stem cells derived from bone marrow stroma into Leydig or adrenocortical cells," *Endocrinology*, vol. 147, no. 9, pp. 4104–4111, 2006.

[18] J. E. Grove, E. Bruscia, and D. S. Krause, "Plasticity of bone marrow-derived stem cells," *Stem Cells*, vol. 22, no. 4, pp. 487–500, 2004.

[19] D. S. Krause, "Plasticity of marrow-derived stem cells," *Gene Therapy*, vol. 9, no. 11, pp. 754–758, 2002.

[20] M. A. Eglitis and É. Mezey, "Hematopoietic cells differentiate into both microglia and macroglia in the brains of adult mice," *Proceedings of the National Academy of Sciences of the United States of America*, vol. 94, no. 8, pp. 4080–4085, 1997.

[21] G. Ferrari, G. Cusella-De Angelis, M. Coletta et al., "Muscle regeneration by bone marrow-derived myogenic progenitors," *Science*, vol. 279, no. 5356, pp. 1528–1530, 1998.

[22] D. C. Hess, W. D. Hill, A. Martin-Studdard, J. Carroll, J. Brailer, and J. Carothers, "Bone marrow as a source of endothelial cells and NeuN-expressing cells after stroke," *Stroke*, vol. 33, no. 5, pp. 1362–1368, 2002.

[23] S. Kale, A. Karihaloo, P. R. Clark, M. Kashgarian, D. S. Krause, and L. G. Cantley, "Bone marrow stem cells contribute to repair of the ischemically injured renal tubule," *Journal of Clinical Investigation*, vol. 112, no. 1, pp. 42–49, 2003.

[24] D. S. Krause, N. D. Theise, M. I. Collector et al., "Multi-organ, multi-lineage engraftment by a single bone marrow-derived stem cell," *Cell*, vol. 105, no. 3, pp. 369–377, 2001.

[25] É. Mezey, S. Key, G. Vogelsang, I. Szalayova, G. David Lange, and B. Crain, "Transplanted bone marrow generates new neurons in human brains," *Proceedings of the National Academy of Sciences of the United States of America*, vol. 100, no. 3, pp. 1364–1369, 2003.

[26] D. Orlic, J. Kajstura, S. Chimenti et al., "Bone marrow cells regenerate infarcted myocardium," *Nature*, vol. 410, no. 6829, pp. 701–705, 2001.

[27] B. E. Petersen, W. C. Bowen, K. D. Patrene et al., "Bone marrow as a potential source of hepatic oval cells," *Science*, vol. 284, no. 5417, pp. 1168–1170, 1999.

[28] M. F. Pittenger, A. M. Mackay, S. C. Beck et al., "Multilineage potential of adult human mesenchymal stem cells," *Science*, vol. 284, no. 5411, pp. 143–147, 1999.

[29] N. D. Theise, M. Nimmakayalu, R. Gardner et al., "Liver from bone marrow in humans," *Hepatology*, vol. 32, no. 1, pp. 11–16, 2000.

[30] E. R. Simpson, M. S. Mahendroo, G. D. Means et al., "Aromatase cytochrome P450, the enzyme responsible for estrogen biosynthesis," *Endocrine Reviews*, vol. 15, no. 3, pp. 342–355, 1994.

[31] S. A. Ross, P. J. McCaffery, U. C. Drager, and L. M. De Luca, "Retinoids in embryonal development," *Physiological Reviews*, vol. 80, no. 3, pp. 1021–1054, 2000.

[32] J. Hua, S. Pan, C. Yang, W. Dong, Z. Dou, and K. S. Sidhu, "Derivation of male germ cell-like lineage from human fetal bone marrow stem cells," *Reproductive BioMedicine Online*, vol. 19, no. 1, pp. 99–105, 2009.

[33] E. Munetsuna, Y. Hojo, M. Hattori et al., "Retinoic acid stimulates 17β-estradiol and testosterone synthesis in rat hippocampal slice cultures," *Endocrinology*, vol. 150, no. 9, pp. 4260–4269, 2009.

[34] J. L. Tilly and B. R. Rueda, "Minireview: stem cell contribution to ovarian development, function, and disease," *Endocrinology*, vol. 149, no. 9, pp. 4307–4311, 2008.

[35] H. J. Lee, K. Selesniemi, Y. Niikura et al., "Bone marrow transplantation generates immature oocytes and rescues long-term fertility in a preclinical mouse model of chemotherapy-induced premature ovarian failure," *Journal of Clinical Oncology*, vol. 25, no. 22, pp. 3198–3204, 2007.

[36] X. Fu, Y. He, C. Xie, and W. Liu, "Bone marrow mesenchymal stem cell transplantation improves ovarian function and structure in rats with chemotherapy-induced ovarian damage," *Cytotherapy*, vol. 10, no. 4, pp. 353–363, 2008.

[37] Y. Kang, M. J. Cheng, and C. J. Xu, "Secretion of oestrogen from murine-induced pluripotent stem cells co-cultured with ovarian granulosa cells in vitro," *Cell Biology International*, vol. 35, no. 9, pp. 871–874, 2011.

20

Membrane Localization of Membrane Type 1 Matrix Metalloproteinase by CD44 Regulates the Activation of Pro-Matrix Metalloproteinase 9 in Osteoclasts

Meenakshi A. Chellaiah and Tao Ma

Department of Oncology and Diagnostic Sciences, Dental School, University of Maryland, Baltimore, MD 21201, USA

Correspondence should be addressed to Meenakshi A. Chellaiah; mchellaiah@umaryland.edu

Academic Editor: Sakae Tanaka

CD44, MT1-MMP, and MMP9 are implicated in the migration of osteoclast and bone resorption. This study was designed to determine the functional relationship between CD44 and MT1-MMP in the activation of pro-MMP9. We used osteoclasts isolated from wild-type and CD44-null mice. Results showed that MT1-MMP is present in multiple forms with a molecular mass ~63, 55, and 45 kDa in the membrane of wild-type osteoclasts. CD44-null osteoclasts demonstrated a 55 kDa active MT1-MMP form in the membrane and conditioned medium. It failed to activate pro-MMP9 because TIMP2 binds and inhibits this MT1-MMP (~55 kDa) in CD44-null osteoclasts. The role of MT1-MMP in the activation of pro-MMP9, CD44 expression, and migration was confirmed by knockdown of MT1-MMP in wild-type osteoclasts. Although knockdown of MMP9 suppressed osteoclast migration, it had no effects on MT1-MMP activity or CD44 expression. These results suggest that CD44 and MT1-MMP are directly or indirectly involved in the regulation of pro-MMP9 activation. Surface expression of CD44, membrane localization of MT1-MMP, and activation of pro-MMP9 are the necessary sequence of events in osteoclast migration.

1. Introduction

Matrix metalloproteinases (MMPs) are a group of endopeptidases that regulate osteoclast migration and bone resorption [1–3]. Proteinases mobilize bone matrix proteins and determine where and when bone resorption should be initiated. MMP9 is predominantly expressed by osteoclasts and osteoclast precursors in adult bone [4]. MMP9 has been proven to be indispensable for the migration of osteoclasts through collagen both in periosteum and developing marrow cavity of primitive long bones [5]. Long bones of MMP9-knockout mice are 10% shorter than bones from wild-type mice [6]. Osteoblasts express MMP2, and bones of MMP2-knockout mice are lower in bone mineral density than those of wild-type mice. Observations in these knockout mice suggest that MMP2 and MMP9 may have a compositional and structural influence, respectively, on the biomechanical properties of whole bones.

MT1-MMP was found to be highly expressed in purified osteoclasts as compared with alveolar macrophages, bone stromal cells, and various other cell types [7]. The localization of MT1-MMP was shown in the sealing zone of osteoclast in vivo. Its distribution suggests that this enzyme modifies the bone surface to facilitate the migration and attachment of osteoclasts as well as to scavenge the resorption lacunae [8]. CD44 is a cell surface molecule, originally identified as a receptor for hyaluronic acid [9] and later found to have affinity to several matrix components including osteopontin, collagen, fibronectin, and matrix metalloproteinases (MMPs) [10–15]. Cell surface colocalization of MT1-MMP and CD44 at the leading edge of migratory cells at lamellipodia suggests the possibility that the MT1-MMP may be involved in the processing of CD44 receptor [16].

We have previously shown that CD44 standard (CD44s) is the most abundantly expressed isoform in osteoclasts [17, 18]. Cao et al. have demonstrated that the tibia of CD44$^{-/-}$ mice was shorter. The cortical bone was thicker and medullary area was smaller [19]. In vitro studies exhibited a significant decrease in bone resorption by osteoclasts from CD44$^{-/-}$

Membrane Localization of Membrane Type 1 Matrix Metalloproteinase by CD44 Regulates the Activation of Pro-Matrix
Metalloproteinase 9 in Osteoclasts

195

mice. A decrease in bone resorption activity of $CD44^{-/-}$ osteoclasts is due to reduced motility [20] which resulted in a mild osteopetrotic phenotype [19]. Surface expression of CD44 can influence signaling pathways that are critical for the activation of MMPs and motility [3, 21, 22].

The proteolytic activities of MMPs are regulated by their endogenous inhibitors, the tissue inhibitor of metalloproteinases (TIMPs) [23, 24]. TIMP2, -3, and -4 are shown as strong inhibitors of MT1-MMP [25, 26]. Increased staining for TIMP2 was also observed in association with increased synthesis of MMP2 and MMP9 [27]. MT1-MMP mediated activation of MMP2 has been shown to require the assistance of TIMP2 on the cell surface [28]. The components of the trimolecular complex MT1-MMP/TIMP2/pro-MMP2 regulate MMP2 activation. It has been shown that increased activation of MT1-MMP/MMP2 complex also activates pro-MMP9 [29]. Although MMP2 is expressed by osteoclasts, the secretory level and activity of MMP2 are significantly lower in the conditioned media of osteoclasts than WT osteoclasts [20]. Although MMP9 has been shown to be involved in the migration of osteoclasts, the mechanism of its activation remains unknown. Recent studies have shown the activation of pro-MMP9 by an MT1-MMP associated protein through RhoA and actin remodeling [30].

Earlier studies from our laboratory showed that CD44 surface expression is regulated by actin remodeling through RhoA activation. CD44 deficiency in osteoclasts increases the secretion of MT1-MMP and reduces activation of pro-MMP9 [20]. We hypothesize that activation of pro-MMP9 requires surface localization of MT1-MMP/CD44s complex. Our studies with CD44-null osteoclasts indicate that CD44 functions as a docking molecule for MT1-MMP. MT1-MMP localized on the membrane may function in the activation of pro-MMP9 as well as proteolytic processing and expression of CD44.

2. Materials and Methods

2.1. Materials. MMP antibody microarray was bought from Ray Biotech Inc. (Norcross, CA., USA, Cat. number: HO149801) Antibodies to CD44, MT1-MMP (goat-polyclonal), TIMP-2, and MMP9 were purchased from Santa Cruz Laboratory. MT1-MMP activation assay kit was bought from Amersham (Cat. number: RPN 2637). MT1-MMP antibody (MAB3329) was also purchased from Chemicon. GAPDH antibody was bought from Abcam Inc (Cambridge, MA., USA). CY2- or CY3-conjugated secondary antibodies were purchased from Jackson ImmunoResearch Laboratories, Inc. (West Grove, PA, USA). MCSF-1 was bought from R&D Systems, Inc. (Minneapolis, MN, USA). All other chemicals were purchased from Sigma (St. Louis, MO, USA).

2.2. Preparation of Osteoclast Precursors from Mice. Wild-type (WT) and CD44-null ($CD44^{-/-}$) mice in a C57/BL6 background were used for osteoclast preparation. $CD44^{-/-}$ mice generated by Dr. Tak W. Mak (Ontario Cancer Institute, Toronto, ON, Canada) were used [32]. WT and $CD44^{-/-}$

mice were generated in the animal facility of the University of Maryland Dental School. Breeding and maintenance were carried out as per the guidelines and approval of the institutional animal care and use committee. WT mice were also bought from Harlan Laboratory on occasion. Osteoclasts were generated in vitro using mouse bone marrow cells as described previously [18].

2.3. RNA Interference-Mediated Silencing of Endogenous Protein. Transfection of small interfering RNA (SiRNA) sequences for targeting endogenous MT1-MMP and MMP9 was carried out using the MIRUS transfection reagent according to the manufacturer's instructions as described previously [33]. From six sequences tested, the SiRNA sequences 5'-GA-AGCCUGGCUACAGCAAUAU-3' reduced the expression of MT1-MMP most efficiently. The 5'-GGUCCAUGCUG-CAGAAAAACU-3' scrambled RNA (ScRNAi) sequence was used as a control. Sense (5'-GGCAUACUUGACCGCUAU-TT-3') and SiRNA (5'-AUAGCGGUACAAGUAGCCTC-3') sequences for MMP9 were made from Ambion, Inc. (Austin, TX, USA). Transfected SiRNA concentrations were 50 nM for MT1-MMP and 40 nM for MMP9. The effect of different SiRNA (Mt1-MMP or MMP9) on the cellular levels of CD44 is shown in supplementary Figure 1 (see Supplementary Material available online at http://dx.doi.org/10.1155/2013/302392). Nontargeting scrambled nucleotides were used at matching concentrations. Cell viability was tested by trypan blue exclusion test, and it was found that treatment with indicated SiRNA or ScRNAi had no significant effect on cell viability.

2.4. Preparation of Total Cellular and Membrane Lysates. Following various treatments, osteoclasts were washed three times with cold PBS and lysed in a radio immune precipitation (RIPA) buffer to make total cellular lysates as described previously [17]. To prepare the cell membrane lysates, osteoclasts from wild-type and $CD44^{-/-}$ mice were washed three times with cold PBS and removed from the plates by gentle scraping with cold PBS. Cells were pelleted and resuspended in 2 mL of a buffer containing 250 mM sucrose, 20 mM Tris-HCl, and 1 mM EDTA pH 7.8. Lysis buffer was supplemented with EDTA-free complete mini-protease-inhibitor-cocktail (1 tablet per 10 mL buffer) immediately before use. Osteoclasts were homogenized in a hand homogenizer with 20–25 strokes. The homogenate was transferred to a 10 mL oak ridge centrifuge tube, and volume was made up to 10 mL close to the top of the tube. Tubes were centrifuged at 41,500 rcf (20,000 rpm) for 10 min at 4°C to pellet membranes. Pellet was washed three times with a buffer containing 100 mM sucrose, 10 mM Tris-HCl pH 7.5, 5 mM $MgCl_2$, and 50 mM NaCl. Subsequently, pellet was lysed in a cold RIPA buffer.

2.5. Preparation of Conditioned Media from WT and $CD44^{-/-}$ Osteoclasts. Mature WT and $CD44^{-/-}$ osteoclast cultures were kept in serum-free medium for 24–36 h. The conditioned media were collected and centrifuged to remove cell debris. Media were concentrated (~20X concentrated) using a Centricon concentrator (Amicon, Beverly, MA, usa) and quantitated as described previously [34].

2.6. Fluorescence-Activated Cell Sorting (FACs) Analysis and Cell Surface Labeling by Biotinylation.
FACs analysis was performed essentially as described previously [17]. Cells were labeled with NHS-biotin according to the manufacturer's guidelines (Pierce, Rockford, IL, USA). Briefly, cells were incubated with 0.5 mg/mL biotin for 30–40 min at 4°C and washed three times with cold PBS. Cells were lysed with RIPA lysis buffer. Equal amounts of protein lysates were used for immunoprecipitation with an antibody to MT1-MMP. The immune complexes were adsorbed onto streptavidin agarose which precipitates the biotinylated MT1-MMP protein from the cell surface as described previously [17]. These complexes were used to determine the activity of membrane associated MT1-MMP enzyme.

2.7. MT1-MMP and MMP9 Activity Assays.
Conditioned medium or immune complexes pulled down with streptavidin agarose were used for MMP-9 (BIOMOL; Cat. Number: AK410) and MT1-MMP activity assay (Amersham; Cat. number: RPN 2637) according to the manufacturer's recommendations and as previously described [34].

2.8. Gelatin Zymography and Reverse Zymography of Conditioned Media from Osteoclasts.
Gelatin zymography of culture media (~25–30 μg protein) or membrane fraction (~40–50 μg protein) was performed as described previously [20, 35]. The reverse zymogram was done to digest the gelatin (2.25%) present in the gels as described previously [36] with minor modifications. Zymogram gels containing 2.25% gelatin were incubated in the conditioned medium collected from WT osteoclasts (20X concentrated; ~50 μg protein) or 150 ng/mL progelatinase A or B for 12–15 h at 37°C. Conditioned medium provides a source of activated gelatinases which are able to degrade gelatin in the gel extensively and more efficiently within 8–10 h at 37°C. After proteolysis, gels were then stained with a staining solution (0.025% Coomassie brilliant blue in 50% methanol and 7% acetic acid) for 1-2 h at RT. Gels were then destained by washing three times in a destaining solution (40% methanol, 10% acetic acid, and 50% distilled water) for 1-2 h at RT. TIMP2 was observed as dark bands against a lighter background.

2.9. Transwell Migration Assay.
Transwell cell migration assay was performed in transwell migration chambers (8-μm pore size; Costar) with osteoclasts from wild-type mice as described previously [37]. Osteoclasts (2×10^4 cells) were added to the upper chamber containing the membrane and allowed to adhere for 2–3 h at 37°C. After the cells attached to the membrane, the culture medium (100 mL) was replaced with the following reagents at the indicated concentrations in the presence of RANKL and mCSF-1: Si and scrambled RNAi for MT1-MMP (50 nM) and MMP9 (40 nM; [38]), GM6001 (15 μM; [20]), and TIMP1 or TIMP2 (100 ng/mL; [39]). Counting of the migrated osteoclasts to the underside of the membrane was performed as described previously [37].

2.10. Immunoprecipitation and Immunoblotting Analysis.
Equal amount of membrane proteins (50 or 100 μg) or conditioned media (25 μg) was used for immunoprecipitation with antibodies of interest. Immune complexes from the immunoprecipitates as well as total cellular lysate and conditioned media proteins were subjected to SDS-PAGE and immunoblotting analyses as described previously {Chellaiah, 1996 366/id}.

2.11. Immunostaining and Gelatin Matrix Degradation Assay.
Matrix degradation assay was done using cross-linked fluorescein isothiocyanate (FITC)-conjugated gelatin matrix-coated cover slips as described previously {Desai, 2008 7105/id}. Osteoclasts cultured on gelatin matrix or glass cover slips were immunostained with MT1-MMP and MMP9 as described previously [40]. Immunostained osteoclasts on cover slips or gelatin matrix (green) (red) were scanned and imaged with a Bio-Rad confocal laser scanning microscope. Images were stored in TIF format and processed by using Photoshop (Adobe Systems, Inc., Mountain View, CA, USA).

2.12. Data Analysis.
All comparisons were made as "% control," which refers to vehicle or scrambled RNAi treated and untreated cells. The other treatment groups in each experiment were normalized to each control value. A value of <0.05 was considered significant. Data presented are means ± SEM of experiments done at different times normalized to intraexperimental control values. For statistical comparisons, analysis of variance (ANOVA) was used with the Bonferonni corrections (Instat for IBM, version 2.0; GraphPad Software, San Diego, CA, USA).

3. Results

3.1. CD44 Knockdown Reduces MT1-MMP Localization in the Membrane of Osteoclasts.
As shown previously, gelatin zymogram analysis demonstrates an increase in the activity of MT1-MMP in the conditioned medium of CD44$^{-/-}$ osteoclasts (Figure 1(a), lane 2) as compared with WT osteoclasts (lane 3) [20]. Activation of pro-MMP9 to active MMP9 is decreased considerably in the conditioned medium of CD44$^{-/-}$ osteoclasts (Figure 1, lane 2). WT osteoclasts exhibited both pro- (92 kDa) and active (84 kDa) MMP9 (Figure 1, lane 3). WT or CD44$^{-/-}$ osteoclasts failed to show detectable levels of MMP2 activity in the zymogram analysis. Reverse zymogram analysis of the lower half of the zymogram gel demonstrated a protein with molecular mass ~24–27 kDa (Figure 1(b)) which was confirmed as TIMP2 in immunoblotting analysis (Figure 1(d)). Secretion of TIMP2 is more in WT osteoclasts (lane 3 in (b) and lane 2 in (d)). Immunoblotting analysis of the conditioned medium (c) with an MT1-MMP antibody confirmed that the increased activity of MT1-MMP in the conditioned media is due to an increase in the secreted or soluble MT1-MMP levels in CD44$^{-/-}$ osteoclasts (Figure 1(c), lane 1).

FACs analysis demonstrated a significant decrease in the surface levels of MT1-MMP in CD44$^{-/-}$ osteoclasts. This suggests that CD44 may have a role in the membrane localization

Membrane Localization of Membrane Type 1 Matrix Metalloproteinase by CD44 Regulates the Activation of Pro-Matrix Metalloproteinase 9 in Osteoclasts

197

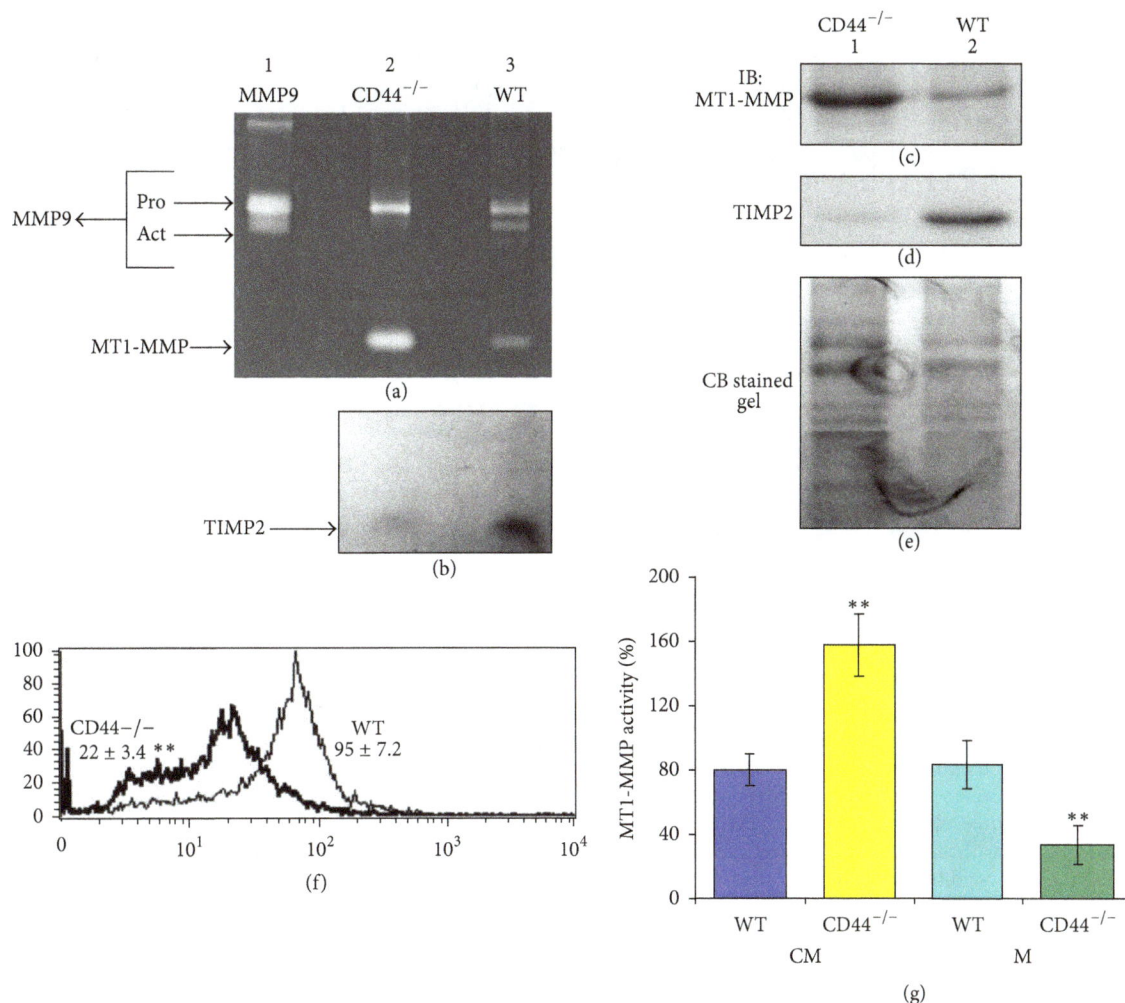

FIGURE 1: CD44 knockout reduces the levels of active MMP9 but increases MT1-MMP in osteoclasts. (a) *Zymogram analysis*. Conditioned media from $CD44^{-/-}$ (lane 2) and WT osteoclasts (lane 3) were analyzed by zymogram analysis. Recombinant MMP9 protein (UBI) was used as an identification marker. (b) *Reverse zymography*. The lower half of the zymogram was cut and incubated with culture media to digest the gelatin in the gel. Subsequently, the gel was stained with Coomassie blue to detect TIMP2 protein as described in the Methods-section. Dark TIMP2 band (indicated by arrows) was observed in the light background. (c–e) *Immunoblotting analyses*. Soluble or secreted levels of MT1-MMP and TIMP 2 in the conditioned media of $CD44^{-/-}$ (lane 1) and WT (lane 2) osteoclasts were determined by immunoblotting analysis with relevant antibody in succession after stripping the blot. Finally the blot was stained with a Coomassie blue stain to determine the amount of protein loaded in each lane (e). (f) and (g) *FACs analysis (f) and MT1-MMP activity assay (g)*. (f) A representative histogram is shown for $CD44^{-/-}$ and WT osteoclasts. Each FACs analysis was performed in quadruplicates. Mean fluorescence intensity of MT1-MMP from two different osteoclast preparations is shown in Figure 1(f). (g) Concentrated conditioned media and MT1-MMP immunoprecipitates pulled down with streptavidin agarose were subjected to the MT1-MMP activity assay to determine the soluble and cell surface (membrane) associated enzyme activity, respectively. $^{**}P < 0.01$ versus WT osteoclasts in (f) and (g). Results shown in (g) are representative of three different experiments with three different osteoclast preparations.

of MT1-MMP (f). An increase in MT1-MMP activity in the conditioned medium of $CD44^{-/-}$ osteoclasts in an in vitro assay (Figure 1(g)) supports the zymogram analysis shown in (a). Furthermore, MT1-MMP activity concurs with the MT1-MMP levels in the conditioned medium (a and c) and membrane (f) in WT and $CD44^{-/-}$ osteoclasts.

3.2. Membrane Localization of MT1-MMP (~63–65 kDa) Protein Is Significantly Reduced in $CD44^{-/-}$ Osteoclasts. MT1-MMP exists in different forms. Therefore, we investigated

whether CD44 would serve to maintain the membrane localization of any of the forms of the MT1-MMP. Membrane lysate fraction (Figure 2, lanes 1 and 3) and conditioned medium (lanes 2 and 4) of WT and $CD44^{-/-}$ osteoclasts were subjected to immunoblotting analysis with an MT1-MMP antibody. Membrane fraction of WT osteoclasts demonstrated latent (~63–65 kDa) and catalytically inactive (~45 kDa) forms of MT1-MMP. Conditioned medium of WT osteoclasts demonstrated a 55 kDa protein band corresponding to the active MT1-MMP form (lane 2) and this band is ~fivefold

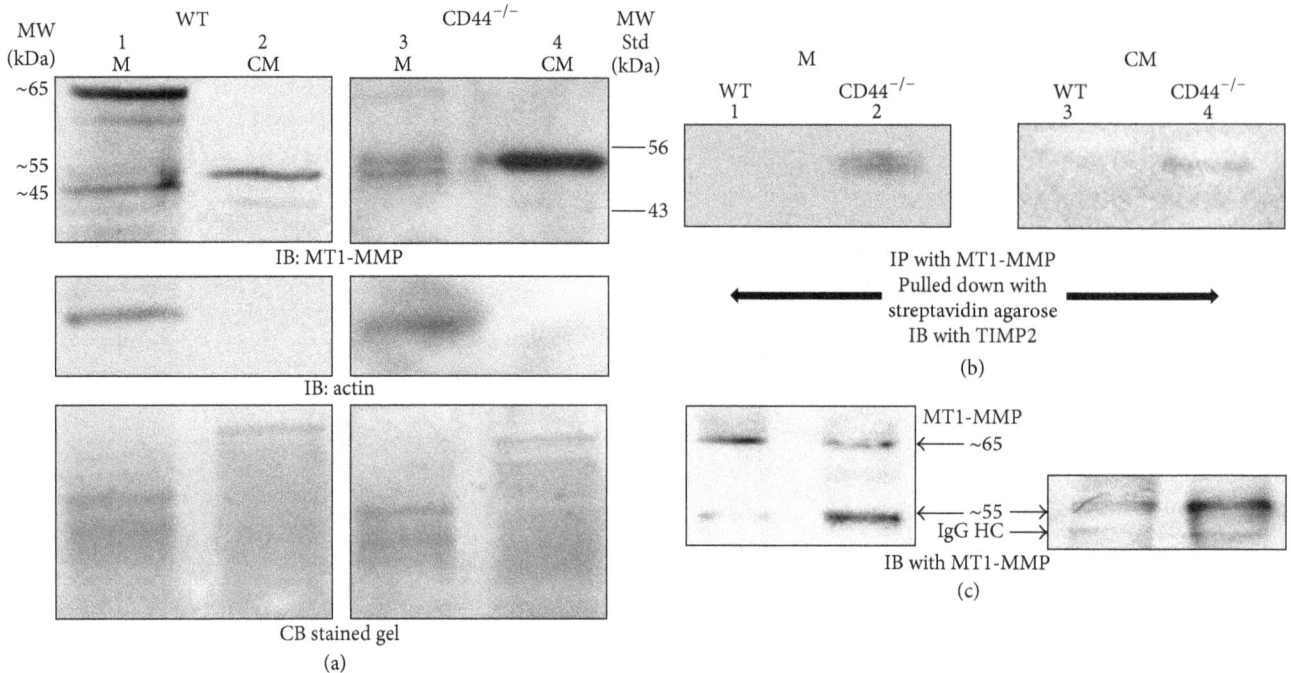

FIGURE 2: CD44 knockdown reduces membrane localization of MT1-MMP (~63 kDa) in osteoclasts. (a) Membrane fraction (M; lanes 1 and 3) and conditioned medium (CM; lanes 2 and 4) of osteoclasts isolated from WT (lanes 1 and 2) and CD44$^{-/-}$ (lanes 3 and 4) mice were analyzed by immunoblotting with an MT1-MMP antibody (top panel). Well-characterized protein standards (indicated on the right side of the figure) were used to estimate the approximate molecular weight of MT1-MMP in the membrane fraction and conditioned media. Estimated approximate molecular masses of MT1-MMP (~65, 55, and 45) are indicated on the left side of the figure. Membranes were stripped successively and blotted with an antibody to actin (middle panels) and stained with Coomassie blue (bottom panels) as loading controls for membrane fraction and conditioned medium, respectively. Data shown are representative of three independent experiments with similar results. (b) and (c) TIMP2 binds active MT1-MMP form (~55 kDa) in the membrane and conditioned medium of CD44$^{-/-}$ osteoclasts. Equal amount of protein from membrane fraction (50 μg; M) and conditioned medium (25 μg; CM) from WT (lanes 1 and 3) and CD44$^{-/-}$ (lanes 2 and 4) was used for immunoprecipitation with an antibody to MT1-MMP and immunoblotted (IB) with an antibody to TIMP2 (top panels in (b)). Membranes were stripped and blotted with an antibody to MT1-MMP ((c); bottom panels). Protein A-HRP was used as secondary antibody as described previously [31]. This analysis provided specific signals corresponding to 55 kDa. Data shown are representative of three independent experiments with similar results.

higher in CD44$^{-/-}$ osteoclasts (lane 4). These osteoclasts also demonstrated the presence of ~55 and ~45 kDa MT1-MMP forms in the membrane fraction (lane 3). However, the level of 55 kDa is considerably lower than the levels observed in the conditioned medium. The presence of extremely weak amount of ~63 kDa MT1-MMP form in the membrane fraction of CD44$^{-/-}$ osteoclasts suggests a role for CD44 in the membrane localization of MT1-MMP. The mechanism of formation of ~55 kDa MT1-MMP needs further elucidation.

3.3. TIMP2 Makes Complex with Membrane-Bound and Secreted Active Form of 55 kDa MT1-MMP.

Previous studies showed that TIMP2 interacts with the membrane-tethered MT1-MMP with its catalytic domain and inhibits its activity [41]. To determine interaction of TIMP2 with membrane-associated MT1-MMP, we used lysates made from WT and CD44$^{-/-}$ osteoclasts surface labeled with NHS-biotin. Lysates were immunoprecipitated with an antibody to MT1-MMP and pulled down with streptavidin agarose. Conditioned medium from these osteoclasts was also used for

immunoprecipitation with an antibody to MT1-MMP. Co-precipitation of TIMP2 with MT1-MMP was observed in the membrane (Figure 2(b), lane 2) and conditioned medium (lane 4) of CD44$^{-/-}$ osteoclasts. Coprecipitation of TIMP2 with MT1-MMP is more in the membrane fraction regardless of the levels of MT1-MMP in the membrane (Figure 2(c)). Membranes shown in Figure 2(b) were stripped and blotted with an antibody to MT1-MMP (Figure 2(c)). As shown in Figure 2(a), 65 kDa protein band corresponding to latent MT1-MMP is more in WT than CD44$^{-/-}$ osteoclasts. Consistently, 55 kDa protein band corresponding to active MT1-MMP form ((b), lane 2) was found in CD44$^{-/-}$ osteoclasts in both membrane ((c), lane 2) and conditioned medium ((c), lane 4). Latent form was found to a lesser extent in WT osteoclasts ((b), lanes 1 and 3).

3.4. MT1-MMP and Not MMP9 Knockdown Reduces CD44 Expression—Immunoblotting Analysis.

Localization of MMPs to the cell membrane through their interaction with cell surface CD44 receptor has been demonstrated in different cell types including osteoclasts [22, 34, 42, 43].

Membrane Localization of Membrane Type 1 Matrix Metalloproteinase by CD44 Regulates the Activation of Pro-Matrix Metalloproteinase 9 in Osteoclasts

199

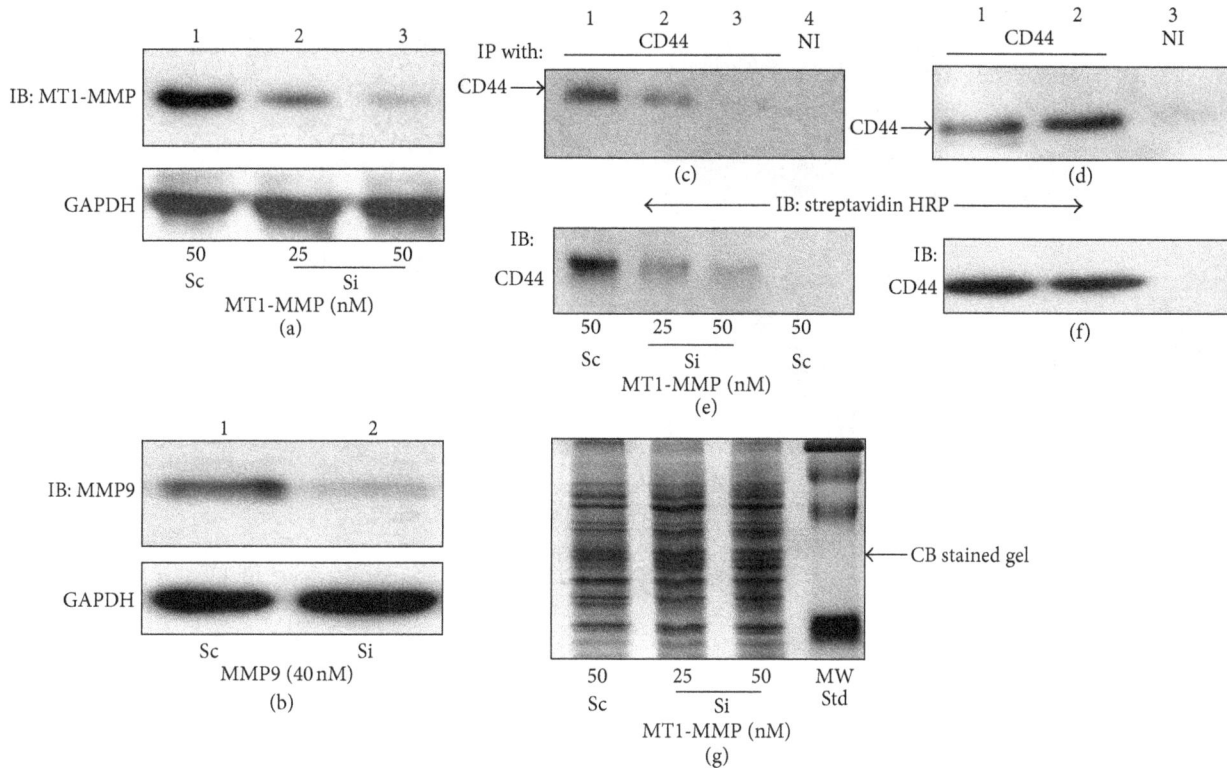

FIGURE 3: MT1-MMP knockdown reduces CD44 surface expression. (a) and (b) Immunoblotting analyses with an antibody to MT1-MMP (a) and MMP9 (b) were performed in total cellular lysates (50 μg protein) made from osteoclasts treated with a scrambled RNAi (Sc; lane 1 in (a) and (b)) or SiRNA sequences to MT1-MMP (lanes 2 and 3 in (a)) and MMP9 (lane 2 in (b)). The blots in (a) and (b) was reprobed with an antibody to GAPDH after stripping (bottom panel in (a) and (b)). GAPDH level was used as a control for loading. ((c) and (d)) Immunoblotting analysis of surface expression of CD44s in osteoclasts treated with a SiRNA (Si) or scrambled RNAi (Sc) to MT1-MMP (c) and MMP9 (d). Equal amounts of proteins (50 μg protein) prepared from osteoclasts treated as indicated in (c) and (d) and surface labeled with NHS-biotin were immunoprecipitated with an antibody to CD44 (lanes 1–3 in (c); lanes 1 and 2 in (d)). Lysates from scrambled RNAi-treated osteoclasts were immunoprecipitated with a nonimmune serum (NI; lane 4 in (c); lane 3 in (d)). Immunoprecipitates were probed with streptavidin-HRP to determine the surface levels of CD44 protein. Blot in (c) and (d) was stripped and reprobed with an antibody to CD44 to determine the total cellular levels of CD44 protein ((e) and (f)). (g) To ensure that equal protein amount was used for immunoprecipitation, duplicate polyacrylamide gels were run with 50 μg total protein and stained with Coomassie blue staining.

The function and molecular association of MT1-MMP and CD44 at the cell surface is not known although CD44$^{-/-}$ osteoclasts demonstrated a decrease in the membrane localization of MT1-MMP. In order to identify whether MT1-MMP or MMP9 activity is necessary for the expression of CD44, osteoclasts from WT mice were transfected with SiRNA to MT1-MMP or MMP9 (Si, Figures 3(a) and 3(b)) and incubated for 48–72 h as described previously [38]. Scrambled RNAi sequences (Sc) transfected cells were used as controls. A dose-dependent decrease in MT1-MMP protein level was observed. A maximum silencing (>80–85%) was observed at 50 nM SiRNA of MT1-MMP (Figure 3(a)) and 40 nM SiRNA of MMP9 (Figure 3(b), lane 2).

Osteoclasts were surface labeled with NHS-biotin and equal amounts of lysate proteins (50 μg) were immunoprecipitated with an antibody to sCD44 or nonimmune IgG (NI; lane 4 in (c) and lane 3 in (d)). Immunoprecipitates were blotted with streptavidin-HRP to determine the surface levels of CD44 ((c) and (d)). Subsequently, the blot was stripped and blotted with an antibody to sCD44 to determine the

total cellular levels of CD44 protein ((e) and (f)). Surface and total cellular levels of sCD44 are reduced in a dose-dependent manner in osteoclasts transfected with MT1-MMP SiRNA (Figures 3(c) and 3(e), lanes 2 and 3) as compared with scrambled RNAi transfected cells ((c) and (e), lane 1). However, knockdown of MMP9 in osteoclasts had no effect on either the surface or total cellular levels of sCD44 ((d) and (e), lane 2). Concurrently, 50 μg of total cellular protein was loaded on a separate gel for Coomassie blue staining (Figure 3(g)) which is used as a loading control.

3.5. MT1-MMP Knockdown Reduces Cellular Levels of CD44 and Migration in Osteoclasts Immunostaining and Transwell Migration Analyses. Consistent with the observations shown in Figure 3(e), immunostaining analysis indeed has shown a significant decrease in total cellular levels of CD44s in MT1-MMP knockdown osteoclasts (Figure 4(b)). Colocalization (yellow) of MT1-MMP (green) and CD44 (red) was observed in the podosomes (indicated by arrows in Figure 4(a)) and membrane of osteoclasts transfected with scrambled RNAi

(a)

(b)

(c)

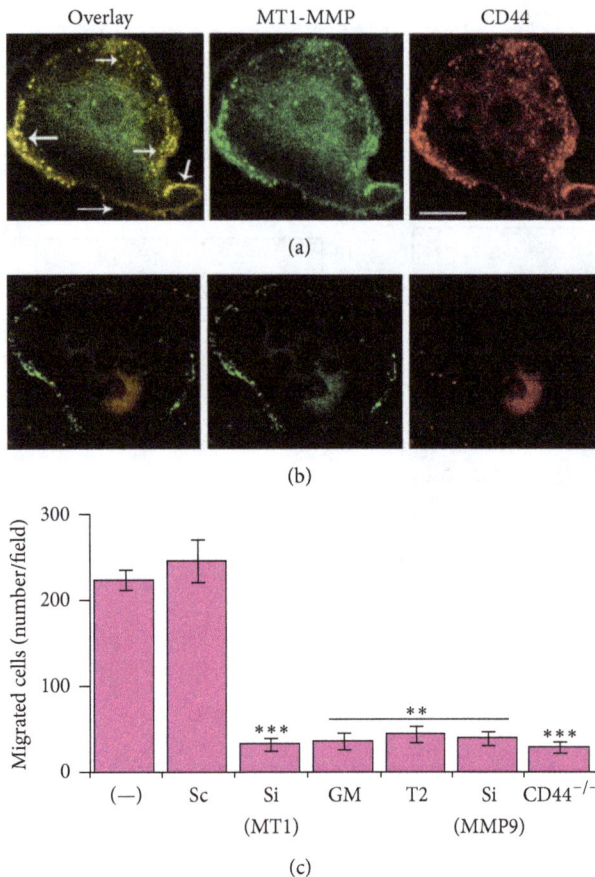

FIGURE 4: MT1-MMP and MMP 9 have roles in osteoclast migration but only MT1-MMP has a role in the expression of CD44. ((a) and (b)) Confocal microscopy analysis of osteoclasts transfected with scrambled RNAi (a) and SiRNA sequences to MT1-MMP. Osteoclasts immunostained for CD44 (red) and MT1-MMP (green) were analyzed by confocal microscopy. Colocalization (yellow) of MT1-MMP and CD44 was observed in the membrane and podosomes of osteoclasts (indicated by arrows in overlay panel). Scale bar: 50 μm. (c) The effects of various treatments on the migration of WT osteoclasts were assessed using transwell migration assay. Osteoclasts from CD44$^{-/-}$ (CD44-null) mice were also used. WT osteoclasts were treated with the following treatments: scrambled RNAi (Sc), SiRNA MT1-MMP or MMP9 (Si), MMP inhibitor GM6001 (GM), and TIMP2. Cells in ~20–25 fields were counted. Values are expressed as mean number of cells migrated/field. The experiment was repeated three times with three different osteoclast preparations. Assay was performed in quadruplicates in each experiment. Data shown are the mean ± SEM of one experiment. **$P < 0.01$ and ***$P < 0.001$ compared with (−) and Sc controls.

(a). A significant decrease in the expression of CD44 in MT1-MMP knockdown and CD44$^{-/-}$ osteoclasts is associated with a significant decrease in the migration of osteoclasts in a transwell migration assay (Figure 4(c)). Migration was also affected by soluble recombinant TIMP2 and a broad spectrum MMP inhibitor GM6001. These data collectively suggest that MT1-MMP regulates the expression of CD44s. Both CD44 and MT1-MMP are involved in the migration of osteoclasts.

3.6. MT1-MMP Knockdown Reduces MMP9 Activity.

The presence of MMP9, TIMP2, and MT1-MMP in the conditioned media of osteoclasts suggests that formation of this complex may regulate the activation of pro-MMP9. Our results show that knockdown of MT1-MMP had a significant inhibitory effect on the activation of pro-MMP9 (Figure 5(a), lane 3). However, MT1-MMP knockdown had no effect on total cellular levels of MMP9 ((c), lane 2). Both pro- and active MMP9 were detected in untransfected ((a), lane 2) and ScRNAi transfected (lane 4) osteoclasts. Coomassie blue stained polyacrylamide gel was used as a loading control (b) for the zymogram gel shown in (a).

3.7. TIMP2 Reduces MT1-MMP and MMP9 Activities in WT Osteoclasts.

TIMP2 is unique among TIMPs because it inhibits a number of MMPs besides its role in the activation. MT1-MMP initiates pro-MMP2 activation in a process that is tightly regulated by the level of TIMP2 [44]. To this end, to assess TIMP2 regulation of MT1-MMP and MMP9 activities, WT osteoclasts were treated with TIMP1 and TIMP2 (100 ng/mL for 20 h at 37°C). Osteoclasts knockdown of MT1-MMP and MMP9 with respective SiRNA was also used. Untreated (−) and scrambled RNAi treated osteoclasts were used as controls. Equal amount of membrane protein was used for MMP9 and MT1-MMP activity assay (Figure 5(d) and (e)). Basal MMP9 or MT1-MMP activity in untreated (−) osteoclasts is considered as 100%. TIMP2 reduces both MMP9 (Figure 5(d)) and MT1-MMP (Figure 5(e)) activities in osteoclasts. Under these conditions, TIMP1 did not inhibit MT1-MMP activity. MT1-MMP knockdown (Si) reduced MMP9 activity (Figure 5(d)) whereas MMP9 knockdown (Si) had no effect on MT1-MMP activity (Figure 5(e)). Consistent with the observation shown in Figure 1(g), MT1-MMP activity is significantly reduced in CD44$^{-/-}$ osteoclasts. These data provide the evidence that MT1-MMP is a potential regulator of MMP9 activity. We show here TIMP2/MT1-MMP complex formation (Figure 2) and suppressed activity of MT1-MMP (Figure 5(e)) in the cell membrane. These findings suggest that TIMP2 is a key determinant of MT1-MMP activity. TIMP2 may inhibit MMP9 through its interaction with MT1-MMP.

3.8. SiRNA to MT1-MMP Reduces Matrix Degradation.

We have previously demonstrated that reducing MMP9 levels by RNA interference inhibited the degradation of matrix but not the formation of podosomes [38]. Since MT1-MMP appears to have a role in the regulation of MMP9 activity, we further determined the ability of MT1-MMP knockdown cells to degrade the gelatin matrix in vitro. First, we determined the possible interaction of MT1-MMP and MMP9 on the cell surface. Immunostaining analysis was performed in osteoclasts not permeablized with Triton X-100 as described previously [17]. Distribution is shown in a representative osteoclast (Figure 6(a)). Punctate codistribution (yellow) of MT1-MMP (green) and MMP9 (red) was observed on the surface of osteoclasts (a).

Membrane Localization of Membrane Type 1 Matrix Metalloproteinase by CD44 Regulates the Activation of Pro-Matrix Metalloproteinase 9 in Osteoclasts

201

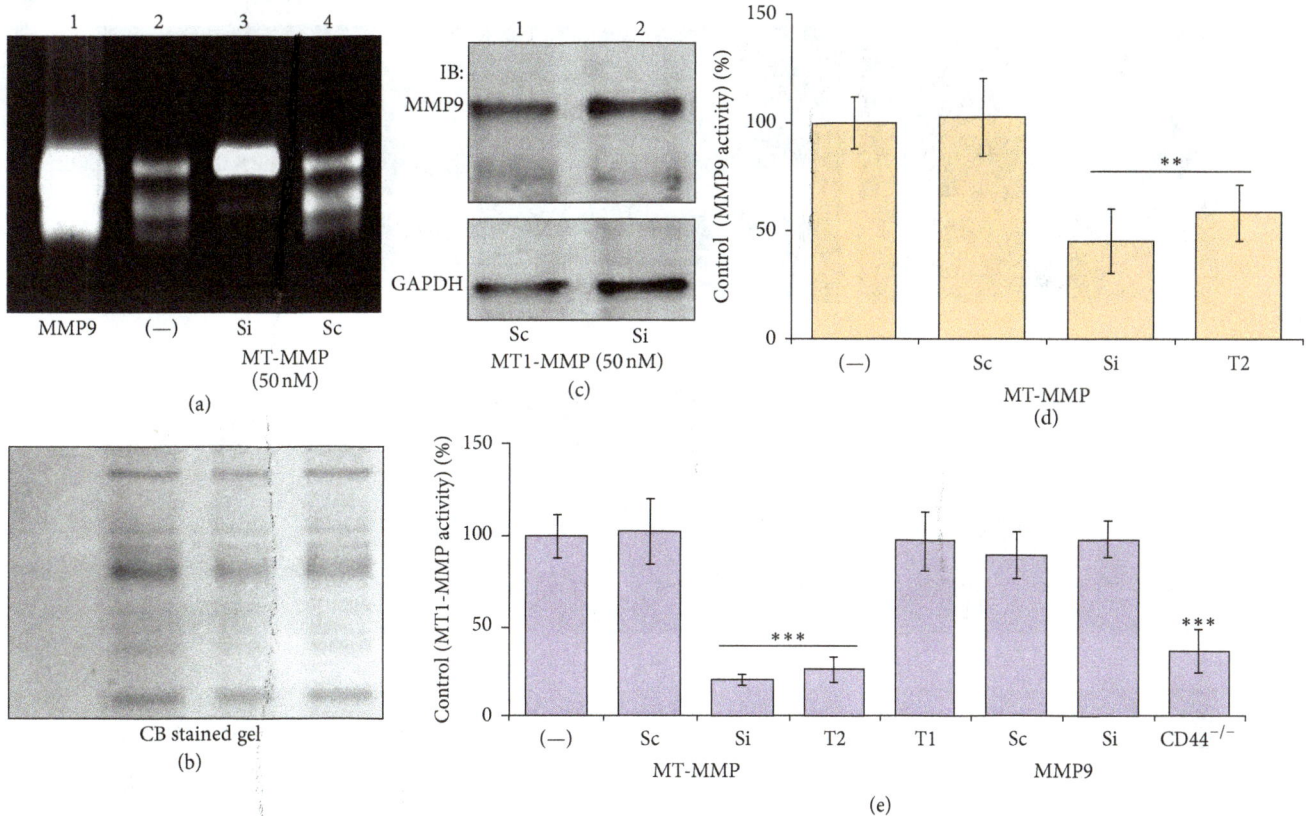

FIGURE 5: MT1-MMP regulates the activity of MMP9. (a–d) The effects of various treatments on MMP9 activity. Equal amounts of membrane proteins (50 μg) from osteoclasts untransfected (−) (lane 2) or transfected with SiRNA (Si, lane 3) and ScRNAi (Sc, lane 4) to MT1-MMP were used for gelatin zymography analysis (a) and MMP9 activity assay in vitro (d). Lysates made from osteoclasts treated with TIMP2 (T2) were also used for MMP9 activity assay in vitro (d). To ensure that equal protein amount was used for zymogram analysis (a), duplicate polyacrylamide gel was run with same amount of protein and stained with Coomassie blue staining (b). Cell extracts from osteoclasts treated with ScRNAi (Sc) and SiRNA (Si) to MT1-MMP were subjected to immunoblotting analysis with an MMP9 antibody to detect total cellular levels of MMP9 ((c); top panel). Stripping and reprobing of the same blot with an antibody to GAPDH was used as loading control (bottom panel in (c)). (e) The effects of various treatments on MT1-MMP activity. MT1-MMP activity was determined in the membrane fraction of osteoclasts subjected to various treatments as indicated as follows: ScRNAi (Sc) and SiRNA (Si) to MT1-MMP and MMP9, TIMP1 (T1), and TIMP2 (T2). Untransfected and CD44-null osteoclasts are indicated as (−) and CD44$^{-/-}$, respectively. Values in (d) and (e) are mean ± SE from three different experiments. $^{**}P < 0.05$ and $^{***}P < 0.001$ compared to controls ((−) and Sc). The results shown are representative of three different experiments.

Next, in order to determine the ability of MT1-MMP knockdown on MMP9 expression/activity and matrix degradation, osteoclasts were cultured on a matrix made of gelatin conjugated with FITC dye (green) as shown previously [38]. After incubation for 3–6 h, surface distribution of MMP9 was determined by immunostaining with an antibody to MMP9. Confocal microscopy analysis demonstrated punctate distribution of MMP9 (Figure 6(b), MMP9) on the surface as shown in (a). Matrix degradation (indicated by asterisks in gelatin panel) was observed underneath and around the osteoclasts ((b); overlay). Osteoclasts transfected with SiRNA against MT1-MMP failed to demonstrate either punctate distribution of MMP9 on the cell surface or degradation of the gelatin matrix (Figure 6(c)). Diffused distribution of MMP9 was observed on the cell surface ((c); MMP9). MT1-MMP knockdown had no effect on the total cellular levels of MMP 9 (Figure 5) but affected distribution and activity of MMP9.

3.9. Reducing the Levels of CD44, MT1-MMP, and MMP9 Attenuates OC Migration but Not Polarization, Sealing Ring Formation, and Bone Resorption. We have previously demonstrated localization of MMP9 and actin in sealing rings of resorbing osteoclasts [20]. We next proceeded to determine if SiRNA to MT1-MMP or MMP9 has the potential to attenuate sealing ring formation and bone resorption. After treating with Sc RNAi and indicated SiRNA for 17–20 h (Figure 7), osteoclasts were cultured on dentine slices for 8–12 h. Cells were stained for actin with rhodamine phalloidin. Diffuse staining of actin is due to intense actin staining in sealing ring and cell membrane. Wild-type osteoclasts treated with a ScRNAi demonstrated convoluted multiple overlapping pits (Figures 7(a) and 7(a′)). Simple resorption pits were observed in CD44$^{-/-}$ osteoclasts and wild-type osteoclasts treated with an SiRNA against MT1-MMP or MMP9 (Figures 7(b)–7(d); 7(b′)–7(d′)). Osteoclast migration, adhesion, polarization,

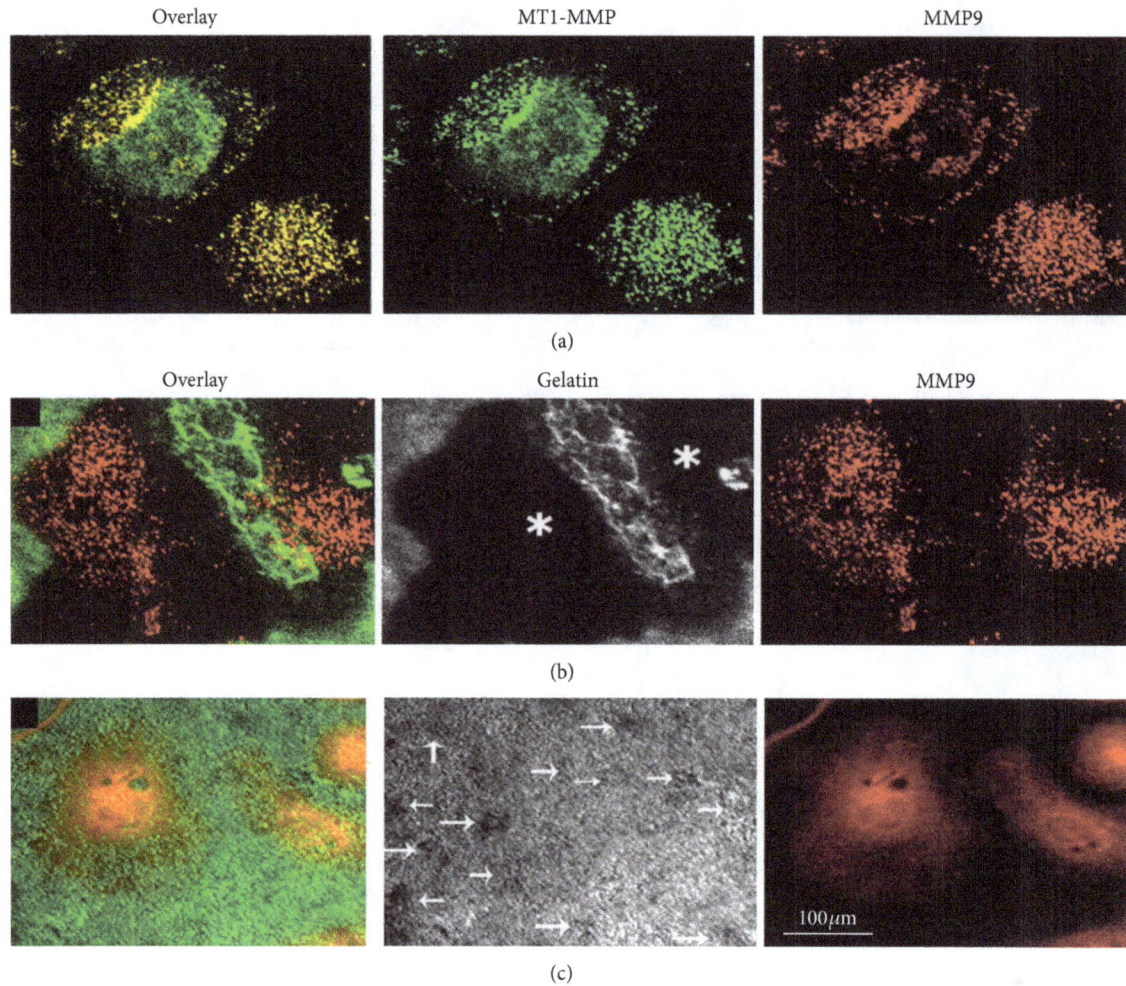

FIGURE 6: MT1-MMP knockdown reduces matrix degradation and punctate localization of MMP9 in osteoclasts. (a) Confocal microscopy analysis of distribution of MT1-MMP and MMP9 in osteoclasts cultured on glass cover slips. Immunostaining was done in nonpermeablized osteoclasts with Triton-X100. Punctate colocalization (yellow) of MT1-MMP (green) and MMP9 (red) was observed on the cell surface. (b) and (c) Confocal microscopy analysis of osteoclasts immunostained with a MMP9 antibody (red) and degradation of gelatin matrix (green). Osteoclasts transfected with scrambled RNAi (b) and SiRNA (c) sequences to MT1-MMP were subjected to gelatin degradation for 6 h. Distribution of MMP9 (red) in nonpermeablized osteoclasts and FITC-gelatin matrix (green) is shown in the overlay panels ((b) and (c)). The results represent one of three experiments performed. Scale bar: 100 μm. Matrix degraded areas are indicated by asterisks in (b) and (c) (gelatin panel).

and resorption are very much coupled processes. Formation of simple pits in these osteoclasts (b–d; b′–d′) suggests that these osteoclasts are not defective in adhesion, polarization, and resorption but defective in migration.

4. Discussion

It has been suggested that MT1-MMP modifies the bone surface to facilitate the migration and attachment of osteoclasts [8]. Immunostaining analysis in osteoclasts demonstrated colocalization of MT1-MMP and MMP9 in osteoclast podosomes. Matrix degradation was observed in areas where podosomes are present [38]. Formation of structurally and functionally linked protease/protein complexes on podosomes dissolves the matrix proteins and paves the way for the migration of osteoclasts. We have previously

demonstrated colocalization of CD44 and MMP9 at the migratory front of osteoclasts [20]. Our previous and present studies in osteoclasts knockdown of MT1-MMP and null for CD44 (Figures 5 and 6) indicate that spatial distribution of MMP9, CD44, and MT1-MMP on the cell surface and podosomes functions in a controlled and coordinated manner to digest extracellular matrix proteins enabling osteoclast migration [20, 38]. However, neither polarization nor bone resorption are affected under these conditions (Figure 7). These observations agree with published roles for MMP9 and MT1-MMP (a.k.a. MMP14) in osteoclast migration and not bone resorption [45].

MT1-MMP has now become one of the best characterized enzymes in the MMP family. However, the role of CD44 in the membrane localization of MT1-MMP is not well understood. Our studies have shown the presence of three forms

Membrane Localization of Membrane Type 1 Matrix Metalloproteinase by CD44 Regulates the Activation of Pro-Matrix Metalloproteinase 9 in Osteoclasts

203

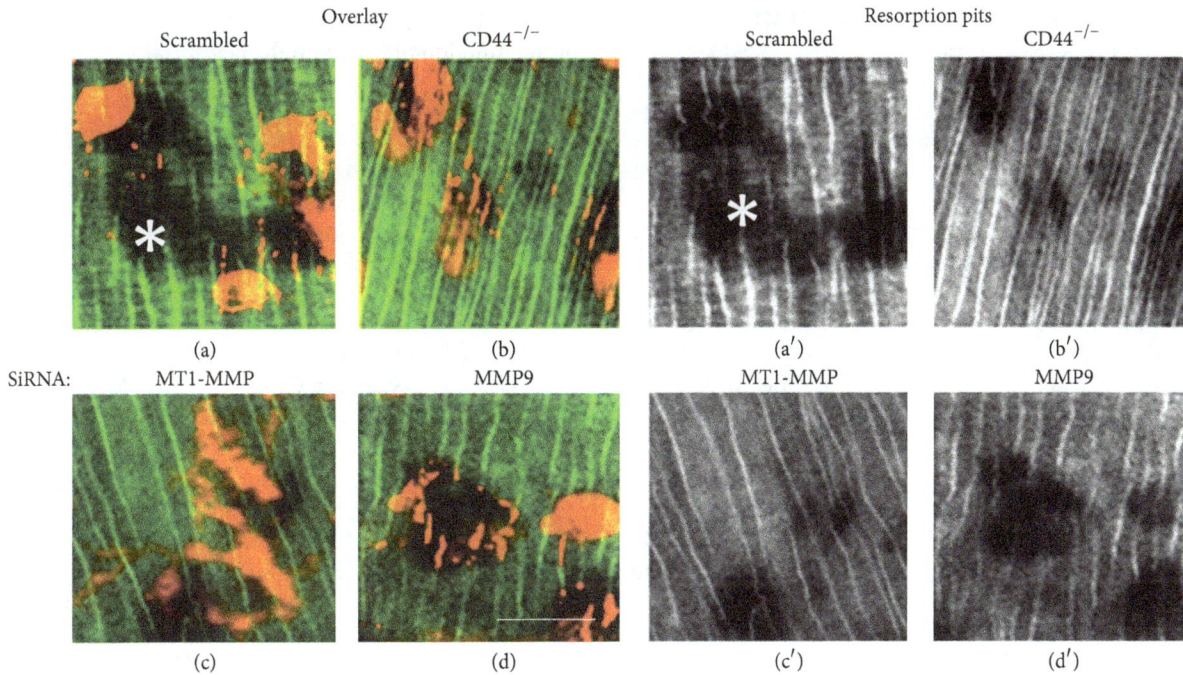

FIGURE 7: Reducing the levels of CD44, MT1-MMP, and MMP9 attenuates OC migration but not polarization, sealing ring formation, and bone resorption. Confocal microscopy analysis of distribution of actin in indicated osteoclasts cultured on dentine slices. Osteoclasts were stained for actin (red; (a–d)). Resorption pits underneath the osteoclasts are shown separately in (a′)–(d′). Dentine slices are shown in green by the reflected light. Experiments were repeated two times. Three dentine slices were used for each condition in both experiments. Results represent one of the two experiments and dentine slices performed with two different osteoclast preparations. Scale bar: 25 μm.

(~63, 55, and 45 kDa) of MT1-MMP in WT osteoclasts. These forms are present in the following order 63 > 45 ≫ 55 kDa in WT osteoclasts. However, CD44$^{-/-}$ osteoclasts displayed the active form of MT1-MMP (55 kDa) as secreted MT1-MMP in the conditioned medium. Our study with osteoclasts from CD44$^{-/-}$ mice provides molecular evidence that CD44 regulates membrane localization of MT1-MMP. The extracellular portion of MT1-MMP has a catalytic domain (CAT) linked to the hemopexin- (HPX-) like domain through the hinge region [13]. MT1-MMP binds to the extracellular portion of CD44 through the hemopexin- (HPX-)like domain and colocalizes in the cells. Disruption of MT1-MMP/CD44 complex by overexpressing the HPX domain resulted in inhibition of the cleavage and shedding of CD44 [46]. It appears that CD44 plays a key role in assembling MT1-MMP to the cell surface. There also is evidence that activation of MMP9 by an MT1-MMP associated protein occurs through activation of RhoA and actin remodeling [30]. We have previously demonstrated that surface expression of CD44 is linked with actin polymerization and CD44 phosphorylation by RhoA activation [20]. As suggested by others CD44 may be one of the key proteins that interact with MT1-MMP and promote cell migration [16, 30].

CD44 was shown to be shed by proteolytic cleavage by MMPs [47–49]. The shedding was shown to be inhibited by the inhibitors of MMPs or serine proteinases [49]. Earlier observations by Okamoto et al. showed a relationship between proteolytic processing of CD44 by MMPs and

transcriptional activation in the nucleus [50]. We have shown here that SiRNA to MT1-MMP reduces total and surface levels of CD44. This indicates that CD44 is a direct target of MT1-MMP in osteoclasts. MT1-MMP and MMP9 are shown to interact with CD44 in different cell systems either directly or indirectly [43, 46, 51, 52]. We have previously demonstrated localization of MMP9 and actin in the sealing ring of resorbing osteoclasts. However, colocalization is negligible [20]. An increase in the secretion of MT1-MMP and a decrease in the active form of MMP9 suggest a possible role for CD44 in the localization of MT1-MMP and MMP9 on the cells surfaces.

Among the MT-MMPs, MT1-MMP is shown to be frequently expressed in highly migratory cells including macrophages, endothelial cells, and invasive cells [7, 16, 53]. We show here that MT1-MMP forms a complex with TIMP2 in CD44$^{-/-}$ osteoclasts. The secreted level and activity of MT1-MMP are more in CD44$^{-/-}$ than WT osteoclasts. However, a significant decrease in the level of TIMP2 was observed in the conditioned medium of CD44$^{-/-}$ osteoclasts as compared with WT osteoclasts. Based on the ratio of MT1-MMP and TIMP2 in the conditioned media, we suggest that the activity of MT1-MMP is tightly regulated by the levels of TIMP2. Low levels of TIMP2 in the conditioned media of CD44$^{-/-}$ osteoclasts correlated with an enhanced activation of MT1-MMP when compared with WT osteoclasts. TIMP2 binds free MT1-MMP at the catalytic site [23]. It has been shown by others that the concentration of 55 kDa MT1-MMP on the cell surface is directly and positively regulated by

TIMP2. In the absence of TIMP2, MT1-MMP undergoes autocatalysis to a 44 kDa form [44].

The levels of 55 kDa MT1-MMP and TIMP2 are considerably lower in the membrane fraction of WT osteoclasts. Autocatalytic product is seen as 45 kDa in the membrane fraction of WT osteoclasts. Latent MT1-MMP binds TIMP2. Immunoblotting analysis of total membrane fraction consistently shows an MT1-MMP protein with MW ~60 kDa (Figure 1(a), lane 1). It is possible that 60 kDa MT1-MMP contributes to the activity shown in Figure 1(g) in WT osteoclasts. This may function as a latent form of MT1-MMP in WT osteoclasts and possibly did not have affinity to TIMP2. As suggested by others [44] it is possible that differences in the molecular weight of latent MT1-MMP in WT and CD44$^{-/-}$ osteoclasts may be due to differential solubility of these forms when NP-40 containing lysis buffer was used. After surface activation, it may be released into the medium. Since the ratio of MT1-MMP to TIMP2 is less, we presume that MT1-MMP is more active in the conditioned medium than in membrane. Reduced activity in the membrane may be due to interaction of TIMP2 with latent MT1-MMP on the cell surface as shown in Figure 2(b) (left panel). It is possible that differences in the activity of MT1-MMP in WT and CD44$^{-/-}$ osteoclasts may be due to a variation in the ratio of MT1-MMP to TIMP2 (Figure 2).

By cross linking experiments, it has been hypothesized that MT1-MMP and TIMP2 form a "receptor" complex that binds MMP2 via its C terminus. High concentrations of TIMP2 inhibit MMP2 activation, because all available MT1-MMP molecules form MT1-MMP/TIMP2 complex. This leads to a condition of no free MT1-MMP molecules to activate MMP2 [41]. MMP2 level is considerably low or not detected at times in the conditioned media of WT or CD44$^{-/-}$ osteoclasts (Figure 1; [20]). MT1-MMP/TIMP2/MMP2 complex formation was not observed in WT osteoclasts (data not shown). Therefore, it is unlikely that MT1-MMP/TIMP2/MMP2 complex is involved in the activation of pro-MMP9 in osteoclasts. MMP3, MMP7, and MMP13 have been shown to activate pro-MMP9 [54, 55]. Although MT1-MMP knockdown regulates pro-MMP9 activation in WT osteoclasts, it is not known whether it directly or indirectly activates pro-MMP9. MT1-MMP knockdown studies suggest MT1-MMP as an upstream regulator of pro-MMP9. To gain further insight into the role of MT1-MMP in the activation of pro-MMP9, future studies should focus on whether MT1-MMP activates pro-MMP9 in cooperation with other MMPs (MMP3, -7, or -13) in osteoclasts.

5. Conclusions

Our data provide molecular evidence that CD44 regulates membrane localization of MT1-MMP. Interestingly, observations in CD44$^{-/-}$ osteoclasts suggest that CD44-TIMP2-MT1-MMP axis regulates MMP9 activity. Surface expression of CD44, localization of MT1-MMP on the cell surface, and activation of pro-MMP9 are the necessary sequence of events in osteoclast migration. Delineating the specific mechanisms that regulate MT1-MMP and consequently MMP9 activity in osteoclasts remains an important goal in our understanding of the function of CD44 on the cell surface.

Abbreviations

MMP:	Matrix metalloproteinase
MT1-MMP:	Membrane type 1 matrix metalloproteinase
HPX:	Hemopexin
CD44:	Cluster of differentiation (also well known as cell surface glycoprotein and adhesion receptor)
CD44s:	Standard CD44
CD44$^{-/-}$ mice:	CD44-null mice
WT mice:	Wild-type mice
TIMP1 or -2:	Tissue inhibitor of metalloproteinase 1 or 2
ECM:	Extracellular matrix
SiRNA:	Small interference RNA
FITC:	Fluorescein isothiocyanate (green-fluorescent fluorescein dye was used)
GM6001:	Broad spectrum MMP inhibitor.

Conflict of Interests

The authors declare no conflict of interests.

Acknowledgments

This work was supported by the National Institute of Health (NIH) Grant AR46292 to Heenakshi A. Chellaiah. Dr. Tak Mak (University of Toronto, Toronto, ON, Canada) is gratefully acknowledged for the kind gift of CD44-null mice, on a C57BL/6 background [32]. The authors gratefully acknowledged the assistance of Dr. Venkatesababa Samanna (Postdoctoral fellow; Georgia State University, Atlanta, GA, USA) in the maintenance of CD44-null colony and osteoclast preparations.

References

[1] X. Yu, P. Collin-Osdoby, and P. Osdoby, "SDF-1 increases recruitment of osteoclast precursors by upregulation of matrix metalloproteinase-9 activity," *Connective Tissue Research*, vol. 44, supplement 1, pp. 79–84, 2003.

[2] L. Blavier and J. M. Delaissé, "Matrix metalloproteinases are obligatory for the migration of preosteoclasts to the developing marrow cavity of primitive long bones," *Journal of Cell Science*, vol. 108, part 12, pp. 3649–3659, 1995.

[3] P. Spessotto, F. M. Rossi, M. Degan et al., "Hyaluronan-CD44 interaction hampers migration of osteoclast-like cells by down-regulating MMP-9," *Journal of Cell Biology*, vol. 158, no. 6, pp. 1133–1144, 2002.

[4] J. S. Nyman, C. C. Lynch, D. S. Perrien et al., "Differential effects between the loss of MMP-2 and MMP-9 on structural and tissue-level properties of bone," *Journal of Bone and Mineral Research*, vol. 26, no. 6, pp. 1252–1260, 2011.

[5] M. T. Engsig, Q.-J. Chen, T. H. Vu et al., "Matrix metalloproteinase 9 and vascular endothelial growth factor are essential

for osteoclast recruitment into developing long bones," *Journal of Cell Biology*, vol. 151, no. 4, pp. 879–889, 2000.

[6] T. H. Vu, J. M. Shipley, G. Bergers et al., "MMP-9/gelatinase B is a key regulator of growth plate angiogenesis and apoptosis of hypetrophic chondrocytes," *Cell*, vol. 93, no. 3, pp. 411–422, 1998.

[7] T. Sato, M. D. C. Ovejero, P. Hou et al., "Identification of the membrane-type matrix metalloproteinase MT1-MMP in osteoclasts," *Journal of Cell Science*, vol. 110, part 5, pp. 589–596, 1997.

[8] K. Irie, E. Tsuruga, Y. Sakakura, T. Muto, and T. Yajima, "Immunohistochemical localization of membrane type 1-matrix metalloproteinase (MT1-MMP) in osteoclasts in vivo," *Tissue and Cell*, vol. 33, no. 5, pp. 478–482, 2001.

[9] J. Lesley and R. Hyman, "CD44 can be activated to function as an hyaluronic acid receptor in normal murine T cells," *European Journal of Immunology*, vol. 22, no. 10, pp. 2719–2723, 1992.

[10] G. F. Weber, S. Ashkar, and H. Cantor, "Interaction between CD44 and osteopontin as a potential basis for metastasis formation," *Proceedings of the Association of American Physicians*, vol. 109, no. 1, pp. 1–9, 1997.

[11] G. N. Thalmann, R. A. Sikes, R. E. Devoll et al., "Osteopontin: possible role in prostate cancer progression," *Clinical Cancer Research*, vol. 5, no. 8, pp. 2271–2277, 1999.

[12] S. Goodison, V. Urquidi, and D. Tarin, "CD44 cell adhesion molecules," *Journal of Clinical Pathology*, vol. 52, no. 4, pp. 189–196, 1999.

[13] M. Seiki, H. Mori, M. Kajita, T. Uekita, and Y. Itoh, "Membrane-type I matrix metalloproteinase and cell migration," *Biochemical Society Symposium*, no. 70, pp. 253–262, 2003.

[14] C. Johnson and Z. S. Galis, "Matrix metalloproteinase-2 and -9 differentially regulate smooth muscle cell migration and cell-mediated collagen organization," *Arteriosclerosis, Thrombosis, and Vascular Biology*, vol. 24, no. 1, pp. 54–60, 2004.

[15] M. A. Chellaiah and K. A. Hruska, "The integrin $\alpha v\beta 3$ and CD44 regulate the actions of osteopontin on osteoclast motility," *Calcified Tissue International*, vol. 72, no. 3, pp. 197–205, 2003.

[16] M. Kajita, Y. Itoh, T. Chiba et al., "Membrane-type 1 matrix metalloproteinase cleaves CD44 and promotes cell migration," *Journal of Cell Biology*, vol. 153, no. 5, pp. 893–904, 2001.

[17] M. A. Chellaiah, N. Kizer, R. Biswas et al., "Osteopontin deficiency produces osteoclast dysfunction due to reduced CD44 surface expression," *Molecular Biology of the Cell*, vol. 14, no. 1, pp. 173–189, 2003.

[18] M. A. Chellaiah, R. S. Biswas, S. R. Rittling, D. T. Denhardt, and K. A. Hruska, "Rho-dependent Rho kinase activation increases CD44 surface expression and bone resorption in osteoclasts," *Journal of Biological Chemistry*, vol. 278, no. 31, pp. 29086–29097, 2003.

[19] J. J. Cao, P. A. Singleton, S. Majumdar et al., "Hyaluronan increases RANKL expression in bone marrow stromal cells through CD44," *Journal of Bone and Mineral Research*, vol. 20, no. 1, pp. 30–40, 2005.

[20] V. Samanna, T. Ma, T. W. Mak, M. Rogers, and M. A. Chellaiah, "Actin polymerization modulates CD44 surface expression, MMP-9 activation, and osteoclast function," *Journal of Cellular Physiology*, vol. 213, no. 3, pp. 710–720, 2007.

[21] I. Abécassis, B. Olofsson, M. Schmid, G. Zalcman, and A. Karniguian, "RhoA induces MMP-9 expression at CD44 lamellipodial focal complexes and promotes HMEC-1 cell invasion," *Experimental Cell Research*, vol. 291, no. 2, pp. 363–376, 2003.

[22] Q. Yu and I. Stamenkovic, "Localization of matrix metalloproteinase 9 to the cell surface provides a mechanism for CD44-mediated tumor invasion," *Genes and Development*, vol. 13, no. 1, pp. 35–48, 1999.

[23] S. Zucker, M. Drews, C. Conner et al., "Tissue inhibitor of metalloproteinase-2 (TIMP-2) binds to the catalytic domain of the cell surface receptor, membrane type 1-matrix metalloproteinase 1 (MT1-MMP)," *Journal of Biological Chemistry*, vol. 273, no. 2, pp. 1216–1222, 1998.

[24] J. Zhong, M. M. C. Gencay, L. Bubendorf et al., "ERK1/2 and p38 MAP kinase control MMP-2, MT1-MMP, and TIMP action and affect cell migration: a comparison between mesothelioma and mesothelial cells," *Journal of Cellular Physiology*, vol. 207, no. 2, pp. 540–552, 2006.

[25] S. Hernandez-Barrantes, Y. Shimura, P. D. Soloway, Q. A. Sang, and R. Fridman, "Differential roles of TIMP-4 and TIMP-2 in pro-MMP-2 activation by MT1-MMP," *Biochemical and Biophysical Research Communications*, vol. 281, no. 1, pp. 126–130, 2001.

[26] H. F. Bigg, C. J. Morrison, G. S. Butler et al., "Tissue inhibitor of metalloproteinases-4 inhibits but does not support the activation of gelatinase a via efficient inhibition of membrane type 1-matrix metalloproteinase," *Cancer Research*, vol. 61, no. 9, pp. 3610–3618, 2001.

[27] G. Dew, G. Murphy, H. Stanton et al., "Localisation of matrix metalloproteinases and TIMP-2 in resorbing mouse bone," *Cell and Tissue Research*, vol. 299, no. 3, pp. 385–394, 2000.

[28] Y. Itoh, A. Takamura, N. Ito et al., "Homophilic complex formation of MT1-MMP facilitates proMMP-2 activation on the cell surface and promotes tumor cell invasion," *EMBO Journal*, vol. 20, no. 17, pp. 4782–4793, 2001.

[29] M. Toth, I. Chvyrkova, M. M. Bernardo, S. Hernandez-Barrantes, and R. Fridman, "Pro-MMP-9 activation by the MT1-MMP/MMP-2 axis and MMP-3: role of TIMP-2 and plasma membranes," *Biochemical and Biophysical Research Communications*, vol. 308, no. 2, pp. 386–395, 2003.

[30] D. Hoshino, T. Tomari, M. Nagano, N. Koshikawa, and M. Seiki, "A novel protein associated with membrane-type1 matrix metalloproteinase binds p27kip1 and regulates RhoA activation, actin remodeling, and matrigel invasion," *Journal of Biological Chemistry*, vol. 284, no. 40, pp. 27315–27326, 2009.

[31] A. Lal, S. R. Haynes, and M. Gorospe, "Clean western blot signals from immunoprecipitated samples," *Molecular and Cellular Probes*, vol. 19, no. 6, pp. 385–388, 2005.

[32] R. Schmits, J. Filmus, N. Gerwin et al., "CD44 regulates hematopoietic progenitor distribution, granuloma formation, and tumorigenicity," *Blood*, vol. 90, no. 6, pp. 2217–2233, 1997.

[33] M. A. Chellaiah, "Regulation of podosomes by integrin $\alpha v\beta 3$ and Rho GTPase-facilitated phosphoinositide signaling," *European Journal of Cell Biology*, vol. 85, no. 3-4, pp. 311–317, 2006.

[34] V. Samanna, H. Wei, D. Ego-Osuala, and M. A. Chellaiah, "Alpha-V-dependent outside-in signaling is required for the regulation of CD44 surface expression, MMP-2 secretion, and cell migration by osteopontin in human melanoma cells," *Experimental Cell Research*, vol. 312, no. 12, pp. 2214–2230, 2006.

[35] B. Desai, M. J. Rogers, and M. A. Chellaiah, "Mechanisms of osteopontin and CD44 as metastatic principles in prostate cancer cells," *Molecular Cancer*, vol. 6, article 18, 2007.

[36] Q. T. Le and N. Katunuma, "Detection of protease inhibitors by a reverse zymography method, performed in a tris(hydroxymethyl)aminomethane-Tricine buffer system," *Analytical Biochemistry*, vol. 324, no. 2, pp. 237–240, 2004.

[37] M. Chellaiah, N. Kizer, M. Silva, U. Alvarez, D. Kwiatkowski, and K. A. Hruska, "Gelsolin deficiency blocks podosome assembly and produces increased bone mass and strength," *Journal of Cell Biology*, vol. 148, no. 4, pp. 665–678, 2000.

[38] B. Desai, T. Ma, and M. A. Chellaiah, "Invadopodia and matrix degradation, a new property of prostate cancer cells during migration and invasion," *Journal of Biological Chemistry*, vol. 283, no. 20, pp. 13856–13866, 2008.

[39] E. I. Deryugina, M. A. Bourdon, G. X. Luo, R. A. Reisfeld, and A. Strongin, "Matrix metalloproteinase-2 activation modulates glioma cell migration," *Journal of Cell Science*, vol. 110, part 19, pp. 2473–2482, 1997.

[40] M. A. Chellaiah, N. Soga, S. Swanson et al., "Rho-A is critical for osteoclast podosome organization, motility, and bone resorption," *Journal of Biological Chemistry*, vol. 275, no. 16, pp. 11993–12002, 2000.

[41] A. Y. Strongin, I. Collier, G. Bannikov, B. L. Marmer, G. A. Grant, and G. I. Goldberg, "Mechanism of cell surface activation of 72-kDa type IV collagenase. Isolation of the activated form of the membrane metalloprotease," *Journal of Biological Chemistry*, vol. 270, no. 10, pp. 5331–5338, 1995.

[42] B. Desai, T. Ma, J. Zhu, and M. A. Chellaiah, "Characterization of the expression of variant and standard CD44 in prostate cancer cells: identification of the possible molecular mechanism of CD44/MMP9 complex formation on the cell surface," *Journal of Cellular Biochemistry*, vol. 108, no. 1, pp. 272–284, 2009.

[43] Q. Yu and I. Stamenkovic, "Cell surface-localized matrix metalloproteinase-9 proteolytically activates TGF-β and promotes tumor invasion and angiogenesis," *Genes and Development*, vol. 14, no. 2, pp. 163–176, 2000.

[44] S. Hernandez-Barrantes, M. Toth, M. M. Bernardo et al., "Binding of active (57 kDa) membrane type 1-matrix metalloproteinase (MT1-MMP) to tissue inhibitor of metalloproteinase (TIMP)-2 regulates MT1-MMP processing and pro-MMP-2 activation," *Journal of Biological Chemistry*, vol. 275, no. 16, pp. 12080–12089, 2000.

[45] P. Hou, T. Troen, M. C. Ovejero et al., "Matrix metalloproteinase-12 (MMP-12) in osteoclasts: new lesson on the involvement of MMPs in bone resorption," *Bone*, vol. 34, no. 1, pp. 37–47, 2004.

[46] N. Suenaga, H. Mori, Y. Itoh, and M. Seiki, "CD44 binding through the hemopexin-like domain is critical for its shedding by membrane-type 1 matrix metalloproteinase," *Oncogene*, vol. 24, no. 5, pp. 859–868, 2005.

[47] M. Goebeler, D. Kaufmann, E.-B. Bröcker, and C. E. Klein, "Migration of highly aggressive melanoma cells on hyaluronic acid is associated with functional changes, increased turnover and shedding of CD44 receptors," *Journal of Cell Science*, vol. 109, part 7, pp. 1957–1964, 1996.

[48] D. Naor, R. V. Sionov, and D. Ish-Shalom, "CD44: structure, function, and association with the malignant process," *Advances in Cancer Research*, vol. 71, pp. 241–319, 1997.

[49] I. Okamoto, Y. Kawano, H. Tsuiki et al., "CD44 cleavage induced by a membrane-associated metalloprotease plays a critical role in tumor cell migration," *Oncogene*, vol. 18, no. 7, pp. 1435–1446, 1999.

[50] I. Okamoto, Y. Kawano, D. Murakami et al., "Proteolytic release of CD44 intracellular domain and its role in the CD44 signaling pathway," *Journal of Cell Biology*, vol. 155, no. 5, pp. 755–762, 2001.

[51] L. Y. Bourguignon, Z. Gunja-Smith, N. Iida et al., "CD44v(3, 8-10) is involved in cytoskeleton-mediated tumor cell migration and matrix metalloproteinase (MMP-9) association in metastatic breast cancer cells," *Journal of Cellular Physiology*, vol. 176, pp. 206–215, 1998.

[52] H. Nakamura, N. Suenaga, K. Taniwaki et al., "Constitutive and induced CD44 shedding by ADAM-like proteases and membrane-type 1 matrix metalloproteinase," *Cancer Research*, vol. 64, no. 3, pp. 876–882, 2004.

[53] K. B. Hotary, I. Yana, F. Sabeh et al., "Matrix metalloproteinases (MMPs) regulate fibrin-invasive activity via MT1-MMP-dependent and -independent processes," *Journal of Experimental Medicine*, vol. 195, no. 3, pp. 295–308, 2002.

[54] R. Fridman, M. Toth, I. Chvyrkova, S. O. Meroueh, and S. Mobashery, "Cell surface association of matrix metalloproteinase-9 (gelatinase B)," *Cancer and Metastasis Reviews*, vol. 22, no. 2-3, pp. 153–166, 2003.

[55] K. C. Nannuru, M. Futakuchi, M. L. Varney, T. M. Vincent, E. G. Marcusson, and R. K. Singh, "Matrix metalloproteinase (MMP)-13 regulates mammary tumor-induced osteolysis by activating MMP9 and transforming growth factor-β signaling at the tumor-bone interface," *Cancer Research*, vol. 70, no. 9, pp. 3494–3504, 2010.

Permissions

The contributors of this book come from diverse backgrounds, making this book a truly international effort. This book will bring forth new frontiers with its revolutionizing research information and detailed analysis of the nascent developments around the world.

We would like to thank all the contributing authors for lending their expertise to make the book truly unique. They have played a crucial role in the development of this book. Without their invaluable contributions this book wouldn't have been possible. They have made vital efforts to compile up to date information on the varied aspects of this subject to make this book a valuable addition to the collection of many professionals and students.

This book was conceptualized with the vision of imparting up-to-date information and advanced data in this field. To ensure the same, a matchless editorial board was set up. Every individual on the board went through rigorous rounds of assessment to prove their worth. After which they invested a large part of their time researching and compiling the most relevant data for our readers. Conferences and sessions were held from time to time between the editorial board and the contributing authors to present the data in the most comprehensible form. The editorial team has worked tirelessly to provide valuable and valid information to help people across the globe.

Every chapter published in this book has been scrutinized by our experts. Their significance has been extensively debated. The topics covered herein carry significant findings which will fuel the growth of the discipline. They may even be implemented as practical applications or may be referred to as a beginning point for another development. Chapters in this book were first published by Hindawi Publishing Corporation; hereby published with permission under the Creative Commons Attribution License or equivalent.

The editorial board has been involved in producing this book since its inception. They have spent rigorous hours researching and exploring the diverse topics which have resulted in the successful publishing of this book. They have passed on their knowledge of decades through this book. To expedite this challenging task, the publisher supported the team at every step. A small team of assistant editors was also appointed to further simplify the editing procedure and attain best results for the readers.

Our editorial team has been hand-picked from every corner of the world. Their multi-ethnicity adds dynamic inputs to the discussions which result in innovative outcomes. These outcomes are then further discussed with the researchers and contributors who give their valuable feedback and opinion regarding the same. The feedback is then collaborated with the researches and they are edited in a comprehensive manner to aid the understanding of the subject.

Apart from the editorial board, the designing team has also invested a significant amount of their time in understanding the subject and creating the most relevant covers. They scrutinized every image to scout for the most suitable representation of the subject and create an appropriate cover for the book.

The publishing team has been involved in this book since its early stages. They were actively engaged in every process, be it collecting the data, connecting with the contributors or procuring relevant information. The team has been an ardent support to the editorial, designing and production team. Their endless efforts to recruit the best for this project, has resulted in the accomplishment of this book. They are a veteran in the field of academics and their pool of knowledge is as vast as their experience in printing. Their expertise and guidance has proved useful at every step. Their uncompromising quality standards have made this book an exceptional effort. Their encouragement from time to time has been an inspiration for everyone.

The publisher and the editorial board hope that this book will prove to be a valuable piece of knowledge for researchers, students, practitioners and scholars across the globe.

List of Contributors

Wayne Carver and Edie C. Goldsmith
Department of Cell Biology and Anatomy, University of South Carolina, School of Medicine, Columbia, SC 29209, USA

Akhilesh Prajapati, Bhavesh Mistry and Sarita Gupta
Department of Biochemistry, Faculty of Science, The Maharaja Sayajirao University of Baroda, Vadodara, Gujarat 390005, India

Sharad Gupta
Ex-assistant Professor karamsad medical college and Gupta Pathological laboratory, Vadodara, Gujarat 390001, India

Swayam Prakash Srivastava, Daisuke Koya and Keizo Kanasaki
Department of Diabetology & Endocrinology, Kanazawa Medical University, Uchinada, Ishikawa 920-0293, Japan

Jung Jin Lim and Hyung Joon Kim
Fertility Center of CHA Gangnam Medical Center, College of Medicine, CHA University, 606-5 Yeoksam dong, Gangnam-gu, Seoul 135-081, Republic of Korea

Kye-Seong Kim
Department of Anatomy and Cell Biology, College of Medicine, Hanyang University, Seoul 133-791, Republic of Korea

Jae Yup Hong
Department of Urology, CHA Bundang Medical Center, CHA University, Seongnam 463-712, Republic of Korea

Dong Ryul Lee
Fertility Center of CHA Gangnam Medical Center, College of Medicine, CHA University, 606-5 Yeoksam dong, Gangnam-gu, Seoul 135-081, Republic of Korea
Department of Biomedical Science, College of Life Science, CHA University, Seoul 135-081, Republic of Korea

Dionysia Lymperatou, Efstathia Giannopoulou, Angelos K. Koutras and Haralabos P. Kalofonos
Clinical Oncology Laboratory, Division of Oncology, Department of Medicine, University of Patras, Patras Medical School, 26504 Rio, Greece

Jared M. Campbell
Discipline of Obstetrics and Gynaecology, School of Paediatrics and Reproductive Health, University of Adelaide, Medical School South, Level 3, Frome Road, Adelaide, SA 5005, Australia
Centre for Stem Cell Research, University of Adelaide, Medical School South, Level 3, Frome Road, Adelaide, SA 5005, Australia

Michelle Lane
Discipline of Obstetrics and Gynaecology, School of Paediatrics and Reproductive Health, University of Adelaide, Medical School South, Level 3, Frome Road, Adelaide, SA 5005, Australia
Repromed, 180 Fullarton Road, Dulwich, SA 5065, Australia

Ivan Vassiliev and Mark B. Nottle
Centre for Stem Cell Research, University of Adelaide, Medical School South, Level 3, Frome Road, Adelaide, SA 5005, Australia
Repromed, 180 Fullarton Road, Dulwich, SA 5065, Australia

Ana A. Q. A. Santos and Gerly A. C. Brito
Department of Morphology, Faculty of Medicine, Federal University of Ceará, Delmiro de Farias, 60416-030 Fortaleza, CE, Brazil

Manuel B. Braga-Neto, Marcelo R. Oliveira and Rosemeire S. Freire
Department of Physiology and Pharmacology, Faculty of Medicine, Federal University of Ceará, 1127 Coronel Nunes de Melo, 60430-270 Fortaleza, CE, Brazil

Eduardo B. Barros, Thiago M. Santiago and Luciana M. Rebelo
Department of Physics, Faculty of Physics, Federal University of Ceará, 922 Campus do Pici, 60455-760 Fortaleza, CE, Brazil

Claudia Mermelstein
Biomedical Sciences Institute, Federal University of Rio de Janeiro, 373 Avenue Carlos Chagas, 21941-902 Rio de Janeiro, RJ, Brazil

Cirle A. Warren and Richard L. Guerrant
Division of Infectious Diseases and International Health, Center for Global Health, University of Virginia, 345 Crispell Drive, Room 2709, Charlottesville, VA 22903, USA

Nicolas Etique, Laurie Verzeaux, Stéphane Dedieu and Hervé Emonard
CNRS FRE 3481 MEDyC (Matrice Extracellulaire et Dynamique Cellulaire), Laboratoire SiRMa (Signalisation et Recepteurs Matriciels), Universite de Reims Champagne-Ardenne (URCA), Moulin de la Housse, Bat. 18, Chemin des Rouliers, BP 1039, 51687 Reims Cedex 2, France

Haitao Li
Department of Orthodontics, College of Dental Medicine, Nova Southeastern University, 3200 S. University Drive, Fort Lauderdale, FL 33328, USA

Amsaveni Ramachandran, Qi Gao, Sriram Ravindran, Yiqiang Song and Anne George
Brodie Tooth Development Genetics & Regenerative Medicine Research Laboratory, Department of Oral Biology (M/C 690), College of Dentistry, University of Illinois at Chicago, Chicago IL 60612, USA

Carla Evans
Department of Orthdontics, College of Dentistry, University of Illinois at Chicago, 801 S. Paulina Street, Chicago IL 60612, USA

Antonio Junior Lepedda, Elisabetta Zinellu, Pierina De Muro, Gabriele Nieddu and Marilena Formato
Dipartimento di Scienze Biomediche, University of Sassari, Via Muroni 25, 07100 Sassari, Italy

Giovanni Andrea Deiana, Laura Fancellu, Piera Canu and Gianpietro Sechi
Dipartimento di Medicina Clinica e Sperimentale, University of Sassari, Viale San Pietro 10, 07100 Sassari, Italy

Daniela Concolino and Simona Sestito
Unita Operativa di Pediatria Universitaria, Azienda Ospedaliera "Pugliese-Ciaccio", Viale Pio X, 88100 Catanzaro, Italy

Vassiliki T. Labropoulou
Hematology Division, Department of Internal Medicine, University Hospital of Patras, 26500 Patras, Greece
Clinical Oncology Laboratory, Division of Oncology, Department of Medicine, University Hospital of Patras, 26500 Patras, Greece

Haralabos P. Kalofonos
Clinical Oncology Laboratory, Division of Oncology, Department of Medicine, University Hospital of Patras, 26500 Patras, Greece

Spyros S. Skandalis, Nikos K. Karamanos and Achilleas D. Theocharis
Laboratory of Biochemistry, Department of Chemistry, University of Patras, 26500 Patras, Greece

Panagiota Ravazoula
Department of Pathology, University Hospital of Patras, 26500 Patras, Greece

Petros Perimenis
Division of Urology, Department of Medicine, University Hospital of Patras, 26500 Patras, Greece

Spyros A. Syggelos
Department of Anatomy, Histology, Embryology, Medical School, University of Patras, 26500 Patras, Greece

Alexios J. Aletras, Ioanna Smirlaki and Spyros S. Skandalis
Laboratory of Biochemistry, Department of Chemistry, University of Patras, 26500 Patras, Greece

Maja Weber, Ilka Knoefler, Ekkehard Schleussner and Udo R. Markert
Placenta-Lab, Department of Obstetrics, University Hospital Jena, Germany
Abteilung fur Geburtshilfe, Placenta-Labor, Universitatsklinikum Jena, Bachstr. 18, 07740 Jena, Germany

Justine S. Fitzgerald
Abteilung fur Geburtshilfe, Placenta-Labor, Universitatsklinikum Jena, Bachstr. 18, 07740 Jena, Germany

Albert H.-L. Chan
Department of Periodontics and Oral Medicine, University of Michigan School of Dentistry, 1011 North University Avenue, Ann Arbor, MI 48109-1078, USA

Rangsiyakorn Lertlam, James P. Simmer and Jan C. C. Hu
Department of Biologic and Materials Sciences, University of Michigan School of Dentistry, 1210 Eisenhower Place, Ann Arbor, MI 48108, USA

Chia-Ning Wang
Department of Biostatistics, University of Michigan School of Public Health, 109 Observatory Street, 1700 SPH I, Ann Arbor, MI 48109-2029, USA

Hong Zhang
Department of Orthodontics, Guanghua School of Stomatology, Sun Yat-Sen University, Guangzhou 510055, China
Dental and Craniofacial Research Institute, School of Dentistry, University of California, Los Angeles, CA, 90095, USA

Xinli Zhang
Department of Oral and Maxillofacial Surgery, Sun Yat-sen Memorial Hospital, Sun Yat-Sen University, Guangzhou 510120, China

Zhiguang Zhang
Department of Oral and Maxillofacial Surgery, Guanghua School of Stomatology, Sun Yat-Sen University, Guangzhou 510055, China

Weiwei Chen
Dental and Craniofacial Research Institute, School of Dentistry, University of California, Los Angeles, CA, 90095, USA
Institute for Medical Biology, College of Life Sciences, South-Central University for Nationalities, Wuhan 430074, China

Jin Nam
Department of Bioengineering, University of California, 900 University Avenue, Riverside, CA 92521, USA
Center for Bioengineering Research, University of California, Riverside, CA 92521, USA

Maricela Maldonado
Department of Bioengineering, University of California, 900 University Avenue, Riverside, CA 92521, USA

J. Liu, H. K. Nam and N. E. Hatch
Department of Orthodontics and Pediatric Dentistry, School of Dentistry, University of Michigan, Ann Arbor, MI 48109-1078, USA

T.-G. Kwon
Department of Oral and Maxillofacial Surgery, School of Dentistry, Kyungpook National University, Jung Gu, Daegu, Republic of Korea

Adele L. Boskey, Lyudmila Spevak and Lyudmila Lukashova
Musculoskeletal Integrity Program, Hospital for Special Surgery, 535 E 70th Street, New York, NY 10021, USA
Department of Endodontics, School of Dental Medicine, University of Pittsburgh, 3501 Terrace Street, Pittsburgh, PA 15261, USA

Kostas Verdelis, Elia Beniash and Xu Yang
Department of Oral Biology, School of Dental Medicine, University of Pittsburgh, 3501 Terrace Street, Pittsburgh, PA 15261, USA

Wayne A. Cabral and Joan C. Marini
Bone & Extracellular Matrix Branch, NIH/ NICHD, Bethesda, MD 20892, USA

Dalei Zhang, Bei Yang, Weiying Zou, Mingdi Xiong, Lei Wu, Jinglei Wang and Ting Zou
Department of Physiology, Medical College of Nanchang University, Nanchang, Jiangxi 330006, China

Junhong Gao
Institute of Acupuncture and Moxibustion, China Academy of Chinese Medical Sciences, Beijing 100700, China

Sifan Xu
Institute of Chinese Minority Traditional Medicine, Minzu University of China, Bejing 100081, China

Xiaying Lu
Department of Physiology, Medical College of Nanchang University, Nanchang, Jiangxi 330006, China
Department of Physiology, Gannan Medical University, Ganzhou, Jiangxi 341000, China

Meenakshi A. Chellaiah and Tao Ma
Department of Oncology and Diagnostic Sciences, Dental School, University of Maryland, Baltimore, MD 21201, USA

www.ingramcontent.com/pod-product-compliance
Lightning Source LLC
Chambersburg PA
CBHW080650200326
41458CB00013B/4797